A Natural History of the
Sonoran Desert

A Natural History of the
Sonoran Desert

ARIZONA-SONORA DESERT MUSEUM

Edited by
Steven J. Phillips & Patricia Wentworth Comus

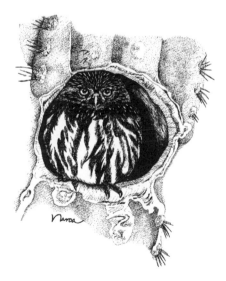

ARIZONA-SONORA DESERT MUSEUM PRESS
Tucson

UNIVERSITY OF CALIFORNIA PRESS
Berkeley   Los Angeles   London

Arizona-Sonora Desert Museum Press
2021 North Kinney Road
Tucson, Arizona 85743
asdmpress@desertmuseum.org

University of California Press
Berkeley and Los Angeles, California

University of California Press, Ltd.
London, England

Library of Congress Cataloging-in-Publication Data
A natural history of the Sonoran Desert / Arizona-Sonora Desert Museum.
p.   cm.
Includes bibliographical references and index.
ISBN 0-520-21980-5 (paper: alk. paper)
I. Natural history—Sonoran Desert.   I. Arizona-Sonora Desert Museum (Tucson, Ariz.)
QH104.5.S58N38   1999
508.791'7—dc21
99-33675
CIP

Book design by Steven J. Phillips
Cover and jacket design by Barbara Jellow
Cover and jacket photographs by Bruce Griffin

Printed in Canada through Spectrum Books, Inc.

10   09   08   07   06   05   04   03
12   11   10   9   8   7   6   5   4

The publisher gratefully acknowledges the financial support
of the family and friends of

David J. Swan

in memory of him and of his love for the Sonoran Desert

Preface

When conservationist and educator William H. Carr envisioned the Arizona-Sonora Desert Museum a half century ago he wanted to create a place where people would pause and reflect on this marvelous land called the Sonoran Desert and learn about its endlessly fascinating ecology. Today, with more than a half million visitors annually, the Desert Museum carries on the work of Bill Carr in ever-expanding ways: world-class exhibits and interpretive programs, educational outreach programs in southern Arizona and northern Mexico, scientific research and conservation symposia, and an award-winning website. Publishing this book is one more way the Desert Museum brings the stories of the desert to new audiences.

To many people, all deserts seem alike: blowing sand, vast expanses of barren land interrupted by the occasional oasis, and perhaps a few lizards and snakes. But the Sonoran Desert, in direct contrast to this perception, is teeming with life. It is the most biologically diverse of the North American deserts, for each of its habitats contains life forms found nowhere else. These endemic plants and animals have developed remarkable strategies for adapting to the rigors of desert climates, harsh to us but hospitable to them.

Today, as more and more people visit the Sonoran Desert and make it their home—a seven-fold increase in population in the 50 years since the Museum was founded—the vulnerability of the desert has never been more evident. Its richness of life is threatened as we dam rivers, pump ground water, and watch housing projects spread out across the desert landscape. It is therefore our hope that this book will inspire you to learn more about this fragile land and step lightly upon its soil.

I would like to express my deep gratitude to the editors and writers, particularly to Steve Phillips, Pat Comus and Dr. Mark Dimmitt, who worked tirelessly to bring this volume to publication. We are also grateful for the help provided by the University of California Press and for the support this project received from the family and friends of David Swan.

Richard H. Daley
Executive Director
Arizona-Sonora Desert Museum

Contents

Acknowledgments

A number of researchers and natural historians took time out of their hectic schedules to review portions of this book. We are extremely grateful for their comments. We, however, accept full responsibility for any errors that remain in the text. Reviewers included Kathleen B. Pigg, Glenn E. Walsberg, Andrew Salywon, Jeremy Buegge, Brian Sullivan, Marc A. Baker, Timothy Tilton, Tom McGarvin, Jon Spencer, Bill Shaw, George Bradley, O. Eugene Maughan, and Jane Cole. Desert Museum docent reviewers included Lois Baker, Ann and Dick Field, Joan Picken, Donna Allen, Linda Gregonis, Pat Hennigan, Barbara Bickel, and Jim Quinlan. Experts on various subjects contributed their thoughts and answered particular questions; they include Allison Alberts, Carol Crosswhite, Lesley Fitzpatrick, Wendy Hodgson, Elizabeth Slauson, Pauline McNaney, Dennis D. Massion, Donald J. Pinkava, Stanley R. Szarek, Ron Tiller, Peter B. Moyle, Dean A. Hendrickson, and David Parsons. Many other people provided insights into the natural history of this region, including Leslie R. Landrum, James W. Cornett, Michael Wilson, Richard Felger, Stephen McLaughlin, Ted Fleming, Judith Bronstein, Mike Pruss, John Phelps, Jim Heffelfinger, Steve Spangle, Bridget Watts, and Linda Laack.

Over the last three years we have called on countless individuals to supply us with source material. Chief among them are Mitzi Frank, Art Needleman, Tom Danton, "Steamer" Lawhead, Johanna Alexander, and Kathy Tudek. Mary Beth Ginter took on the task of translating Spanish vernacular names of plants into English. Peter Kresan and Harry Casey came through with a couple of difficult photo needs. Desert Museum associate director Nancy Laney was always willing to help wherever needed, with everything from reviewing contracts to hunting down historic photographs.

We are also indebted to the following Desert Museum docents and volunteers for their many and varied contributions: Mary Chris Carbonaro, Daphne Stevens, Karen Dunin-Wasowicz, Peggy Larson, Amy Pate, Helen Roubcek, Carole Bonhorst, Kassandra Chizek, Herb Elins, Jeanne Rosengren, Nancy Hannan, Brendan Craughwell, Daron Overpeck, Jim Sullivan, Bill Taylor, Fred Grossman, Nancy Cole, and Aaron Gilbreath.

Special gratitude is owed to volunteer illustrator Jeff Martin who spent countless hours on his exquisitely detailed arthropod drawings; to Tony Burgess who somehow carved time out of his relentless work routine to visit the Museum's director of natural history and share his keen knowledge of plant physiology and ecology; and to Edward F. Anderson who dropped everything to take on the prickly task of reviewing the Cactaceae section. A very special thanks to former Desert Museum education director Carol Cochran who championed the idea of producing this volume and who helped get it off the ground in 1996.

Our thanks to the ever cheerful and always helpful staff at the University of California Press, especially Doris Kretschmer, Barbara Jellow, and Tony Crouch.

A huge debt of gratitude to the family and friends of David Swan and the Guido A. and Elizabeth H. Binda Foundation. Without their support this book may never have made the long and expensive journey into print.

Finally, a warm note of thanks to our two loved and loving spouses, Christina McNearney-Phillips and Louis F. Comus, Jr., who endured panic attacks, bouts of grouchiness, canceled vacations, and lost weekends, never complaining and always supportive. Thanks, Tina. Thanks, Lou.

Welcome to the Sonoran Desert

Gary Paul Nabhan

Imagine crossing a threshold—not into someone's home but into another world. Imagine that world to be inhabited by creatures with names as wild as Gila monster, chuckwalla, vinagaroon, boojum and devil's claw. Close your eyes, and smell a world filled with the fragrances of night-blooming cactus flowers, sacred datura, and the aromatic oils of creosote bush released into the air after the first summer rains. Listen hard, and hear the distant calls of Cactus Wrens, cicadas, Scaled Quail, Curve-billed Thrashers and spadefoot toads. Then open your eyes again, and see that the world you've entered into is swarming with leafcutter ants, carpenter bees, hummingbirds, and kangaroo rats.

You have not entered into someone else's home, but one which for a day, a year, or an entire lifetime, may be your own: the Sonoran Desert. It is a homeland that rambles over some 100,000 square miles (260,000 km²) in two countries and five states. It is home to 60 species of mammals, more than 350 kinds of birds, 20 amphibians, 100 or so reptiles, and 30 native freshwater fish. Well over 2000 native species of plants occur within the Sonoran Desert proper, of the 5000 or so which occur in Sonora, Arizona, Baja California, and adjacent southern California. This region is also home to at least seventeen indigenous cultures, as well as many others which have adopted it: Latino and Anglo, Chinese and Chicano, Arabic and African in origin.

Not far from the upland edge of this desert is a museum which serves as its threshold—an entry point through which the richness of the Sonoran Desert is revealed. For roughly a half century, the Arizona-Sonora Desert Museum has introduced more people to the flora, fauna, and habitats of this unique region than has any other medium of environmental education. Today, over a half million visitors come through its doors each year; several hundred thousand more receive its message through its publications, through its web site and annual conservation conferences, and through "The Desert Speaks," a TV collaboration with The Nature Conservancy and the University of Arizona. In short, the Museum's staff have done much to discover the living riches of this region, and have explored the many ways to celebrate and elucidate the value of its habitats to the whole of human society.

When the Museum staff began to work on this natural history of our region, our initial intent was to distill into a single publication the essential stories that we had gathered together in our various training courses over the decades. We had trained thousands of docents (volunteer interpreters) to help us tell visitors about the creatures and characters, the scenes and the scenery, the ecological processes and plots around which these stories revolve. We wanted to capsulize what we had all learned in this ecological theater, so that others would be aware of the wonderful evolutionary play going on all around us.

As we began to edit older writings and drawings we had used over the years, we realized how many of those stories had changed since their last telling, how innovative research by our own staff and by others had shifted the plots, and added new characters, both heroes and villains. We began to write an entirely fresh compilation of these character profiles and essential vignettes as new means to explain how a desert really works. As we read one another's chapters and discussed them, we realized how richly complex and startling our

new understanding of the desert's patterns had become. Rather than portraying the Sonoran Desert as a stark, biologically impoverished landscape where all creatures compete "tooth and claw" for scarce resources, we were surprised by how many *mutualisms*—interactions between species which benefit each other—occur all around us. We were in awe of the implications of recently published inventories which demonstrate that deserts—not rainforests—are richest in pollinator diversity and perhaps in reptile diversity as well. We were also saddened to learn how rapidly this diversity is disappearing, as uninformed people who believe the Sonoran Desert to be a wasteland unknowingly turn it into one.

This book, then, is a fresh look at the ecological and cultural patterns which shape the richest, most complex desert in all of the Americas. If you have the chance to cross the threshold, to let its creatures and cacti speak directly to you, leaf through this book before you go and after you return. Read it by campfire light at night, or when you get stuck below a rocky overhang that summer day when a sudden downpour forces you to head for cover. Savor it as you sit beneath a saguaro cactus, eating the succulent fruit, hearing the breeze make music as it blows through the cactus spines—a living thumb piano. And use this book as a wellspring for reflection, to remind all of us human mortals that there are other lives on the face of this earth which enrich our own.

Biomes & Communities of the Sonoran Desert Region

Mark A. Dimmitt

World travelers can scarcely help but notice the great diversity of landscapes on this planet. That diversity is as much due to the vegetation as to the landforms. Closer observation reveals a mind boggling diversity of plants and animals in these different places. People who travel in the American Southwest or on the western side of any other continent near 30° latitude will see dramatic changes within distances of only a few miles.

In apparent contrast, some widely distant parts of planet Earth show certain broad scale similarities. The vegetation of the Mediterranean coast of Europe looks remarkably similar to the chaparral of Southern California, though no two plant species occur in both places (except some introduced weeds). Both the diversity and the similarities have the same underlying causes, namely the interactions between geography and the global climate machine which governs the biosphere on both grand and microscopic scales.

The Sonoran Desert region has a great variety of both species and habitats, the latter ranging from extremely hot, arid desert to semiarid tropical forest to frigid subalpine meadows. Our focus is on the Sonoran Desert in the heart of the region, but to understand it we need to know something about the other habitats that border it. These adjacent geographical features and biological communities exert profound, complex influences on the desert itself.

Ecologists who study nature on a global scale recognize a few basic, widespread classes of habitats that are easily identified by their dominant plant *life forms*, which are basic categories based on general appearance, for example, tree, shrub, annual, succulent, and so on. Such global scale habitats are called *biomes*, and are determined primarily by the climatic factors of temperature and rainfall. These factors are in turn determined by latitude, elevation, and wind patterns. Biome classification is based

3

on vegetation because plants, being generally immobile, are the most obvious and easily recognizable components of a biological community. In addition, plants are more definitive of their biomes because, since they are rooted in place, they must be adapted to that specific environment. Plants, therefore, are often *endemic* (occurring only in the named area) to a biome or smaller community. Biomes do contain characteristic animal life as well, including many endemic insects and other invertebrates. Most vertebrates, however, are more mobile and rather few species are restricted to a single habitat.

All of the world's biomes occur in the Sonoran Desert region. This tremendous diversity in a fairly small area is due to two influences. For one thing, this region is on the west side of a continent near 30° North latitude, a position where several biomes typically occur in close proximity (a phenomenon explained later in this chapter). Secondly, our great topographical relief creates the cold, wet climates that allow northern biomes to occur farther south than they would ordinarily.

It is important to recognize that biomes and most other biological classifications are largely subjective concepts—an attempt to make sense of the nearly incomprehensible diversity of nature. In addition, their boundaries are rarely distinct. Wherever two biological communities or biomes meet there is usually a zone of intergradation which is sometimes very wide. For these reasons classifications differ among classifiers. For example, some biologists recognize thornscrub as a separate biome while others call it an *ecotone* (transition zone) between desert

and tropical forest. Some combine tropical and temperate forests into the same biome based simply on vegetation height and density. The biomes as defined here are so distinctive that you should be able to place any terrestrial habitat on the planet within one of them at a glance (see plate 1).

Biomes are subdivided into a hierarchy of smaller categories, defined by the particular species that inhabit them. There are many classification systems and the categories have many names. We use the general terms *biotic community, biological community*, or simply, *community*. The names used here for the communities are mostly those of Brown and Lowe (1982).

TUNDRA is the most poleward and highest-elevation biome and is characterized by extremely cold winters. The dominant plant life forms are ground-hugging woody shrubs and perennial herbs. Intense cold excludes trees and succulents and the growing season is too short for annuals.

Temperatures become warmer at lower latitudes (toward the equator) at the same elevation. But an increase in elevation at a given latitude has the same climatic effect as does traveling toward a pole: temperature decreases. So a climate that supports tundra, like that in the arctic, can be found on high mountains all the way to the equator. The other cold biomes in both hemispheres also extend toward the equator where sufficiently high elevations meet their climatic requirements. The San Francisco Peaks near Flagstaff, Arizona, rise to 12,600 feet (almost 3900 m). These are the only mountains in our region that extend above timberline (about 11,200 feet, 3400 m elevation in Arizona). There, only forty-five miles (72 km) from the saguaros of the Sonoran Desert, is a small area of alpine tundra that includes some of the same plant species that occur in the arctic tundra of Alaska (see plate 2).

CONIFEROUS FOREST (also known by its Russian name, *taiga*) is dominated by cone-bearing trees, especially pines, firs, and spruces in the northern hemisphere. Many conifers are adapted to cold only a little less severe than in tundra. Tree height ranges from a few feet (a couple of meters) near the tundra boundary or at timberline to over 300 feet (90 m) in more temperate latitudes. Some coniferous woodlands extend into subtropical climates, for example, in the southeastern United States.

In our region coniferous forest occurs in the higher mountain ranges, mostly to the north and east of the Sonoran Desert. Our most widespread coniferous community is Petran Montane (Rocky Mountain) forest, the dominant vegetation of the cold-temperate Rocky Mountains. Its elevation increases southward into Mexico until it is pushed off the tops of the mountains by excessive aridity and warmth. In the mountains west of the Sonoran Desert are isolated islands of Sierran (as in Sierra Nevada) coniferous forest, characterized by different conifer species.

TEMPERATE DECIDUOUS FOREST is characterized by dense stands of broadleaf trees that drop their foliage in winter. The winters are milder than those of most conifer-dominated climates, though still too cold for plant growth. Summers are typically warm and humid. There are more species

than in the two more poleward biomes. The herbaceous perennial life form is well-represented along with the trees and shrubs. Although pure temperate deciduous forest is rare in our region, it is represented by scattered aspen groves and ribbons of riparian trees.

The foothills and lower mountain slopes east of the Sonoran Desert are wooded with oaks and pines, a mixture of coniferous forest and temperate deciduous forest tree types. The oaks, however, are mostly evergreen species; they are not deciduous except during severe droughts. This Madrean evergreen woodland (also called Mexican oak-pine woodland) is a warm-temperate community of the Sierra Madre Occidental. It extends as far north as central Arizona, where it is squeezed out by the cool-temperate Rocky Mountain forests above it and the more arid grassland and desert below. (Though its official name is woodland, in its southern part it's actually a forest; i.e., the tree canopies overlap.) This is a semiarid community which experiences a dry season in spring (see plate 3).

In the Sonoran Desert region tundra, coniferous forest, and temperate deciduous forest are restricted to mountains that rise well above the intervening basins. In southeastern Arizona and northeastern Sonora there is a gap between the massive Rocky Mountains and the Sierra Madre Occidental. The mountain ranges in this gap are distinct entities separated by intervening valleys. The cool, moist communities on their upper elevations are isolated from one another by "seas" of hot, arid habitat. Because this isolation is analogous to oceanic islands, the terms "mountain islands" and "sky islands" have been coined for these and similar ranges.

GRASSLAND is a semiarid biome characterized by warm, humid summers with moderate rain and cold, dry winters. (The central valley of California is an exception; it is a winter-rainfall grassland at a lower than typical elevation.) Grass is the dominant life form; scores of species form a nearly continuous cover over large areas. Other well-represented life forms are annuals and *geophytes* (herbaceous perennials such as bulbs that die to the ground each year). Populations of trees, shrubs, and succulents are kept at low levels by periodic fires during the dry season.

Most of the grasslands in the western states are intermediate between the true prairies of the American Midwest and deserts. They are called semi-desert or desert grasslands. (Again the California grasslands are an exception. They are heavily influenced by the unique California floristic province and not much by the Midwest prairies.) Compared with prairie grassland, the grasses in desert grassland are shorter, less dense, and are more frequently interspersed with desert shrubs and succulents. Desert grassland or chaparral borders the northern Sonoran Desert on the east (see plate 4).

CHAPARRAL is a semiarid biome that occurs on the west coast of every continent between about 30° and 40° North latitude. This smallest biome is unique for its Mediterranean climate: mild, moist winters and hot, dry summers. Mature chaparral consists almost solely of woody evergreen shrubs with small leathery leaves. The numerous species form impenetrable thickets from five to eight feet (1.5 to 2.5 m)

tall. During the long dry summers the typically resinous foliage and dry woody stems become explosively combustible.

Wildfires raze large areas to ash-covered earth every few decades. Fires are not harmful to this community; they are in fact necessary for maintaining its vigor. Following fires the bare ground is briefly colonized by a large number of annual species, but the land is soon reclaimed by the shrubs which sprout from seeds or root crowns. Trees and succulents are rare life forms in chaparral because they are more vulnerable to destruction by the very hot fires.

This young biome evolved from early Tertiary tropical forest during the Pliocene and Pleistocene. The uplift of the great mountain ranges of western North America blocked the summer monsoon moisture from reaching the far west, creating a summer dry season (see chapter on "Deep History of the Sonoran Desert").

The main area of chaparral occurs west of the coast, transverse, and peninsular ranges and is called Californian chaparral (see plate 5). Disjunct patches of chaparral occur inland of these ranges and are called interior chaparral. Interior chaparral differs in having only a few species; it is often comprised almost entirely of manzanita (two species of *Arctostaphylos*) and shrub live oak (*Quercus dumosa*). Interior chaparral also receives substantial summer rainfall, though the plants do not respond to it.

California chaparral borders the western edge of the Sonoran Desert in California and northern Baja California, and interior chaparral is scattered along the desert's northeastern edge where it meets the Mogollon Rim of Arizona. Interior chaparral also occurs in isolated patches on the lower slopes of some mountain islands.

DESERT is the driest biome, its vegetation is determined solely by the extreme aridity. Temperature and seasonality of rainfall determines the specific vegetation and fauna, but all desert vegetation looks more or less similar; most plants are widely spaced and have small or absent leaves (see plate 6). A detailed discussion of deserts follows this section.

THORNSCRUB is intermediate between the desert and tropical forest biomes. The vegetation consists largely of short trees, ten to twenty feet (3-6 m) tall, and shrubs, with cacti also being common in the "New World" communities. It is generally more dense and taller than desert vegetation, and many species are thorny. Annuals and herbaceous perennials are abundant, and vines—a primarily tropical life form— are well represented. During the dry season most perennial plants are drought-deciduous (as opposed to plants of more temperate regions which are cold-deciduous). In contrast, the rainy season, though short, is moderate and dependable and the vegetation grows lush. The climate is nearly frost-free, so temperature is not limiting; the vegetation is determined by the alternating dry and wet seasons (see plate 7).

TROPICAL FOREST is determined by the absence of freezing temperatures and the occurrence of ample rainfall for at least part of the year. Some tropical forests have a dry season, while tropical rain forest is never stressed for water. Tropical deciduous forests have a dry season lasting from three to nine months, during which time many of the plants become deciduous. Many of the tree species flower during the winter-spring dry season while leafless. In the rainy season the dense vegetation grows luxuriantly and forms closed canopies of foliage. The upper canopy ranges from fifteen to thirty feet (4.5 to 9 m) above ground in dry forests, to 150 feet (45 m) in lowland rain forests. Almost all life forms are represented, though annuals are nearly absent from rainforest. Flowering *epiphytes* (plants that grow on other plants or rocks but are not parasitic) are almost completely restricted to tropical habitats, and are a major component of wet tropical floras (see plate 8).

To the south, the Sonoran Desert merges almost imperceptibly into thornscrub in central Sonora, and thornscrub in turn merges with the northern limit of tropical deciduous forest in the southern tip of that state. A major proportion of the Sonoran Desert's biota evolved from ancestors in these tropical biomes; examples are noted in the species accounts.

RIPARIAN COMMUNITIES

Riparian communities are not biomes. Though they could be considered iso-lated ribbons of deciduous forest, they are better viewed as a unique habitat type. They occur within any biome wherever there is perennial water near the surface. The term *riparian* specifi-cally refers to the zones along the banks of rivers; however, it is also applied to the shoreline communities along slow or nonflowing waters such as marshes and lakes.

The drier the surrounding habitat, the more distinct is the riparian zone. In the desert or grassland a flowing

Biomes and Communities

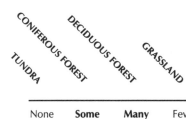

	TUNDRA	CONIFEROUS FOREST	DECIDUOUS FOREST	GRASSLAND	DESERT	THORNSCRUB	TROPICAL FOREST
TREES	None	**Some**	**Many**	Few	None to Some	Many	**Many**
WOODY SHRUBS	**Few**	Some	Many	Some	**Many**	**Many**	Many
SUBSHRUBS	Few?	Few	Some	Some	Many	Many	Some
ANNUALS	None	None	Some	Many	**Many**	Some	Some
HERBACEOUS PERENNIALS	**Some**	Some	Many	Many	Many	Some	Some
GRASSES	Few	Some	Many	**Many**	Some	Some	Some
SUCCULENTS	None	Few	Few	Some	**None to Many**	**Many**	Some
VINES	None	Few	Few	Few	None to Some	Many	Many
FLOWERING EPIPHYTES	None	None	None	None	None	Some	**Some to Many**

(**Boldface** signifies dominant life form)

stream supports a conspicuous oasis with forests and wildlife that would not otherwise occur in the area. The available water also augments populations of more arid-adapted species in the adjacent habitat.

Riparian zones are so different at different latitudes and elevations that they should be thought of as several communities with similar physical characteristics, primarily their dependence on perennial water. Montane streams support alder and aspen, while at lower elevations there are cottonwoods and sycamores. In tropical deciduous forests a riparian zone may be visually indistinguishable during the wet season because the overall appearance of stream-bank and hillside trees is similar, though the species may be different. But in the dry season most of the slope vegetation is deciduous, while tropical riparian species are typically evergreen.

Some ecologists broaden the concept of riparian communities to include the banks of dry washes in deserts. A wash in the Lower Colorado River Valley with its woodland of palo verdes (*Cercidium* spp.), ironwoods (*Olneya tesota*), and desert willows (*Chilopsis linearis*) is clearly distinct from the surrounding creosote bush flats. These dry washes occupy less than five percent of the area of this subdivision of the Sonoran Desert, but support ninety percent of its bird life. This concentration of life is the result of the greater availability of water, even though the wash may carry surface water for only a few hours a year. Desert drainageways should be labeled "dry riparian" or "desert riparian" to avoid confusion with wetter habitats that have surface water all or most of the year.

Dry riparian habitats share most of their defining characteristics with traditional "wet" riparian habitats. They are chronically disturbed, unstable sites where water and nutrients are harvested and concentrated from larger areas (watersheds). Finally, they are corridors for dispersal of plants (seeds) and animals (see plate 9).

What Is a Desert?

Although many people visualize deserts as dry, desolate wastelands, the term actually defines a wide spectrum of landscapes and plant and animal population densities. The Sonoran Desert does have seas of sand and expanses of desert pavement that are nearly devoid of visible life, but most of it is more reminiscent of a sparse woodland savanna.

The common denominator of all deserts is extreme aridity—water is freely available only for short periods following rains. Desert is often defined as a place that receives less than ten inches (250 mm) of annual average rainfall, but this definition is inadequate. For example, the Pacific coast of northern Baja California and the north slope of Alaska both receive less, but those places are vegetated with chaparral and tundra, respectively. An accurate measure of aridity must compare rainfall (abbreviated P for precipitation) with potential water loss through

evaporation and transpiration (the loss of water from leaves). Potential evapotranspiration (abbreviated PET, the water that would be lost from evaporation and transpiration if water were present to evaporate) is difficult to measure accurately, but is crudely estimated to be sixty percent of pan evaporation (the water that evaporates from a wide pan of water exposed to the weather). Pan evaporation varies severalfold within a local area depending on slope and exposure to wind, so it is applicable only to the specific site where it is measured. Tucson receives an average of twelve inches (305 mm) of rain a year, while the pan evaporation is about 100 inches (254 cm). In other words, the climate of Tucson could evaporate eight times more water per year than is supplied by rain, a pan evaporation to precipitation ratio of 8:1. Using the sixty percent estimate for PET, Tucson's PET/P ratio is 4.3; climatologists classify areas with ratios higher than 3.0 as semiarid. This moisture deficit presents a significant challenge to the biota, but is not large compared to that of hyperarid deserts such as that around Yuma, Arizona which has a PET/P ratio of 30, or the interior Sahara Desert's 600.

A concise nontechnical definition of a desert is "a place where water is severely limiting to life most of the time." (Without the word "severely" the phrase defines semiarid habitats such as grassland, chaparral, and tropical deciduous forest.) Though desert plants and animals must cope with scarce water, the common perception that they are struggling to survive is grossly inaccurate. The native biota are adapted to and usually thrive under these conditions and, in fact, most of them require an arid environment for survival. Look at it this way: if a desert received much more rain, it wouldn't be a desert. A different, wetter, biome would replace it. Thus an alternative and more positive definition might be: "**A desert is a biological community in which most of the indigenous plants and animals are adapted to chronic aridity and periodic, extreme droughts, and in which these conditions are necessary to maintain the community's structure.**" (The desert biome requires chronic aridity, but not all of its component species do.)

WHY ARE DESERTS SO DRY?

The low rainfall typical of deserts is more easily understood if one knows a little about the basics of global climate. Atmospheric thermodynamics is an extremely complicated field, but the basic rules are simple. First, hot air rises and cool air sinks. Second, rising air expands and cools, while sinking air compresses and becomes warmer. Third, warmer air can hold more water vapor than cooler air. These three natural phenomena plus the sun's heat determine where rain falls on the planet.

The sun shines almost vertically on the equatorial belt year round, but it shines on the polar regions at a shallow angle and only in the summer of each respective pole. There are two consequences. A beam of sunlight ten square feet (1 m²) shines on about ten square feet of Earth's surface at the equator at noon, but it covers more than twice that area near the poles. The sun's light and heat are thus less concentrated at higher latitudes. In addition, at the

equator the sunlight travels straight down through the atmosphere, but near the poles it travels through much more air where more of the light is reflected, absorbed, or scattered and less reaches the ground. This is why the equator is hot and the poles are cold.

Because of the great quantity of heat delivered to the equatorial belt, it is a zone of warm, rising air. It absorbs much water vapor from the oceans and land vegetation. As this air rises it cools. Eventually it reaches saturation (dewpoint temperature) and water vapor condenses into clouds and often falls as rain. So the equatorial region is both hot and wet.

The equatorial air rises, then spreads horizontally at high elevations to the north and south. Eventually the now cool air sinks and flows along the surface to replace the rising air at the equator, forming a circulation cell. It tends to sink at about 30° north and 30° south latitude. (These two zones were called the "horse latitudes" by mariners. Before motorpower, sailing vessels could get becalmed in these latitudes for weeks at a time. To reserve precious water for themselves, the crew threw horses and other livestock overboard; other ships would encounter the floating carcasses.) As the air sinks it warms by compression, and because there is no source of evaporating water, it becomes drier with increasing temperature. Not only can sinking air not produce rain, but when it reaches the ground it absorbs water from the soil and vegetation, creating even more arid conditions.

The horse latitude zones of sinking air are not continuous belts. The combination of the Earth's rotation and the interaction between land masses and oceans creates stable high pressure zones (sinking air) over the oceans west of the continents. The resulting aridity is reinforced by the cold ocean currents that also occur on western coasts at this latitude; the cold water further inhibits the potential for rising air currents that are necessary to make rain. Thus on the west edge of every large land mass there is a hyperarid area near 30° latitude called a *horse latitude desert*. Despite the proximity of the oceans, the high pressure zone is so strong over the Atacama and Sahara deserts that decades may pass without rain.

cools and drops most of its moisture on the windward slope. On the leeward side it descends, warms, and dries. At latitudes that have a prevailing wind direction, rain shadow deserts are created on mountains' lee sides.

Aridity is the primary attribute of deserts, but it also generates several other characteristics of deserts. In addition to being meager, desert precipitation is also highly variable and unpredictable. The more arid the desert, the more variable is its rainfall. The average annual precipitation is a poor predictor of the rainfall in a given

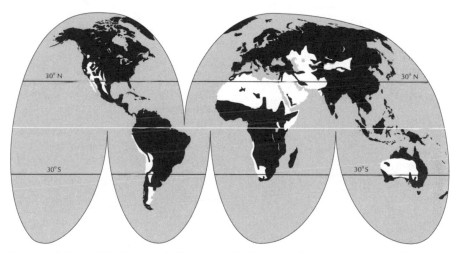

Deserts of the world (white areas). Horse latitude deserts are those on the western edges of all the continents near 30° north and south latitude. The rest are rain shadow deserts. The difference between the two types can be seen in South America, where the Andes Mountains stretch the entire length of the continent near the west coast. North of 30° the trade winds blow from the northeast, causing a rain shadow desert on the west or coastal side of the Andes. South of 30° the easterlies blow onshore and the rain shadow desert is on the eastern, inland, side of the Andes. Near the 30th parallel there is no prevailing wind; the stable high pressure zone creates horse latitude deserts on both sides of the Andes Mountains.

Deserts are also caused by rain shadow effects wherever there are mountains and prevailing winds. Where wind encounters a mountain, it is forced up and over. As it rises, it

year. For example, Yuma, Arizona has an average annual rainfall of three and one-half inches (90 mm), but in most years it receives less, sometimes none at all. When the stable weather pattern that enforces aridity breaks down occasionally, Yuma may receive two or three times its annual average, sometimes in a single storm.

Desert temperatures vary widely both daily and seasonally. The dry, transparent air and cloudless skies transmit maximal solar energy to the ground where much of it is absorbed and converted to heat; the temperature rises dramatically. At night the same conditions permit most of this heat to be radiated to the sky, and the temperature plummets. (Water vapor, either as humidity or cloud cover, reflects infrared heat and slows heat loss.) Daily temperature variation can be more than 50°F (28°C). The same conditions create great seasonal fluctuation. High-elevation deserts that have 100°F (38°C) days in summer can experience nights below 0°F (-18°C) in winter.

Besides the heat it creates, the intense sunlight in arid lands is itself a challenge. The ultraviolet radiation can damage animals' retinas, cause skin cancer, and destroy vital plant molecules such as chlorophyll. Desert organisms have evolved a variety of adaptations to avoid getting too much sun.

THE NORTH AMERICAN DESERTS

North America has four major deserts: Great Basin, Mohave, Chihuahuan and Sonoran. All but the Sonoran Desert have cold winters. Freezing temperatures are even more limiting to plant life than is aridity, so colder deserts are poorer in both species and life forms, especially succulents.

The Great Basin Desert (plate 10)

is both the highest-elevation and north-ernmost of the four and has very cold winters. The seasonal distribution of precipitation varies with latitude, but temperatures limit the growing season

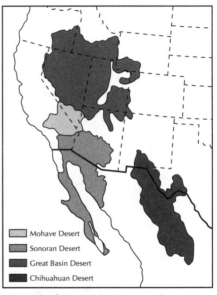

Mohave Desert
Sonoran Desert
Great Basin Desert
Chihuahuan Desert

The four North American deserts

to the summer. Vegetation is dominated by a few species of low, small-leafed shrubs; there are almost no trees or succulents and not many annuals. The *indicator plant* (the most common or conspicuous one used to identify an area) is big sagebrush (*Artemisia tridentata*), which often grows in nearly pure stands over huge vistas. (Such cold shrub/deserts in the "Old World" are called *steppes.*)

The Mohave Desert (plate 11) is characterized largely by its winter rainy season. Hard freezes are common but not as severe as in the Great Basin Desert. The perennial vegetation is composed mostly of low shrubs; annuals carpet the ground in wet years. There are many species of these two life forms, but few succulents and trees grow there. The only common tree species is the characteristic joshua tree (*Yucca brevifolia*), an *arborescent* (treelike) yucca that forms extensive woodlands above 3000 feet (900 m) elevation.

Though the Chihuahuan Desert (plate 12) is the southernmost, it lies at a fairly high elevation and is not protected by any barrier from arctic air masses, so hard winter freezes are common. Its vegetation consists of many species of low shrubs, leaf succulents, and small cacti. Trees are rare. Rainfall is predominantly in the summer, but in the northern end there is occasionally enough winter rain to support massive blooms of spring annuals. The Chihuahuan Desert is unexpectedly rich in species despite the winter cold.

THE SONORAN DESERT

The Sonoran Desert as currently defined covers approximately 100,000 square miles (260,000 sq. km) and includes much of the state of Sonora, Mexico, most of the southern half of Arizona, southeastern California, most of the Baja California peninsula, and the islands of the Gulf of California. Its southern third straddles 30° north latitude and is a horse latitude desert; the rest is rain shadow desert. It is lush in comparison to most other deserts. The visually dominant elements of the landscape are two life forms that distinguish the Sonoran Desert from the other North American deserts: legume trees and large columnar cacti. This desert also supports many other life forms, encompassing a rich spectrum of

some 2000 species of plants, over 550 species of vertebrates, and unknown thousands of invertebrate species.

The amount and seasonality of rainfall are defining characteristics of the Sonoran Desert. Much of the area has a bi-seasonal rainfall pattern, though even during the rainy seasons most days are sunny. From December to March frontal storms originating in the North Pacific occasionally bring widespread, gentle rain to the northwestern two-thirds. From July to mid-September, the summer monsoon brings surges of wet tropical air and localized deluges in the form of violent thunderstorms to the southeastern two-thirds. So distinct are the characters of the two types of rainfall that Sonoran residents have different Spanish terms for them—the winter rains are *equipatas* (derived from the Yaqui-Mayo word for rain, *quepa*), the summer rains are *las aguas* ("the waters" in Spanish).

The Sonoran Desert prominently differs from the other three North American deserts in having mild winters. Most of the area rarely experiences frost, and the biota are partly tropical in origin. Many of the perennial plants and animals are derived from ancestors in the tropical thornscrub to the south, their life cycles attuned to the brief summer rainy season. The winter rains, when ample, support great populations of annuals (which make up nearly half of the species of our plants). Some of the plants and animals are opportunistic, growing or reproducing after significant rainfall in any season (see the chapter "Deep History of the Sonoran Desert" for more details on its evolution).

SUBDIVISIONS OF THE SONORAN DESERT

Forrest Shreve was the first person to define the Sonoran Desert by dividing it into seven subdivisions, based on the diverse and distinctive vegetation found here. One of Shreve's subdivisions (the Foothills of Sonora) has since been reclassified as Foothills Thornscrub, a non-desert biome.

Lower Colorado River Valley

Named for its location surrounding the lower Colorado River in parts of four states, this is the largest, hottest, and driest subdivision. It challenges the Mohave Desert's Death Valley as the hottest and driest place in North America. Summer highs may exceed 120°F (49°C), with surface temperatures approaching 180°F (82°C). The intense solar radiation from cloudless skies on most days and the very low humidity suck the life-sustaining water from plants, water that cannot be replaced from the parched mineral soil. Annual rainfall in the driest sites averages less than three inches (76 mm), and some localities have gone thirty-six months with no rain. Even so, life exists here, abundantly in the rare wet years.

The terrain consists mostly of broad, flat valleys with widely-scattered, small mountain ranges of almost barren rock. There are also seas of loose sand and the spectacular Pinacate volcanic field (see plate 14). The valleys are dominated by low shrubs, primarily creosote bush (*L. tridentata*) and white bursage (*Ambrosia dumosa*). These are the two most drought-tolerant perennial plants in North America, but in the driest areas of this subdivision even they are restricted to drainageways. Trees grow only along the larger washes. The mountains support a wider variety of shrubs and cacti, but the density is still very sparse. Columnar cacti, one of the indicators of the Sonoran Desert, are rare (virtually absent in California) and are restricted to valley floors. Annual species comprise over half the flora, up to ninety percent at the driest sites; they are mostly winter growing species and appear in large numbers only in wet years (see plate 13).

This is the only part of the Sonoran Desert that extends into California, where most residents call it the Colorado Desert. North of a sagging line between the Coachella Valley (Palm Springs) and Needles, California, it merges almost imperceptibly into the lower Mohave Desert.

Arizona Upland

This northeastern subdivision is the highest and coldest part of the Sonoran Desert. Located in south-central Arizona and northern Sonora, the terrain contains numerous mountain ranges, and valleys narrower than those of the Lower Colorado River Valley subdivision. Trees are common on rocky slopes as well as drainageways, and saguaros grow on slopes above the cold valley floors. This community is also called the "saguaro-palo verde forest." It is the only subdivision that experiences frequent hard winter frosts, so many species of the lower elevation and more southerly subdivisions cannot survive here. Nevertheless it is a rich area. The small range that is the Desert Museum's

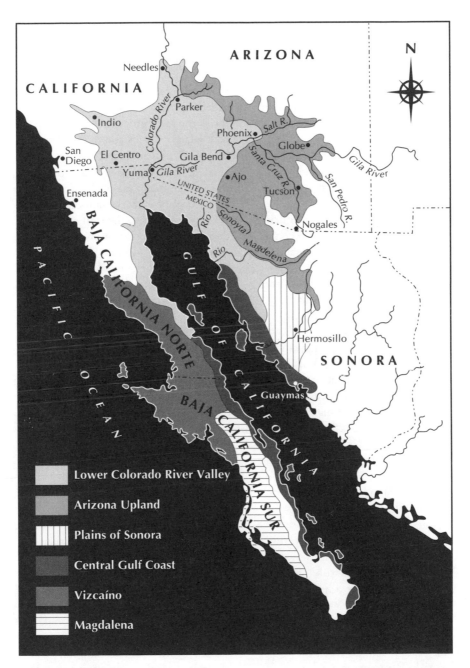

Subdivisions of the Sonoran Desert. The six subdivisions reflect the biological diversity of this large desert and the fact that it has been intensively studied. Each subdivision has a different climate, topography, and vegetation.

home, the Tucson Mountains, has about 630 taxonomically distinct kinds of plants. This richness is partly explained by the two equal rainy seasons which total twelve inches (305 mm) per year on average. The hilly terrain provides a multitude of microhabitats on north and south slopes and deep, shaded canyons. The proximity to chaparral, woodland, and grassland communities contributes still more species to the flora (see plate 15).

Biologists are increasingly concluding that the Arizona Upland's climate, vegetation density, and biodiversity resemble thornscrub more than desert. Don't be surprised if this subdivision is reclassified in the near future.

Tucson is the only major city located in Arizona Upland, although much of metro Phoenix's parks and land above 2000 feet (600 m) in elevation share its characteristics. Residents who have moved to this area from temperate climates often complain about the lack of seasons. Actually Arizona Upland has five seasons which, though more subtle than the traditional temperate four, are distinct if one learns what to look for.

The following description is for Tucson, but is fairly applicable to the rest of Arizona Upland and to the eastern one-half of the Lower Colorado River Valley subdivision as well. The seasons are a little later at higher latitudes and elevations, earlier at lower ones. The monsoon is later and more sporadic farther west; in some years it fails to reach the Colorado River.

SUMMER MONSOON or summer rainy season *(early July to mid-September).* In local native tradition, the year begins with the most dramatic weather event of the region—the often abrupt arrival of the summer rains (plate 16). A tropical air mass brings humidity and moderates the temperatures from June's extremes; frequent thunderstorms occur; this is the main growing season for many of the larger shrubs and trees. (*Monsoon* is derived from an Arabic word for "season," and was applied to a wind that changes directions seasonally. Be aware that it does not refer to rain or storms per se, but rather to the shift of wind direction which brings moist air that can generate storms—in our case, a southerly wind in July. The word is often misused, even by some weather reporters.)

A sixth season, late summer, lasting from mid August through September, is sometimes added; this is a hot and dry period after the monsoon ends— nonexistent in some years.

AUTUMN *(October & November).* Warm temperatures; low humidity; little rain; few species in flower, but the growing season for winter annuals begins if there is enough rain. Late summer and autumn occasionally receive heavy rains from the remains of Pacific hurricanes (tropical storms).

WINTER *(December & January; sometimes February).* Mostly sunny, mild days, with intermittent storms that bring wind, rain, and cool-to-cold temperatures; February often warm and dry, more spring-like (see plate 17).

SPRING *(early to late February through April).* Mild temperatures; little rain; often windy; one of two flowering seasons; winter annuals may start blooming in February in warm, wet years (see plate 18).

FORESUMMER *(May & June).* High temperatures; very low humidity; no rain in most years; May is very warm and often windy; June is hot and usually calm. There is little biological activity except for the flowering and fruiting of saguaro and desert ironwood. Most plants and many animals are dormant until the rains arrive (see plate 19).

Plains of Sonora

This small region of central Sonora is a series of very broad valleys between widely separated ranges. It supports denser vegetation than does Arizona Upland because there is more rain (with summer rain dominant) and the soils are generally deeper and finer. It contains most of the same species as Arizona Upland, plus some tropical elements, because frost is less frequent and less severe. There are abundant legume trees, especially mesquite, and relatively few columnar cacti. The few hills in this region support islands of thornscrub. Most of this subdivision has been converted to agriculture in the last few decades.

If Arizona Upland is reclassified as thornscrub, the wetter Plains of Sonora subdivision would also have to be reclassified from desert to thornscrub.

Central Gulf Coast

The Central Gulf Coast occupies a strip along both sides of the Gulf of California. Extreme aridity dictates the distinctive appearance of this subdivision. It straddles the horse latitude belt, and desert vegetation grows right to the seashore. Small shrubs are nearly absent; their shallow root systems and lack of water storage cannot sustain them through the droughts which commonly last for several years. Dominating the vegetation are large stem-succulents, particularly the massive cardón (*Pachycereus pringlei*, a giant relative of the saguaro), and trees such as palo verde, tree ocotillo (*Fouquieria diguetii* and *F. macdougalii*), ironwood, elephant tree (*Bursera* spp.), and limberbush (*Jatropha* spp.); the trees are leafless most of the time. The average annual rainfall of less than five inches (125 mm) occurs mostly in summer, though not dependably enough to call it a rainy season. A year with no rain is not rare (see plate 20).

Vizcaino

The Vizcaino subdivision is on the Pacific side of the Baja California peninsula. Though rainfall is very low, cool, humid sea breezes with frequent fog ameliorate the aridity. Winter rain predominates and averages less than five inches (125 mm). This subdivision contains some of the most bizarre plants and eerily beautiful landscapes in the world. There are fields of huge, sculpted white granite boulders or black lava cliffs that shelter botanical apparitions such as boojums (*Fouquieria columnaris*), twisted and swollen Baja elephant trees (*Pachycormus discolor*), sixty-foot (18 m) tall cardones, strangler figs (*Ficus petiolaris* ssp. *palmeri*) that grow on rocks, and blue palm trees (*Brahea armata*). In stark contrast, the coastal Vizcaino Plain is a flat, cool, fog desert of shrubs barely a foot tall, with occasional mass blooms of annual species (see plates 21 and 26).

Magdalena

Located in coastal Baja California south of the Vizcaino, Magdalena is similar in appearance to the Vizcaino but the species are somewhat different. Most of its meager rainfall comes in summer and the aridity is modified by Pacific breezes. The bleak coastal Magdalena Plain's only conspicuous endemic plant is the weird creeping devil cactus (*Stenocereus eruca*), but inland the rocky slopes are rich and dense with trees, succulent shrubs, and cacti (see plate 22).

Foothills of Sonora

This was Shreve's seventh subdivision of the Sonoran Desert. It has since been reclassified as foothills thornscrub community and is no longer considered part of the desert biome because of its greater rainfall, taller trees and cacti, and denser vegetation.

Shreve's delineation of the Sonoran Desert's boundary and subdivisions are the most widely accepted. There are at least five other major attempts to define this area with dramatically differing boundaries. One version excludes most of Baja California from the Sonoran Desert. Another includes the Mohave as part of the Sonoran Desert. (Indeed, it is difficult to distinguish the two along the currently accepted boundary.) These differences of interpretation reflect the great diversity of geography and biota found here.

The discussion is not yet over. Time will determine whether Arizona Upland and the Plains of Sonora will remain parts of the Sonoran Desert or be reclassified as thornscrub. Whatever they are called, all of these regions are fascinating places for nature lovers, whether they are classified as scientists or tourists.

Additional Readings

Brown, David E., ed. *Biotic Communities: Southwestern United States and Northwestern Mexico.* Salt Lake City: University of Utah Press, 1994.

Dunbier, Roger. *The Sonoran Desert: Its Geography, Economy, and People.* Tucson: University of Arizona Press, 1968.

Lowe, Charles H. *Arizona's Natural Environment: Landscapes and Habitats.* University of Arizona Press, 1964.

Shreve, Forrst. *Vegetation of the Sonoran Desert.* Washington, D.C.: Carnegie Institution of Washington, 1951.

Shreve, Forrest and Ira L Wiggins. *Vegetation and Flora of the Sonoran Desert.* Stanford: Stanford University Press, 1964.

Walter, Heinrich. *Ecology of Tropical and Subtropical Vegetation.* New York: Van Nostrand Reinhold Co., 1971.

Sonoran Desert
Natural Events Calendar

Roseann Beggy Hanson & Jonathan Hanson

Winter

December, January, early February

➤ December

Average high: 65°F (18.3°C)
Average low: 39°F (3.9°C)
Relative humidity at 5 am: 61%
Relative humidity at 5 pm: 34%
Normal rainfall: .94″ (23.9mm)

December is a mild month, with mostly cool but sunny days and only a few nights dropping below freezing, and possibly a few days of light winter rains.

Flora

Fruits of desert mistletoe, Christmas cactus, and netleaf hackberry trees are ripening, providing food for many birds and mammals. If winter rains begin, many shrubs such as brittlebush, creosote, and ocotillos will sprout bright new leaves, but many trees, such as mesquites, palo verdes, and sycamores, will drop their leaves as the temperatures dip below freezing.

Fauna

Birds in the Mimic-Thrush family, such as mockingbirds and Curve-billed Thrashers, begin to establish mating territories. They "map" territories by singing their famous copycat songs from the tops of trees, poles, or fences. Cactus Wrens begin to build their breeding nests; they also build separate roosting nests—some pairs will build three or more nests. Anna's Hummingbirds may also breed this month. The males can be heard singing their squeaky songs around feeders.

Nature Watching Tips for December

Tohono Chul Park, in northwest Tucson at the corner of Oracle and Ina roads, is a great place to watch Cactus Wrens building nests in cholla cacti. For information on the park, call (520) 575-8468.

➤ January

Average high: 64°F (17.8°C)
Average low: 38°F (3.3°C)
Relative humidity at 5 am: 62%
Relative humidity at 5 pm: 32%
Normal rainfall: .86″ (21.8mm)
Normal snowfall: .31″ (7.9mm)

In January, deserts will experience freezing temperatures half a dozen times, while up in the mountains it will freeze nearly every night and a foot of snow may fall. Desert days are mostly clear and pleasant, ranging from cool to warm; sometimes cold rains will arrive from the northwest.

Flora

At lower elevations, Frémont cottonwoods begin to sprout new leaves and open blossoms. If rain has been falling, many annuals and grasses will be sprouting new green growth.

Fauna

Mockingbirds, Curve-billed Thrashers, Cactus Wrens, and packrats (white-throated woodrats) begin their breeding seasons. Male Phainopeplas—glossy black crested birds with white wing patches—perch conspicuously in palo verde or mesquite trees that are well-endowed with desert mistletoe berries, an important food source. They perform fluttery flight-displays to attract females. Mule deer breeding season, called the "rut," is in full swing. Males engage in violent-looking but seldom injurious sparring with their antlers. Mountain lions feed well this month in the mountain foothills, where distracted male deer make easier prey.

Nature Watching Tips for January

Dozens of Phainopeplas usually make a great show in tall palo verde and mesquite trees along Highway 286, about ten miles (16 km) south of Three Points (Robles Junction).

➤ February

Average high: 67.5°F (19.7°C)
Average low: 40.2°F (4.6°C)
Relative humidity at 5 am: 59%
Relative humidity at 5 pm: 27%
Normal rainfall: .63″ (16.0mm)

Winter is loosening its grip on the deserts; only a couple of nights may dip below freezing. Warm days tempt us later in the month with thoughts of spring—days in the 80s are possible—but cold snaps are still more probable.

Flora

Blooming shrubs may include chuparosa, samota, and desert mock-orange. Later in the month, if the fall and winter rains were generous, desert wildflowers may begin their show with such species as Mexican gold poppies, lupines, and owl-clover. In mountain canyons, deciduous trees such as alders and walnuts may bloom before they send out new leaves.

Fauna

Costa's Hummingbirds join Anna's Hummingbirds in establishing breeding territories around backyard feeders or blooming chuparosa shrubs. Male Costa's Hummingbirds display for females and mark their territories with a distinctive "zing" call. Rust-colored hummers that show up at feeders around mid-month are Rufous Hummingbirds, which are migrating through the Southwest. Gila Woodpeckers hammer away at tree trunks or even metal pipes around buildings; they are marking their territories using sound. In the cool evenings or pre-dawn mornings, Great Horned Owls can be heard calling softly, usually in duos, as they begin their breeding season. Pipevine swallowtail butterflies are common fliers this month.

Nature Watching Tips for February

Early wildflowers begin to bloom in Organ Pipe Cactus National Monument; especially beautiful is the elusive ajo lily. For information call (520) 387-6849. For early blooms around Tucson, explore the south-facing rocky slopes of the Tucson Mountains, in Tucson Mountain Park, or Saguaro National Park, especially near the Desert Museum. Popular February and March wildflower viewing areas include Picacho Peak State Park, Pinal Pioneer Parkway between Oracle Junction and Florence, Catalina State Park northwest of Tucson, and the Arivaca-Ruby loop road off I-19 (note: Ruby Road is rough but usually passable for low-clearance cars). Phoenix area hot spots include Bartlett Lake, Echo Canyon, the Desert Botanical Garden, Usury Mountain Park, and South Mountain Park.

Spring

Late February, March, April

➢ March

Average high: 71.9°F (22.2°C)
Average low: 43.8°F (6.6°C)
Relative humidity at 5 am: 53%
Relative humidity at 5 pm: 23%
Normal rainfall: .71″ (18.0mm)

Spring begins in earnest in the desert, with warm and sunny days and cool nights; days are warming in the high country—enough to begin melting snow and filling creeks; we hear that magical sound of running water.

Flora

March is wildflower month. Look for dozens of wildflower species, including globe mallows, penstemons, evening primroses, desert marigolds, blue dicks, gilias, bladderpods, dock, chia, desert hyacinths, and many more. Many shrubs bloom as well, including desert lavender, hop bush, brittlebush, and Mormon tea.

Fauna

Many animals are breeding or preparing to breed. Elf Owls arrive from wintering grounds in Mexico to breed in saguaro-mesquite desert; males arrive first and try to win females with their distinctive barking call as they perch in cavities in saguaro cacti or other trees. Burrowing Owls and Barn Owls, which are also found in desert or even urban areas, also breed this month. Migratory songbirds begin to arrive, either to breed or to rest on their way to northern breeding grounds; riparian areas are especially good places to see them. Desert tortoises and desert box turtles emerge from burrows and begin to mate. Turkey vultures move back to southern Arizona for the summer.

Nature Watching Tips for March

Spring bird migration is an exciting time to head out to riparian corridors, where many songbirds rest en route to northern breeding grounds or stop to breed here in southern Arizona. Good bets include the Anza Trail along the Santa Cruz River near Tubac, Sabino Canyon in northeast Tucson, Ciénega Creek east of Tucson, Hassayampa River Preserve northwest of Phoenix, Boyce Thompson Southwestern Arboretum east of Apache Junction, and the San Pedro National Riparian Conservation Area near Sierra Vista. The intersection of Baseline and Salome roads, approximately fifty miles (80 km) west of Phoenix, is a wonderful place to look for Bendire's, Sage, Crissal and Le Conte's Thrashers through the end of March. Also, contact the Tucson Audubon Society at (520) 629-0510 and the Audubon Society of Maracopa County at (602) 631-9761 about their birding tours.

➤ April

Average high: 80.5°F (26.9°C)
Average low: 50.1°F (10.1°C)
Relative humidity at 5 am: 42%
Relative humidity at 5 pm: 16%
Normal rainfall: .31" (7.9mm)

Dryness begins to settle on the desert, with just a few days of possible sprinkles. Spring weather arrives in the mountains, with days in the 60s (>15°C) and only half the nights dropping below freezing.

Flora

The desert "bean" trees (legumes such as blue palo verdes, catclaw acacias, and mesquites) begin to open yellow to creamy blooms, and the cacti begin to bloom as well, with a few of the prickly pears, chollas, and hedgehogs starting off. By the end of the month, some saguaro cacti will open their big, white flowers. Brittlebush may still bloom; look for iron-cross blister beetles— they are black, yellow and red with black cross patterns on their backs— feeding on the brittlebush blossoms.

Fauna

Bird migration continues. Summer's hawks, including Swainson's, Zone-tailed and Black Hawks, begin to arrive and get busy finding mates for the summer breeding season. In mountain canyons around Tucson many hummingbirds arrive and breed, including Broad-billed, Black-chinned and Magnificent. White-winged Doves also return and fill the late-spring air with the signature summer call, "Who-cooks-for-you?" As the days lengthen and warm up, reptiles become more visible—time to watch for rattlesnakes (although remember they can be out any month of the year). Desert iguanas, lesser earless lizards and western whiptail lizards begin breeding. Bobcats, coyotes and foxes are having litters. And butterfly activity picks up; look for great blue hairstreak, hackberry, skipper, blue, and queen butterflies.

Nature Watching Tips for April

A walk up Romero Canyon Trail in Catalina State Park may yield some late-spring wildflowers and good butterfly activity in late morning and throughout warm afternoons. Before the trail climbs the western slopes of the Santa Catalina Mountains, it crosses through a mature mesquite *bosque* (Spanish for forest) where you can look for songbirds, hummingbirds, and White-winged Doves. For information on the park, call (520) 628-5798. April is also a good time for people in the Phoenix area to look for warblers, flycatchers and Bullock's Orioles in the Forest Service picnic areas on Bush Highway along the Salt River.

Foresummer Drought

May, June

➤ May

Average high: 88.8°F (31.6°C)
Average low: 57.4°F (14.1°C)
Relative humidity at 5 am: 34%
Relative humidity at 5 pm: 13%
Normal rainfall: .15″ (3.8mm)

May will likely see the first day over 100°F (38°F) in the desert; most days will be clear, dry and hot. Many animals, humans included, begin to retreat to the mountains, where balmy days and cold but not freezing nights beckon.

Flora

Many species of Cactaceae bloom nocturnally, including saguaros, senitas, organ pipes, and queens-of-the-night, also known as night-blooming cereus. Delicate lavender blossoms open on desert ironwood and smoke trees. Desert spoon and soaptree yuccas put up tall, woody bloomstalks with white flowerettes; desert spoon is dioecious, with male or female flowers. Many red, trumpet-shaped flowers are blooming in mountain canyons as hummingbirds become more numerous and continue breeding; among the most spectacular blossoms are those of the coral bean.

Fauna

Female nectar-feeding bats, many of which are pregnant, migrate from

Mexico into desert areas where nocturnally blooming plants are flowering. The two species are Mexican long-tongued and lesser long-nosed, the latter of which are endangered; they give birth in colonial maternity caves. Gila monster eggs, laid ten months ago, begin to hatch; the young lizards are perfectly formed miniature versions of their venomous parents and immediately fend for themselves. In mountain canyons, red-spotted toads are mating, filling the nights with their loud trills.

Nature Watching Tips for May

Balmy evenings around full moon are a great time to watch bats and moths visit cactus flowers. Don't forget to watch for rattlesnakes and scorpions where you step or sit. Visit Saguaro National Park's Tucson Mountains or Rincon Mountains units, which are open to foot traffic in the evenings. Call (520) 733-5158 (Tucson Mtns. unit) or (520) 733-5153 (Rincon Mtns. unit). For people in the Phoenix area, enjoy an evening at the Desert Botanical Garden. Call (480) 481-8134 for a rundown of activities.

➤ June

Average high: 98.5°F (36.9°C)
Average low: 67.3°F (19.6°C)
Relative humidity at 5 am: 32%
Relative humidity at 5 pm: 13%
Normal rainfall: .24″ (6.1mm)

Famously hot and dry characterizes June in the desert, with lots of days over 100°F (38°C)—and a few as high as 110°F (43°C) or more—and often not a drop of moisture. The high country will remain relatively cool, in the 80s even when the mercury climbs over the century mark down below.

Flora

Saguaro cactus fruits ripen, split open, and fall to the ground; many birds, insects, and mammals feed on them. Bean pods on mesquites, palo verdes, and catclaw acacias are ripening , as are jojoba seeds. If winter or early spring rains were plentiful, sacred datura may bloom. Organ pipe cactus continue to open their pale lavender blooms.

Fauna

Many snakes bear live young or lay eggs this month, including gopher snakes, common kingsnakes, Sonoran whipsnakes, and western diamondback and tiger rattlesnakes. Days are filled with the buzzing of male cicadas, also known as "cactus dodgers," as the insects try to attract mates. Lesser Nighthawks fill the warm nights with their unique trilling calls. This is a good month to see hawks that breed in riparian areas with tall cottonwoods or sycamore trees; look for Gray, Black and Zone-tailed Hawks, and Mississippi Kites.

Nature Watching Tips for June

Traditionally, many Tohono O'odham, the Desert People, make saguaro fruit-gathering excursions beginning this month. The Desert Museum hosts several saguaro fruit-harvesting workshops, including field trips. For information call (520) 883-3025. During the summer months the Desert Museum is open late on Saturday nights and offers some enjoyable evening programs. For details call (520) 883-1380.

Summer Monsoon

July, August, Early September

➤ July

Average high: 98.4°F (36.9°C)
Average low: 73.6°F (23.1°C)
Relative humidity at 5 am: 57%
Relative humidity at 5 pm: 28%
Normal rainfall: 2.54" (64.5mm)

Summer rains arrive, bringing welcome relief from the hot and dry days of May and June. Locally we call them "monsoons," which is a slight misnomer since the term refers to a seasonal shift in winds, bringing wet and dry periods to a region. Creeks run again, and a second springtime begins with the abundant rain.

Flora

Summer rains produce a second round of wildflowers, including summer poppies, devil's claw, and morning glories, as well as blooms of woody plants, and also some of the agaves. Riparian canyons, especially those close to the Mexican border, are lush, hot, humid and full of life.

Fauna

Summer rains trigger a second breeding season for many animals, from insects to the birds and mammals that feed on the insects. Many butterflies emerge or arrive with the rains; look for monarchs, sulphurs, queens, fritillaries, and two-tailed swallowtails. Giant four-inch-long palo verde beetles, which as nymphs feed on the roots of their namesake host plants, emerge to mate and lay eggs. Amphibians such as the spadefoots, Sonoran green toads, and red-spotted toads begin their short and frenzied reproductive cycles in the shallow rain puddles throughout the region. The nectar-feeding bats and their new young begin to move south, following the blooms of agaves. Look for swirling swarms of winged leaf-cutter and harvester ants the morning after heavy rain; these are new queens and males which will mate and establish new colonies.

Nature Watching Tips for July

Evening drives or walks along back-country roads will reveal an orgy of amphibian and reptile activity in the hot and humid monsoon season. The roads throughout Avra Valley west of Tucson are favorites among herpetophiles.

➤ August

Average high: 96.2°F (35.7°C)
Average low: 72°F (22.2°C)
Relative humidity at 5 am: 65%
Relative humidity at 5 pm: 33%
Normal rainfall: 2.03" (51.6mm)

Summer rains continue throughout the month, with dramatic lightning and thunderstorms. Temperatures remain high, although they are a little less hot than in July. Mornings tend to be clear, while the storms build and break sometime after noon.

Flora

Prickly pear cactus fruit begins to ripen. Many birds, mammals and insects will feed on them. Blooms continue on plants such as barrel cacti, asters, four o'clocks, buffalo gourds and ground cherries. The little orange fruits of desert hackberry shrubs ripen this month; look for Empress Leilia hackberry butterflies in the foliage.

Fauna

Young regal horned lizards continue to hatch this month. Late-summer through fall bird migrations get underway, with species that spent the summer in the north arriving here for the winter or passing through on their way farther south. Hummingbirds such as Rufous and Allen's begin to migrate through, feeding on summer-season blossoms. In some years, snout butterflies become so numerous that they can clog automobile radiators on rural roads.

Nature Watching Tips for August

There is a dramatic increase in hummingbird activity this month, as many species begin "tanking up" fat reserves for migration, and other species arrive here on their way south; check out established feeders at mid-elevation canyons such as Madera Canyon in the Santa Rita Mountains near Green Valley. Call Coronado National Forest at (520) 281-2296.

➤ September

Average high: 93.5°F (34.2°C)
Average low: 67.3°F (19.6°C)
Relative humidity at 5 am: 55%
Relative humidity at 5 pm: 27%
Normal rainfall: 1.34″ (34.0mm)

Rains continue in the first few days of September, but by the end of the month it will seem like a repeat of May: dry and hot. Fall may begin to creep into mountain canyons and the higher elevations, as nighttime temperatures drop closer and closer to freezing.

Flora

Blooming continues in the deserts and desert grasslands; asters and sunflowers are especially lush. Also, look for the golden yellow blooms of snake-weed, turpentinebush, goldeneyes, and telegraph plants. Sticker and burr season begins as plants put out their seeds in ways that ensure good transportation away from the parents; canyon ragweed, with its almond-sized, fishhook-spined fruits, are a particularly noticeable fruiting plant for hikers. If rains were sufficient, a good crop of prickly pear cactus fruit should be very ripe. Bright pink piles of coyote scat on backroads and trails are dead giveaways to the omnivorous nature of those desert canines. Seep-willow shrubs bloom in washes and rocky canyon bottoms.

Fauna

Fall bird migrations reach their peak; ponds, such as golf course hazards or wastewater treatment plants, are excellent stopovers for waterfowl, and some will stay for the winter. Turkey Vultures, Western Kingbirds, and many species of hawks congregate as they prepare to move south. Desert bighorn sheep are breeding. Common butterflies include gray hairstreaks, funereal duskwings, and painted ladies.

Nature Watching Tips for September

If the summer rains were good and the corresponding blooms of milkweed are abundant, large numbers of beautiful monarch butterflies may be seen in low- to mid-elevation canyons. One good place for butterfly watching is Brown Canyon, in the Baboquivari Mountains. Call the U.S. Fish and Wildlife Service's Buenos Aires National Wildlife Refuge at (520) 823-4251. White Tanks on the northwest side of Phoenix is another very good place to do some butterfly watching.

Fall

Late September, October, November

➢ **October**

> Average high: 84.1°F (28.9°C)
> Average low: 56.5°F (13.6°C)
> Relative humidity at 5 am: 53%
> Relative humidity at 5 pm: 25%
> Normal rainfall: .79″ (20.1mm)

Summer hangs on, with warm-to-hot days but cooling nights. It will freeze and most likely snow in the higher elevations. Some years might see unseasonable heavy rains this month, although this is uncommon.

Flora

Allergy sufferers bemoan the blooming of desert broom, a shrub that colonizes disturbed areas. However, broom flowers are a favorite of hundreds of butterflies, bees, wasps, and beetles. Many plants are fruiting, including barrel cacti, soapberry trees, desert hackberries, and wolfberries. Fall colors splash through canyons.

Fauna

Resident desert birds, or those that spend their winters here, are gorging on the many plant fruits. Most snakes head to winter burrows this month. One exception is the rosy boa, which lives in the warmer western deserts and will give birth this month and next. (Although most snakes remain fairly inactive in the colder months, keep in mind that they can emerge any time of year if enough warm days pass to sufficiently rouse them.) Wintering hawks arrive; common raptors include Northern Harriers, Rough-legged and Ferruginous Hawks, Kestrels, Merlins, and Prairie Falcons. Turkey Vultures are mostly gone, but loud groups of Common Ravens replace them on carrion patrol.

Nature Watching Tips for October

Butterfly expert Robert Michael Pyle once remarked that he's never seen so many butterfly species on one plant at one time than when observing a flowering desert broom (he saw more than forty-five species). Settle down near one with a pair of binoculars and enjoy great purple hairstreaks, snouts, many different sulphurs, swallowtails, checkerspots and more, as well as bees, wasps, beetles, and flies. Look for desert broom in washes or in disturbed areas, such as around new housing developments.

➤ November

> Average high: 72.5°F (22.5°C)
> Average low: 45.2°F (7.2°C)
> Relative humidity at 5 am: 54%
> Relative humidity at 5 pm: 28%
> Normal rainfall: .59″ (15.0mm)

Balmy weather finally begins to settle on the deserts, while winter grips the mountains. Expect mostly dry weather, although well over half a foot of snow may fall up high. Storms from the northwest may bring cold, even freezing, temperatures to the deserts as well.

Flora

Fall colors snake across low- and mid-elevation canyons where sycamores, cottonwoods, ashes, and walnuts begin to give up their leaves. Desert broom seeds take to the air in cloudy puffs that look like snow. Desert mistletoe berries begin to form. If many branch ends of mesquite trees begin to brown up and die, chances are mesquite girdler beetles were at work. The females bore a trough around, or "girdle," a twig and then lay their eggs in the soon-to-die tips.

Fauna

Anna's and Costa's hummingbirds are the most common hummingbirds at feeders now; both will breed in the winter. Some Costa's Hummingbirds leave southern Arizona for the winter, returning in early spring to breed before the foresummer drought sets in. Although reptiles are mostly inactive, larger desert mammals such as bobcats, coyotes, badgers, and gray and kit foxes will remain active throughout the winter. Male desert mule deer rub the velvet off their antlers, which grew throughout the summer. Soon they will begin sparring with each other as they prepare for breeding competitions in the winter months.

Nature Watching Tips for November

Fall in the desert grasslands is subtle but beautiful—golden grasses, flocks of twittering winter sparrows, and raptors floating high above make for special days in the well-preserved grasslands in the Altar Valley southwest of Tucson (Highway 286 runs through its heart) and the San Rafael Valley (southeast of Sonoita).

Weather values are from Tucson International Airport. Source: *Arizona Climate: The First Hundred Years* (Sellers, Hill, and Sanderson-Rae, 1985)

Nature Watching in the Sonoran Desert Region

Roseann Beggy Hanson

Native desert people, such as the Tohono O'odham and Pima Bajo, observed their surroundings closely. Just as you might watch for price reductions and new product arrivals at supermarkets, they watched the movements of animals and the fruiting of plants, the abundance or paucity of rain, the rise and fall of temperatures. They knew that in the mountain canyons when the coatis formed large troops and ventured to lower elevations, fall was arriving with cooler temperatures and the promise of acorns, hackberries, and wolfberries. Or that when a badger goes digging for squirrels or kangaroo rats, a clever coyote often hovers nearby, waiting for the hunted rodent to run out another hole.

Observing animals in their natural habitats is as rewarding as it can be challenging. Planning, patience, knowledge of habitats and animals, and a little luck are what you need to successfully "hunt" wildlife with binoculars or camera. This chapter offers a little help, with suggested hot spots for observing wildlife and flora of the Sonoran Desert region, and some how-to tips as well.

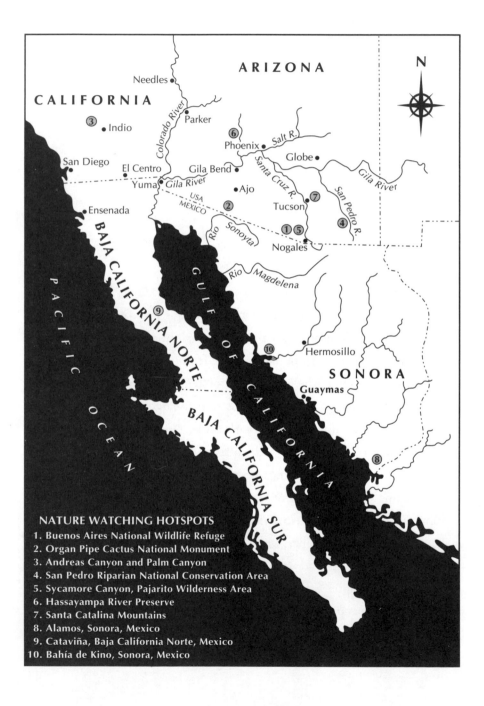

NATURE WATCHING HOTSPOTS
1. Buenos Aires National Wildlife Refuge
2. Organ Pipe Cactus National Monument
3. Andreas Canyon and Palm Canyon
4. San Pedro Riparian National Conservation Area
5. Sycamore Canyon, Pajarito Wilderness Area
6. Hassayampa River Preserve
7. Santa Catalina Mountains
8. Alamos, Sonora, Mexico
9. Cataviña, Baja California Norte, Mexico
10. Bahía de Kino, Sonora, Mexico

Ten Sonoran Desert
Nature Watching Hot Spots

1 Buenos Aires National Wildlife Refuge

Desert grassland with riparian corridors, ciénegas and Madrean oak woodland

This diverse refuge, under management of the U.S. Fish and Wildlife Service, is fifty miles (80 km) southwest of Tucson near the Mexican border and includes 116,000 acres of desert grassland, riparian woodlands, ciénegas, and oak woodlands (3200 to 4600 feet (975 to1402 m). Wildlife-watching is great year-round, featuring over 317 species of birds and other animals including unusual wildlife such as Masked Bobwhite, pronghorn, Gray Hawks, Underwood's mastiff bats; even jaguars have been sighted within the refuge. In spring and fall, the migratory bird show is spectacular, featuring hawks, hummingbirds (12 species recorded), warblers, and many other songbirds. The heart of the vast refuge is a broad grassland valley flanked by riparian areas at Arivaca Creek and Ciénega on the east and, to the west, sycamore-lined Brown Canyon below Baboquivari Peak. The entire refuge is open to the public; call refuge headquarters for information about Brown Canyon, which is open by guided tour only. Over 100 miles (160 km) of rugged dirt roads are open; there are fifteen miles (24 km) of walking and hiking trails. (520) 823-4251.

2 Organ Pipe Cactus National Monument

Arizona Upland and Lower Colorado River Valley subdivisions of the Sonoran Desert

This 516 square mile (1336 km²) National Park Service preserve 140 miles (225 km) southwest of Tucson, along the Mexico-Arizona border, features two distinct subdivisions of the Sonoran Desert amid jagged volcanic mountain ranges (approximately 1000 to 4800 feet, 305 to 1463 m). Most of the habitat is classic cactus-palo verde, Arizona Upland desertscrub, with lusher ribbons of desert ironwood and mesquite trees along the many large, dry washes; along the southern and western edges of the monument are creosote bush and bursage-dominated Lower Colorado River Valley communities, and in a small hollow called Senita Basin the landscape is dotted with senita cacti, ashy limberbushes, and elephant trees. The namesake organpipe cacti, which offer lovely white blooms in May and June, are near their northern limit in the monument (see plate 23). Fall through early spring are the best seasons for temperate weather; in early spring Rufous Hummingbirds migrate through following the scarlet ocotillo blossoms and the stunning white ajo lilies may bloom. Late-spring through summer offer the best cactus blooms and wildlife viewing opportunities, but also hot weather. There are five miles

(8 km) of maintained hiking trails, many miles of backcountry trails and two scenic, graded, unpaved drives, 21 and 53 miles (34 to 85 km) long. (520) 387-6849.

3   Andreas Canyon and Palm Canyon, California

Desert fan palm oasis

The desert fan palm oasis is a unique habitat to the Sononran Desert region. Almost all of the known 158 oases occur in the Lower Colorado River Valley subdivision. Desert fan palms are the most massive palms in North America, especially when their skirts of dead leaves or "petticoats" are intact. The palms typically live in rocky canyons with permanent springs, creating shady, cool oases attractive to people and wildlife. In a good year a single tree can produce 350,000 fruits (350 pounds)—food for early native people and animals. In California most of the desert fan populations on the western edge of the Sonoran Desert are similar genetically and were likely established from seeds in coyote scat. Cahuilla Indians regularly set the palm groves on fire to clean out the underbrush, reduce insects, and increase fruit production. As in all riparian habitats, water and nutrients are concentrated, increasing productivity and diversity of animals and plants. Often desert animals are seen within a few feet of species more typical of other environments such as riparian, woodland, or chaparral habitats. Andreas and Palm canyons are well-developed, mature desert fan palm oases that are easily visited from Palm Springs. (800) 790-3398.

4   San Pedro Riparian National Conservation Area

Riparian corridor

The BLM-administered San Pedro Riparian National Conservation Area is one of the most important preserves in North America, protecting thirty-six miles (58 km) of this rare undammed river, a vital corridor for millions of neotropical migratory birds. Over 380 species of birds and eighty-two species of mammals have been recorded within the cottonwood, willow and thornscrub corridor. Recently, reintroduction of once-native beavers has been proposed. The San Pedro RNCA is about ten miles (16 km) east of Sierra Vista, Arizona, and is easily accessed at the historic San Pedro House, a hub for several riverside trails. Birdwatching opportunities are excellent: Green Kingfishers, Yellow-billed Cuckoos, Tropical Kingbirds and Gray Hawks are favorites in spring and summer. (520) 458-3559.

5   Sycamore Canyon, Pajarita Wilderness Area

Riparian corridor in desertscrub–Madrean evergreen woodland interface
(See plate 24)

The north-south border canyons of southern Arizona are natural corridors through which tropical and montane (Sierra Madrean) Mexican species funnel into the United States. The results of the invasion are typical in canyons such as Sycamore—a wealth of locally rare species such as Elegant Trogons, Five-striped Sparrows, brown vine snakes, ball moss

Hummingbird Bats

In the Chiricahua and Huachuca mountains of southern Arizona, hummingbird enthusiasts have discovered nocturnal interlopers among their feathered friends. Nectar-feeding bats that feed on blooming agave and columnar cacti are also adept at locating bird feeders filled with sugar water. In many areas, flocks of bats arrive to drain feeders just as the birdwatchers are going to bed. Upon waking, avian naturalists were initially amazed at their drained feeders, looking for leaks or other nocturnal culprits such as ringtails or raccoons. But diligent overnight observations revealed the real story.

Nectar bats echolocate to maneuver around obstacles and to find their way on pitch dark nights, but their skills in this regard are not as finely tuned as those species which pursue flying prey. Most nectar bats rely heavily on their excellent sense of smell and upon vision to locate nectar-producing flowers. The sweet aroma of sugar-water-filled feeders proved an easy and reliable target for hungry bats.

Bats usually arrive at feeders well past sundown and feed in flocks, hitting a single feeder several times up until midnight. Two to half-a-dozen bats may circle a feeder in a well-choreographed "holding pattern," while one at a time they peel off to dip in for a drink. Watching bats at feeders is not as easy as watching hummingbirds. The bats are much quicker and do not often hover, but their loud dove-like wing beats are unmistakable and often announce their presence long before their dusky shadows flit rapidly in and out of sight.

In many areas of southern Arizona, nectar bats are relying heavily on hummingbird feeders. Scientists are not sure of the behavioral and health effects of this artificial nectar source on the bats. Like most animals—humans included—the bats readily take advantage of an easy, reliable food source. But after gorging on "candy" for a while , it is quite possible that they seek out the more nutritionally complete flower nectar and pollen that make up essential nutrients in their diets. Meanwhile, the presence of bats at hummingbird feeders provides unprecedented encounters with one of America's most rarely seen and most interesting bat species.

— Janet Tyburec

(the only epiphytic flowering plant in Arizona, in the Bromeliad or "pineapple" family), a rock fern called *Asplenium exiguum*, and the only wild Arizona population of Sonora chub (see page 521), a small native fish. Coyotes, Mexican opossums, coatis, bobcats, and mountain lions are sometimes seen. The creek, the headwaters of Sonora's Río de la Concepción, flows in Sycamore Canyon year-round, though there are dry stretches during certain times of the year; the six-mile-long trail, which ends at the border fence, crosses the creek many times. Coronado National Forest is the land management agency for this canyon. (520) 281-2296.

6 Hassayampa River Preserve, Arizona Nature Conservancy

Arizona Upland, riparian corridor with cottonwood-willow forest and lake communities

This exquisite Nature Conservancy property, preserving some five miles (8 km) of one of central Arizona's last perennial streams and a lake habitat, lies at the northern reaches of the Sonoran Desert. The well-preserved cottonwood-willow gallery (Frémont cottonwood and Goodding willow) is one of the most threatened forest types in North America. Along the river live such species as Gilbert's skink, Zone-tailed Hawk, Mississippi Kite, Yellow-billed Cuckoo, Willow Flycatcher, mule deer, javelina, and ringtail; over 230 species of birds have been recorded at the preserve. At Palm Lake, a four-acre pond-and-marsh habitat, live five species of rare desert fish: bonytail chub, Colorado River squawfish, razorback sucker, Gila topminnow, and desert pupfish. The Hassayampa River Preserve, approximately 2000 feet (610 m) elevation, is about 60 miles (97 km) northwest of Phoenix and offers excellent nature-watching opportunities year-round. (520) 684-2772.

7 Santa Catalina Mountains

Arizona Upland through mixed-conifer forest

The twenty-six-mile drive up the Santa Catalina Mountains, which flank Tucson's north side, is like driving from Mexico to Canada in just a few hours. Beginning at Milepost 0, the foothills are covered in lush vegetation typical of the Arizona Upland subdivision of the Sonoran Desert—a great spot to see saguaro cacti, Curve-billed Thrashers, and Gambel's Quail. Five miles (8 km) up the road is Molino Basin; along the way you pass desert grassland, with yuccas, ocotillos and Black-chinned Sparrows, and a riparian woodland, with tall Arizona sycamores, walnut trees, black-necked garter snakes, and canyon treefrogs. In Molino Basin, oak woodland begins with three species of oaks, alligator juniper, border piñon, Acorn Woodpeckers, and Mexican Jays. At Milepost 12 is a beautiful oak-pine woodland with tall ponderosa pines, silverleaf, blue, Arizona and Gambel's oaks, Pygmy Nuthatches, and gray squirrels. After that the road climbs to over 8000 feet (2440 m) through mixed coniferous forest, with ponderosa pines, white firs, golden aspens, Abert's squirrels, and black bears. Near the top of the mountain is the small community of Summerhaven, where several restaurants offer good food; here one often sees Blue-throated and Broad-tailed Hummingbirds. Coronado National Forest is the land agency. (520) 749-8700.

8 Alamos, Sonora, Mexico

Tropical deciduous forest, tropical riparian sabino forest, Madrean pine-oak forest

The pueblo of Alamos is a charming Spanish colonial town immersed in memories of an elegant and rich past, with cobblestone streets, vast old *mansiones*, vibrant gardens of colorful tropical plants, a palm-studded gazebo plaza in front of the 260 year old La Purisima Concepción cathedral (see plate 25), and a classical musical festival in January. Alamos is located

430 miles (695 km) south of the U.S.-Mexico border, a day's drive from Tucson on good roads. Excellent hotels, restaurants, and guides are available. For the naturalist, Alamos is the gateway to the New World tropics, the northernmost opportunity to experience the tropical dry season and the diversity of the lush summer monsoon forest—rampant vines, lianas, and epiphytes, tropical animals including boa constrictors, brown vine snakes, Lilac-crowned and White-fronted Amazon Parrots, Black-throated Magpie Jays, and other tropical species. Trails from 1300 to 4260 feet (400 to 1300 m) elevation in the nearby Sierra de Alamos provide access to tropical deciduous forest (see plate 8), oak woodland, and pine-oak forest. The Río Cuchujaquí (see page 35), just to the southeast, is a scenic tropical river lined by very large sabinos, or Mexican bald cypress, with indigo snakes, Tiger Herons, and more than 730 species of plants. In 1995, the Sierra de Alamos-Arroyo Río Cuchujaquí was declared a federally protected area for fauna and flora by the Mexican government.

9 Cataviña, Baja California Norte, Mexico

Vizcaíno subdivision of the Sonoran Desert

Driving down the Baja peninsula on Highway I from Tijuana, one skirts the Pacific Coast until El Rosario, where the highway turns east and begins to cross the rocky backbone of Baja. The deserts of Baja begin here, in the strangely beautiful Vizcaíno subdivision of the Sonoran Desert.

Around a tiny wayside stop called Cataviña spread some of the most scenic of Baja's landscapes. Giant granite boulders are strewn across the sandy landscape of *cirio* ("boojums"), palo adán (*Fouquieria macdougalii*), and the cardón cacti, cholla, pitaya, and garambullo, or old man cactus (see plate 21). Just north of Cataviña a perennial creek crosses the highway; look here for lush blue fan palms and many songbirds taking advantage of the oasis (see plate 26). The Cochimi Indians used to live in this area, hunting the elusive desert bighorn sheep. Sometimes fog rolls in from the Pacific, lending an eerie air to this arid land.

10 Bahía Kino, Sonora, Mexico

Central Gulf Coast subdivision of the Sonoran Desert

A little over 150 miles (241 km) south of the Arizona border is the sleepy Mexican fishing and holiday town of Bahía Kino, or Kino Bay, on the Sea of Cortez. This picturesque bay is at the eastern edge of the sea's many midriff islands, a biological treasure trove where fin, pilot and orca whales, giant manta rays and several species of porpoises and dolphins may be seen. North of Kino lies some of the most beautiful desert in North America—the Central Gulf Coast subdivision of the Sonoran Desert—with giant cardón cacti, two species of elephant tree, and a stand of boojum trees a short distance up the coast (near Puerto Libertad). Migratory songbirds and shorebirds are some of the wildlife that make this area so special.

A Few More Great Places to Explore

➤ Cactus Forest Loop Trail, Saguaro National Park, East

An easy hiking trail through dense saguaro cactus and palo verde-mesquite forest, the five-mile Cactus Forest Trail showcases lots of resident desert birds such as Gila Woodpeckers, Cactus Wrens, Curve-billed Thrashers, and Phainopeplas and offers a chance to see how saguaros grow under "nurse" plants. (520) 733-5153.

➤ Hugh Norris Trail and King Canyon, Saguaro National Park, Tucson Mountains

The King Canyon Trail begins along a large desert wash (across from the Desert Museum) where desert dwellers such as coyotes, javelina and mule deer are often seen. It ascends gradually through beautiful saguaro cactus and palo verde dominated desert, joining the Hugh Norris Trail to gain the summit of Wasson Peak with its glorious views of the Tucson Basin, Avra, and Altar Valleys, and Rincon, Santa Catalina, and Santa Rita Mountains. Of botanical interest is the remnant desert grassland and interior chaparral atop the peak. Length: 3.5 miles (5.6 km) from the King Canyon trailhead to Wasson Peak. (520) 733-5158. (See plate 27.)

➤ Garden Canyon, Huachuca Mountains

The Huachuca Mountains' proximity to Mexico make for excellent opportunities to see such southern species as Elegant Trogons, Montezuma Quail, and Spotted Owls. Just a few minutes to the southeast of Sierra Vista, this canyon shelters habitats from desert grassland up through oak woodland and oak-pine woodland, with a montane riparian community running through all. The area is especially good for butterfly watching and wildflower viewing in September, and is also very rich in ferns. Access (usually permissible) is through Fort Huachuca (520) 533-7085.

➤ Aravaipa Canyon, Arizona Nature Conservancy and Bureau of Land Management

Rare raptors such as Mississippi Kites, Zone-tailed, Gray, and Black Hawks call this cottonwood and willow lined riparian canyon home, as do tanagers, coatis, bobcats, mountain lions and native fishes. Access is through either the western end (Mammoth) or the eastern end (Klondyke) by permit only. The small effort needed to obtain a permit is more than offset by the solitude this system affords. (520) 348-4400. (See plate 28.)

➤ Sulphur Springs Valley

Sulphur Springs Valley is near Willcox, Arizona. Don't miss the wintertime spectacle of tens of thousands of Sandhill Cranes and Snow Geese. January is the best time to see these beautiful birds, with tours and seminars available during the festival. (520) 384-2272.

A "Talk" with Coyote

I often wander around in the desert by myself and have been very fortunate in seeing lots of wildlife and in observing many of the miracles that happen every day in nature. One of my favorite memories is of an encounter with Coyote.

I was walking up a lushly vegetated wash at dawn one summer morning. A coyote started ambling across the wash, then suddenly realized I was there. He began to trot away, so I sat down on the sand. That immediately aroused his interest since it's not a typical reaction from people, and it's also a very non-threatening gesture. So the coyote stopped and sat down, watching me. I simply sat, waiting for him to make the next move. He did. He lay down and put his head on his paws, still watching and assessing me. I lay down resting my head on my hands, staring into those intelligent brown eyes. After a minute he rolled on his side, so of course I did too. We spent several wonderful minutes changing positions and rolling around.

Suddenly Coyote sat up cocking his ears around. He glanced at me one last time as though to say goodbye, then moved off into the brush, just before a horseback rider came into view down the wash. It was an incredible, magical feeling staring into the soul of Coyote and finding myself judged worthy of a few private moments of play.

— Pinau Merlin

➤ Palm Canyon, Kofa National Wildlife Refuge

The jagged Kofa Mountains thrust straight out of the flat desert pan 63 miles (101 km) north of Yuma, Arizona. Much of this 665,000-acre refuge is designated wilderness and is closed to motorized travel. One of the exceptions is the nine-mile rocky dirt road up to the mouth of Palm Canyon, where there are some fifty native palms tucked away in nearly inaccessible side canyons. The refuge is home to hundreds of bighorn sheep, five species of rattlesnakes, and many other desert animals. Fall through spring are the best times to visit. Camping is available along the dirt road. (520) 783-7861.

➤ South Mountain Park

This largest municipal park in the world rises in southern Phoenix. High slopes and isolated valleys are home to coyotes, Gila monsters, rock squirrels, javelinas, and a variety of desert plants, including elephant tree, *Bursera microphylla*, here at its northern limit. Ancient petroglyphs chipped into rock varnish suggest that the mountain was sacred to prehistoric peoples. (602) 262-7275.

➤ Kitt Peak, Quinlan Mountains

Kitt Peak, located 50 miles (80 km) west of Tucson, is the home of the National Optical Astronomy Observatories' viewing complex. The peak, which is accessible via a paved all-weather road, rises from the desert floor to beautiful evergreen woodland at almost 7000 feet (2130 m). On your trip to the summit look for coatis (a resident troop frequents the parking lot), birds such as Acorn Woodpeckers, Mexican Jays, Spotted Towhees (in summer), and the cryptic-colored mountain spiny lizard, which might be seen even in winter because of its ability to regulate its body temperature and escape into deep crevices. (520) 318-8726.

> Superstition Wilderness

Saguaro-palo verde forests, chaparral, pinyon, juniper, and oak woodlands, and even small pockets of Ponderosa pine, are all part of the rugged Superstition Wilderness landscape. The wilderness, roughly forty to sixty-four miles (64-100 km) east of Phoenix, also includes scenic and geological features, cliff dwellings, and a rich history. The less-visited eastern half offers the best wildlife viewing. (602) 610-3300.

> Cañon Nacapule,
San Carlos-Guaymas,
Sonora, Mexico

Just inland from the Mexican resort town of San Carlos is the magical Nacapule Canyon, an enclave of tropical deciduous forest at the southern edge of the Sonoran Desert. Several species of fig trees, including rock and strangler figs (*nacapule* is Spanish for *Ficus pertusa*), and tropical palm trees thrive in the oasis; Mexican boa constrictors have also been found there.

Nature-Watching Tips

Too often we charge off down a trail, binoculars in hand, optimistically scanning for wildlife—then we end up wondering why we didn't see very much despite the great distances we covered. One of the best tips for increasing the chances for seeing wildlife is to slow down. Or better yet, stop. Pick a spot to hide, such as under a tree or between some boulders, get comfortable, and put your patience in gear. You'll be surprised how much wildlife will wander by.

Tips for Mammal and Bird Watching

- Dress in neutral-colored clothing that blends in well with natural colors, don't wear any scents, and stay downwind of your viewing area; if you need sun or insect protection, wear long sleeves, pants and a hat.

- Choose a spot where two habitat types converge—a riparian corridor and a grassland, or desertscrub and a wash, for example.

- Head out at dawn and at dusk, the times most animals are active in the Sonoran Desert.

- Move your binoculars slowly, and study areas through them thoroughly for signs of animals—a tail, an ear, a leg, a twitch.

- Train yourself to look up—such as for deer scrapes on saplings or signs of browsing on twig-ends—and down—for tracks, burrows, a pile of feathers, or shredded pine cone scales, all of which indicate animal activity.

Tips for Watching Butterflies (and Other Arthropods)

- Butterflies and most other arthropods won't become active until the sun warms things up to about 65°F (18°C). Midday is best.

- In dry seasons, search out areas of moist earth such as pond edges, wash seeps or muddy spots on trails (especially where there is dappled sunlight-and-shade) where many species congregate in what lepidopterists call "puddle parties."

- Blooming plant species that are especially attractive to butterflies include desert broom (*Baccharis sarathroides*), and seep willow (*Baccharis salicifolia*), milkweeds (*Asclepius* spp.), windmills (*Allionia incarnata*), groundsels (*Senecio* spp.), and bee brush (*Aloysia* spp.) to name but a few. In summer, desert hackberry (*Celtis pallida*) is an excellent butterfly "magnet."

- During the warm months, carefully search grass, shrubs and trees for any of the thousands of spiders, insects, and insect larvae. For example, bagworms—a moth larva— might be found foraging on *Acacia* spp. or *Mimosa* spp. Or, look carefully at mesquite trees for signs of mesquite girdler beetles.

Tips for Reptile and Amphibian Watching

- In the Sonoran Desert region, many reptiles and amphibians are most active in summer, especially after rains and in the evenings when it cools off. Many lizards and snakes, however, can be viewed throughout the year, assuming the ambient temperature is to their liking (below 65° F; 18°C is generally too cold). Roadside ditch pools are very good amphibian habitats.

- Go rural road cruising after sunset, since many reptiles and amphibians bask on the warm surfaces; the best roads are dark-colored, curbless, and have lots of natural vegetation on the margins. Drive slowly or walk and be careful of traffic. National park roads that are closed to traffic at night are ideal for evening walks.

Observe But Don't Participate

When exploring for plants or sitting and enjoying the wildlife, it's important to remain an observer and resist becoming a participant. Using food to bait animals in for a better photograph, or handling baby animals, is harmful to the animals in the long run. Approaching nests or burrows too closely can cause an adult to flee, leaving the young exposed to predation. In sensitive habitats, such as along stream banks or in desert areas with cryptobiotic ("with hidden life" or "living") soil, care is necessary to avoid trampling delicate growth or causing erosion. Also, avoid staying near *tinajas* (water holes or tanks) and other solitary water sources for long as you may be keeping animals from a life-sustaining resource.

Binoculars can greatly enhance your nature experiences by bringing wildlife nearer without your having to get close. Use them to get a better look at birds, butterflies, mammals, and reptiles.

Going South

If you venture into Mexico, you will need a few extra travel items. Many American automobile insurance companies offer coverage up to fifty miles (80 km) into Mexico, but most of the best places to explore are a hundred or more miles across the border. In Tucson and Phoenix, in border towns such as Nogales and Lukeville, and at the car permit stop at the border, you can purchase Mexican insurance for your automobile (look in the Yellow Pages under "insurance" or "automobile insurance, Mexico"). Take your car registration, passport or voter identification, and photo IDs. Depending on where you are going in Mexico, you may or may not need a personal travel visa and a permit for your automobile. It is important that you consult with your Mexican insurance agent about your destination. If you are a member, AAA offers excellent Mexico maps and tips for Mexico travel.

Along the tranquil Río Cuchujaqui, just southeast of Alamos, Sonora, Mexico

Additional Readings

Arizona-Sonora Desert Museum. *Tucson Mountains Trail Guide.* Tucson: Arizona-Sonora Desert Museum Press, 1995.
———. *Mount Lemmon Road Guide.* Tucson: Arizona-Sonora Desert Museum Press, 1995.
Carr, John N. *Arizona Wildlife Viewing Guide.* Helena, MT: Falcon Press, 1992.
Hanson, Roseann Beggy and Jonathan Hanson. *Southern Arizona Nature Almanac.* Boulder: Pruett Publishing Company, 1996.
Tucson Audubon Society. *Davis and Russell's Finding Birds in Southeast Arizona.* Tucson: Tucson Audubon Society, 1995.

The Arizona Public Lands Information Center coordinates information from all U.S. and Arizona agencies that are concerned with public lands and is an excellent source for maps and books: 222 N. Central Avenue, Phoenix, AZ 85004. (602) 417-9300; www.publiclands-usa.com/html/home.html.

Desert Storms

Mrill Ingram

It has been said that weather in the Sonoran Desert is a story of monotonous, cloudless days, interrupted by catastrophic exceptions. Whether catastrophic or not, those exceptions are a major part of our story here: how the Sonoran Desert region gets its rain and how this pattern of rainfall influences life in the desert—wild and human.

PATTERNS OF RAIN

Generally speaking, the Sonoran Desert averages only three to fifteen inches (76 to 400 mm) of rain a year. In the Arizona Upland subdivision of the Sonoran Desert, rain falls about equally in two rainy seasons—a winter one in December and January, and a summer one in July through early September. August, September and December are the region's wettest months; May and June are the driest. While there are local variations depending on elevation and proximity to mountains, this pattern basically holds for the entire area.

Rainfall here is infrequent and undependable. The most salient feature of rainfall is not so much its rarity, but its variability, or capriciousness, to put it in terms of human personality. The rain does not fall in even patterns. Sometimes rainfall over a summer will be recorded in small showery increments, but often the rain falls in a few large storms. And while not "normal," it isn't unusual for a single storm to produce fifty percent more rain than typically falls in a whole year. Yuma, Arizona, for example, is one of the driest places on earth, averaging about 3½ inches (89 mm) a year. Yet deluges in the past have dumped over four inches (100 mm) in a single day.

Why It Doesn't Rain Much in the Sonoran Desert

This area's climate is, in a word, dry. Ringed by mountains that keep the rain away for much of the year, the Sonoran

Desert quietly bakes. Moist air moving east from the Pacific Ocean is forced to rise over the Cascades and Sierra Nevada, cooling as it rises. Since cool air cannot retain as much water vapor as warmer air can, the excess water precipitates. Moisture blowing in from the ocean is effectively drained, and the air that moves down the ranges' eastern slopes is usually so dry it cannot produce any more rain. This phenomenon, the *rain shadow effect*, describes such aridity on the inland side of coastal mountains.

30° latitude (north and south)—where many of the planet's deserts lie (see map on page 11). As this equatorial air descends upon the Sonoran Desert region, it creates a very stable and warm atmosphere. The rains stay away until the high pressure system weakens, allowing moisture to slip into the region.

At least two things are necessary for rain: a source of moisture and a delivery system for that moisture. The Sonoran Desert's main source for summer and winter rain is the Pacific Ocean. Moist air that makes it to the

 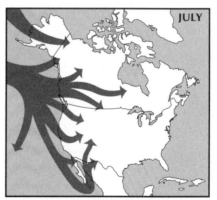

The arrows in these two diagrams indicate large-scale air cirulation patterns as they flow from the ocean across western North America, bringing moisture with them. Typical circulation patterns for January and July are shown.

Redrawn from Katherine K. Hirschboeck, "Climate and Floods," *National Water Summary 1988-89—Floods and Droughts: Hydrology*, U.S.G.S. Water-Supply Paper 2375 (Washington, D.C., 1991)

The dependability of our warm, cloudless, often windless, days is primarily the result of what's called the North Pacific high pressure zone, and is related to the circulation of the earth's atmosphere. (See the chapter "Biomes and Communities of the Sonoran Desert Region" for more information.) Air that heats up at the equator rises, as warm air does, and moves poleward, until it cools enough to sink at roughly

Sonoran Desert region, which lies between 23° and 35° latitude, arrives on the back of strong westerlies— winds that are part of a belt of eastward-moving winds that circle the earth from about 30° to 60° latitude. Independent air streams within this belt curl around and shift their flows depending on the season and other atmospheric changes. The patterns of rain change accordingly.

Summer Rains — The Big Tease

The word *monsoon* refers to a system of winds that changes seasonally, bringing wet and dry periods to a region. In the Sonoran Desert, the summer monsoon consists of winds from the sea flowing inland to fill the partial vacuum created by rising continental air warmed by the summer sun. These winds bring moisture. The belt of westerlies shifts north in the summertime so that the Pacific high sits around 40° latitude, allowing moist air from the Pacific off of Baja California to move into the region.

These westward-moving winds actually circle around into the area and often reach the Sonoran Desert as southeasterly winds. This is one reason people in the past assumed that monsoon moisture comes from the Gulf of Mexico. Recent studies, however, lead many meteorologists to believe that most moisture from the southeast is drained by the 6500 foot (1980 m) Mexican Sierra Madre and so doesn't reach the Sonoran Desert. This issue is yet unresolved.

Once Pacific moisture reaches our area, usually in July, the increased humidity means we really begin to feel the heat. If June has proceeded as usual,

---

Pipe Dreams

The excitement of a summer rain reminds us that water is precious in this arid environment. Long periods of drought are the norm, and desert plants and animals are adapted to water scarcity. But what about the millions of people now living in the burgeoning cities of the Sonoran Desert region? From 1990 to 1997 alone, Phoenix grew by 22 percent and its metropolitan population topped 2,600,000. With per capita water use exceeding 300 gallons per day in Phoenix, the aridity of the desert seems to impose little restraint on these modern desert dwellers. New golf courses appear to spring up every week in our desert cities (Phoenix and Tucson alone had more than 200 public and private golf courses in 1998). This, despite the fact that each 18-hole golf course uses an average of 185 million gallons of water annually—the equivalent of 3500 single family homes!

How can the desert supply all this water? The answer is that it can't. Rainfall replenishes only a fraction of the fossil groundwater withdrawn every year from ancient aquifers laid down thousands of years ago. (The aquifer underlying central Tucson has fallen by more than 200 feet in the last 50 years.) To supplement depleting groundwater supplies, distant river basins are diverted through massive water projects to transport water into this arid region.

By mining fossil groundwater and importing river water from more humid climates, the residents of the Sonoran Desert have buffered themselves from the reality of their arid environment. This dependence on imported and non-renewable sources of water must be addressed in planning for the long-term sustainability of our desert cities.

the desert surface is very hot, causing the moist air moving in to expand and rise. The hot air rising off the desert floor moves upward in great columns called *thermals*, which can be three to five miles (5 to 8 km) in diameter. Broader areas of cooler air separate the thermal columns, which is why thunderheads can be so wildely scattered. The creation of thermals can be a violent business, and local updrafts can move at over fifty feet (15 m) per second. The strong convection upwards is usually matched by strong downdrafts which kick up sand and dust as they hit the land. Above, the air cools as it rises, until at about 17,000 feet (5200 m) the moisture freezes. A growing thunderhead can tower 40,000 feet (12,000 m) or more, with the whole top containing a raging snowstorm—a strange concept to a person broiling at ground level.

And even with all the buildup, it is not at all uncommon to have a "frustrated" thunderstorm. Towering cumulus clouds sweep across valley floors, whirling skirts of wind and dust, and throwing lightning bolts. Yet all the rain can evaporate before reaching the ground. This creates one of the more awesome desert sights: *virga*—the trailing vaporous streams of rain that hang from a thunderhead with frayed ends drying in the layer of hot air over the desert's surface.

The rain that does reach the desert floor in a summer thunderstorm typically does so with great vigor. Although the dry desert can absorb substantial amounts of water, much of the rain rolls off the hard-baked ground. Sheets of water wash across the land, filling arroyos and riverbeds in minutes, the flow carrying along sand, rocks, and plants, carving new stream channels and eroding stream banks. This runoff is a critical resource for desert life, whether it is providing a temporary pool for a desert spadefoot (*Scaphiopus* spp.), a cool spell and source of groundwater recharge for urban desert dwellers, or irrigation for a Tohono O'odham squash field.

Winter Rain — A More Lasting Affair

If summer monsoons are torrid affairs—never predictable in terms of the next tempestuous rendezvous— winter storms resemble somewhat more stable relationships. While winter precipitation is, in fact, as variable as summer rain, the precipitation is often more predictable, since storm tracks can become established. For example, during the winter months, the westerlies shift south to about 35° latitude and the major storm track brings winter storms off the Pacific to the northwest and into the Great Plains region. This usually produces no more than partly cloudy skies and strong winds in the northern part of the Sonoran Desert.

Occasionally, however, a trough of low pressure forms over the western United States, causing the prevailing flow to push storms further south along the west coast, sometimes as far as San Francisco, and then across the mountains to the Sonoran Desert. Meteorologists can detect the storm approaching the coast and warn desert dwellers days ahead of time that a storm has entered California and will soon reach the Sonoran Desert region. Even the mountains can't keep all the moisture away. Although these storms are embedded in fast-moving air currents and don't usually linger more than a day or two, they are important sources of gentle, soaking rain. (Desert dwellers in Sonora, Mexico typically call winter rains *las equipatas*, or "little packages" of rain, in contrast to summer rains—*las aguas*, "the waters.") Once the airflow pattern is established it tends to persist, so that several storms will follow one another over the course of several weeks. This frequently also means that when it rains in Tucson, it is dry in Seattle and vice versa.

Tropical Rain — One Heck of a Date

Another manner in which rain comes to the Sonoran Desert is by tropical cyclones, which originate in the eastern part of the North Pacific, usually in the early fall. These giant storms have established some of the all-time records of monthly precipitation in the Sonoran Desert region. The Spanish word *chubasco* is frequently used by Sonoran Desert dwellers to refer to these tropical storms. ("Chubasco" is more generally defined as any extremely violent storm.) Although infrequent, these storms are memorable.

Consider, for example, some statistics from the flood of 1983. About 10,000 people were displaced. Water, mud and debris severely damaged or destroyed over 1300 homes; 1700 received lesser damage. Many people who fled from their homes were cut off from help because roads, bridges, phone lines, and electric lines were washed away. Interstate 10, the main link between Phoenix and Tucson, was washed out at the Gila River, and twenty other main highways were closed. Nine people drowned trying to cross flooded washes; four others were killed when aircraft got caught in downbursts and crashed.

These large storms begin out at sea, and, as they churn over Baja California, the storms pick up additional energy from the warm waters of the upper Gulf of California. They reach the Sonoran Desert region with renewed energy. The Yuma area is frequently hardest hit, occasionally receiving its whole annual allotment of precipitation in a matter of hours. Even when the storm remains at sea it can still produce heavy rains in the desert. In 1970 and 1983, the tropical storms Norma and Octave pounded the Pacific side of Baja California. Moisture moved up into the Sonoran Desert region from the south, met a cold front moving into the area from the north and caused tremendous flooding across the area. Tropical storms are a normal part of the weather pattern, and they have visited the Sonoran Desert region once or twice per decade in recent times.

El Niño in the Sonoran Desert

El Niño generally refers to a naturally occurring, unpredictable condition in which warm water "pools" in the western Pacific. This occurs when trade winds, which typically keep the ocean water circulating, weaken. The pool of warm water drifts eastward toward South America, causing atmospheric pressure gradients along the equator to weaken, and trade winds to diminish even more. Changes in the ocean temperatures reinforce changes in atmospheric circulation, and the two

Flood damage along Tucson's Rillito River, October 1983.

sets of changes intensify and drive each other, though neither one is clearly the initiator of El Niño.

One result of these "chicken-and-egg" changes is that the powerful tropical Pacific storms begin to form farther east than usual, and the jet stream over the northern Pacific Ocean is invigorated and pulled farther south. More moisture and more storms are thus carried to the southwestern U.S. and northern Mexico. El Niño events increase the likelihood and severity of winter storms in the Sonoran Desert region. They can also increase the chance of tropical storms from the eastern Pacific. Floods have occurred more often in many Arizona rivers during El Niño events than in other years. We also typically see more winter days with more rainfall, while the northwestern U.S. typically sees fewer days with high precipitation. El Niño events usually last for several seasons. Typically, during the spring, the seasonal cycle reasserts itself, and the tropical ocean cools back to normal temperatures. Sometimes the warm El Niño events give way to unusually cold sea-surface tempera-tures, a condition called *La Niña.* The effects of the El Niño and La Niña on global climate are, in part, mirror images of each other, and drought is a common occurrence in the Sonoran Desert region during a La Niña event.

Patterns of Life

What do these different types of storms mean for life in the desert? How does wildlife cope with shifts from drought to flood within hours and then perhaps back to no rain again for months? How plants, animals, and people in the Sonoran Desert respond to the rainfall's variability is a particularly critical aspect of survival in the desert.

As we've seen, rain falls in the northeastern Sonoran Desert generally in a "bimodal pattern"—that is, twice a year: thunderstorms in the summer, and larger, gentler storms in the winter. These rainy times are not predictable, and a season's worth of rain can fall in a day or drizzle in over a month, or rain may not come at all. Given these conditions, survival in the desert has required adaptations—even, as we shall see, on the part of the modern urban dweller.

There are perhaps three things that characterize a good desert denizen: knowing how to wait, knowing how to hold onto what rain does fall, and knowing how to get down to business when opportunities arise. Consider the patience, resourcefulness, and speedy sex of spadefoot toads. Cued by the vibration of rain or thunder, spadefoot toads emerge from interments of ten to eleven months. Taking advantage of the temporary ponds from the rains, spadefoots pursue breeding with all the intensity of creatures denied for most of the year. After months of waiting for opportunity to knock, they begin the next generation of spadefoots within a day's time. "Patience" also characterizes the Gila monster, which does absolutely nothing for nearly nine months of a year. Mornings in April, May, and June will find the large lizard seeking bird eggs and baby quail, but most of the time it waits out the hot, dry days, living off the fat stored in its expandable plump tail.

The variability of rainfall is reflected in reproductive cycles. Many desert plants and animals do not automatically attempt to reproduce every year, but wait until sufficient rain has fallen to make the investment of energy worthwhile. Spadefoots will not emerge without a heavy enough rain to fill the temporary ponds, giving them time to mate and the tadpoles time to develop. Gambel's Quail and Rufous-winged Sparrows will not nest unless sufficient rain has fallen to support insect life and fruit development that will in turn supply baby birds with food. Brittlebush plants (*Encelia farinosa*) bloom generously most years, but hold back their yellow flowers in times of drought.

Many desert plants exemplify the ability to hold on to what one has. Barrel cactus (*Cylindrocactus* and *Ferocactus* spp.), saguaro cactus (*Carnegiea gigantea*), and succulents like agaves (*Agave* spp.) are well known for their ability to store gallons of life-sustaining moisture within. Saguaros grow visibly plump after a wet monsoon season. The extensive, shallow roots of these plants help them capture much of the ephemeral moisture before it evaporates. The intense competition for water when it is available is one reason desert plants are so well dispersed over the landscape. Other plants, such as mesquite (*Prosopis* spp.), push their roots many, many feet deep to tap into underground moisture.

Lifegiving Desert Oases

Riparian areas are the precious gems of the desert. These year-round streams or springs are the incubators of desert life. More than eighty-five percent of desert animals depend upon riparian areas for some phase of their life cycle.

Sonoran Desert riparian areas are sustained by the rain that falls in the mountains and foothills and migrates into the region's alluvial valleys and aquifers. As overpumping of groundwater occurs, the hydrologically-connected riparian areas are degraded and eventually lost. Cottonwood-willow forests that once lined many rivers in the Sonoran Desert, including the Salt, Gila, Santa Cruz, and Rillito rivers, have disappeared as groundwater pumping and surface water diversions disrupted flows and lowered the water table below the plants' root zone.

It is estimated that more than ninety percent of Arizona's riparian areas have been lost in the past century, and many of the remaining areas are imperiled by surrounding urban, agricultural and mining development. The spectacular San Pedro riparian corridor in southeastern Arizona, for example, home to an estimated 400 species of birds, 83 mammal species, and 47 reptile and amphibian species, faces a bleak future because of groundwater pumping in nearby Sierra Vista, and Fort Huachuca, and by burgeoning housing developments near the river.

Efforts to protect the San Pedro and the other few remaining riparian treasures will require concerted and dedicated action on a number of fronts to prevent their demise.

People and Rain

Desert survival has required that human life also adapt to scarce and variable rainfall. Traditional Tohono O'odham farmers plant tepary beans, squash, corn, melons, and other crops bi-annually, in order to take advantage of the rhythm of summer and winter rains. Their fields are designed to catch water washing across the land after storms, often channeling it to areas that have been prepared for planting. A single summer or winter rain can make or break a harvest, and some years, the fields are not planted at all. The Tohono O'odham also scatter their desert plots widely among several washes, in order to maximize the chances that even scattered thunderstorms will soak at least one field.

Perhaps as a reflection of their sensitivity to the vagaries of Sonoran Desert rainfall, the Tohono O'odham seem to dislike jumping to any conclusions about the weather. Linguist William Pilcher noted that the Tohono O'odham avoid any assumption that rain will fall for sure: ". . . it is my impression that (they) abhor the idea of making definite statements. I am still in doubt as to how close a rain storm must be before one may properly say *t'o tju* (it is going to rain on us), rather than *tki' o tju:ks* (it looks like it may be going to rain on us)." Life-giving rain, upon which Tohono O'odham have traditionally been utterly dependent, is not taken for granted, and when it falls, is considered good fortune.

Many modern farmers of large-scale commercial crops, in comparison, have so divorced themselves from the natural rhythm of the desert that they actually dislike rain. Cotton, for example, is grown solely on irrigation water. To increase their yields, some farmers "stress" their plants slightly by withholding water, and the plants put out bigger blooms as a result. Any natural rainfall during this period is seen, therefore, as interfering with the growers' management plans.

Contemporary city dwellers, likewise, are mostly buffered from the vagaries of desert precipitation by modern technologies such as irrigation, cooling systems, and bridges. Pumped water, swimming pools, lawns, and air conditioning allow city people to live comfortably through the desert's hottest times. Is modern urban society therefore immune to the unpredictability of desert rain?

The changeable nature of desert rainfall has required adaptation, even on the part of the city dweller. This is particularly evident with regard to the paradoxical hazard of arid lands—flooding. The natural desert tendency to flood is exacerbated in cities. Over paved surfaces like roads, rain water will move more than eight times as fast as it could in a wooded area. And in urban areas, the proportion of runoff, that is, the water which flows over the surface rather than sinking into it, is about four times that of undeveloped desert. So, not only is more runoff moving much faster in a city, it has less opportunity to soak into the ground. This problem is worsened as more earth disappears under asphalt and pavement, and as more people live closer to flood plains and arroyos.

While Tohono O'odham farmers and desert spadefoots welcome floodwaters, a typical urban response has been irritation and, especially in the face of violent storms, fear for human life and safety. For decades, as cities in the region grew at tremendous speed, the urban response to the threat of flooding was to pour concrete. Natural drainages were widened, straightened, channeled, and lined in concrete. Water that fell over a large urban area quickly flowed into concrete ditches and rushed away. Millions of dollars went toward engineering projects including huge storm drains to accommodate the hundred-year floods that inexplicably seemed to occur every seven years.

Yet, in spite of the investments in flood control, when the big, bad storms hit, such as that causing the flood of 1983, human engineering has repeatedly met its match. In fact, the "structural" approach often serves to worsen flooding and other problems. Concrete ditches move floodwaters away fast—so fast that unlined channels downstream suffer worse erosion as they are hit with more and faster-moving water. The water quality of this runoff also presents a problem, since rain collected from streets, parking lots, and buildings carries sediments, pollutants from cars, and nutrients from fertilizers used in landscaping. Additionally, a cement-lined wash cannot serve as a recharge route to the underground aquifer, an important natural function of desert washes. And a wide cement ditch detracts from the aesthetic appeal of a neighborhood.

The Value of an Urban Wash

Over the past several decades, city dwellers have gradually realized that engineering cannot completely remove their vulnerability to the threat of floods. In addition, some people have also come to view runoff from desert rains not as a nuisance to be guided away, but rather as a resource that can support desert wildlife and recharge underground aquifers. Engineering projects are still a major aspect of flood control plans, but Tucson, Phoenix, Scottsdale, and other cities, have begun to look to the desert's natural drainage system as part of the solution.

A newcomer may not even notice the network of arroyos around a city, or think much of the dry beds natives call "rivers." But for desert city dwellers the arroyos are assets. Contemporary city planning has prohibited construction in flood plains and encouraged the development of extensive river park systems. These parks provide attractive recreation areas for city dwellers, and also buffers during a flood. Small, unlined washes through urban neighborhoods offer pathways and cool, green spots sheltering native plants and urban wildlife such as quail, roadrunners, javelina and coyotes. The unlined urban washes also allow rainwater to soak into and recharge underground aquifers.

City people in the desert may never respond to the arrival of the first summer monsoon with the enthusiasm of the spadefoot toad, but there is a sense of joy as the first drops fall. The changes in city flood-control decisions indicate an ethic of urban desert living that welcomes desert wildlife along arroyos, acknowledges vulnerability to desert storms, and seeks answers, at least in part, in the desert's natural flow. These plans are expensive, yet they enjoy real political and economic support. So when the rains come and the washes begin to run, perhaps more people are pausing to watch the arroyo vegetation turn green, breathe in the smell of wet earth, and wonder at the marvelous event of a desert storm.

Additional Readings

Carr, Jerry E. *National Water Summary 1988-89—Floods and Droughts: Hydrology*, Water-Supply Paper 2375. Washington, D.C.: U.S. Geological Survey, 1991.

Olin, George. *House in the Sun: A Natural History of the Sonoran Desert.* Tucson: Southwest Parks and Monuments Association, 1994.

Sellers, William D., Richard H. Hill and Margaret Sanderson-Rae. *Arizona Climate: The First Hundred Years.* Tucson: University of Arizona, 1985.

Smallwood, J. B., Jr., ed. *Water in the West.* Manhattan, KS: Sunflower University Press, 1983.

The website http://geochange.er.usgs.gov/sw/changes/natural/ has information on a number of topics related to climate change and land use in the Southwest.

Desert Air and Light

David Wentworth Lazaroff

ocks, rattlesnakes, roadrunners, coyotes, cacti—all familiar and tangible parts of that great and complicated whole we call the Sonoran Desert. But the desert is also made of less substantial things, and these, too, contribute to its special character. Among these elusive ingredients are many subtle and mysterious phenomena involving air and light.

These aren't matters to ponder indoors! Let's take a drive in my Volkswagen bus. You can sit up front.

Phantom Water

It's 2:00 p.m. on a very hot April afternoon, and we've just pulled out of the entrance to the Arizona-Sonora Desert Museum, west of Tucson, Arizona. Far ahead of us a puddle seems to be continuously evaporating off the sun-baked pavement, its receding edge matching its speed to ours. Even when I floor the accelerator we can't catch up to the water, but somehow the cars passing our Volkswagen easily do. They drive into it without a splash, and they seem to be reflected in it, upside down.

Of course, real water doesn't behave this way. The retreating puddle is just the most familiar form of that often misunderstood phenomenon of air and light, the *mirage*.

The puddle mirage (which can also be seen on warm days in more temperate climates) starts with simple physics. A shallow layer of sizzling air lies on the surface of the hot pavement. When light from the sky encounters this superheated layer it's bent, or *refracted*, upwards toward our eyes. The effect is very much as though a mirror were laid flat on the road. We see an image of the bright sky, and even upside down images of cars and cacti.

We can't catch up to the water because light entering the hot air layer close to us isn't refracted upwards steeply enough to reach our eyes. The mirage looks like a puddle because when we see sky and automobiles

The puddle mirage. Light from the sky is refracted upward by a layer of superheated air on the pavement. Light entering the superheated layer close to the bus isn't bent upward sharply enough to reach the driver's eyes.

apparently reflected off the ground, our brains insist on interpreting the scene as something familiar. In everyday experience, the most common reflective object we see on the ground is a body of water.

Is the puddle mirage real or is it an illusion? It's actually a little of both—straightforward physics plus the workings of the human mind. Add a little dehydration, which we almost always experience on hot days like this, and perhaps we're even more inclined to see water where it isn't!

Ground-level layers of hot air aren't restricted to pavement. Under the right conditions they can conjure up larger "ponds" and "lakes" in the open desert. And these layers have important consequences for living things. Many spring annual wildflowers begin their lives in the fall as rosettes of leaves spread flat on the desert floor, where they bask in the thin layer of warmer air on cool, sunny days. When the weather heats up in the spring a plant may lift these leaves off the ground, or the rosette may die back as the stem grows upward, and cooler, higher leaves take over the duties of photosynthesis.

Air temperatures can drop so rapidly in the first few centimeters above the soil that long legs can be a real advantage for a small desert animal. When they find themselves on hot ground, many desert lizards stretch out their legs full length and lift their bodies as high as they can. Some small day-active animals escape the natural oven near the ground by climbing or flying to higher perches, and of course many simply seek shade. Ground-hugging desert creatures live in a world very different from the one you and I experience at the "higher altitudes" of human existence.

Atmospheric Uprisings

By now we've turned west onto the main highway toward Kitt Peak and the Tohono O'odham Nation. Cruising in our VW at a breathtaking forty-five miles per hour, we can't easily see the "heat waves" that are rising all around us above the desert floor. Heat has expanded the air layer at ground level, making it less dense and lighter than the cooler air above it. Everywhere bubbles of air are rising like hot air balloons—but without the balloons.

We can't see these ascending bubbles directly, but light passing though them is refracted in randomly-changing directions, causing distant objects to ripple and dance—an effect known as

When Light Needs a Brake Adjustment

Why does light bend when it meets a hot ground-level layer of air, as in the puddle mirage?

The well-known "speed of light," about 186,000 miles per second (300,000 kilometers per second), is actually light's speed limit. Light zips along that fast only in an absolute vacuum. When light travels under less ideal conditions, such as through ordinary air, it's slower. In effect air puts on the brakes, and the denser the air is, the more forcefully the brakes are applied.

Imagine our Volkswagen bus is a light wave. It's been shrunk to matchbox size and turned on its right side, and it's hurtling obliquely downward toward the ground. Because we're driving through air the brakes are dragging, and the speedometer reads just under light's speed limit—a white-knuckle situation if there ever was one!

At the last instant before impact our right (lower) wheels enter the superheated layer of air just above the pavement. The hot air has expanded, so it's less dense than the cooler air above it. Suddenly the brakes on the right wheels aren't dragging as strongly as those on the left. The right side of the bus starts to move faster and the vehicle pulls to the left, away from the ground. Saved by refraction!

Refraction is simply the changing of a wave's direction when different parts of it move at different speeds. Any kind of wave. Sound waves in the atmosphere travel faster in the warmer air at lower altitudes than in the cooler air higher up, so they tend to bend upward, like light in the puddle mirage. That's why sometimes you can't hear the thunder from a distant monsoon storm, even though you can see the lightning: the sound passes above your head!

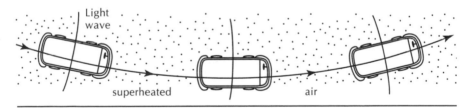

The part of the light wave in the superheated air near the ground travels faster than the part in the cooler, denser air above it. As a result the wave veers upward, the way a bus with bad brakes pulls to one side.

shimmer or *atmospheric boil.* Telephoto shots through shimmering desert air are a staple of western movies, evoking an impression of heat for comfortable viewers in climate-controlled theaters.

High above the ground the rising bubbles can be quite large. These are the *thermals* described in the chapter

"Desert Storms." Thermals, too, are usually invisible, but there are sometimes clues to their whereabouts. A hawk wheeling in the sky may be saving energy by riding the rising air of a thermal. Soaring Turkey Vultures gain another benefit besides lift. Unlike most birds, they have an excellent sense of smell, and occasionally the ascending air carries with it the tempting fragrance of decaying flesh!

Toward the western side of the valley two *dust devils* are parading slowly across the desert floor. A dust devil, which looks something like a miniature tornado, is a special kind of thermal. Like people, some thermals are better organized than others, and dust devils are thermals of the best organized kind.

The exact conditions that create these grit-charged whirlwinds are somewhat mysterious. Dust devils form most frequently around mid-day or in the early afternoon, when solar heating is most intense. As heated air rises above a surface "hot spot," nearby air spirals inward and upward to take its place—like water whirling into a sink drain, only upside down. The rotation of the air (or water) accelerates as it approaches the center of the action, as a spinning skater speeds up when he pulls in his arms.

The analogy between a dust devil and a draining sink is apt for another reason. An endlessly repeated "factoid," easily refuted by any skeptic who puts it to the test, holds that draining sink water rotates in opposite directions north and south of the equator, thanks to the *Coriolis effect*. The Coriolis effect, a consequence of the earth's rotation, does indeed cause hurricanes to spin in opposite directions in the northern and southern hemispheres, but it has negligible influence on systems as small as sinks.

Or as small as dust devils. Not surprisingly, the sink drain myth has extended itself to these whirling dervishes, which are commonly believed to spin one way in Australia and another way in Arizona. In fact, no matter where on earth dust devils appear, about half turn clockwise and the other half counter-clockwise, in stubborn disregard of what they're "supposed" to do.

Dust devils are much less powerful than tornadoes, but they are capable of doing damage. A very large one could conceivably knock over a tall, flat-sided vehicle like ours. No dust devil could lift Dorothy (or Toto) off the ground, but it's not hard to imagine one levitating a lizard.

Do dust devils have any other effects on living things? No doubt they sometimes disperse the seeds of desert plants over longer than usual distances, and they help spread the fungal spores that cause the disease coccidioidomycosis, better known as valley fever. Perhaps more important, dust devils raise tiny soil particles high into the air, where they may drift hundreds of miles. Airborne dust is a major factor in the formation of the limy desert soil layers called *caliche*, as well as clay-rich *argillic* layers, both of which have profound effects on desert vegetation, as explained in the chapter "Desert Soils."

Floating dust has interesting effects on the desert sky, too. More on that farther down the road.

Got Those Bouncing Photon Blues

Speaking of the desert sky, why is it so blue? For that matter, why is the sky blue anywhere? Oddly enough, we owe the answer to the latter question largely to a trio of scientists who worked in the foggy climate of the British Isles.

In the late 19th century the physicists John Tyndall and Lord Rayleigh showed that the sky is blue because of the way sunlight interacts with air. Most ordinary visible light passes through the atmosphere relatively undisturbed, but occasionally a light particle—a *photon*—runs into an air molecule and bounces off it, a process called *scattering*. The light we see in the sky is sunlight that has been scattered off air molecules. But why is it blue instead of white?

About two centuries before Tyndall's and Rayleigh's investigations the great Sir Isaac Newton had demonstrated that ordinary white sunlight is a mixture of all the colors of the rainbow, from red through violet. Tyndall and Rayleigh showed that how strongly light is scattered by air molecules depends on its color (more precisely, on its wavelength). Light from the blue-violet end of the spectrum is much more likely to bounce off an air molecule than is light from the red-orange end. As a result, most reddish light travels through the atmosphere more or less unimpeded, but enough bluish light is scattered into our eyes to make the sky appear blue.

At least it appears blue to us. Other creatures may see it differently. Invisible to us, the radiation lying just beyond the violet end of the spectrum—the *ultraviolet*, or UV—is scattered even more than the violet. The desert sky is so bright with UV around midday that an exposed surface receives about as much of it scattered from the sky as directly from the sun—a good reason for humans to wear broad-brimmed hats! Unlike us, hummingbirds and honey bees can see ultraviolet light. What color might the sky look to them?

Sonoran Desert skies are such a deep blue (to human eyes) because desert air is unusually pure, that is, compared to the air above many other places on the planet; it's relatively free of the tiny floating particles and droplets called *aerosols*. Aerosols come in a wide range of sizes, and the larger ones reduce the blueness of the sky. Unlike air molecules, they scatter light of all colors about equally. As a result, they seem to fill the sky with white light, diluting the blue.

Desert air has so few of these large aerosols partly because it's so dry. In more humid climates water vapor condenses on microscopic airborne particles, forming tiny droplets that we see as hazes and fogs. This is especially true in coastal areas, where tiny salt crystals from evaporating ocean spray are especially good at capturing water vapor and creating water droplets. In

invisible							invisible
infrared	red	orange	yellow	green	blue	violet	ultraviolet

The visible spectrum, as seen in a rainbow, or more often nowadays in the surface of a compact disk. Also shown are the invisible infrared and ultraviolet radiation just off the ends of the visible spectrum.

fact, morning fog is a routine occurrence in parts of the Sonoran Desert along the western coast of Baja California, as described in the chapter "Biomes and Communities."

Of course, even inland desert air is far from aerosol free. Beautiful water droplet hazes can form in the Sonoran Desert overnight after rainstorms, but they often evaporate quickly after sunrise. On spring days like this one the desert sky can be noticeably brightened by airborne pollen. And, unfortunately, even Sonoran Desert skies can be sullied by particulate pollution from cities and factories. But the most characteristic desert aerosol is dust. A major dust storm can make the car in front of you disappear from view, but small everyday dust particles have more benign effects, as we'll see.

Deceptive Distances

As we drive westward onto the Tohono O'odham Nation we can see desert mountain ranges receding into the distance in every direction. Nearby slopes show the warm colors of volcanic rocks and blooming palo verde trees, but more distant mountains seem bluish and washed out. That's because air between us and any object in the landscape scatters blue light into our eyes. The more distant the object, the more air intervenes, and the greater is the bluish tint and the loss of contrast.

This effect, called *aerial perspective*, is well known among landscape painters and photographers, who exploit it to create the impression of depth. It's one of the cues we all use unconsciously to gauge distance in the outdoors. In regions where the atmosphere is thick with aerosols, aerial perspective is strong and obvious, but here in the desert it's often quite subtle. More than one desert traveler has suffered dire consequences because the mountain with the next water hole was much farther away than it looked!

Blue skies, blue mountains—and blue birds. Among the oaks and telescope domes on the summit of Kitt Peak (now disappearing in our rear-view mirror), Mexican jays are preening their plumage. The blue in their feathers isn't a pigment; it's caused by the scattering of sunlight off tiny structures in the feathers—the same process that makes the sky blue. Scattering even creates the blue in the open eye of the astronomer awakened by the scolding jay outside her window!

Speaking of sleep, it looks like you could use a siesta, too. That's another familiar effect caused by warm desert air. . . .

Left-over Light

I see that last pothole woke you up. We've just left the Tohono O'odham Nation. I've turned onto this dirt road so that we can park among the creosote bushes and enjoy the closing act in the day's drama of air and light—that unparalleled boon to the photographic industry—a desert sunset.

Balanced on the jagged horizon, the sun has turned the orange color of a desert globemallow blossom. Believe it or not, it looks that way because of the same phenomenon that makes skies, jays, and eyes blue: light scattering. At the end of the day, sunlight travels a much longer path through the atmosphere to reach us than it does at noon.

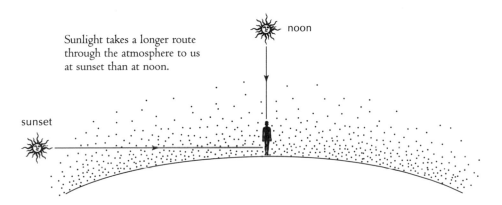

Sunlight takes a longer route through the atmosphere to us at sunset than at noon.

noon

sunset

Along the way so much violet, blue, and even green light is "scattered out" that mostly red, orange, and yellow light gets through. In other words, the reddish light we're seeing now is merely the sunlight that's left over after helping to create blue skies for travelers farther west.

The same purity of the air that makes desert skies so blue also helps make desert sunsets so beautiful. There usually aren't enough large aerosols here to quench the light of the setting sun or to reduce visibility. However, smaller aerosols, mainly in the form of airborne dust, can actually intensify the colors of a sunset.

Unlike larger aerosols, which, as you'll remember, scatter light of all colors about equally, very small particles scatter bluer light preferentially, much as air molecules do. They brighten the blue sky and make sunsets even redder. It's because of these tiny particles that desert sunsets are usually more colorful than sunrises. Breezes and dust devils increase the dustiness of air by the end of the day, but some of the dust settles out during the calm of the night. (When you see a dust devil, think of it as a sunset in the making!)

Now the sun has dropped below the horizon, and we need to turn 180 degrees to watch the next important event. A blue-gray band is rising slowly above the eastern horizon. This is the *earth's shadow*—literally the shadow of our planet on its own atmosphere. From the point of view of any hawks still soaring in this darkened region of air, the sun is below the horizon. Just above the shadow is a band of reddish light. Hawks soaring up there can still see the setting sun. We see the reddish band because reddened sunset light is being scattered back to us by dust in the air.

The curtain of night rises more quickly, then merges into the darkening sky. If this were a cloudy evening, these few minutes would be the most spectacular—a time of shifting shadows and rapidly-changing hues, as fiery sunlight finds its way through openings between the shifting clouds.

Part of the fun of watching a desert sunset a few months from now will be following the changing appearance of a monsoon thundercloud. As the sun slips farther below the horizon, the up-tilting shadow of the

The earth's shadow and the colors of a thundercloud just after sunset. The bottom of the cloud (1) is in darkness. The middle of the cloud (2) is lit by sunlight reddened by scattering in the lower atmosphere. The top of the cloud (3) is white; it's lit by sunlight that has passed only through the rarefied upper air.

earth swallows the cloud from the bottom up. Just above the rising shadow the cloud is bathed in the warm light of the setting sun, but for many minutes the top of the cloud remains bright white, because at that great altitude the air is too thin to appreciably dim and redden the sunlight. Then, just before it's extinguished, the summit turns pink as it's lit by sunlight that has passed close to the earth's surface far to the west.

Out of the Oven, Into the Refrigerator

As twilight colors fade we can feel the coolness of night settling over the desert. The sun-warmed ground is cooling by radiating its heat into the sky. Most of the radiation lies in the *infrared*, or IR, just beyond the red end of the visible spectrum. (See the illustration on page 33.) While we can't see infrared, in a sense rattlesnakes can. On nights like this they use sensitive IR detectors on their heads to find the warm bodies of rodents in the dark. You can read about that in the "Rattlesnakes" chapter.

(Watch your step out here, by the way.)

As the ground cools it in turn chills the air just above it. Gradually, the layer of hot air that seethed above the sun-baked soil during the day will be replaced by a quieter layer of cold air. Unlike warm air, which is light and floats upward, cold air is dense and flows downhill. (You can prove it. Stand in front of your refrigerator in your bare feet and open the door.)

Later tonight cool air will slide downward off the slopes of nearby mountain ranges and collect here on the valley floor. It will pool especially in low places, which will be noticeably colder by morning than places only a few meters higher. This *cold air drainage* is most apparent on clear winter nights. You do not want to lay down your sleeping bag on the bottom of an arroyo in January!

Sometimes the lake of cool air that gathers overnight in a desert basin causes morning mirages very different from the puddle mirage we saw this afternoon. Under these conditions light is refracted downward instead of upward, and, thanks to a peculiar atmospheric astigmatism, distant objects can seem to stretch out vertically, creating illusory cliffs and towers. If there happens to be a city in the basin, pollutants may be trapped in the cold air. This situation is often called a *temperature inversion* because the air's temperature rises with increasing height, instead of falling, as is usually the case.

Night-time radiative cooling and cold air drainage can pose a real threat to some desert plants, especially on the cooler northeastern edge of the Sonoran Desert. Frost sensitive plants like desert ironwoods may be confined to mountain foothills from which cold air drains away on winter nights. Radiative cooling is one reason why young saguaros are more likely to survive under a nurse plant. At night the overarching branches of a tree or shrub are slightly warmed by the infrared radiation from the ground. The branches then reradiate some of the heat back downwards, keeping the young cactus above freezing through a chilly night.

Radiative cooling operates well only when the desert sky is clear. Clouds work like nurse plants on a larger scale—they, too, intercept upward-traveling infrared and reradiate it. Even invisible water vapor absorbs infrared, so there's less cooling when there's more moisture in the air. In fact, water vapor is the most important gas responsible for the famous *greenhouse effect*, which keeps most of the earth, Sonoran Desert included, from resembling Antarctica.

Dry, cloudless skies let sunlight in during the day and heat out at night— that's why in a desert there's often such a great temperature difference between day and night.

The sky is dark now. Stars are twinkling overhead, and even more toward the horizon. (The scintillation is caused by random refraction by moving air of varying temperatures—an effect much like the shimmer we saw this afternoon.) Later tonight the atmosphere will settle down, the "seeing" will improve, and astronomers back at Kitt Peak will train their telescopes upward through the clear window of the desert sky. Perhaps they'll take a look at that red planet up there, where dust devils whistle across a rock-strewn world so dry that the Sonoran Desert seems like a rainforest by comparison.

Alas, there are some deserts where even a Volkswagen can't go!

Additional Readings

Bohren, Craig F. *Clouds in a Glass of Beer: Simple Experiments in Atmospheric Physics.* New York: John Wiley and Sons, 1987.

Gedzelman, Stanley David. *The Science and Wonders of the Atmosphere.* New York: John Wiley and Sons, 1980.

Meinel, Aden and Marjorie. *Sunsets, Twilights, and Evening Skies.* Cambridge: Cambridge University Press, 1983.

Minnaert, Marcel. *Light and Color in the Outdoors.* New York: Springer-Verlag, 1993.

Schaefer, Vincent J., and John A. Day. *A Field Guide to the Atmosphere.* Boston: Houghton Mifflin, 1981.

The Deep History of the Sonoran Desert

Thomas R. Van Devender

The Sonoran Desert is considered to be the most "tropical" of the North American deserts. Its climate is virtually frost-free, and summer rainfall comes from the tropical oceans. The Sonoran Desert's structurally diverse vegetation, which includes columnar cacti and leguminous trees, certainly differs from those of the shrub-dominated Great Basin, Mohave and Chihuahuan deserts. It has both geographic and biologic connections with more tropical communities. In a single day's travel, naturalists can begin in the oak woodlands and desert grasslands in southern Arizona along Interstate 19, and travel through the various desertscrub, thornscrub, and tropical deciduous forest habitats along México 15 in Sonora, experiencing the remarkable transition from temperate zone communities to the New World tropics. This vegetational gradient gives us a sense of the Sonoran Desert's connection with the tropics, but does not really explain it. The explanation lies in the tropical roots of the Sonoran Desert, deep in its evolutionary history.

A WALK THROUGH TIME

The Sonoran Desert that we see today, with its characteristic assemblages of plants and animals is quite recent, at least in terms of geologic time. In fact, it and the other North American deserts are among the youngest biotic communities on the continent. Although some Sonoran species evolved in ancestral seasonally-dry tropical communities, the development of the unique regional climates and the evolution of characteristic desert-adapted plants and animals are thought to have combined to form the Sonoran Desert by about 8 million years ago (mya) in the late Miocene. Similar conditions developed many times subsequently as global climates changed, with the Sonoran Desert continually expanding, contracting, and redefining itself. The most recent expansion of the Sonoran

Desert into its modern area in Arizona and California occurred only 9000 years ago, with the modern communities of plants and animals developing 4500 years later. This chapter is a walk through time examining the conditions that led to the development of the Sonoran Desert and exploring what shaped its dynamic history.

The Paleocene (66.4 to 57.8 mya)

In the Paleocene epoch, soon after the extinction of the dinosaurs (65 mya), most of North America was covered with temperate evergreen and tropical rainforests. There was little regional variation. The warm climates promoted humid forests with strong Asian affinities; primitive ferns (*Anemia* spp.), cycads (*Diön, Zamia* spp.), and palms grew as far north as Alaska. The flowering plants (angiosperms), whose spectacular evolutionary radiation began in the Late Cretaceous, became increasingly important in the forests, displacing archaic cycads, conifers, and tree ferns. The earliest indisputably recognizable fossil grasses were found in sediments dating to about 58 mya. These were broad-leaved forest grasses ancestral to the modern bamboos.

The Eocene (57.8 to 36.6 mya)

In the Eocene, deciduous trees became increasingly common, providing the first evidence of a dry season. The landscape now included tropical deciduous forests, in which trees dropped their leaves in response to drought. During these periods, sunlight passed

through the leafless canopy and heated and dried the surface of the ground. Many new species of plants and animals evolved, adapting to these new heat and moisture regimes. The origins of cacti and other succulents likely occurred during the Eocene in dry tropical forests. Fossils of an alligator (*Allognathosuchus*), a softshell turtle (*Trionyx*), a primitive tortoise (*Geochelone*), a primitive monitor lizard (*Varanidae*), a ground boa (*Boidae*), and many small mammals from Ellesmere Island in northeastern Canada, then at 78° latitude, indicate that the world was very warm and that plants and animals freely traversed a land bridge between North America and Europe. The results were dramatic shifts in the biota as more advanced forms displaced archaic ones.

The Oligocene (36.6 to 23.7 mya)

By the Oligocene, grasses, including the important taxonomic groups in arid lands, had achieved relatively modern diversity. Unfortunately, most of this dramatic radiation in the grasses, one of the most important plant families, was not captured in the fossil record. *Gopherus*, the genus of the modern desert and gopher tortoises (*G. agassizii* and *G. polyphemus*), appeared in the Oligocene. Modern genera of lizards in the Oligocene fauna were skinks (*Eumeces*) and beaded lizard or Gila monster (*Heloderma*). The snake fauna was dominated by small ground boas related to the living rubber boa (*Charina bottae*) and desert rosy boa (*Lichanura trivirgata*) of western North America and the sand boas (*Eryx* spp.)

Geologic Time Scale

EON	ERA	PERIOD		EPOCH	TIME SPAN
Phanerozoic	Cenozoic	Quaternary		Holocene	11,000yrs - today
				Pleistocene	1.8mya* - 11,000
		Tertiary	Neogene	Pliocene	5.3 - 1.8mya
				Miocene	23.7 - 5.3mya
			Paleogene	Oligocene	36.6 - 23.7mya
				Eocene	57.8 - 36.6mya
				Paleocene	66.4 - 57.8mya
	Mesozoic	Cretaceous			144 - 66.4 mya
		Jurassic			208 - 144 mya
		Triassic			245 - 208 mya
	Paleozoic	Permian			286 - 245 mya
		Carbon-iferous	Pennsylvanian		320 - 286mya
			Mississippian		360 - 320mya
		Devonian			408 - 360mya
		Silurian			438 - 408mya
		Ordovician			505 - 438mya
		Cambrian			570 - 505 mya
Proterozoic	Grouped as Precambrian				2500 - 570 mya
Archean					3800 - 2500 mya
Hadean					4500 - 3800 mya

(mya = million years ago)

* The age of the beginning of the Pleistocene is in dispute

of Africa. The lizards common today in the Sonoran Desert—Iguanidae and Teidae, (that is, iguanas and their relatives, and whiptails)—and the common snakes of the Sonoran Desert today—Colubridae and Viperidae (colubrid snakes, for example, bull snakes, kingsnakes; and pitvipers) were uncommon or absent at these early dates.

The Miocene (23.7 to 5.3 mya)

A series of enormous volcanic eruptions from the middle Oligocene to the middle Miocene (about 30 to 15 mya) changed the climates and established the modern biogeographic provinces of North America. (See the chapter "The Geologic Origin of the Sonoran Desert" for a more complete discussion of these processes and the terminology used to describe them.) The Rocky Mountains were uplifted to new heights by the accumulation of over a kilometer and a half of volcanic rock. A kilometer (.6 mile) thick layer of rhyolitic ash fell in the Sierra Madre Occidental in northwestern Mexico—on top of a kilometer of early Tertiary (the geologic period that includes the Paleocene, Eocene, Oligocene, Miocene, and Pliocene epochs) andesites. As regional uplift pushed them even higher, the mountains interrupted the upper flow of the atmosphere for the first time. Tropical moisture from both the Pacific Ocean and the Gulf of Mexico was blocked from the mid-continent, drying out the modern Great Plains and the Mexican Plateau. Harsher climates segregated drought- and cold-tolerant species into new environmentally-limited biomes, including tundra, conifer forests and grasslands, and restricted them along elevational and latitudinal environmental gradients. The Miocene also was a time when major evolutionary radiations began in many of today's successful groups including composites (plants in the sunflower family), grasses, toads, iguanid and teid lizards, colubrid snakes and pitvipers.

The "Miocene Revolution"

After the rise of the Sierra Madres and the resulting changes in climate, tropical forests were found only in the lowlands along the coasts of Mexico and Central America. There, newly evolved species joined archaic ones in the biome known as tropical deciduous forest. And, along the lower, drier edges of tropical deciduous forest evolved a new biome—thornscrub. Thornscrub looks very much like a transitional state between tropical deciduous forest and Sonoran Desert: its vegetation is shorter and sparser than tropical forest and it does not require much moisture. During the Miocene, thornscrub may well have been the regional vegetation in drier areas to the north that are now Sonoran Desert. Thornscrub, in fact, may be the ancestral biome of many Sonoran Desert plants and animals.

The Sonoran Desert itself came about during a drying trend in the middle Miocene (15 to 8 mya). Much of the desert's vegetation, however, predates the Sonoran Desert itself, having evolved in climates that also required adaptations to aridity. For example, guayacán (*Guaiacum coulteri*), organpipe cactus (*Stenocereus thurberi*), palo brea

(*Cercidium praecox*), senita (*Lophocereus schottii*), and tree ocotillo (*Fouquieria macdougalii*) likely evolved in thornscrub. Other plants such as desert ironwood (*Olneya tesota*), foothills palo verde (*Cercidium microphyllum*), and saguaro (*Carnegiea gigantea*) evolved along with the Sonoran Desert.

Another important chapter in the history of the Sonoran Desert concerns the formation of the Baja California peninsula. Before 12 million years ago, much of the land which is now Baja California was part of the Mexican mainland. Activity along the San Andreas fault caused the Gulf of California to open, and several large chunks broke away from the mainland and drifted in splendid isolation northwestward. The timing of the formation of Baja California is controversial, with some estimates as recent as 5 to 6 million years ago. These islands were populated with tropical plants and animals which soon evolved into regional endemics, including the boojum tree or cirio (*Fouquieria columnaris*). Eventually the islands combined and joined California to form the Baja peninsula. Today, the modern islands on both sides of the peninsula in the Pacific Ocean and the Gulf of California, many of which are geologically young, are the most active evolutionary arenas in the Sonoran Desert Region, and have many endemic species.

The Pliocene (5.3 to 1.8 mya)

During the latest Miocene and early Pliocene, geological forces again altered landscapes and climate regimes, causing a reversal to more tropical climates.

Sea level rose enough that the Gulf of California expanded into the Los Angeles area of southern California. A fossil skull of an iguana (*Pumilia novaceki*), a primitive relative of the tropical green iguana (*Iguana iguana*) that today occurs no farther north than southern Sinaloa, was found in sediments 2.5 to 4.3 million years old in southern California. With tropical circulation patterns enhanced by warmer oceans, tropical forests of western Mexico likely expanded, reaching farther north than they do today in Sonora; likewise, the Sonoran Desert in Arizona and California extended further, perhaps as far as southern Nevada.

The Pleistocene (1.8 mya to today)

The warmth of the Pliocene ended abruptly at the beginning of the Pleistocene about 1.8 million years ago, as the Earth entered a new climatic era that far surpassed the middle Miocene in cool, continental conditions. Traditionally, four ice ages or glacial periods were recognized, based on terrestrial sedimentary deposits in North America, and these were widely correlated with conditions in Europe and South America. However, recent studies of sediment cores from the ocean floors record fifteen to twenty glacial periods in the Pleistocene. Ice ages were about ten times longer than interglacials (the warm periods between ice ages) which lasted 10,000 to 20,000 years. Officially, the end of the Pleistocene was defined as the beginning of the Holocene 10,000 years ago, based on changes in sediments in European lakes. Today we understand that the Holocene is the present interglacial period and that the cyclic environmental fluctuations of the Pleistocene likely have not ended.

In the last glacial period (the Wisconsin), the massive Laurentide continental glacier covered most of Canada and extended as far south as New York and Ohio. As much as two miles (3 km) of ice covered the Great Lakes and New York City. Boreal forest with spruce and jack pine moved southward displacing the mixed deciduous forests of the eastern United States. Mountain glaciers covered the tops of the Rocky Mountains and the Sierra Nevada in the western United States and the Sierra Madre del Sur in south-central Mexico. Now-dry playa lakes in the Great Basin were full. Enough water was tied up in ice on land to lower sea level about 425 feet (130 m).

During the last half of this glacial period (from 45,000 to 11,000 years ago), plant remains in ancient packrat (*Neotoma* spp.) middens document the expansion of woodland trees and shrubs into areas that had been desert. Woodlands with singleleaf pinyon (*Pinus monophylla*), junipers (*Juniperus* spp.), shrub live oak (*Quercus turbinella*), and joshua tree (*Yucca brevifolia*) were widespread in the present Arizona Upland subdivision of the Sonoran Desert. Ice age climates with greater winter rainfall from the Pacific Ocean and reduced summer monsoonal rainfall from the tropical oceans favored woody cool-season shrubs related to plants living farther north, rather than to the summer-rainfall trees, shrubs and cacti of tropical forests and sub-tropical deserts. The isolated chaparral communities in central Arizona, mostly in a northwest-southeast band below the Mogollon Rim, are relicts of ice-age chaparral connections with California. Many species are shared between California and Arizona chaparral, including shrub live oak. The Arizona black rattlesnake (*Crotalus viridis cerberus*) in Arizona chaparral is essentially a dark form of the southern Pacific rattlesnake (*C. v. helleri*); it is more distantly related to the other four subspecies of western rattlesnakes in Arizona.

Warm desert communities dominated by creosote bush (*Larrea tridentata*) were restricted to below 1100 feet (300 m) elevation in the Lower Colorado River Valley in the Sonoran Desert and in the southern Chihuahuan Desert. Although brittlebush and saguaro returned to Arizona soon after the beginning of the present interglacial (the Holocene) about 11,000 years ago, the Sonoran Desert did not re-form until about 9000 years ago, as the last displaced woodland plants retreated upslope. Relatively modern community composition was not achieved until about 4500 years ago when foothills palo verde, desert ironwood and organpipe cactus arrived from their retreats to the south and to low elevations. However, the modern assemblages that we recognize as the Sonoran Desert communities must have recurred many times during the Pleistocene interglacials, only to retreat to warmer climates as ice age climates returned. Modern desert communities have been present for only about five percent of the 2.4 million years of the Pleistocene, while ice age woodlands in the desert lowlands persisted for about ninety percent of this period.

Tropical Interglacials

Surprisingly, the vertebrate fossil record suggests that some interglacial climates were more tropical during the Holocene. El Golfo de Santa Clara is near the mouth of the Colorado River in northwestern Sonora. Early Pleistocene (1.8 mya) fossils reflect a climate that was frost free, with much greater rainfall in the warm season, and with higher humidity than today. Greater summer rainfall would suggest that tropical oceans were warmer than they are today, in contrast to most of the Pleistocene when ocean waters were colder. The fauna included such mammals as antelope, bear, camels (dromedaries and llamas), cats, horses, proboscidians, and a tapir (*Tapirus*). The giant anteater, capybara (*Neochoerus*), and ground sloths in the fauna were members of ten families of mammals that immigrated into North America in the late Pliocene or early Pleistocene after the opening of the Panamanian land bridge during the Great American Interchange. In contrast, the imperial mammoth (*Mammuthus imperator*), a hyena (*Chasmoporthetes johnstoni*), and jaguar (*Felis onca*), were Eurasian immigrants. The nearest populations of giant anteater are 1800 miles (3000 km) to the southeast in the humid, tropical lowlands of Central America! As for many large mammals, the modern distribution may not accurately reflect their physiological range limits because of human predation in the last 11,000 years. Other fossils in the fauna include the Sonoran Desert toad (*Bufo alvarius*), slider "turtle" (*Trachemys scripta*), boa constrictor (*Constrictor constrictor*), and the large extinct California beaver

(*Castor* cf. *C. californicus*). The Sonoran Desert toad is a regional endemic, while the slider and boa constrictor occur today in Sonora in wetter, more tropical areas to the southeast. Although the El Golfo area is today part of the hyperarid Gran Desierto, the delta of the Colorado River was historically a very wet area that supported extensive cottonwood (*Populus fremontii*) gallery forests with abundant beaver. There is even a December 1827 account of a large spotted cat (likely a jaguar) that entered James Ohio Pattie's camp on the Colorado River south of Yuma to feed on drying beaver skins.

Rancho La Brisca is in a riparian stream canyon north of Cucurpé, fifty-four miles (90 km) south of the Arizona boundary in Sonora. A pocket of 150,000-year-old ciénega sediments yielded abundant Sonoran mudturtle (*Kinosternon sonoriense*), fish, and other small vertebrate fossils associated with bison (*Bison*). The presence of the sabinal frog (*Leptodactylus melanonotus*) 144 miles (240 km) north of the northernmost extant population on the Río Yaqui indicates that the climate of a late Pleistocene interglacial was also more tropical than at the site today.

Sonoran Desert Mammoths?

A few years ago, a Mayo Indian found a very large bone in the bank of the arroyo behind his house in Teachive, a village in coastal thornscrub in southern Sonora. For him, as it has been for others who have discovered fossil mammoth bones throughout North America and Europe for centuries, it was puzzling and perhaps frightening. What animal could be so much larger than a deer or a cow? Why has no one seen these monsters? Many a legend was born to explain them and their disappearance.

Today we know that about 11,000 years ago, nearly two-thirds of the large mammals of North America went extinct. Common, widespread grazers, including horses and mammoths, disappeared at the very time that spruce and pine retreated and grasslands expanded from Arizona to Canada. Paul S. Martin of the University of Arizona forcefully presented the case that big game hunters caused widespread extinctions within a few hundred years after their entry into North America from Siberia via the Bering Strait. The theory of "overkill" of "naive" large mammals is controversial, and some suggest changes in climate may have caused the extinctions. However, the paleobotanic record gives no evidence of climatic changes severe enough to have resulted in the extinction of so many large animals over such a broad, diverse area. The well-preserved plant remains in packrat middens provide additional insights. A species could respond to a major climatic change by (1) adapting genetically (speciation), (2) becoming extinct, or (3) adjusting its geographical distribution. At the beginning of the Holocene, the last glacial/interglacialclimatic shift, there are essentially no records of speciation or extinction in plants or small animals. Most simply shifted their geographic and elevational ranges. Moreover, woodland plants survived in desert lowlands for several thousand years after the megafaunal extinctions and before the expansion of the Sonoran Desert. The biotic communities of North America have had

fewer large herbivores in the last 11,000 years than at any time in the last 20 million years! The impacts of these herbivores on tropical deciduous forest, thornscrub, and the Sonoran Desert flora were undoubtedly profound—but we will never fully understand the ecological roles of these missing animals.

THE ARCTIC CONNECTION

Tropical communities with their great species diversity, outrageous morphological adaptations, and mixtures of archaic and advanced species have been important evolutionary arenas. However, the evolutionary mechanisms are not so clear, considering that the climatic fluctuations that isolate populations and stimulate the evolution of new species are more intense at high latitudes. Recent paleomagnetic dating of fossil-bearing sediments in northern Canada indicate that some plants appeared there as much as 18 million years earlier, and some mammals two to four million years earlier than at lower latitudes. If the relative ages are not the result of a fragmentary fossil record, they may indicate that important biotic innovations evolved in the mild "tropical" Arctic climate with its months-long polar day-night cycle, and then the species moved southward into the tropics.

Fossil records of tropical plants and animals from Arctic latitudes with six-month-long nights not only reflect very warm global climates but raise questions about how these organisms survived the dark. Today reptiles spend the cold winters in hibernation and hot, dry periods in estivation. Deciduous plants shed their leaves for long periods, either triggered by the onset of cold temperatures in temperate latitudes or the beginning of the dry season in tropical latitudes. The Arctic fossils suggest that deciduousness in plants and hibernation and estivation in reptiles could have arisen as responses to the polar night, and later shifted to these other stimuli.

WHOSE CHILD IS THIS?

Although the perception of the Sonoran Desert as "tropical" is based partly on the presence of columnar cacti, there are climatic, physical, and deep historical connections as well. As discussed above, the region likely supported tropical deciduous forest from the Eocene to the early Miocene and then thornscrub later in the Miocene. The desert biota of this region are rich in endemics, most of which evolved in more tropical communities prior to the Sonoran Desert itself. Many Sonoran species reach their southern limits in thornscrub—the structural, biotic, and historical link to tropical deciduous forest—not in the tropical deciduous forests of southern Sonora. With the exception of organpipe cacti and a few others, the paucity of desert species shared with the tropical deciduous forests challenges the popular idea that "tropical deciduous forest is the mother of the Sonoran Desert."

Thornscrub could more accurately be called the "mother" of the Sonoran Desert and tropical deciduous forest its "grandmother" or "great aunt"!

Additional Readings

Bahre, Conrad J. *A Legacy of Change. Historic Human Impact on Vegetation in the Arizona Borderlands.* Tucson: University of Arizona Press, 1991.

Betancourt, Julio L., Thomas R. Van Devender, and Paul S. Martin. *Packrat Middens. The Last 40,000 Years of Biotic Change.* Tucson: University of Arizona Press, 1990.

Davis, Goode P., Jr. *Man and Wildlife in Arizona: The American Exploration Period 1824-1865.* Phoenix: Arizona Game and Fish Department, 1982.

Grayson, Donald K. *The Desert's Past. A Natural Prehistory of the Great Basin.* Washington, D.C.: Smithsonian Institution Press, 1993.

Hastings, James R. and Raymond M. Turner. *The Changing Mile.* Tucson: University of Arizona Press, 1965.

Imbrie, John and Kathryn P. Imbrie. *Ice Ages. Solving the Mystery.* Cambridge: Harvard University Press, 1986.

Kurtén, Bjorn and Elaine Anderson. *Pleistocene Mammals of North America.* New York: Columbia University Press, 1980.

Martin, Paul S. and Richard G. Klein, eds., *Quaternary Extinctions: A Prehistoric Revolution.* Tucson: University of Arizona Press, 1989.

The Geologic Origin of the Sonoran Desert

Robert Scarborough

Geology challenges the human imagination. First of all, there is the notion of geologic time. We humans think in terms of lifetimes or of centuries. What can 10,000 years possibly mean to us, let alone 65,000 or 70 million? Space, in geologic terms, is an equally difficult notion, though we don't often recognize the difficulty. We often wonder what the land we stand on was like in times past. What was here, we ask, when dinosaurs roamed, or when mountains were new? Though we may imagine different landforms, different vegetation, we probably still imagine "here" as a definite, and permanent, spot on the globe, a place with constant longitude and latitude, a place whose changes we could trace through time, perhaps by reading the record in the strata beneath our feet. But geology mocks our notion of permanence. Geology deals with continents that drift, collide and re-form, with rivers and oceans that appear and disappear,

with mountain ranges whose battered remnants have been carried away and now lie buried on some other continent. Geology warns us not to be too literal as we imagine the history of our planet.

Geologic Interactions with the Living Community

Geology, the interpretation of earth and life history, encompasses much more than the study of sterile rock masses. Continents, oceans, the climate, our atmosphere, and all life have co-evolved on this planet in a complex, interwoven web. Inorganic environmental changes occur, and all life forms must adapt quickly, in terms of geologic time.

There are many ways the earth may influence a local ecosystem. The least obvious is slow continental drifting across lines of latitude or longitude, which affects circulation patterns in the oceans, storm tracks, mean temperatures, and the timing and duration of seasons.

Mountain chains appear near coastlines for various geologic reasons, setting up *orographic* (mountain-induced) cooling of rising moist air masses to form coastal fog deserts and rain shadow deserts on the protected sides, such as coastal Baja California and the hyper-dry Mohave Desert, respectively. Upland canyons, piedmonts, and mountaintops create new ecological niches, sites of adaptations and evolutionary change. The Sonoran Desert and nearby mountain islands exhibit nearly two miles of vertical relief, from sea-level deserts to mountaintops at 9500 feet (2900 m) that harbor subalpine spruce-fir forests, cool enough to have supported semi-permanent ice masses on shady north slopes during the Pleistocene (the past two million years). Then as mountains slowly erode to flatlands, the climate pattern changes.

As climates and habitats change, plant and animal species either adapt, migrate to more favorable ground, or become extinct. Migratory routes are often determined by geologic processes. For example, climate dictates that a river be perennial or intermittent, and that a lake expand or dry up. New mountains produce new rivers. These changes may block or expedite migration for a terrestrial animal while serving as barriers or corridors for an aquatic one.

Ancient life affects later geologic and climatic conditions. Biologically-produced gases (oxygen, carbon dioxide, methane, nitrous oxide) maintain a chemically reactive atmosphere that in turn influences rates of rock weathering, the nature of sedimentary deposits, and the content of gases in the atmosphere.

Desert soils, highly variable in their water-holding capability, salinity, and alkalinity determine the kinds of plants that will survive on them. Some desert plants, for example, are well-adapted to soils that would be toxic to other plants. (See the chapter "Desert Soils.")

The Topography of the Sonoran Desert and Adjacent Lands

The Sonoran Desert lies in a region of the West called the Basin and Range geologic province. This curious country consists of broad, low-elevation valleys rimmed by long, thin, parallel mountain ranges, which extend from northern Mexico across much of Arizona, California, Utah, and Nevada, northward to the southern plains of Idaho. (See the map, page 73.) Normally dry streams in each valley either connect to a major through-flowing river, such as the Gila or Salt rivers (see plate 9), or else drain into a valley's internal low spot where a salt-encrusted playa forms. California's Imperial Valley, and several other valleys within the American Soutwest, exhibit this internal drainage.

Topography is an important influence upon the unique climate of the Sonoran Desert, since topographic barriers direct, confine, or block moist air masses. We can see such a confinement for the Sonoran Desert on the map on page 73, which resembles a shallow bathtub, breached at its south end where it receives Pacific moisture, and at its north end where it joins the rain-starved Mohave Desert.

Let us consider the topography of the areas on the periphery of the

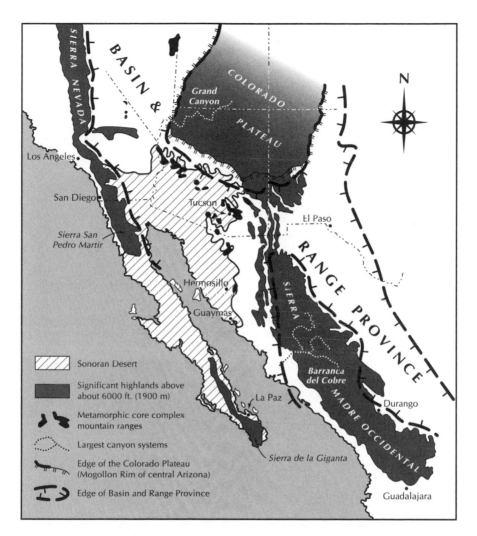

This map shows the Basin and Range geologic province, the highlands that confine the Sonoran Desert, and the largest canyon systems of the region. Also shown is an unusual series of mountains in the region called "metamorphic core complexes."

Sonoran Desert, since these lands both help define and affect the desert itself.

Considerable high country exists west of the Sonoran Desert in Baja California's Sierra San Pedro Martir and in southern California's coastal Laguna Mountains, loosely joined to the imposing Sierra Nevada farther

north through the Palm Springs region. The backbone rock of this country is vast masses of Cretaceous-aged granites, 140 to 80 million years old (see plate 26).

The Sonoran Desert is bounded to the northeast by a mile-high escarpment called the Mogollon Rim, which forms the distinctive southern edge of the Colorado Plateau province. The Colorado Plateau extends north across Utah and western Colorado and consists of a grand "pancake" pile of sedimentary rocks of diverse age (from 30 million to 1200 years old), exposed over a wide area, but most famously in the walls of Arizona's Grand Canyon National Park.

The eastern edge of the Sonoran Desert in southeastern Arizona consists of a honeycomb series of high valleys and mountain ranges, including the Pinaleno and Chiricahua Mountains. Mountaintops range from 3000 feet (915 m) in the west to 10,000 feet (3050 m) in the east. The elevations of valley bottoms rise from sea level near Yuma to 5000 feet (1525 m) in southeast Arizona, where deserts are replaced by grassland valleys. Since rising air cools, annual precipitation and wintertime cold extremes intensify to the east, causing the desert, with its frost-sensitive plants, to gradually give way to grassland.

The Sonoran Desert's southeastern edge is defined by Mexico's Sierra Madre Occidental, a tall, mountainous accumulation of 30-million-year-old volcanic rocks, which stand exposed in the cliffs of the pine-covered country of Barranca del Cobre (Copper Canyon) of Sonora-Chihuahua.

Formation of Basin and Range

The recent geologic history of the Sonoran Desert includes an event unique in all the world, one that tore the country apart. This "Basin and Range disturbance" was the culmination of several events that have taken place over the last 40 million years. Before this geologic onslaught there is good evidence the region stood as an extensive upland, devoid of today's mountains and smoothly connected to all the surrounding highlands (see map on on page 73 and plate 32). Starting with this flatter land, several interconnected geologic actions produced the modern Sonoran Desert landscape.

VOLCANISM AND REGIONAL ARCHING

A visitor to Basin and Range country 40 to 20 million years ago might have been alerted to one of the deep-seated geologic events by noting numerous active volcanic centers in the region. These volcanic areas occasionally produced tremendous explosions, and left behind extensive volcanic flows (see plates 23 and 24). Some of these volcanoes then collapsed, forming large circular basins called *calderas*, as seen in the Chiricahua and Superstition mountains. Other centers, such as the Ajo, Kofa, Galiuro, and Gila mountains, ejected ash-flow materials from long, thin fissure vents.

Exciting as this surface activity was, far more impressive events were occurring below. Intense heat rising into the crust was hot enough to entirely melt and soften portions of the lower continental crust into a viscous fluid, similar in consistency to molasses taken out of a refrigerator. This heating became important when the Pacific Coast became somehow attached or glued to the edge of the Pacific Ocean tectonic plate, which was at the time beginning to move northwest relative to the main continent. This movement applied a stretching force to the region. Basin and Range crust, being hot and fluid, could

Desert Pupfish

How strange that many isolated springs of the Sonoran Desert contain several varieties of small cyprinid fish, the desert pupfish. These little, unlikely desert survivors can tolerate mildly saline and very warm waters. Their dispersal across much of Basin and Range country, even into totally isolated valleys, must have occurred during the Pleistocene when the Colorado River system flowed more vigorously, allowing them to explore all the back alleys of this aquatic kingdom. Then when drier times came, their habitats shrank back to only the perennial springs they inhabit today.

At least one variety of pupfish has gone extinct with the introduction of game fish. This survival story, like so many others, continues to unfold as the Sonoran Desert evolves.

not resist this force and so began to stretch apart in a giant geo-taffy pull.

Think of what happens as you bite into a caramel candy coated with hard chocolate: the fluid caramel stretches while the brittle coating shatters. In this way Basin and Range crust began breaking up, resulting in tremendous disruptions. Early pull-apart action 25 to 20 million years ago was localized along the line of mountains (shown as metamorphic core complexes on the map, page 73), where tremendous heat from beneath was concentrated. (See illustration A on page 76.) This heated zone across Arizona responded to the pull-apart force by forming a huge fault zone, along which all the land west to the Pacific Coast was pulled away. Several nearly flat faults, such as the Catalina detachment fault shown on page 76 B, accommodated this motion. Once uncovered by the faulting, the fluid granite rocks of these "metamorphic core complex" mountains rose or arched up further into an aligned series of "pimples," due to their heat and buoyancy. Rocks above the detachment faults moved a considerable distance to the west, perhaps ten miles or farther, from the arched terrain. These spectacular processes probably lasted for a few million years.

THE ORIGIN OF MODERN VALLEYS

Following this activity, the stretching action quickly spread across most of the entire region, which would eventually become the modern Basin and Range country. By about 12 million years ago, the entire substrata of Basin and Range country was involved with the expanded taffy-pull, stretching out some thirty to eighty percent more than its original width, while the brittle crust above shattered into hundreds of long, thin segments. Narrower alternating segments tended to sink into the taffy, while alternating wider slices maintained more of their old heights. (See illustration C on page 76.) Virtually all mountains of the region were simultaneously born in this way; this also explains the semi-parallel trend of the region's mountains and valleys—they are perpendicular to the direction of stretching.

By about 8 million years ago, the pull-apart action stopped, the thinned crust cooled, and Basin and Range

A. Heating from Beneath, Arching (30 million years ago)

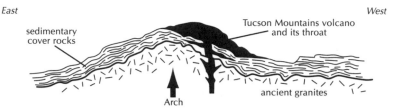

East   *West*

- sedimentary cover rocks
- Tucson Mountains volcano and its throat
- Arch
- ancient granites

B. Volcanism and Detachment Fault (25 million years ago)

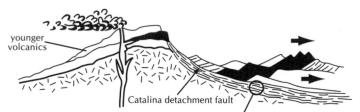

- younger volcanics
- Catalina detachment fault

granite deforms into layered gneiss during movement along the Catalina detachment fault

C. Basin and Range Faulting (12 to 6 million years ago)

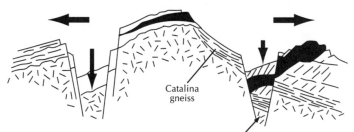

- Catalina gneiss

block under Tucson valley drops 10,000 feet or more

D. Today–Basins Filled with Sand, Gravel and Clay

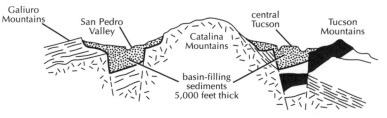

- Galiuro Mountains
- San Pedro Valley
- Catalina Mountains
- central Tucson
- Tucson Mountains
- basin-filling sediments 5,000 feet thick

Geologic cross-sections through the Tucson area illustrating the recent geologic history of the Sonoran Desert

Aerial photo of Baboquivari Mountains showing typical Basin and Range structure: long, thin, rugged ranges isolated between broad, wide valleys.

mountains and valleys stabilized. Since that time the chief geologic activity has been the movement of rock debris off mountains and into adjacent valleys through stream action, as seen in illustration D. Most Basin and Range valleys are filled with 5000 feet (1525 m) or more of gravel, sand, and clay beds, the geologic containers for our desert aquifers. Basin and Range country is unique—no other region of similar origin is identified on the planet. Plates 4, 6, 9, and 13 show the end result of Basin and Range formation and the long period of subsequent erosion.

Near the end of major Basin and Range formation, about 6 million years ago, a severe sideways ripping action began along the Pacific coastline; this continues in our own time. The crack at the edge of the ripped-off land is called the San Andreas fault; it is responsible for the separation of the Baja California peninsula from mainland Mexico and for the opening of the Gulf of California. (Refer to the seismicity map on page 78.) The northward dragging action is related to that which originally ripped apart Basin and Range. These days the San Andreas fault allows for the transport of all land on its west side (the Baja California peninsula and parts of western California) toward the north in a very irregular fashion at the startling speed of a few inches per year.

Earthquakes

Except for the major influence of San Andreas fault activity, the Sonoran Desert is seismically quiet, with notice-

Plot of ten years (January 1, 1984 to December 31, 1994) of seismicity of all magnitude earthquakes from the Preliminary Determination Epicenters catalogue compiled by the National Earthquake Information Center. The San Andreas fault, the focus for many earthquakes shown on this map, runs up the Gulf of California and northward along California's coastal region. Plotted by Mark Tinker and Susan Beck at the University of Arizona.

able earthquakes felt less than once per few decades. However, slight readjustments to changing conditions do occur. Ground ruptures associated with prehistoric earthquakes are known from near the towns of Green Valley and Gila Bend in Arizona.

The last major earthquake in the Tucson region was felt on May 3, 1887, at 2:15 pm. That afternoon the earth's surface ruptured with an estimated magnitude of 7.2 on the Richter scale at a place some twenty miles (38 km) south of Douglas, Arizona, near the village of Bavispe, Sonora (see plate 30). It was responsible for fifty-one deaths in Mexico, while cracking plaster walls in Tucson and El Paso, knocking over an adobe wall at the Spanish cemetery at Mission San Xavier del Bac near Tucson, and stopping large pendulum clocks in Phoenix. Tucson residents reported large clouds composed of dust and forest fire smoke rising above the crest of the Santa Catalina Mountains and first thought it to be an erupting volcano. Apparently the fires were ignited by falling boulders crashing together and causing sparks on the dry hillsides. Tohono O'odham residents at the village of Pan Tak below Kitt Peak reported a massive rock fall. Geysers of water shot up from the flood plain of the San Pedro River, while other streams and springs throughout the region either dried up or initiated flow. An estimated 50,000 cattle died that year in the San Pedro River valley as a direct result of wildfires or later starvation. A line of recent small earthquakes from that fault zone appears on the map on page 78.

Young Volcanism

There have been two contrasting styles of volcanism in the Sonoran Desert in recent geologic time. The intense volcanic episode mentioned above produced rhyolites (light-colored volcanic rocks, relatively rich in silica, aluminum, potassium and sodium), created as the western edge of the North American continent moved over the Pacific ocean floor. The main mass of the Tucson Mountains is composed of rhyolite produced during an earlier rhyolite volcanic episode some 70 million years ago. (See illustration on page 76.) In contrast, younger volcanic rocks formed during Basin and Range time (the last 10 million years) are called basalt (dark-colored volcanic rock, rich in iron, magnesium and calcium); basalt formed beneath the continent during pull-apart actions and rose along deep cracks. Rhyolite volcanoes tend to explode violently, like Mount St. Helens or Krakatoa. Basaltic eruptions are non-explosive; they produce lavas with a consistency of fifty-weight motor oil, which spread quickly across valley floors.

There are three basalt volcanic fields in the Sonoran Desert, all formed within the past four million years. One of these, the Pinacate field, lies just north of Rocky Point, Sonora, and has rightfully become an international showcase of natural history. The field contains a central 4000-foot tall (1220 m) stratified volcano composed of multiple lava flows and ash layers, surrounded by approximately 400 outlying single-eruption basalt cinder cones and flows (see plate 14), and ten unusual steam-blast explosion

Sonoran Desert's Fossil Legacy

Life's drama is abundantly preserved in the rock record of the region. Arizona's oldest rocks—1800 to 1400 million years of age—have not been adequately searched for remains of microscopic life. The oldest stratified rocks of the Sonoran Desert date from about 1200 million years ago, and contain horizons where mats and small mushroom-shaped colonies of algae once grew in protected aquatic habitats. The main threats to the survival of these organisms were the enormous Precambrian ocean tides.

Paleozoic shales and limestones (570 to 240 mya) contain occasional remains of trilobites, shark and fish teeth, crinoids and corals, bryozoans, conodonts, clams, brachiopods, oysters and a variety of cephlapods (a class of mollusks, which includes the octopus and squid). We envision these creatures' homes as shallow tropical ocean waters with coral reefs, lagoons and inlets, reminiscent of the Bahamas-Florida-Mississippi Delta region.

Mesozoic fossil beds (150 to 90 mya) represent a regional trend toward terrestrial conditions as the land rose and drained. River floodplain deposits of Jurassic or Cretaceous age in the Tucson Mountains contain tracks of lizards found with rare fossils of horsetails (Equisitum spp.) and petrified wood. Cretaceous beds contain clams, sharks, marine reptiles like the mosasaur (an aquatic monitor lizard), and turtles. Cretaceous low-elevation coniferous forests were resplendent with cycad and ginkgo trees, through which glided flying reptiles. Late Cretaceous strata of the northern Santa Rita Mountains contain a remarkable fossil record including the titanic long-necked sauropods, horned and duckbill dinosaurs, and some of Arizona's oldest fossil mammals. All these lived along large river floodplains and shores of ancient inland lakes, sharing territories with crocodiles and lizards.

Cenozoic deposits contain a mammal-dominated fauna that inhabited a land reminiscent of a lush East African savanna. Earlier forms included ancestral horses, giant rhino-like titanotheres, and oreodonts (ancestors to peccaries and camels). By late Miocene and Pliocene time (10 to 2 mya) grasses and grazers became widespread. Pleistocene fauna of the last 2 million years included camels, herds of bison and near-modern horses, mastodons, imperial mammoths, giant ground sloths, wolves, lions, giant beavers, and short-faced bears. North America's first people left dart points imbedded in fossil remains of some of these animals at sites near former springs. The modern Sonoran Desert ecosystem seems a distant cousin to the ancient environments of the region.

---

craters, called *diatremes*, some with diameters in excess of a mile (see photo, page 81). These unusual craters owe their explosive origin to an encounter of the rising magma with water-saturated sediments, which adds the force of steam blasts to the normal volcanic fountain.

The Pleistocene Climate

The Pleistocene Period, the last two million years, is noted for its glaciations and worldwide flip-flopping climate changes. These climate shifts have left marks upon the Sonoran Desert. Though glaciers in Arizona mountains were confined to elevations above 9000

Aerial view of Pinacate volcanic field near Rocky Point, Sonora, Mexico. The circular feature in the upper right is McDougal Crater. With a diameter of 6000 feet (1830 m) it is the largest crater in the field. The rugged mountain in the left foreground is composed of Precambrian granites. Volcanic activity has taken place here during the past 4 million years; the field is considered dormant, but not extinct.

feet (2740 m), vigorous stream runoff during the first million years removed much soil and debris from mountain slopes and deposited it in many large fan-like deposits below the mouths of the larger canyons. (See photo, page 83.) Then the streams in the larger valleys of the region began to downcut their channels to lower levels, which then sidecut, causing the formation of several sets of flat-topped terraces above modern flood plains (see plate 3). The last permanent high-elevation ice masses were rapidly melting 14,000 years ago, and the regional climate was becoming drier and modern (interglacial) by 12,000 years ago. This last climate change marks the birth of the modern Sonoran Desert ecosystem. Climatologists cannot predict future climate shifts.

Landforms

The Sonoran Desert contains a characteristic series of landforms, shown in the illustration on page 82. Sparse regional rainfalls tend to lack the force to move sediments very far from the mountains. However, rare heavy rains produce torrents of mud, rocks, and vegetation that cascade rapidly down steep narrow canyons in the mountains. This debris flow spreads out at the fronts of the mountains into cone-shaped masses called *alluvial fans*. (See photograph on page 83.) When neighboring alluvial fans coalesce along a mountain front, the resulting landform is a *bajada* (bah-HAH-dah). The term bajada is generally reserved for those areas where obvious alluvial

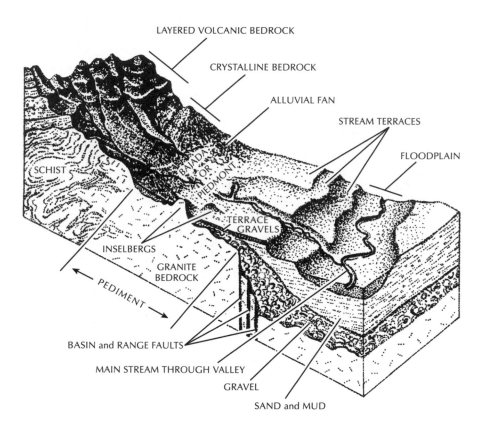

Labels in illustration:

LAYERED VOLCANIC BEDROCK

CRYSTALLINE BEDROCK

ALLUVIAL FAN

STREAM TERRACES

FLOODPLAIN

BAJADA OR PIEDMONT

SCHIST

TERRACE GRAVELS

INSELBERGS

GRANITE BEDROCK

PEDIMENT

BASIN and RANGE FAULTS

MAIN STREAM THROUGH VALLEY

GRAVEL

SAND and MUD

Typical Sonoran Desert landforms

fans line the mountain front, while *piedmont* is used in situations where alluvial fan shapes are not obvious.

Beneath the bajada lies an important hidden feature of desert geology: the *pediment*. Pediments are buried shoulders of mountain rocks that extend from the edge of the exposed mountain some miles toward the valley center, where they contact the buried Basin and Range fault, beyond which lies thick valley alluvium (gravel, sand, silt, and clay). Pediments form as the mountain front is worn back with time by all the streams exiting the mountain front; then the shoulder is buried by a thin layer of gravel as the valley fills with alluvium. Their presence, though invisible, is very important for human development, since the main valley aquifer—often a mile thick—is confined to the centers of the valleys (see illustration). Water wells drilled into the pediment often do not yield sufficient water for even a single residence. Isolated small hills near mountains, called *inselbergs*, are exposed rock masses that have not worn away; they are a sure sign of the pediment's presence.

Major valleys contain one or more main stream channels that are normally dry. Floodplains are strips of flat land adjacent to the channel that in former times were subject to flooding. But since the 1890s, river floods have tended to incise and widen the channels, so that the floodwaters do not flow out onto the floodplains, except locally.

Alluvial fan. Boulder debris in front of the Gila (Sheep) Mountains near Yuma. Large rocks like these are transported in massive debris flows down steep canyons, then spread out along the mountain front during times of rare, intense rains.

This post-1890s channel enlargement is part of a regional trend throughout the West called "arroyo cutting," likely caused by a combination of factors, including increased cattle grazing following development of regional railroads in 1882, devegetation of hillsides by the mining industry for mine timbers and coke, and a possible unrecognized, subtle climate shift. Local governments take a risk in stabilizing channel enbankments with soil cementation. Haphazard bank stabilization increases channel erosion (bank caving) and floodplain inundation downstream of the protected reaches. This is because cement-lined channel walls prevent infiltration and force more water down the channel.

In the mountains, *balanced boulders* form from certain rock types, such as granite and thickly-layered sandstone or volcanic ash. The rocks weather down along cracks or joints and tend to form spires or irregular columns. Rounding happens as corners weather faster than sides, just as the cube of ice in your iced tea melts into a sphere. If a flat (horizontal) set of joints is also present, weathering along this base eventually forms a rounded balanced boulder. Boulders are eventually shaken free by earthquakes or uneven weathering and litter the nearby ground. These same kinds of rocks, when more fracture-free, may weather into large, spectacular domes that develop concentric rounded joints just like a layered onion, from which segments of layers separate, exposing a cone-shaped core. Balanced boulders can be seen along the Gates Pass road in the Tucson Mountains, along the Mt. Lemmon Highway at Geology Vista, in Texas Canyon along I-10 east of Benson, and on Camelback Mountain (Echo Canyon) in Phoenix. Spectacular conical granite domes with "onion-peel" structures may be seen along the high ridges on the west side of the Santa Catalina Mountains above Catalina State Park.

Mountains composed of volcanic layers weather into rugged tablelands cut by sharp canyons, such as in the Ajo, Tumacacori, and Kofa Mountains (see the photo below). The sharp eye may spot a vertical cylinder of resistant rock weathering away from a volcanic cliff; often this is a volcanic neck or plug, which was formed in the feeder vent for a volcanic flow. There are good examples of volcanic necks in the Superstition, Tucson, and Ajo mountains.

Sand dune fields are common in the Sonoran Desert. Dune fields occupy downwind portions of valleys where wind-dispersed sand has accumulated. Sands for the Gran Desierto of northwestern Sonora (plate 29), the Algodones dune field east of El Centro, and the Mohawk Valley field east of Yuma all derive large quantities of sand from the Colorado River delta. The vast Cactus Plain dune field near Parker derives its sand from the old shoreline sands of the Colorado River.

Cliffs of volcanic tuffs exposed along the western front of the Kofa Mountains near Oak Canyon in western Arizona. The volcanic eruption that created these mountains was but one of hundreds that occurred throughout the Southwest between 40 and 20 million years ago.

Desert pavement is a sparsely vegetated desert flatland totally covered with a single layer of desert-varnished rocks. *Desert varnish* is a black, shiny coating on the exposed surfaces of undisturbed rocks. (See plate 31; also the chapter "Desert Soils" for a discussion of the causes of these phenomena as they have come to be understood in recent years.)

Many other minor dune fields are found throughout the region, their sands derived from local river flood plains. The Gran Desierto contains examples of star dunes, with several radiating sharply-crested sand ridges coming off a high point (photo on next page).

Maybe the only way to fully comprehend the geologic processes that

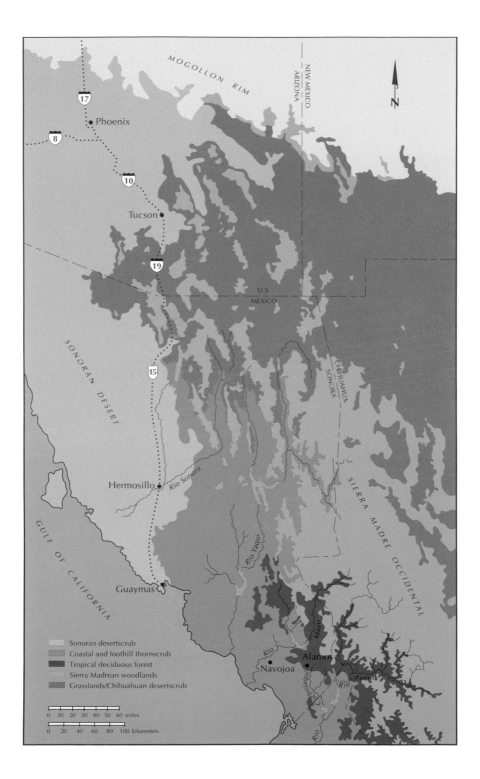

MOGOLLON RIM

NEW MEXICO
ARIZONA

N

17

Phoenix

8

10

Tucson

19

U.S.
MEXICO

SONORAN DESERT

15

CHIHUAHUA
SONORA

Hermosillo   Rio Sonora

GULF OF CALIFORNIA

Rio Yaqui

SIERRA MADRE OCCIDENTAL

Guaymas

Sonoran desertscrub
Coastal and foothill thornscrub
Tropical deciduous forest
Sierra Madrean woodlands
Grasslands/Chihuahuan desertscrub

Rio Mayo

Alamos

Rio   Navojoa

Rio Fuerte

Rio Chihuahua

Rio

0 10 20 30 40 50 60 miles

0   20   40   60   80 100 kilometers

PLATE I (previous page) Biotic
Communities of the Sonoran Desert
region. All of the world's major vegetation
types (biomes) are found in this region.
The southernmost tundra in North
America occurs near Flagstaff, Arizona
(just north of the area covered by this
map). The northern limit of tropical
forest extends into southern Sonora. All
of the others exist somewhere in-between.
See pages 3-4.

PLATE 2 (above) Tundra biome, represented here by arctic tundra in the Alaska
National Wildlife Refuge. Alpine tundra
occurs at high elevations as far south as
Flagstaff, Arizona. See pages 4-5. (Photo
by Carol Cochran)

PLATE 3 Coniferous Forest and Temperate
Deciduous Forest biome components are
intermixed in the middle elevations of
the Southwest. This Mexican oak-pine
woodland community in southeastern
Arizona's Pedregosa Mountains is also
called Madrean evergreen woodland.
See pages 5 and 81. (Photo by Mark
Dimmitt)

PLATE 4 Grassland biome in the San
Rafael Valley, southeastern Arizona. True
prairie grassland such as this is rare in our
region. More common is a blend of grassland and desert vegetation called desert
(or semidesert) grassland. Near here is the
headwater cienega of the Santa Cruz River
which flows through Tucson on its way to
the Gila River. See pages 6 and 270.
(Photo by Thomas R. Van Devender)

PLATE 5 Chaparral biome in San Diego County, California. It is the only biome with mild wet winters and hot dry summers, called a Mediterranean climate. These shrubs form a continuous, nearly impenetrable cover eight to twelve feet (2.4-3.7 m) tall. See pages 6-7. (Photo by Thomas R. Van Devender)

PLATE 6 Desert biome, Sonoran Desert
(Lower Colorado River Valley subdivision)
in southwestern Arizona. The Tinajas
Altas Mountains are in the background.
The average annual rainfall at this site
is less than four inches (100 mm). See
page 7. (Photo by Thomas R. Van
Devender)

PLATE 7 Thornscrub biome, coastal thornscrub in southern Sonora. The foreground is a cleared road shoulder; undisturbed thornscrub often consists of impenetrable tangles of mostly spiny vegetation. See page 7. (Photo by Thomas R. Van Devender)

Plate 8 (facing page) Tropical Forest biome near Alamos, Sonora. This tropical deciduous forest receives about thirty inches (760 mm) of rain during the three summer months. By the end of the dry season most of the woody plants are leafless and only the cacti are green. See pages 7 and 34-35. (Photo by Mark Dimmitt)

PLATE 9 (above) Dry Riparian Habitat in the Lower Colorado River Valley subdivision of the Sonoran Desert. In this very arid region trees can grow only in drainages that collect runoff from occasional rains. This desert wash cuts a water gap through tilted volcanic rocks between the Growler Mountains (right) and the Puerto Blanco Mountains (left) of Organ Pipe Cactus National Monument, Arizona. A true riparian zone is visible as a greener ribbon in the canyon, where groundwater is forced near the surface in the narrow gap. See pages 8-9 and 72. (Photo by Peter Kresan)

PLATE 10 Great Basin Desert in eastern Oregon. This desert's valley floors are typically dominated by big sagebrush, the gray shrubs in this photo, which are the indicator species of the Great Basin Desert. Other shrub species are few and there are few other life forms of plants in this very cold desert. See page 12. (Photo by Thomas R. Van Devender)

PLATE 11 Mohave Desert east of Searchlight, Nevada. The large plants are joshua trees, the indicator species of this desert, but they occur only in the this desert's higher elevations. The lower-elevation portions of the Mohave Desert are difficult to distinguish from the adjacent Sonoran Desert. See pages 12-13. (Photo by Thomas R. Van Devender)

PLATE 12 Chihuahuan Desert in southwestern Texas. There is no single indicator plant; this desert is characterized by a large diversity of shrubs and especially small cacti and other succulents. Trees are almost absent. The mountain is limestone, a very abundant rock in this region. See page 13. (Photo by Thomas R. Van Devender)

PLATE 13 Sonoran Desert, Lower Colorado River Valley subdivision, Cargo Muchacho Mountains in southeastern California. It is so arid here that most plants grow in drainages that collect runoff. In this photo shrubs are growing in small drainages and trees, one of the Sonoran Desert's indicator life forms, line the wash in the background. See plate 9 for an aerial view. See also page 14. (Photo by Mark Dimmitt)

PLATE 14 (above) Sonoran Desert, Lower Colorado River Valley subdivision, Mayo cinder cone in the Pinacate volcanic field in northwestern Sonora. This region at the head of the Gulf of California is the arid core of the Sonoran Desert. One crater near here received no rain for seven years. There are several types of volcanoes in the Pinacate. See pages 14 and 79-80. (Photo by Mark Dimmitt)

PLATE 15 (facing page) Sonoran Desert, Arizona Upland subdivision, a valley in the Tucson Mountains. Most of this sub-division consists of more hilly terrain. Columnar cacti (saguaros) and legume trees (foothill palo verdes—most of the background vegetation) are indicator plants of the Sonoran Desert. See pages 15-16, 184-193 and 232-233. (Photo by Steve Phillips)

PLATE 16 Summer rainy season in
Arizona Upland. From July through mid
September localized thunderstorms bring
brief but violent downpours that make up
half the annual rainfall in this subdivision.
See pages 16, 25-27 and 149. (Photo by
Jim Honcoop)

PLATE 17 Winter in Arizona Upland, snow on the Tucson Mountains. Winters here are typically mild with occasional storms that bring the other half of the annual precipitation, usually as rain. About twice per decade a snow blankets the desert landscape. See pages 16 and 19-20. (Photo by Jim Honcoop)

PLATE 18 (facing page) Spring in Arizona Upland, Picacho Peak State Park, Arizona. Spring is the greater of two flowering seasons that result from the two rainy seasons. Lupines (blue flowers) are annuals that appear only in wetter than average years. Shrubs such as brittlebush (yellow flowers) bloom in most years. Ocotillo produces its red flowers annually without fail. See pages 16, 21-23 and 147-148. (Photo by Jim Honcoop)

PLATE 19 (above) Foresummer in Arizona Upland. The months of May and June are hot and dry. Most of the plants are dormant and awaiting the summer rains, but saguaro cacti use their stored water to flower and fruit at this time. See pages 16-17, 23-24 and 184-193. (Photo by Jim Honcoop)

PLATE 20 Sonoran Desert, Central Gulf
Coast subdivision near Loreto, Baja
California Sur. Rain rarely falls in this
subdivision despite the adjacent ocean.
See page 17. (Photo by John Wiens)

PLATE 21 Sonoran Desert, Vizcaino subdivision in the Cataviña boulder field in Baja California. This stunning landscape is home to several of the world's most bizarre plants, including the sometimes contorted boojum. See pages 17, 35 and 240-241. (Photo by Thomas R. Van Devender)

PLATE 22 Sonoran Desert, Magdalena subdivision near Ciudad Constitución, Baja California Sur. Half of this subdivision is a rich "woodland" of succulent trees and cacti. The other half is a flat plain with very short, sparse vegetation. The Magdalena Plain has one outstanding plant, the creeping devil cactus. See pages 17-18 and 198-199. (Photo by David Seibert)

PLATE 23 Organ Pipe Cactus National Monument, Arizona. The eastern portion is a biologically rich example of Arizona Upland. The portion of the Ajo Mountains shown here are composed of rhyolite tuffs from explosive volcanic eruptions. See pages 31-32, 74-75 and 194. (Photo by Mark Dimmitt)

PLATE 24 (facing page) Sycamore Canyon, Pajarito Mountains, Arizona. This wet canyon on the Mexican border has more than forty plants that are found nowhere else in Arizona. Wildlife is similarly rich. The canyon, which cuts through ancient volcanic flows, is a magnet to birds and birders; it also harbors rare mammals, fish, and reptiles. See pages 32-33 and 74-75. (Photo by Kim Duffek)

PLATE 25 (above) Alamos, Sonora, Mexico. A charming Spanish colonial town surrounded by tropical deciduous forest. This relaxed tourist paradise offers good hotels, restaurants, culture, and a comfortable base from which to explore the surrounding natural beauty. See pages 34-35. (Photo by Steve Phillips)

PLATE 26 (above) A broad palm canyon in the Cataviña boulder field, Baja California. Much of Baja California is underlain by a large granitic batholith. Where it is exposed on the surface the huge boulders add extra character to a landscape already famous for its strange plants. This perennial oasis is also a great birding spot. See pages 17, 35 and 165-167. (Photo by Thomas R. Van Devender)

PLATE 27 (facing page) Saguaro National Park, Tucson Mountains, Arizona. This small mountain range harbors about 630 species of flowering plants. Half of them are annuals that appear only in wet years. Most of the animal groups have not been enumerated, but are also diverse. The Park has several scenic drives and hiking trails. See pages 36 and 150. (Photo by Mark Dimmitt)

PLATE 28 (facing page) Aravaipa Canyon, southeastern Arizona. From either end of this gash through the Galiuro Mountains, the canyon provides an easy walk along a shallow riffle-stream lined with cotton-woods and willows. The canyon walls are clothed with Sonoran Desert vegetation. Side canyons present many tiny vignettes such as this fern-framed rivulet. Wildlife is abundant. See pages 7-9 and 36. (Photo by Pinau Merlin)

PLATE 29 (above) Gran Desierto, Sonora. This huge sand sea—the largest in North America—partially originates in the rocks of the Colorado Plateau as far away as Utah and Wyoming. The gravel weathered from the rocks was transported down the Colorado River and further reduced to sand, which then blew here from the river delta to the northwest. Wind transport of sand was favored during the Ice Ages when sea level was lowered by as much as 300 feet (100 meters). See pages 84 and 87-88. (Photo by Mark Dimmitt)

PLATE 30 Aerial View of the Pitaycachi
fault line (arrows) southeast of Douglas,
Arizona, in Sonora. This fault was responsible for the magnitude 7.2 earthquake in
May 1887 which killed fifty-one people
in Sonora. The Sonoran Desert region
contains thousands of inactive faults;
active faults are rare. See page 79.
(Photo by Robert Scarborough)

PLATE 31 Desert pavement, Pinacate volcanic region, Sonora. Desert pavements occur only in the drier parts of the Sonoran Desert. Flat pavements such as this one covered with tightly-packed, small stones took at least several tens of thousands of years to develop. The dark, shiny patina on the rocks is deposited by bacteria, which live on this inhospitable surface. See pages 84 and 94-97. (Photo by Robert Scarborough)

PLATE 32 Wilderness of Rocks, Santa
Catalina Mountains. Current geological
thinking suggests prior to Basin and Range
breakup of the country some ten million
years ago most of the Sonoran Desert
region lay as an extensive flat highland,
connected to adjacent lands such as the
Colorado Plateau province, and was much
more tree-covered than the desert. Modern
mountaintops are the only remnant of this
old highland. See page 74. (Photo by
Robert Scarborough)

A star dune field along an isolated set of hills, northwest of Rocky Point, Sonora, Mexico. Sand for this dune field is blown in by wind from coastal wave action along the Gulf of California.

have shaped the Sonoran Desert is to immerse oneself in the study of the region's natural history. Still, hints of profound relations are everywhere around us: storm tracks funneled by Basin and Range topography, which ultimately define the region's ecological limits; the very different past worlds represented in the fossil record; hard granite rocks torn apart by small lichen colonies; the wide variety of landforms and rock types. Also, we must remind ourselves that the true character of the land is much more than what's on the surface—the often-neglected third dimension is vital, as miners, soil engineers, and water well drillers know. All these effects of the past are interwoven into a tapestry of cause and effect on a grand and wonderous scale.

Additional Readings

Chronic, Halka. *Roadside Geology of Arizona*. Missoula, Montana: Mountain Press Publishing Company, 1983.

Harris, Stephen L. *Agents of Chaos*. Missoula, Montana: Mountain Press Publishing Company, 1990.

Nations, Dale and Edmund Stump. *Geology of Arizona*. Dubuque, Iowa: Kendall/Hunt Publishing Company, 1983.

McPhee, John. *Basin and Range*. New York: Noonday Press, 1990.

Sharp, Robert P. *Geology Field Guide to Southern California*. Dubuque, Iowa: Kendall/Hunt Publishing Company, revised edition 1976.

Sheldon, John. *Geology Illustrated*. San Francisco: W.H. Freeman & Company, 1966.

Sykes, Godfrey. *The Colorado Delta*. American Geographical Society Special Publication No. 19, edited by W.L.G. Joerg. Port Washington, N.Y.: Kennikat Press, original pub. 1937, reissued 1970.

Desert Soils

Joseph R. McAuliffe

I grew up in Nebraska and had a narrow view of "soil" for the first part of my life; soil was the deep, dark loamy stuff that supported the bountiful agriculture of the Corn Belt. When I first came to the Sonoran Desert nearly twenty years ago, I hardly considered that anything deserving the label "soil" existed beneath the desert's gravelly and rocky surface. I've learned so much since then; the soils found throughout the Sonoran Desert are far more varied and complex than any I studied as a college student in the Midwest.

Desert soils are downright unusual! They vary tremendously in texture; many are sandy and gravelly, while others contain layers of sticky clay, or even rock-hard, white limy layers. Desert soils may be gray-colored, brown, or even brick red. In the more arid parts of the Sonoran Desert, surfaces of some soils are covered by a layer of small stones that can be as tightly interlocked as pieces of an ancient Roman mosaic, and are coated with dark, shiny rock varnish. Many of these diverse features of desert soils have taken thousands of years or more to form. Characteristics of these soils also greatly affect, and are greatly affected by, desert organisms.

Soil Formation

The varied geological terrains of the Sonoran Desert provide many different kinds of *parent materials* in which soils form. Gravelly or stony *alluvial fans* that spill out of mountain drainages into adjacent basins cover much of the face of the Sonoran Desert (see the chapter "The Geologic Origin of the Sonoran Desert"). The sediments transported all the way to the floors of these basins are usually much finer— sands, silts and clays. The mountains themselves possess various rock types, slopes, and exposures that offer a com- plex array of different soil-forming environments. The monstrous heaps of wind-blown sand in the dune fields of the Gran Desierto in northwestern Sonora (plate 29) and the Cactus Plain east of Parker, Arizona, provide yet another kind of soil parent material.

Soils initially form within the physical matrix of the parent material. Over time, the composition and location of substances within the developing soil change and move around, altering the soil's characteristics. Water is responsible for most of this work of alteration and transportation of materials in soils, even in deserts. Physical and chemical weathering reduces the sizes of coarse particles. Chemical reactions convert some minerals contained in the parent materials into clay minerals. Other chemical reactions paint the soil's colors. Wind-blown dust and solids dissolved or suspended in raindrops slowly add new materials to the desert's surface, and water moves some of these materials downward into the soil. Living things residing in and on a soil further influence the soil's development. All of these processes take time; desert soils that have formed in parent materials that were deposited long, long ago differ considerably from those that have formed in younger deposits.

You can readily observe the effect of time on the degree of soil development in alluvial fans deposited at different times throughout the region. Pleistocene deposits (2.4 million to 11,000 years old) may cover the land surface in some places, especially in areas nearer the mountains, while elsewhere, geologically younger Holocene deposits (less than 11,000 years old) are present. The alluvial deposits on the southwest side of the Tortolita Mountains near Tucson provide an example. Water in Wild Burro Wash transported large volumes of gravelly to rocky alluvium out of Wild Burro Canyon and deposited these materials in alluvial fans on the broad, gently sloping *piedmont* (see the photograph and matching landscape diagram on the next page). Some of these materials were deposited perhaps more than 100,000 years ago, and the surfaces of some of these ancient alluvial fans, although somewhat eroded, still persist. Beneath these old land surfaces are soils that have developed thick, red-colored clayey layers underlain by accumulations of white *caliche* (defined and discussed below). In contrast, soils in nearby, considerably younger Holocene deposits lack the bright red coloration and accumulations of clay and lime. Soils in both the Pleistocene and Holocene deposits started out with the same kind of parent materials: gravelly to rocky, granitic alluvium almost totally devoid of clay and lime. The great contrast in the amount of time over which the soils have developed is responsible for the differences in soil characteristics.

Soil Layers:
Clay, Colors, and Caliche

Many soils of the piedmonts and basin floors of the Sonoran Desert start out as deposits of gravelly or stony alluvium which have fairly uniform characteristics throughout. With the passage of time, however, pronounced horizontal layers called *soil horizons* develop. You can often easily see these horizons in older, well-developed soils because they differ from each other in color. You may also detect them by changes in texture and other

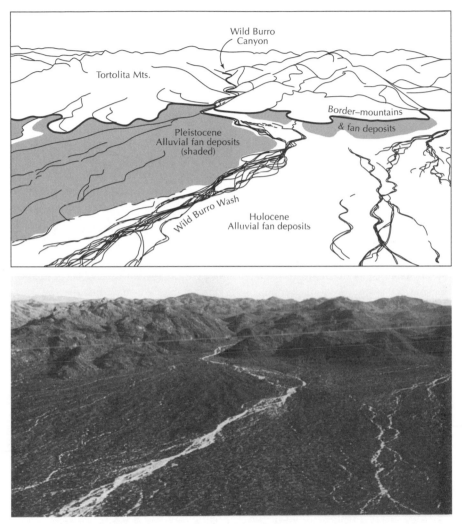

Alluvial fans deposited by Wild Burro Wash on the southwest side of the Tortolita Mountains near Tucson, Arizona. (Photo by Kyle House)

characteristics. Soil horizons form in arid and semi-arid regions through the downward transport of materials by water and the accumulation of various materials at characteristic depths. The depth at which materials accumulate depends mostly on how much precipitation is received (greater amounts of precipitation soaking into the soil transport materials to greater depths) and on the kind of material that is being transported.

CALCIC HORIZONS Many desert soils contain prominent, whitish layers called *calcic horizons*. These are accumulations of calcium carbonate, the same material found in chalk, concrete, and agricultural lime. In the Sonoran Desert, the tops of these horizons are typically less than twenty to forty inches (50 to 100 cm) below the soil surface. Calcic horizons may be very thin (six inches; 15 cm) in some soils and contain only small amounts of calcium carbonate. In other soils, these horizons may be very thick (greater than three feet; 1 m) and strongly cemented. These nearly impenetrable, cemented layers, or *petrocalcic* horizons, are commonly called *caliche*.

Calcic horizons are most typically unique features of soils of arid and semi-arid regions and are usually absent from parts of the world receiving much precipitation. The amount of precipitation that infiltrates into the soil is the most important factor that determines the depth to which calcium carbonate is transported and accumulates. In relatively moist parts of the Sonoran Desert, such as areas near Tucson, Arizona, where annual precipitation averages ten inches (25 cm) or more, calcium carbonate tends to accumulate at depths exceeding ten inches (25 cm). In extremely arid regions, though, such as the area around Yuma which annually receives four inches (10 cm) or less precipitation, calcium carbonate may accumulate within a few centimeters of the surface, or even at the surface. In contrast to desert regions, areas such as the eastern United States receive so much rainfall that calcium carbonate never accumulates in pronounced soil horizons, because it is readily leached from the soil and flushed into the groundwater.

Thick, strongly-cemented calcic horizons take a long time to form. They start as thin, patchy coats of whitish

calcium carbonate on the lower surfaces of pebbles and small stones. In fine-grained parent materials, such as dune sand, that lack coarse materials, calcium carbonate first appears as thin, white, thread-like accumulations where small roots have extracted soil water and caused the calcium carbonate to precipitate. These weakly-developed calcic horizons can form within a few thousand years (see the illustration on page 91). Accumulation of more calcium carbonate eventually produces thicker, continuous coatings on pebbles and stones or pronounced whitish nodules in fine-grained parent materials. Eventually, additional accumulation of calcium carbonate fills the soil interstices between pebbles or nodules and the calcic horizon becomes plugged, greatly restricting the downward movement of water. Once this occurs, calcium carbonate may continue to accumulate on the top of the calcic horizon in hard, cemented layers and may literally engulf and obscure overlying soil horizons in the process (see the photo on page 92). It takes many tens to hundreds of thousands of years for such strongly-developed calcic horizons to form. Sometimes hard, whitish caliche becomes exposed on the surfaces of very old soils when erosion removes overlying, less erosion-resistant soil horizons. These partly eroded soils are very common throughout the Sonoran Desert and are called *truncated* soils.

For a long time the source of the tremendous amounts of calcium carbonate contained in calcic horizons of desert soils was a mystery. An obvious suggestion was that the original parent materials provided the necessary

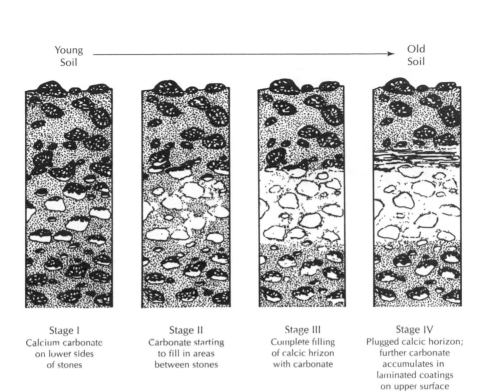

Young Soil → Old Soil

Stage I
Calcium carbonate
on lower sides
of stones

Stage II
Carbonate starting
to fill in areas
between stones

Stage III
Complete filling
of calcic hrizon
with carbonate

Stage IV
Plugged calcic horizon;
further carbonate
accumulates in
laminated coatings
on upper surface

Stages of development of calcic horizons

calcium. This explanation is plausible for parent materials containing limestone and other calcium-rich rocks. However, thick calcic horizons are also found in soils that have formed in parent materials that contain little or no calcium, such as alluvium derived from many igneous rocks or non-calcareous sedimentary and metamorphic rocks. In soils developed in these parent materials, the calcium clearly comes from somewhere else. Another commonly held belief is that groundwater deposited these limy soil layers. Yet calcic horizons are found even in soils where the depth to groundwater exceeds 330 feet (100 m)—even in days before extensive groundwater pumping—and where there is clear evidence that groundwater was far from

the surface even during less arid conditions of the last glacial episode more than 11,000 years ago.

The answer to the mystery was literally "blowing in the wind." Atmospheric additions of calcium contained in dust and precipitation are the predominant sources of the calcium contained in calcic horizons of most desert soils. Dust inputs have been measured in desert environments throughout the American Southwest. In a detailed study that spanned a decade in the northern Chihuahaun Desert near Las Cruces, New Mexico, dust traps positioned about thirty-five inches (90 cm) above the soil surface collected an average of more than one ounce (23 g) of dust per square meter surface area per year. The calcium contained in this

Piece of strongly cemented calcic horizon ("caliche") removed from soil excavation showing laminations on upper surface.

amount of dust was enough to form the equivalent of about a half gram of calcium carbonate per square meter of ground surface in one year's time.

In addition to dust, precipitation also contains dissolved calcium. In the Las Cruces area, two to three times the amount of calcium contained in the deposited dust is delivered to the ground surface dissolved in precipitation. Water moving downward into the soil carries dissolved calcium. Plant roots and other organisms living in the soil respire, producing carbon dioxide which, dissolved in soil water, provides the rest of the chemical building blocks needed to form calcium carbonate. Calcium carbonate precipitates out of solution once the soil begins to dry. Over thousands and thousands of years, the small annual inputs of calcium in dust and precipitation add up to very large accumulations in soil calcic horizons.

CLAY-RICH HORIZONS Some desert soils look like a layered cake with one or more clayey, reddish-brown horizons directly above the white calcic horizon. These clay-rich layers are called *argillic horizons*, formed over long period of time when clay particles suspended in

water are carried downward into the soil and accumulate. Like calcic horizons, the strength of development of argillic horizons depends strongly on soil age. Desert soils formed in materials deposited during the Holocene (less than 11,000 years old) usually lack argillic horizons. Pleistocene deposits that are tens to hundreds of thousands of years old often contain pronounced argillic horizons. Argillic horizons are typically located above calcic horizons, because minute clay particles suspended (but not dissolved) in water are not transported as deeply as is dissolved calcium carbonate (see the photograph on page 93).

In some of the moister parts of the Sonoran Desert, such as areas around Tucson that annually receive an average of ten inches (250 mm) or more precipitation, argillic horizons of some soils may be more than one and one-half feet (½ m) thick and parts may contain more than fifty percent clay. Periods of abundant rainfall can turn these clay-rich layers into sticky traps that can easily mire a vehicle. In contrast, during the dry season, these clay-rich horizons become as hard as adobe brick. When they are dry, strongly developed argillic horizons often have a pronounced prismatic or blocky structure that is a product of repeated swelling and shrinking of the clays.

In more arid parts of the Sonoran Desert, argillic horizons are also present in many soils, but they are not as pronounced; they are much thinner and contain less clay than those found in moister regions.

Argillic horizons do not form in certain kinds of parent materials in arid and semi-arid environments. Soils that have formed in calcium carbonate-rich parent materials, such as limestone or exposed calcic horizons of ancient, truncated soils, typically lack argillic horizons. The presence of calcium carbonate apparently causes clay particles to clump together in a way that prevents them from being dispersed in water, thereby inhibiting the downward movement and accumulation of the clays into argillic horizons.

produce this color. For example, biotite and hornblende are common iron-bearing minerals in parent materials derived from many kinds of igneous and metamorphic rocks. Weathering of these minerals creates new minerals, including the rust-colored iron oxide compounds. As these compounds accumulate, they paint the soil with increasingly strong reddish hues. Very young soils lack accumulations of

Roadcut in Scottsdale, Arizona on a soil developed in a Pleistocene-aged alluvial fan deposit of granitic gravel. The white scale bar is 1 m long and scaled in decimeters. The whitish calcic horizon is positioned below a reddened, clay-enriched argillic horizon.

SOIL COLORS Desert soils come in a variety of colors. Some soils have the same pale, brownish color from top to bottom, but others may be layered with browns, reds, pinks, and whites. Argillic horizons of many older soils in the Sonoran Desert are a distinct, rusty brick red. The weathering (oxidation) and accumulation of iron-bearing minerals contained in the soil

these iron oxides and are typically pale brown or yellowish-brown. In contrast, soils that have been developing for many tens to hundreds of thousands of years on Pleistocene deposits frequently contain soils with strongly reddened horizons.

The degree of reddening that develops within a soil depends strongly on the iron content of minerals in the parent material and on how rapidly the parent materials weather. For example, some granitic rocks with abundant biotite weather very rapidly, producing strongly reddened soils. In some parent materials, the presence of extremely iron-rich minerals such as magnetite or hematite further intensifies red soil colors. Less intense reddening develops in soils derived from parent materials that are relatively poor in iron minerals and are more resistant to weathering. Some soils never develop pronounced red coloration. In deserts, only the slightest reddening of soil horizons occurs in calcareous parent materials derived from limestone, even in old soils. The predominant colors of these soils are light brownish-grays and whites of strongly-developed calcic horizons.

Soil Surfaces: Desert Pavements

Large, flat areas devoid of vegetation and covered by a layer of tightly packed small stones are conspicuous features of extremely arid landscapes. These desert *pavements* are rare or absent in the moister parts of the Sonoran Desert, but become increasingly pronounced in the driest parts (see the photo at the top of page 95 and plate 31).

Some of the most extensive and well-developed areas of desert pavements occur on stony alluvial fan deposits flanking the rugged, low mountains in the extremely arid lower Colorado River Valley. Geologically young deposits (Holocene-aged, less than 11,000 years) lack the flat-surfaced pavements. Surfaces of these young deposits are typically cluttered with large stones and rocks irregularly piled in elevated bars; these low bars are separated by intervening swales. This stony jumble of bars and swales is the topographic imprint of the surface's creation by the powerful tumult of moving water laden with rocky debris. Over time, though, this imprint disappears as the vertical relief of these coarse, rocky deposits is leveled out, eventually forming the flat pavement of small stones. The best-developed pavements are those that have formed over the passage of several tens of thousands to a few hundreds of thousands of years.

Research conducted within the last fifteen years in the Mohave Desert by a team of soil scientists and geologists (L.D. McFadden, S.G. Wells, and M.J. Jercinovich) provides a detailed picture of how desert pavements form on rocky parent materials such as these. Physical weathering of the large rocks on the surface produces the smaller stones that eventually form the pavement surface. These smaller stones tend to accumulate in topographic lows on the original, uneven surface. Special soil characteristics found directly beneath pavements provide clues to an additional process involved in creation of pavements. If you carefully remove the layer of stones from a pavement surface, you will find a distinct, fine-grained soil horizon called a *vesicular A* (or *Av*) *horizon* (see the photograph at the bottom of page 95). The name "vesicular" refers to the many vesicles or large pores found throughout the horizon. "A" denotes its position as the topmost mineral layer of soil. The Av horizon is typically a few centimeters (about an inch) thick, and contains mostly silts and

Closeup of tightly packed stones in a desert pavement. The dark stones are varnish-covered pieces of volcanic rhyolite; the white pieces are quartz on which varnish usually does not form.

clays; it lacks coarse materials, even though small stones of the pavement cover the Av horizon and rocky materials occur in the soil underlying it.

The origin of the fine-grained Av horizon is an important key to understanding how the overlying flat-topped pavement develops. Examination of the minerals contained in the Av horizon at one site in California demonstrated that the materials in this horizon did not originate from the weathering of the rocky parent materials. Instead, dust deposited on the stony surface is the source of the silts and clays of the Av horizon. These fine-grained materials accumulate beneath a layer of surface stones, separating these stones from the rest of the underlying rocky materials. Over time, the further accumulation of fine-textured materials in the Av horizon literally lifts the mono-layer of stones of the pavement

and levels the surface (see the illustration on page 96).

Prior to this work, it was commonly believed that most desert pavements originated through the selective erosion of fine materials from the surface by either wind or water, a process called *deflation.* However, such a process cannot

Side view of a piece of the fine-textured Av horizon removed from beneath a desert pavement showing the air-filled vesicles.

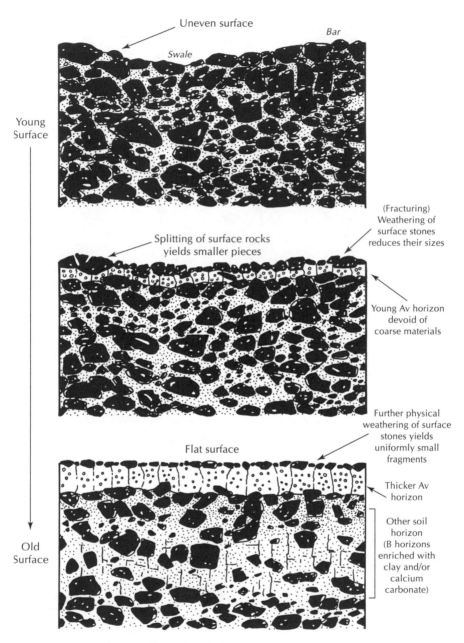

Young Surface

Uneven surface

Swale

Bar

(Fracturing) Weathering of surface stones reduces their sizes

Splitting of surface rocks yields smaller pieces

Young Av horizon devoid of coarse materials

Further physical weathering of surface stones yields uniformly small fragments

Flat surface

Thicker Av horizon

Other soil horizon (B horizons enriched with clay and/or calcium carbonate)

Old Surface

Formation of desert pavement.

explain the development of pavements in stony parent materials that initially lacked fine-grained materials (such as many coarse-grained alluvial fan deposits or areas of exposed bedrock, including basalt flows). Nor can it explain the presence of the fine-grained Av horizon that separates the surface pavement from underlying rocky materials. In coarse parent materials, atmospheric *additions* of fine materials (rather than their selective removal) and incorporation of these materials into the Av horizon below a layer of stones are responsible for creation of the pavement surface.

More rarely, however, stone pavements can also be created by a process of deflation in certain environments where the original parent material consisted of small stones and pebbles mixed with abundant fine-grained sands, such as in beach deposits that ring ancient lake beds in some parts of the region. In such cases, the selective removal of the sand by wind and/or water leaves behind a lag of pebbles that resists further deflation. Once the surface lag stabilizes the fine-grained deposit, the pebbles act as a dust trap and airborne materials accumulate to eventually form a silt- and clay-rich Av horizon beneath the pebble pavement in the same manner as that which occurs on rocky parent materials. These kinds of pavements exist in a limited number of locations in the Sonoran Desert region, but are less common by far than pavements that developed on coarse, rocky parent materials as described previously.

ROCK VARNISH Desert pavements are frequently very dark-colored; in many cases they are nearly black. *Rock varnish* (frequently called *desert varnish*) on the stone surfaces provides this dark complexion, despite the rock's internal color. The glossy coatings of desert varnish on stones are very thin, at most a few hundredths of a millimeter thick—

about the thickness of a sheet of paper. These thin, lustrous coatings contain a variety of constituents. Clay minerals typically form about three-quarters of the bulk of the varnish and manganese oxides impart the dark color. Many other minerals are present in trace amounts. Desert varnish also contains organic matter, apparently derived from microbial activity.

How rock varnish forms is poorly understood. Many of the mineral ingredients of varnish, including clays and manganese, are probably derived from airborne materials that settle on rock surfaces. Bacteria residing on the rock surface may play a major role in concentrating and cementing these materials to form the glossy coatings. Laboratory studies have shown that rock varnish gives off considerable carbon dioxide when moistened, indicating bacterial respiration. However, bacteria are generally absent from the shiny exposed surfaces of varnish, indicating that they reside within and beneath the microscopic varnish layers.

The formation of varnish may actually be a means by which these microbes protect themselves in the exposed, extreme environment of a rock surface in the desert. Interestingly, the manganese oxides in rock varnish very effectively block the transmission of ultraviolet radiation. Perhaps the rock-dwelling microbes manufacture their own manganese-formula sun-screen!

Rock varnish forms very slowly. Surfaces of some rocks, including many coarse-grained granitic rocks, rarely sport thick coats of varnish because they weather and erode faster than varnish can form. But if rock surfaces resist weathering, varnish coatings become increasingly thick and dark with the passage of time. Rock varnish therefore provides geologists with a valuable tool for determining relative ages of different alluvial fan deposits. Stone surfaces in older deposits are generally covered by thicker, darker coats of varnish. In some cases, the thickest, darkest coatings of varnish found on older deposits may have been accumulating for many tens of thousands to over 100,000 years.

Ancient inhabitants of the Sonoran Desert used varnished desert pavements as dark-colored canvases on which they rendered gigantic artistic impressions. By removing the dark varnished stones and exposing the underlying light-colored soil, prehistoric peoples created fantastic images of human figures, animals, and abstract forms. Many of these *intaglios* or *geoglyphs* have been discovered in the vicinity of the lower Colorado River Valley near Blythe, California and Ehrenberg, Arizona. Intaglios of the Sonoran Desert are frequently ten yards (10 m) or so long. The largest yet discovered, located about seventy-five miles (120 km) west of Phoenix, is a human figure nearly the length of a football field. These gigantic works of landscape art may have been created well before European colonization of the Americas. We don't know why the intaglios were created or the purposes they may have served. However, according to legends of indigenous peoples who occupied the area in historic times, some of the giant, human-like figures represent gods or other supernatural beings. The large, skyward-looking images, although usually difficult to recognize from the ground, certainly would be apparent to ancient peoples' deities in

Intaglio (geoglyph) found northeast of Quartsite, Arizona.

skies above. Inhabitants of other desert regions of the world also created intaglios long ago on the surfaces of stone pavements, perhaps for similar reasons. One of the best known examples is located on extensive alluvial fan deposits in the Peruvian coastal desert near Nazca. These intaglios include long, straight lines that are fifteen to twenty miles (24-32 km) long, geometric forms, and images of many kinds of animals.

Rock surfaces covered with dark desert varnish provided prehistoric desert dwellers with another medium for artistic expression. Petroglyphs are designs created by chipping away the surface varnish on large rock and boulder surfaces, exposing lighter-colored rock beneath. These are much, much smaller than the giant intaglios, but similarly include diverse shapes, geometric designs, and human and animal forms.

Darkly-varnished desert pavements take so long to form and are extremely sensitive to disturbance. The intaglios created by ancient peoples can last for centuries. So will the uninspiring and less aesthetically appealing tracks so thoughtlessly created in our time by drivers of off-road vehicles.

Soils and Desert Life

Soils of the desert teem with living things. Plant life ranging from single-celled cyanobacteria to giant saguaro reside on and in the soil. Tunneling termites and burrowing mammals turn over the soil. The activity of some organisms is very ephemeral and occurs in a short span of time after rains. For many others, the soil provides protection from environmental extremes, permitting year-round activity. The survival, growth, and reproduction of living things depends considerably on soil characteristics. Conversely, the activities of organisms living on and in the soil significantly influence the characteristics of soils.

SOIL CRUSTS Some of the most abundant organisms that inhabit the soil are ones that we scarcely notice. Very small organisms, including cyanobacteria, algae, lichens, mosses, and liverworts, form living crusts on many desert soils. Scientists call these coverings *microphytic crusts*. Depending on the environment, they may be a mix of many of the above kinds of photosynthetic organisms. Individual lichens, mosses, and liverworts are visible to the naked eye; others like the cyanobacteria are micro-

scopic, but their presence *en masse* is revealed by the dark color on the soil surface. In some places, the living microphytic crust covers more of the ground surface than do canopies of shrubs, small trees, cacti and other "macrophytes."

Different soil microenvironments support different kinds of microphytic crusts. In the Sonoran Desert, larger lichens, mosses, and liverworts are more common on slopes with northern exposures; these retain moisture for longer periods after rainfall. In some fine-grained soils on the floors of basins in the hottest, driest parts of the Sonoran Desert, the dark-colored soil surface is due to the presence of untold millions of cyanobacteria or "blue-green algae." These dark cyanobacterial crusts lie dormant most of the time but are physiologically "awakened" when the soil surface is wetted. After a rain, these organisms typically remain active for only a day or two before the soil surface again dries.

Microphytic soil crusts significantly affect soils. Cyanobacteria create sticky materials that bind soil particles together. Threadlike structures called *hyphae*, produced by the symbiotic fungi of lichens, knit the soil together, making it more resistant to erosion by the intense splashes of thunderstorm raindrops. Cyanobacteria can fix atmospheric nitrogen, thereby adding nitrogen to the soil in a form that potentially can be used by larger plants.

Like desert pavements, these living crusts can easily be destroyed by human activities. Mechanical disturbance by recreational vehicles poses a significant threat in all desert regions of the

American Southwest. Excessive livestock trampling can also greatly damage microphytic crusts. Once destroyed, recovery of some kinds of microphytic crusts can be very slow, taking decades to perhaps a century or more.

SOILS AND ANIMALS Soil conditions directly affect many kinds of desert animals. For burrowing animals, the choice of a place to excavate a living shelter may depend on soil texture. For example, in the Mohave Desert, desert tortoises apparently require soils that are loose enough to be excavated, but firm enough so that the burrows will not collapse. Consequently, in areas containing both extremely sandy soils versus loamy soils with higher clay content, tortoises tend to construct their burrows in the loamy soils. Different kinds of soil horizons may physically impede burrowing activities of some animals. For example, the mound-like burrows of Merriam's kangaroo rat (*Dipodomys merriami*) frequently are found beneath the canopy of a large creosote bush (*Larrea tridentata*) in sandy to loamy soils that lack substantial development of argillic or calcic horizons. However, on soils with extremely shallow, strongly cemented calcic horizons, this species of kangaroo rat is usually absent or rare, probably because of the difficulty of excavating burrows in these soils.

Soil texture may also affect some animals by controlling the availability of water. Two similar species of large, seed harvester ants of the Sonoran Desert (*Pogonomyrmex rugosus* and *Messor pergandei*) tend to occupy soils that differ slightly in clay content. *P. rugosus* is usually found in soils with higher clay contents than those occupied by *Messor pergandei.* The reasons for this separation are not well understood, but perhaps the influence of soil texture on the infiltration and availability of water affect the two species in different ways.

Animals can profoundly affect characteristics of desert soils. The animals that dig, wriggle, tunnel and burrow through desert soils range in size from microscopic mites and nematodes to badgers and coyotes. Their activities move soil around and cycle nutrients among the realms of "animal, vegetable, and mineral."

The soil directly beneath perennial plant canopies is enriched with organic matter and nutrients, compared to bare areas between plants. Beneath plant canopies, organic materials are buried by the burrowing activities of mammals, and dead plant roots provide abundant organic matter on which many small animals feed. In this case, ecological wealth generates more wealth, and these "fertile islands" foster considerably higher densities and increased activity of small animals living on and in the soil.

Two groups of very small animals, arthropods and nematodes, are abundant in desert soils, especially in the fertile islands beneath plant canopies. Comparatively speaking, some of these animals are relatively large, such as the .04 to .06 inch (1-2 mm) long springtails (order Collembola), small insects that reside in leaf litter. Myriads of others, including many kinds of mites and nematodes, are extremely small, requiring a microscope to see them.

Different species of soil mites play a variety of ecological roles. Many are scavengers that feed on decomposing plant and animal material. Others are herbivores that pierce cells of plant roots with their mouthparts to drain the contents. Some are carnivores that feed on other small arthropods, other mites, and nematodes. Soil nematodes, or microscopic roundworms, are equally varied in the ecological roles they play. Some consume bacteria and other microbes; others subsist on fungi or plants or are omnivorous; some are micropredators. In turn, nematodes are the prey of other small animals, such as predaceous mites. One team of scientists estimated that the top twelve inches (30 cm) of soil at a site in the Mohave Desert contained slightly over one million nematodes per square meter (10.7 ft²). However, this tremendous number of microscopic roundworms amounted to less than .0035 ounce (¹/₁₀ gram) dry weight. Other surveys indicate similar abundances of soil nematodes in the Sonoran Desert. The minute soil arthropods and nematodes, together with the plant foods and organic matter they consume, form complex food webs of producers, prey, and predators. The operation of this web continuously recycles mineral nutrients between living and non-living components of the desert ecosystem.

Larger animals that tunnel and burrow in the soil, including termites and many kinds of rodents, are biological bulldozers. Their excavations can significantly affect characteristics of desert soils. The excavation and turnover of soil by many seed-eating rodents, such as Merriam's kangaroo rat, creates more porous soils into which water more easily infiltrates. The increased soil permeability created by the excavations of some kinds of kangaroo rats may actually benefit creosote bush plants growing on the mound-like burrow systems. In the warm deserts of the American Southwest, termites are perhaps the greatest earth-movers. Subterranean termites transport a tremendous amount of relatively deep soil materials to the surface when they construct their mud-covered tunnels on the soil surfaces and on the cellulose-containing foods they consume. In this way, desert termites accomplish the same kind of ecological role as do earthworms in soils of moister regions. In the Chihuahuan Desert of southern New Mexico, a group of ecologists estimated that termites annually brought over 1760 pounds (800 kg) of soil materials per two and one-half acres (1 hectare) to the soil surface. Termite activity in some parts of the Sonoran Desert is probably comparable. This tremendous turnover of the soil can greatly affect the formation of soil horizons and the distribution of clay, calcium carbonate and other materials. In southern New Mexico, this turnover of soil by termites has apparently thoroughly mixed soil horizons of some older soils, obliterating distinct zones of clay accumulation in argillic horizons.

SOILS AND PLANTS The availability of water usually poses the single greatest limitation to plant life in deserts. Although desert environments are defined by the amount of precipitation received, the amount of moisture that is available for plants to use is modified by features of the soil. Subtle changes in permeability and texture affect how much precipitation is either absorbed by the soil or lost to runoff, and how deeply water infiltrates into the soil.

The different soil horizons discussed earlier affect infiltration and depth of water storage. Relatively large quantities of water can soak into the surface of a sandy-gravelly soil that lacks horizon development. A significant fraction of rainfall may infiltrate to perhaps a meter in such a deep, permeable soil. Water stored deeply is protected from evaporation to the atmosphere. In contrast, clayey argillic horizons of well-developed soils can hold much more water than sandy layers. As a consequence, clayey soil layers absorb much water as it moves downward, greatly limiting deeper infiltration of water. Shallow soil moisture is more readily lost to evaporation. The different kinds of soil water reservoirs (deep versus shallow) vary in their persistence throughout the year. Various perennial plant species respond differently to these contrasting soil moisture conditions. Some species are better suited to soils that foster deep infiltration; others have ways to cope with the extreme seasonal fluctuations of shallow soil moisture.

For example, in the Arizona Upland subdivision of the Sonoran Desert where average annual precipitation generally exceeds eight inches (20 cm), soils on Pleistocene-aged alluvial fan surfaces that have strongly developed argillic horizons are covered by perennial vegetation that is very different from the vegetation on nearby Holocene surfaces or on erosionally truncated soils. Creosote bush is typically the dominant plant species on soils lacking argillic horizons on Holocene fan surfaces. The relatively deep root system of this plant enables it to extract water from deep in the soil. Although creosote

bush also contains abundant shallow roots, the supply of the more constant (although typically meager), deeper supplies of water allows this evergreen shrub to remain active throughout the entire year, even during extended rainless periods when shallow soil moisture has largely disappeared. Creosote bush is also often the dominant species on erosionally truncated soils with shallow, calcic horizons. Some deep water is available in these soils due to cracks and fissures within the otherwise hard, impenetrable layers of caliche. In contrast, creosote bush is often rare or even absent from soils with thick, strongly developed argillic horizons. The principal occupants of these soils are usually triangleleaf bursage (*Ambrosia deltoidea*), and cacti, including staghorn and buckhorn chollas (*Opuntia acanthocarpa* and *O. versicolor*). These plants have relatively shallow roots and are capable of quickly using shallow soil water when it is briefly available and then surviving lengthy periods when it is not. Triangleleaf bursage survives during long dry periods by shedding its leaves and becoming dormant (a condition called *drought dormancy*). In the cacti, water stored in succulent tissues enables the plants to continue photosynthetic activity past the time when water no longer can be extracted by roots.

In the extremely arid parts of the Sonoran Desert, some soil horizons may reduce the infiltration of water to the extent that no perennial plants can survive. Beneath stone pavements, the silts and clays of Av horizons can soak up a considerable amount of moisture, preventing water from passing to deeper parts of the soil. Even in the case of the occasional storm that delivers large

Vegetation dominated by triangleleaf bursage and buckhorn cholla on soil with well-developed argillic horizon on Pleistocene-aged alluvial fan deposit. Harquahala Mountains, Arizona.

amounts of rain, once the fine-textured Av horizon is saturated, very little additional water can soak into the soil surface and the abundant rainfall is quickly lost as runoff. The shallow moisture below the pavement then quickly evaporates, preventing occupancy of the surface by perennial plants. The shallow infiltration of water into pavement soils also creates another soil condition that makes some desert pavements extremely inhospitable to many plants. Because water seldom, if ever, infiltrates far beneath the surface, large quantities of salts, including sodium chloride, tend to accumulate near the surface. In the vicinity of Yuma, Arizona, salt contents in some soils directly beneath varnished pavements have been measured and are about thirty times higher than salt content of nearby soils lacking pavements. Most common perennial plants

of the Sonoran Desert cannot tolerate such salt concentrations, especially when combined with the scarcity of water.

The large amount of runoff from desert pavements concentrates in shallow runnels and washes cut into the pavement surface. In these places where the stone pavement and Av horizon are absent, water has a chance to soak into the soil, sustaining sinuous lines of creosote bush, bursage, ocotillo, palo verde, and desert ironwood that weave across the otherwise barren pavements.

Throughout the expanses of the Sonoran Desert in Arizona, California, Sonora, and Baja California, you will see many places where the character of the vegetation abruptly changes. Differences in soils, often subtle, but sometimes pronounced, and the way soil characteristics influence soil water, are responsible for much of this compositional complexity in the Sonoran Desert.

Selected References

Bassett, C.A. "Lonely Giants." *Arizona Highways* 65: 38-45 (1989).

Chew, R.M. and W.G. Whitford. "A Long-term Positive Effect of Kangaroo Rats (*Dipodomys spectabilis*) on Creosotebushes (*Larrea tridentata*)." *Journal of Arid Environments* 22: 375-386 (1992).

Drake, N.A., M.T. Heydeman and K.H. White. "Distribution and Formation of Rock Varnish in Southern Tunisia." *Earth Surface Processes and Landforms* 18: 31-41 (1993).

Gile, L.H. "Causes of Soil Boundaries in an Arid Region: I. Age and Parent Materials." *Soil Science Society of America Proceedings* 39: 316-323 (1975)

————. "Causes of Soil Boundaries in an Arid Region: II. Dissection, Moisture, and Faunal Activity." *Soil Science Society of America Proceedings* 39: 324-330 (1975).

Gile, L.H., J. W. Hawley and R.B. Grossman. "Soils and Geomorphology in the Basin and Range Area of Southern New Mexico," *Guidebook to the Desert Project. Memoir 39*, Socorro, New Mexico: New Mexico Bureau of Mines and Mineral Resources, 1981.

Gile, L.H. and R.B. Grossman. "Morphology of the Argillic Horizon in Desert Soils of Southern New Mexico." *Soil Science* 106: 6-15 (1968).

Gile, L.H., F.F. Peterson and R.B. Grossman. "Morphological and Genetic Sequences of Carbonate Accumulation in Desert Soils." *Soil Science* 101: 347-360 (1966).

Johnson, R.A. "Soil Texture as an Influence on the Distribution of the Desert Seed-harvester Ants *Pogonomyrmex rugosus* and *Messor pergandei*." *Oecologia* 89: 118-124 (1992).

McAuliffe, J.R. "The Sonoran Desert: Landscape Complexity and Ecological Diversity." *Ecology of Sonoran Desert Plants and Plant Communities*. R. Robichaux, ed. Tucson: University of Arizona Press, 1999.

————. "Landscape Evolution, Soil Formation, and Ecological Patterns and Processes in Sonoran Desert Bajadas." *Ecological Monographs* 64: 111-148 (1994).

McFadden, L.D., S.G. Wells and M.J. Jercinovich. "Influences of Eolian and Pedogenic Processes on the Origin and Evolution of Desert Pavements." *Geology* 15: 504-508 (1987).

Peterson, F.F. "Landforms of the Basin and Range Province Defined for Soil Survey." *Nevada Agricultural Experiment Station Technical Bulletin* 28 (1981).

Polis, G.A., ed. *The Ecology of Desert Communities*. Tucson: University of Arizona Press, 1991.

VandenDolder, E.M. "Rock Varnish and Desert Pavement Provide Geological and Archaeological Records." *Arizona Geology* 22(1): 1, 4-7 (1992).

Weide, D. L., ed. *Soils and Quaternary Geology of the Southwestern United States*. GSA Special Paper 203. Boulder: Geological Society of America, 1985.

West, N.E. "Structure and Function of Microphytic Soil Crusts in Wildland Ecosystems of Arid to Semi-arid Regions." *Advances in Ecological Research* 20: 179-223 (1990).

Human Ecology
of the Sonoran Desert

Thomas E. Sheridan

*I*n 1922, Aldo Leopold and his brother canoed through the delta of the Colorado River. They hunted quail and geese, watched bobcats swat mullet from driftwood logs, and dreamed of *el tigre* (jaguar), whose "personality pervaded the wilderness," even though they never saw any of the big cats. "For all we could tell, the Delta had lain forgotten since Hernando de Alarcón landed there in 1540," Leopold mused. "When we camped on the estuary which is said to have harbored his ships, we had not for weeks seen a man or a cow, an axe-cut or a fence" (Leopold 1949, 141). "On the map the Delta was bisected by the river," the famed conservationist goes on to say, "but in fact the river was nowhere and everywhere, for he could not decide which of a hundred green lagoons offered the most pleasant and least speedy path to the Gulf. So he traveled them all and so did we. He divided and rejoined, he twisted and turned, he meandered in awesome

jungles, he all but ran in circles, he dallied with lovely groves, he got lost and was glad of it, and so were we. For the last word in procrastination, go travel with a river reluctant to lose his freedom in the sea."

By the time Leopold wrote those words in the 1940s, he knew he was writing an elegy, not a paean. The river mighty enough to support a jungle in the desert had already lost its freedom, not to the Gulf of California into which it emptied, but to California farmers and the City of Los Angeles. Beginning in the 1890s, Anglo-American promoters and government engineers strove to break the Colorado to the new Western order. Their first attempts nearly triggered a geological catastrophe, when floods in 1905 sent the Colorado roiling down a canal with no headgate and turned the Salton Sink into the Salton Sea. In 1936, however, a white wall of more than three million cubic yards of concrete rising 726 feet against black rock halted the river in its tracks. Erected to prevent floods and to provide hydro-electric power, Hoover Dam turned the Colorado into a tame ditch for the last 300 miles of its course to the sea. The Colorado and its tributaries, along with the other major rivers that brought water to the Sonoran Desert, such as the Yaqui and the Mayo, became ghosts of the past, victims of the twentieth century, carcasses of sand whose lifeblood had been diverted into cotton fields, copper mines, and vast, sprawling cities.

The Native Americans

Leopold foresaw the transformation and never revisited the delta. "I am told the green lagoons now raise cantaloupes. If so, they should not lack flavor," he wrote. "Man always kills the thing he loves, and so we the pioneers have killed our wilderness" (Leopold 1949, 148).

The passion of those words rang with recrimination. But they also reflected a deeply American romanti-cism as well. Leopold felt that he was witnessing the death of wilderness in the Southwest. Like most Anglo-American newcomers, however, he underestimated the impact that native peoples had already visited upon the Southwestern landscape. Much of the Sonoran Desert, after all, had been homeland to American Indians for at least 12,000 years. Their transforma-tions were more subtle, and in most cases more benign, but human groups have shaped the flora and fauna of the Sonoran Desert, including the Colorado delta, for millenia.

Geoscientist Paul Martin believes that Paleoindians armed with stone-tipped spears overhunted the great Pleistocene mammals of North America and helped drive them to extinction. His "Pleistocene overkill" theory is controversial, but prehistoric Indian societies clearly had an impact upon local animal populations.

Analyzing animal bones from hun-dreds of Hohokam sites in central and southern Arizona, zooarchaeologist Christine Szuter traces a decline in artiodactyls (deer, bighorn sheep, and pronghorn antelope) and an increase in rodents and in lagomorphs (cottontails and jackrabbits) as Hohokam settle-ments grew larger and more sedentary. In other words, the longer Hohokam lived in an area, the more they hunted out the big game and relied upon rabbits and rodents for animal protein. And as they cleared desert vegetation for firewood and fields, they also harvested more jack rabbits, which fled predation by running across flats, and fewer cot-tontails, whose instincts told them to dash for cover that was no longer there.

The Hohokam were sophisticated desert farmers who built ballcourts, platform mounds, and the largest system of irrigation canals in pre-Columbian North America. But even hunters, gatherers, and fisherfolk manip-ulated the distribution of plants and animals. The Comcáac, or Seri Indians, live on the coast of Sonora in one of the driest landscapes on the continent. Seris never cultivated domestic plants or raised domestic animals except dogs, but they did carry certain species of wild plants and animals with them as they moved across the desert and sea. According to ethnobiologist Gary Nabhan and his colleagues at the Arizona-Sonora Desert Museum, their sojourns expanded the range of at least five of the forty-nine species of terrestrial reptiles in Seri territory beyond areas where they naturally occurred.

A case in point is the story of the piebald chuckwallas of Isla Alcatraz in Bahía Kino. These large lizards exhibit traits from three different species— *Sauromalus varius*, from San Esteban Island, *S. ater* from the Sonoran mainland, and *S. hispidus* from the western midriff islands. During the earlier part of this century, when Asian demand for the swim bladders of totoabas (*Totoaba macdonaldi*, the largest member of the croaker family) triggered a fishing boom in the Gulf of California, Seri fishermen released chuckwallas and iguanas as survival foods on islands like Alcatraz. On Alcatraz, at least, the chuckwallas interbred to create a larger, meatier reptile. These hybrids are the result of cultural, not natural, processes of biogeographic distribution. Piebald chuckwallas may not rival Hoover Dam, but they do represent human-induced changes in the Sonoran Desert.

Contrary to romantics like ecologist Stanwyn Shetler, pre-Columbian America was not a "pristine natural kingdom" where "the native people were transparent in the landscape, living as natural elements of the ecosphere" (Shetler 1991, 226).

The most intensive way pre-Columbian Native Americans transformed their environments was through agriculture. Archaeologists are finding evidence that so-called "Archaic" peoples were growing maize (corn) at least 3000 years ago in well-watered areas like the Tucson Basin. Then came pinto and tepary beans, gourds, squash, cotton, and a host of other plants including amaranth and devil's claw. The Cocopas cultivated Sonoran panic grass in muddy sloughs along the Colorado delta. They, like the Quechan, Mojaves, Yoemem (Yaquis), and Yoremem (Mayos), practiced flood-plain-recession agriculture, planting their crops as floodwaters receded. The Hohokam and their successors, the Akimel O'odham (Upper Pimas), on the other hand, dug canals to divert water from Sonoran Desert rivers onto their fields. Hohokam canal systems along the Salt and Gila rivers snaked across the desert floor for nearly 100 miles (160 km) in the Florence area and for 125 to 315 miles (200 - 500 km) in the Phoenix Basin. Hohokam farmers did not use all sections of these canal systems at any one time. Nonetheless, they still irrigated between 30,000 and 60,000 acres (12,000 - 24,000 ha) in the Phoenix Basin alone.

Hohokam farmers also constructed ditches and brush weirs along alluvial fans to divert runoff onto their fields after summer rains. This form of agriculture, sometimes called *ak-chin* among Tohono O'odham in southern Arizona and temporal among *mestizos* (people of mixed Hispanic and Indian ancestry) in rural Sonora, is still being practiced today. North of Tucson, however, the Hohokam developed an enormously labor-intensive type of agriculture that did not survive into the historic period. Archaeologists Paul and Suzanne Fish and their colleagues at the Arizona State Museum discovered more than 42,000 rock piles in association with contour terraces and checkdams on the western slopes of the Tortilita Mountains. They also found huge roasting pits containing charred fragments of agave. The rock piles protected young agave plants from predation by rodents and conserved moisture by reducing evaporation

The Sonoran Desert's prehistoric Hohokam were sophisticated desert farmers who built the largest system of irrigation canals in pre-Columbian North America. This section of canal was exposed during the 1964-1965 excavations at Snaketown, located northwest of the present-day town of Sacaton, Arizona.

around their bases. During the twelfth and thirteenth centuries, more than 100,000 agaves may have been simultaneously growing in these rock pile fields.

The Arrival of the Europeans

The Marana community that cultivated those agaves abandoned the northern Tucson Basin in the mid-fourteenth century. About the same time, in the winter of 1358-59 A.D., a massive flood roared down the Salt and Verde rivers, washing out canals and washing away fields in the Salt River Valley. The flood was followed by two decades of drought and more floods in the early

1380s. Hohokam communities along the Salt may never have recovered from those climatic calamities.

Other Hohokam communities along the Gila River survived into the fifteenth century. By the time the first Europeans settled in the region in the late 1600s, however, Hohokam civilization had collapsed. Some archaeologists speculate that centuries of highly mineralized irrigation water may have saturated Hohokam fields with salts until they could no longer produce crops. Others argue that increasing political conflict may have caused Hohokam society to implode. Pima creation narratives describe the ancestors of the O'odham emerging from beneath the earth to destroy "great houses" ruled by powerful priest-chiefs along the Gila River. The conquest begins with Casa Grande and ends with Pueblo Grande, governed by the priest-chief Yellow Buzzard. Many archaeologists believe that the O'odham are descendants of the Hohokam. O'odham creation narratives, in contrast, portray themselves as conquerors.

Regardless of how or why Hohokam civilization disintegrated, the peoples of the northern Sonoran Desert were living much simpler lives when Jesuit missionary Eusebio Francisco Kino and his companions encountered them. We know much less about the pre-Columbian peoples of Sonora because so little archaeological research has been done south of the international border. At the time of initial European contact in the sixteenth century, large populations of Cahita-speakers—the ancestors of the Mayos and Yaquis—lived along the Mayo and lower Yaqui rivers. Opatas and Eudeves dominated the Sonora River valley and the upper Yaqui and its tributaries. So-called Pimas Bajos, or Lower Pimas, occupied a wide crescent between the Opatas and the Yaquis. All of these peoples spoke Uto-Aztecan languages and relied upon agriculture for at least part of their diet.

Once the Europeans arrived, however, one of the greatest ecological revolutions in the history of the world transformed both lives and landscapes. Biological historian Alfred Crosby calls this revolution the Columbian Exchange—that flow of genes, microbes, plants, and animals between the "Old World" of Eurasia and Africa and the "New World" of the Americas. The first revolutionary wave

may even have preceded the Europeans themselves: Indians infected with Eurasian diseases like measles, influenza, and smallpox may have unsuspectingly unleashed contagions as they traveled up ancient trade routes from central Mexico. Some archaeologists and ethnohistorians even contend that these diseases may have contributed to the demise of Hohokam civilization itself.

We may never know exactly when the first epidemic spread death and devastation among the Indians of the Sonoran Desert. What we do know is this: thousands of Yaquis and Mayos were perishing when the first Jesuit missionaries ventured into Cahita territory in the early 1600s. The pattern repeated itself again and again at five- to eight-year intervals as Spanish settlers pushed northward into the Opatería and the Pimería Alta. Epidemiologically virgin populations had no genetic resistance or cultural mechanisms to resist the microbial onslaught. Like native peoples all over the Americas, the number of Indians in the Sonoran Desert declined by as much as ninety-five percent over the next two centuries.

For the Indians who endured, however, the Europeans brought new crops, new tools, and new animals that revolutionized their economies and their means of transportation. Winter wheat filled an empty niche in their agricultural cycle because it could be planted in November, when frosts at higher elevations in the Sonoran Desert would have withered corn, beans, or squash. Mules and oxen enabled them to cultivate their fields with wooden or iron plows, intensifying their reliance upon agriculture. Cattle, sheep, and goats allowed them to convert non-edible forbs and grasses into beef, mutton, cheese, milk, leather, and tallow. Horses expanded their ranges and shrank distances. A Euro-American agropastoralist economy—irrigation agriculture along the floodplain, animal husbandry in the uplands—supplemented, complemented, and slowly replaced digging-stick agriculture and wild food gathering.

Cattle, horses, and mules had other consequences as well. In November 1697, Kino and his frequent traveling companion Juan Mateo Manje visited Sobaipuri Pimas along the Babocómari River in southeastern Arizona. They found O'odham performing a circular dance around a tall pole dangling nine scalps. Sixteen Janos and Jocome raiders had tried to run off their small herd of livestock—livestock they had received from the Jesuit missionary. Kino's gifts of cattle and horses had made O'odham along the San Pedro River targets of Jocome, Janos, and Apache raiding. Because of those raids, O'odham along the San Pedro eventually had to retreat to the Santa Cruz River Valley, while Pimas along the Gila River organized themselves into larger villages with a standing army of 1000 men—nearly one-fourth their total population. The introduction of Old World livestock into the Sonoran Desert triggered a pattern of raiding and retaliation between O'odham and Apaches that lasted for 200 years.

Spaniards and Mexicans

That same pattern made mission Indians and Hispanic frontiersmen allies of one another, as their guerrilla warfare with the Apaches intensified. Spaniards and their mixed-blood descendants were a minority in the Sonoran Desert until the mid- to late-eighteenth century. That meant that even though they exploited Indian labor and encroached upon Indian land, they also depended upon Yaquis, Mayos, Opatas, and Pimas for their very existence on a dangerous frontier.

Some groups, such as the Yaquis, kept Hispanic colonists from penetrating their territory until the late 1800s. Among the Opatas and Lower Pimas, on the other hand, Spaniards, *mestizos* (Spanish-Indian), *coyotes* (mestizo and Indian), and mulattos settled in or near mission communities despite the protests of the missionaries. *Mestizaje*—racial mixture—weakened ethnic boundaries, as competition for arable land and irrigation water increased in Sonoran Desert river valleys. Those river valleys—the Sonora, the Bavispe, the Santa Cruz—became riparian oases of Hispanic civilization in the desert.

Stock raising was the most land-extensive Euro-American transformation of the Sonoran Desert. Cattle, horses, goats, and sheep searched for forage from river floodplains to mountain crests. In more settled areas like central Sonora, overgrazing became endemic during the Spanish colonial period. The presidio (military garrison) and town of Pitic (modern Hermosillo) alone ran 5000 head of cattle, 3422 sheep, 435 goats, 2138 horses, and 367 mules in 1804 (Radding 1997, 218). But Apache hostilities prevented ranchers from expanding beyond the Santa Cruz Valley in southern Arizona. During the 1820s and 1830s, the Mexican government issued nine land grants in southern Arizona. All were largely abandoned by the 1840s. Hispanic stock raising never was sustained enough to have a widespread impact upon the southern Arizona landscape except along the Santa Cruz.

Apache attacks also checked the northward expansion of mining, the most land-intensive industry in Hispanic Sonora. In less vulnerable areas, however, mining lured thousands of Indians and non-Indians alike from strike to strike. There were two types of mining communities in the region. The discovery of silver in 1683 south of the Río Mayo led to the development of Alamos, a city of wealth and social stratification based upon the enormous capital investment required to extract and process silver ore. Cieneguilla in northwestern Sonora, on the other hand, was a desert boom town where *gambucinos* (prospectors) dry-winnowed alluvial deposits for particles of gold. Vein-mining operations like that of Alamos depended upon large, stable labor forces organized into hierarchies of occupations. Placers like Cieneguilla attracted restless, mobile congregations of Spaniards, mestizos, Yaquis, Opatas, and Pimas, most of whom worked for themselves. No other economic activity so thoroughly rearranged social relationships on New Spain's and then Mexico's northern frontier.

The same could be said about mining's impact upon local and regional environments. Most placers faded into the desert, however, while silver-mining districts like Alamos or San Juan Bautista in the Río Moctezuma Valley continued to generate voracious demands for charcoal, timber, firewood, tallow, salt, and mercury for decades and even centuries. To supply these mining districts, stock ranches also prolifer-

ated, creating what geographer Robert West (1949) calls the "ranch-mine settlement complex." Complex webs of economic relationships linked farflung mining communities with merchant and imperial capital in Mexico City and Madrid. Meanwhile, hillsides were denuded, streams diverted, water tables polluted, and vegetation communities irrevocably changed.

Anglo-Americans

Those social and ecological patterns replicated themselves after the United States wrested away more than half of Mexico's national territory during the Mexican War of 1845-1848. Gold and silver lured Anglo-Americans and Europeans to the northern Sonoran Desert—first along the Colorado River as placer booms like the one at La Paz flared and faded, then in the mountains along the Hassayampa River, where the town of Wickenburg sprang up to reduce and mill silver ore. Precious metals were the only commodities worth the enormous transportation costs on such an isolated and dangerous frontier.

And Arizona was a frontier for the first three decades of its existence as a part of the United States. From the arrival of the Europeans until the 1880s, the Arizona-Sonora borderlands were a frontier in the most basic sense of the term—contested ground where no single tribe, empire, or nation held uncontested sway. Because Apaches, Yavapais, and River Yumans resisted European and Euro-American conquest so successfully for two centuries, Euro-American impact upon the desert environment was intermittent rather than sustained.

That all changed in the 1880s, when the Era of Extraction began. In 1877, the Southern Pacific Railroad reached Fort Yuma on the Colorado River. Three years later, after its largely Chinese crews had laid tracks across some of the hottest, driest terrain in North America, the railroad steamed into Tucson. At the gala celebration on March 20, 1880, Mexican intellectual Carlos Velasco raised a toast to the "irresistible torrent of civilization and prosperity" that would follow the steel rails. Six years later, when Geronimo surrendered to General Nelson Miles for the final time, the frontier came to an end. Suddenly, Arizona and Sonora were safe for global capital, which poured in from the eastern United States, California, and the British Isles, as the Southern Pacific and other railways extended their arteries of commerce across deserts and mountains. Both Arizona and Sonora became extractive colonies of the industrial world, their natural resources ripped from the ground and shipped somewhere else for finishing, processing, and consuming. In Arizona, this was the era of the "Three C's," when cattle, copper, and cotton dominated the economy.

The first major extractive industry to explode across the landscape was stock raising. In 1870, there were perhaps 38,000 head of cattle in the Arizona Territory. By the early 1890s, there were about 1,500,000 head of cattle and more than a million sheep, many of which had been shipped into the territory by rail. Stock raising expanded south of the border as well, as investors like Colonel William Henry Greene put together vast ranches on the grasslands of northern Sonora. Pascual

Encinas even drove his cattle onto the coastal plains west of Hermosillo, establishing his famous Costa Rica Ranch at Siete Cerros on the edge of Seri territory. Encinas tried to train the Seris as laborers and offered them religious instruction. But as more ranchers moved onto their desert hunting and gathering grounds, the Comcáac responded by killing cattle for food. By the late nineteenth century, the "Encinas War" erupted, pitting cowboys armed with repeating rifles against Seris armed with bows and arrows. The Comcáac were decimated. Sonora's march to the sea began.

The second major extractive industry—copper mining—depended even more heavily upon the railroads. Unlike gold and silver, copper was an industrial rather than a precious metal. The evolution of the industry therefore became the triumph of technological innovation over declining grades of ore. Staggering amounts of earth had to be moved to extract a metal that eventually constituted less than one percent of its ore bodies, and that earth had to be moved from mine to smelter in railroad cars, not in freight wagons or on the backs of mules. In copper districts like Bisbee, where Phelps Dodge Corporation's Copper Queen reigned supreme, thousands of miles of shafts and tunnels burrowed underground. And while most of these early districts were in the uplands fringing the Sonoran Desert, Phelps Dodge and other giants eventually chewed into the desert as well, particularly after open-pit mining became feasible. At Ajo, Mammoth, Twin Buttes, and Silverbell, giant holes begat gargantuan slag heaps, which rose above the desert floor like pyramids erected in honor of the Electrical Age.

An 1890s photograph of Bisbee, Arizona. The Copper Queen mine is left center.

Water Control and the Transformation of the Desert

The third major extractive industry—agriculture—led to the ultimate transformation of the Sonoran Desert. For 3000 years, farmers had cultivated their crops along those few stretches of the desert where surface water flowed near arable land. In Sonora, the major agricultural areas were located in the river valleys of the *zona serrana*, the mountainous central and eastern portions of the state. Generally trending northeast-southwest, successive mountain ranges and river valleys cradled Opata, Eudeve, and Lower Pima communities. During the Spanish colonial period, the serrana attracted Spaniards and their mestizo descendants as well, who raised wheat, fruits, vegetables, and sugar cane along its cottonwood-shaded floodplains.

After Mexico won its independence from Spain in 1821, however, Guaymas became Sonora's most important port of entry, and a strong commercial axis between Guaymas and Hermosillo (formerly Pitic)—the gateway to the serrana—developed. Merchants and military officials cast covetous eyes on the rich coastal floodplains of western and southern Sonora, particularly the Yaqui and Mayo river valleys. The Yaquis fought a bloody war of resistance but by the late nineteenth century, the *Hiakim*, or Yaqui homeland, had been occupied by the Mexican army. The Sonoran state government under Governor Rafael Izábal then decided to eliminate the Yaquis once and for all by deporting thousands of Yoemem to the Oaxaca Valley or to henequén plantations in the Yucatán. It was cultural genocide in the service of "order and progress."

Speculators soon descended upon the Yaqui Valley with grand plans to irrigate the coastal plains. In 1890, the Mexican government granted Carlos Conant Maldonado 300,000 hectares (one hectare, or ha = 2.47 acres) along the Río Yaqui, 100,000 ha along the Río Mayo, and 100,000 ha along the Río Fuerte in northern Sinaloa in return for surveying the area and building canals along each river. Conant's Sonora and Sinaloa Irrigation Company, incorporated in New Jersey with U.S. capital, soon went bankrupt, but the Richardson Construction Company purchased Conant's grant in 1906. In exchange for selling 400-hectare blocks of land and supplying irrigation water to colonists, most of them from the United States or Europe, the Companía Constructora Richardson, S.A. received exclusive right to sixty-five percent of the Río Yaqui's flow for ninety-nine years.

The Richardson brothers had big plans to build storage dams on the Yaqui to generate electric power and furnish water to a network of canals capable of irrigating 300,000 hectares. The Mexican Revolution and World War I destroyed their enterprise but the dream of transforming the Yaqui Valley into a vast grid of irrigated agribusiness bore fruit in 1952, when the Mexican government completed Alvaro Obregón Dam at Oviachi forty miles away. Along with dams upriver, the Oviachi reservoir controlled flow along the lower Río Yaqui and eventually channeled its water into three major canals that irrigate nearly 600,000 acres in the Yaqui Valley. Ciudad Obregón, a city of more than 500,000 people, arose to service the largest irrigation district in Sonora. Recognizing the Yaqui Valley's impor-

tance, the Rockefeller Foundation established a wheat-breeding station on the outskirts of Obregón under the direction of Dr. Norman Borlaug. This station became one of the hearths of the Green Revolution, that controversial program that dramatically increased wheat production—all of it dependent upon high inputs of chemical fertilizers and pesticides—around the world.

North of the Yaqui Valley, advances in pump technology after World War II allowed other coastal irrigation districts to bulldoze desert plains and convert them into wheat and cotton fields. The largest was the Costa de Hermosillo where, at its height, 887 pump-powered wells regurgitated water onto more than 100,000 hectares. But discharge exceeded recharge by 250 percent. As water tables plummeted and salt water intruded from the Gulf of California, the Mexican government finally stepped in and halved the amount of water that could be pumped. Many fields were abandoned. Other farmers switched from relatively low-value crops like cotton to high-value, high-risk crops like brandy grapes, citrus, garbanzo beans, and vegetables destined for U.S. markets.

Because of these developments, Sonora's demographic, political, and economic center of gravity shifted from the *serrana* to the coast during the twentieth century. Dam-building and groundwater pumping enabled capital-intensive agricultural districts to plow under great mesquite *bosques* (forests) west of Hermosillo and desert ironwood plains around Caborca. Those twin pillars of modern water control also allowed older cities like Hermosillo (about 800,000 population) to expand, and entirely new cities like Cuidad Obregón to spring up like an industrial flower south of the dry channel of the Río Yaqui. But whether these flowers are perennial or ephemeral remains to be seen. Hermosillo already experiences severe water shortages during dry seasons and dry years. Groundwater districts like Caborca and the Costa de Hermosillo are contracting painfully as aquifers drop and pumping costs escalate. Even the Yaqui Valley with its huge reservoirs faces an uncertain future as the North American Free Trade Agreement (NAFTA) reshapes Mexican agriculture. Most farmers in the valley now plant wheat in the winter and soybeans in the summer. As subsidies are removed and trade barriers lowered, many question whether Sonoran producers can compete with Canadian and U.S. farmers dry-farming the same crops.

In Arizona, the future of agriculture is held hostage to urban growth. During the early twentieth century, the newly created Reclamation Service (precursor of the U.S. Bureau of Reclamation) erected Roosevelt Dam in 1903 on the Salt River east of Phoenix and turned the Salt River Valley into one of the largest agricultural centers in the Southwest. And when a British embargo on long-staple industrial cotton during World War I triggered Arizona's cotton boom, commercial agriculture spread south across the saltbush and creosote bush flats between Phoenix and Tucson. Arizona became one of the leading cotton producers in the world.

But World War II and the postwar boom thrust Arizona from the Era of Extraction into the Era of Transformation, turning an overwhelmingly rural state into an overwhelmingly urban one. Thousands of acres of citrus and cotton sprouted subdivisions and malls as Phoenix and its satellites sprawled into a metropolis of more than 2,500,000 people by 1995. Metro Tucson approached 750,000. By the time the Central Arizona Project (CAP)—a farmer's dream since the 1920s—reached Maricopa, Pinal, and Pima counties, many farmers could not afford its water. The CAP became one more bargaining chip in the water game, that escalating contest that pitted relentlessly expanding cities against farmers, miners, and Indian nations.

Visions and Nightmares

Water has always been the ultimate limiting factor on human society in the Sonoran Desert. Until the late nineteenth century, people largely relied upon surface flow, adapting to rivers rather than making the rivers adapt to them. This century, however, dams have domesticated all major rivers in the region while pumps have mined groundwater aquifers far beyond recharge. Arizona's 1980 Groundwater Management Act mandates that "safe-yield" (when discharge does not exceed recharge) be reached by the year 2025 in four Active Management Areas which together are home to eighty percent of Arizona's population. Certain areas of the Sonoran Desert may indeed see population growth slowed or halted because they are running out of water.

But other areas like metropolitan Phoenix will spread unchecked as long as they can wrest more water away from farmers and Indian nations. For much of its history as a part of the United

The twentieth century has seen the transformation of Arizona from an overwhelmingly rural state to an overwhelmingly urban one, as seen in these two photos of Tucson: about 1900 (top) and 1999 (bottom). Both photos taken at the intersection of Scott and Congress.

States, Arizona has suffered from a bad case of "California envy," battling its more powerful western neighbor for Colorado River water, yet emulating its explosive growth. Unless it consciously decides to restrict growth, however, Arizona will become the southern California of the twenty-first century. Metropolitan Phoenix will embrace five to seven million people. Tucson will reach 1,500,000 and learn to guzzle the CAP water it has twice refused to drink in the 1990s. *Ambos* Nogales (two cities of the same name separated by the international boundary) may resemble El Paso-Júarez if free trade and the *maquiladora* (assembly plant) program keep on luring millions of Mexicans to the border. And since most of Arizona and Sonora's population now live in the Sonoran Desert, the urban assault will only intensify.

Meanwhile the Columbian Exchange continues to rearrange the countryside. Since 1942, the number of non-native plants in Arizona alone has risen from 190 to approximately 330 species. During the 1930s, agents of the Soil Conservation Service promoted a South African grass called Lehmann lovegrass to control erosion. Today it covers more than 400,000 acres of Arizona. Beginning in the 1960s, Sonoran range scientists introduced another African grass—buffelgrass—to increase forage production. More than one million acres of desert and subtropical thornscrub have now been scraped away to plant this exotic. Ecologist Tony Burgess of Columbia University's Biosphere II calls such proliferation the "Africanization of the Sonoran Desert."

Aldo Leopold may have under-estimated past peoples' manipulation of the Southwestern landscape. He would have shuddered at how utterly we have transformed the Sonoran Desert since he and his brother drifted along the green lagoons.

References

Leopold, Aldo. *A Sand County Almanac and Sketches Here and There.* London: Oxford University Press, 1949.

Radding, Cynthia. *Wandering Peoples: Colonialism, Ethnic Spaces, and Ecological Frontiers in Northwestern Mexico, 1700-1850.* Durham: Duke University Press,1997.

Shetler, Stanwyn. "Three Faces of Eden." *Seeds of Change: A Quincentennial Commemoration.* Herman J. Viola and Carolyn Margolis, eds. Washington, D.C.: Smithsonian Institution Press, 1991.

West, Robert. "The Mining Community in Northern New Spain: The Parral Mining District." *Ibero-Americana.* Berkeley: University of California Press, 1949.

Additional Reading

Bahre, Conrad. *A Legacy of Change: Historic Human Impact on Vegetation of the Arizona Borderlands.* Tucson: University of Arizona Press, 1991.

Bowden, Charles. *Killing the Hidden Waters.* Austin: University of Texas Press, 1977.

Fradkin, Philip. *A River No More: The Colorado River and the West.* Tucson: University of Arizona Press, 1984.

Nabhan, Gary. *Gathering the Desert.* Tucson: University of Arizona Press, 1985.

Reisner, Marc. *Cadillac Desert: The American West and Its Disappearing Water.* New York: Penguin Books, 1986.

Sheridan, Thomas. *Arizona: A History.* Tucson: University of Arizona Press, 1995.

A Sense of Our Place in the Sonoran Desert

Following is a selection of books from *Desert Stories: A Reader's Guide to the Sonoran Borderlands*, an Arizona-Sonora Desert Museum Sense of Place Project publication.

> ➤ CONTEMPORARY SCENE

Burckhalter, David. *La Vida Norteña: Photographs of Sonora, Mexico.* Albuquerque: University of New Mexico Press, 1998.

Miller, Tom. *On the Border: Portraits of America's Southwestern Frontier.* NY: Harper & Row, 1981.

Nabhan, Gary. *Desert Legends: Re-Storying the Sonoran Borderlands.* New York: Henry Holt, 1994.

Weisman, Alan. *La Frontera: The United States Border with Mexico.* Tucson: University of Arizona Press, 1986.

Yetman, David. *Sonora: An Intimate Geography.* Alburquerque: Univ. of New Mexico Press, 1996.

> ➤ HISTORY

Clarke, Asa Bement. *Travels in Mexico and California.* Texas A&M University Press, 1988 (originally published in 1852).

Lumholtz, Carl. *New Trails in Mexico: An Account of One Year's Exploration in North-Western Sonora, Mexico and South-Western Arizona, 1909-1910.* Tucson: University of Arizona Press, 1990.

Martinez, Oscar. *Border People.* Tucson: Unversity of Arizona Press, 1994.

Pfefferkorn, Ignaz. *Sonora: A Description of the Province.* Tucson: Univ. of Arizona Press, 1989.

> ➤ NATIVE AMERICAN CULTURES

Fontana, Bernard. *Of Earth and Little Rain: The Papago Indians.* Flagstaff: Northland Press, 1981.

Griffith, James S. *Beliefs and Holy Places: A Spiritual Geography of the Pimería Alta.* Tucson: University of Arizona Press, 1992.

Kelley, Jane Holden. *Yaqui Women: Contemporary Life Histories.* Lincoln: University of Nebraska Press, 1978.

Sheridan, Thomas E. and Nancy J. Parezo. *Paths of Life: American Indians of the Southwest and Northern Mexico.* Tucson: University of Arizona Press, 1996.

Underhill, Ruth. *Singing for Power.* Tucson: University of Arizona Press, 1993. (Originally published in 1938).

> ➤ NATURAL HISTORY AND THE ENVIRONMENT

Abbey, Edward. *Cactus Country.* New York: Time-Life, 1973.

Alcock, John. *Sonoran Desert Summer.* Tucson: University of Arizona Press, 1990.

Bowden, Charles. *Sonoran Desert.* New York: Harry N. Abrams, 1992.

Hayden, Julian D.. *The Sierra Pinacate.* Tucson: Southwest Center Series, University of Arizona, 1998.

Hornaday, William T. *Camp-fires on Desert and Lava.* Tucson: University of Arizona Press, 1983. (First published in 1908).

Krutch, Joseph Wood. *Desert Year.* Tucson: University of Arizona Press, 1985.

Nabhan, Gary. *The Desert Smells Like Rain.* San Francisco: North Point Press, 1981.

West, Robert C. *Sonora: Its Geographic Personality.* University of Texas Press, 1993.

Biodiversity: The Variety of Life that Sustains Our Own

Gary Paul Nabhan

There is a place in the Sonoran Desert borderlands which, more than any other I know, capsulizes what the term *diversity* has come to mean to both natural and social scientists alike. The place is a desert oasis known as Quitobaquito, centered on a spring-fed wetland at the base of some cactus-stippled hills that lie smack dab on the U.S.-Mexico border. Whenever I walk around there, I am astounded by the curious juxtapositions of water-loving and drought-tolerating plants, of micro-moths wedded to single senita cacti, and hummingbirds that have traveled hundreds of miles to visit ocotillos, of prehistoric potsherds of ancient Patayan and Hohokam cultures side by side with broken glass fragments left by O'odham, Anglo- and Hispanic-American cultures.

Walk down from its ridges of granite, schist, and gneiss, and you will see organpipe cactus growing within a few yards of arrowweed, cattails, and bulrushes immersed in silty, saline sediments. The oasis has its own peculiar population of desert pupfish in artesian springs just a stone's throw from the spot where a native caper tree makes its only appearance in the United States. The tree itself is the only known food source for the pierid butterfly that is restricted in range to the Sonoran Desert proper.

More than 270 plant species, over a hundred bird species, and innumerable insects find Quitobaquito to be a moist harbor on the edge of a sea of sand and cinder. Not far to the west of this oasis, there are volcanic ridges that have frequently suffered consecutive years without measurable rainfall, and their impoverished plant and animal communities reflect that.

Quitobaquito is naturally diverse, but its diversity has also been enhanced rather than permanently harmed by centuries of human occupation. Prehistoric Hohokam and Patayan, historic Tohono O'odham, Hia c-ed O'odham, Apache, Cucupa, and Pai Pai visited

Quitobaquito for food and drink long before European missionaries first arrived there in 1698. Since that time, a stream of residents from O'odham, Mexican, Jewish, and Mormon families have excavated ponds and irrigation ditches, transplanting shade and fruit trees alongside them. They intentionally introduced useful plants, and accidentally brought along weedy camp-followers, adding some fifty plant species to Quitobaquito over the centuries. Native birds and mammals have also been affected by human presence there, and some increased in number during the days of O'odham farming downstream from the springs. All in all, Quitobaquito's history demonstrates that the desert's cultural diversity has not necessarily been antithetical to its biological diversity; the two are historically intertwined.

In fact, the Sonoran Desert is a showcase for understanding the curious interactions between cultural and biological diversity. There are at least seventeen extant indigenous cultures that each has its own brand of land management traditions, as well as the dominant Anglo- and Hispanic-American cultures which have brought other land ethics, technologies, and strategies for managing desert lands into the region. While some cultural communities such as the Seri were formerly considered passive recipients of whatever biodiversity occurred in their homeland, we now know that they actively dispersed and managed populations of chuckwallas, spiny-tailed iguanas, and columnar cacti. Floodwater farmers such as the Tohono O'odham and Opata dammed and diverted intermittent watercourses, planted Mesoamerican

crops, and developed their own domesticated crops from devil's claw, tepary beans, and Sonoran panic grass. Anglo- and Hispanic-American farmers and ranchers initiated other plant and animal introductions, and dammed rivers on a much larger scale. Each of these cultures has interacted with native and exotic species at different levels of intensity, including them in their economies, stories, and songs. From an O'odham rainmaking song that echoes the sound of spadefoots, to the Western ballad "Tumblin' Tumbleweeds" written in Tucson over a half century ago, native and invasive species have populated our oral and written traditions as curses, cures, and resources.

Technically speaking, this stuff we call diversity eludes one single definition. For starters, however, *biodiversity* (short for biological diversity) can be generally thought of as the "variety of life on earth." Scientists use this term when discussing the richness of life forms and the heterogeneity of habitats found within or among particular regions. Biodiversity in this sense is often indicated by the relative richness of species in one habitat versus another. Thus it is fair to say that riparian gallery forests of cottonwoods and willows along desert rivers typically support more *avian biodiversity*—a greater number of bird species—than do adjacent uplands covered with desertscrub vegetation. Similarly, there is greater biodiversity in flowering vines in the moist tropical forests of southern Mexico than there is in the Sonoran Desert of northern Mexico.

It is worth noting, however, that ecologists such as E.O. Wilson first coined the term biodiversity to signify something far more complex than the mere number of species (termed *species richness*) found in any given area. Usually ecologists also consider the number of individuals within each species when they assess diversity or heterogeneity. An area where one desert wildflower such as the California poppy dominates eight other species is considered to be less diverse than an area with the same eight species where the numbers of each are more evenly distributed. As Kent Redford of The Nature Conservancy has recently explained, "A species-focused approach to biodiversity has proved limiting for a number of reasons.... [The] use of just species as a measure of biodiversity has resulted in conservation efforts focusing on relatively few ecosystems while other threatened ones are highly ignored. Species do not exist in a vacuum, and any definition of biodiversity must include the ecological complexes in which organisms naturally occur and the ways they interact with each other and with their surroundings."

The integrity of biodiversity can be teased apart into the following components. Although each of them may be separated out by scientists for study, they do not truly exist "apart" from one another.

ECOSYSTEM DIVERSITY: the variety of landscapes found together within any region, and the ways in which their biotic communities interact with a shared physical environment, such as a watershed or coastal plain. A landscape interspersed with native desert vegetation, oasis-like cienegas, riparian woodlands, and croplands is more diverse than one covered entirely by one crop

such as cotton. The Colorado River Delta was once a stellar example of ecosystem diversity, displaying a breath-taking mixture of riparian gallery forests, closed-canopy mesquite bosques, saltgrass flats, backwater sloughs, rivers, ponds, and Indian fields. Much of it is now dead, except for the hypersaline wetlands known as the Cienega de Santa Clara.

BIOTIC COMMUNITY DIVERSITY: the richness of plants, animals, and microbes found together within any single landscape mosaic; such a mosaic can range in scope from the regional to the watershed level. This richness can be shaped by a variety of factors, ranging from the age of the vegetation to land use to soil salinity and fertility. For example, the number of species on well-drained, ungrazed desert mountain slopes covered by columnar cacti, ancient desert ironwoods, and spring wildflowers is greater than that on an alkali flat grazed by goats, where only saltbush, saltgrass, and seepweed may grow. The Rincon Mountains east of Tucson demonstrate a gradient of communities, each with its own diversity, as they rise from desertscrub to xeric woodlands, and coniferous forests.

INTERACTION DIVERSITY: the complexity of interactions within any particular habitat, such as the relationships between plant and pollinator, seeds and their dispersers, and symbiotic bacteria and their legumes. A pine-oak woodland in Arizona's "sky islands" harbors more interspecific interactions than does an even-aged pine plantation. Ramsey Canyon in the Huachuca Mountains showcases such interaction diversity, with over a dozen humming-birds, as well as bats, bees, and butter-flies visiting its myriad summer flowers.

SPECIES DIVERSITY: the richness of living species found at local, ecosystem, or regional scales. A well-managed desert grassland hosts more species than can be found in a buffelgrass pasture intentionally planted to provide livestock forage without consideration of wildlife needs. Quitobaquito, discussed above, is as fine an example of localized species diversity as we have anywhere in the binational Southwest.

GENETIC DIVERSITY: the heritable variation within and between closely-related species. A canyon with six species of wild out-crossing beans contains more genetic variation than does a field of a single highly-bred hybrid bean. Indian fields in southern Sonora demonstrate this concept, for their squashes hybridize with weedy fieldside gourds, and their cultivated chile peppers are inflamed by genetic exchange with wild chiltepines.

All of these components of biodiversity ensure some form of environmental stability to the inhabitants of a particular place. A landscape with high ecosystem diversity is not as vulnerable to property-damaging floods as a bladed landscape is, for a mix of desert grassland and wetlands serves to buffer downstream inhabitants from rapid inundations. A diverse biotic community is less likely to be ravaged by chestnut blight or spruce budworm than a tree plantation can be. A cactus forest with diverse species of native, wild bees is less vulnerable to fruit crop failure than are

orchards or croplands that are exclusively dependent upon the non-native honeybees. A desert grassland with multiple species of grasses and legumes cannot be as easily depleted of its fertility and then eroded as can one with a single kind of pasture grass sucking all available nutrients out of the ground. And finally, a Pima Indian garden intercropped with many different kinds of vegetable varieties will not succumb to white flies or other pests as easily as will an expansive, irrigated lettuce field in the Imperial Valley.

In short, more of "nature's services"—the economic contributions offered by intact ecosystems—are possible when we manage these ecosystems to safeguard or restore their biodiversity, and not allow it to be depleted. Recent estimates by environmental economists suggest that the dollar value of the services such as flood protection and air purification provided by the world's intact wild ecosystems averages thirty-three trillion dollars per year, compared to the eighteen trillion dollar Gross National Product of all nations' human-made products.

The message is clear: when a mosaic of biotic communities is saved together and kept healthy within a larger landscape, few endangered species fall between the cracks and succumb to extinction processes. In contrast, a small wildlife sanctuary designed to save a single species often fails to achieve its goal, for the other organisms which that species ultimately needs in its presence have been ignored or eliminated. Not only do humans benefit from the conservation of large wild-lands landscapes, but many other species do as well.

How does this play out in our Sonoran Desert region?

Ask most people to characterize life in the desert and few will think to mention the word "diversity" as part of their thumbnail sketch of this place. Most of us keep in our heads those pictures of bleak, barren, blowing sandscapes when we hear the word "desert."

The Sonoran Desert does contain one major sea of sand, as well as a long corridor of coastal dunes along the Gulf of California, but even these are seasonally lush with unique and thriving life forms. As one spends more time in a range of Sonoran Desert habitats, one is constantly surprised by how many plants and animals are harbored here.

Travel out of Sonoran Desert vegetation into the higher mountain ranges held within the region and even more astonishing levels of biodiversity can be found. In fact, the "sky islands" of southeastern Arizona and adjacent Sonora are now recognized by the International Union for the Conservation of Nature as one of the great centers of plant diversity north of the tropics.

When we compare our desert with others, the contrast is striking. Overall, the Sonoran Desert has the greatest diversity of plant growth forms—architectural strategies for dealing with heat and drought—of any desert in the world. From giant cacti to sand-loving underground root parasites, some seventeen different growth forms coexist within the region. Often, as many as ten complementary architectural strategies will be found together, allowing many life forms to coexist in the same patch of desert.

Biodiversity in the desert is often measured on a scale that would not be used in the tropical rainforest. Desert ecologists have found twenty kinds of wildflowers growing together in a single square yard (.84 m²), while a single tropical tree might take up the same amount of space. On an acre (.4 ha) of cactus forest in the Tucson Basin, seventy-five to 100 species of native plants share the space that three mangrove shrubs might cover in swamp along a tropical coast. These levels of diversity are a far cry from the "bleak and barren" stereotype, and it may well be that the Sonoran Desert region is more diverse than other arid zones of comparable size.

Consider for example, the flora of the Tucson Mountains, which Arizona-Sonora Desert Museum research scientists recently inventoried with a number of their colleagues. In an area of less than forty square miles (100 km²), this botany team encountered over 630 plant species—as rich a local assortment of plants as any desert flora we know. This small area contains roughly one-sixth of the Sonoran Desert's entire plant diversity. It is disproportionately rich relative to its size, its paucity of surface water, and its elevational range.

Such a diversity of wildflowers and blossoming trees attracts a diversity of wildlife as well. In the Sonoran Desert area within a thirty mile radius of Tucson, you can find between 1000 and 1200 twig- and ground-nesting native bees (all of them virtually "stingless"). As the Desert Museum's research associate Stephen Buchmann wryly notes, "this may mean that the Sonoran Desert region is the richest bee real estate anywhere in the world—the entire North American continent has only 5000 native bee species."

Desert wildflowers attract more than bees. Southern Arizona receives visits from more hummingbird species—seventeen in all—than anywhere else in the U.S. Other pollinator groups, such as butterflies and moths, are well-represented in the region as well. Single canyons near the Arizona-Sonora border may harbor as many as 100 to 120 butterfly species, and moth species may number five to ten times higher than that in the same habitats. When all pollinating organisms breeding or passing through here are counted, it may be that the greater Sonoran Desert has as large a pollinator fauna as any bioregion in the world.

This region is also rich in small mammals and reptiles. Some eighty-six species of mammals have ranges centered within the San Pedro National Riparian Area alone, a record unsurpassed by any natural landscape of comparable size in the U.S.; the area contains half of all mammal species in the binational Sonoran Desert. At least ninety-six species of reptiles are endemic to the Sonoran Desert—found here and nowhere else in the world.

Why is such diversity present in a land of little rain? For starters, our bimodal rainfall pattern brings out completely different suites of wildflowers and their attendant insects at different times of the year. In addition, we benefit from a more gradual transition between tropical nature and desert nature than does the Chihuahuan Desert on the other side of the Sierra Madre—many tropically-derived life forms reach their northernmost limits in the Sonoran Desert due to its relatively frost-free climes. Of course, tropic rainforests are much more diverse in the total number of species they have throughout their biome, in part because of their ages and their high energy budgets. However, there may be more turnover in species from place to place in the Sonoran Desert than in some tropical vegetation types. That is to say, many desert plants and insects are "micro-areal"—occurring only within a 100 by 100 mile spots on the map. Particularly in Baja California, there are extremely high levels of endemism, including some 552 plants unique to the peninsula.

Nevertheless, it remains true that the highest levels of local diversity in this desert region occur where water accumulates. Some of the highest breeding bird densities recorded anywhere in the world come from riparian forests along the Verde and San Pedro river floodplains. More than 450 kinds of birds have historically nested or migrated along the Colorado, San Pedro, and Santa Cruz rivers. And yet, if riparian habitats were among our richest, what have we lost with the removal of cottonwoods from ninety percent of their former habitat in Arizona? Ornithologists cannot name a single Sonoran Desert bird that has gone extinct with riparian habitat loss, but many of the eighty species of birds dependent on these riparian forests have locally declined in abundance. A single desert riparian mammal—Merriam's mesquite mouse—is now extinct due to the loss of riparian habitat at the hand of groundwater pumping, arroyo cutting, and overgrazing. Mexican wolves and black bears that formerly frequented our river valleys are among those mammals no longer found in the Sonoran Desert proper.

Conservation International has estimated that as much as sixty percent of the entire Sonoran Desert surface is no longer covered with native vegetation but is dominated by the 380-some alien species introduced to the region by humans and their livestock. Alien plants such as buffelgrass now cover more than 1,400,000 acres of the region, at the expense of both native plants and animals. Tamarisk trees choke out native willow and cottonwood seedlings. Invasive weeds such as Johnson grass and Sahara mustard have taken over much of certain wildlife sanctuaries and parks in the desert, outcompeting rare native species. Other invasive species such as Africanized bees and cowbirds also compete with the native fauna. Biological invasions are now rated among the top ten threats to the integrity of Sonoran Desert ecosystems, whereas a half century ago they hardly concerned ecologists working in the region. These invaders somehow reach even the most remote stretches of the desert, to the point of being ubiquitous.

The wholesale replacement of natives by aliens is enough of a problem, but desert biodiversity has been even more profoundly affected by habitat fragmentation—the fracturing of large tracts of desert into pieces so small that they cannot sustain the interactions among plant, pollinator, and seed disperser. Such fragmentation does not necessarily lead to immediate extinctions, just declines—there is a time lag before a species' loss of interactions with others leads to complete reproductive failure. Fragmentation caused by urbanization is now considered the number-one threat to the biodiversity

of the region and is not expected to diminish during our lifetimes. The population of Arizona's Maricopa County in the year 2025 is expected to be two and a half times what it was in 1995, and similar growth rates are anticipated along the entire desert coastline of the Sea of Cortez.

In a sense, humans are making the Sonoran Desert much more like the old (and erroneous) stereotype of a barren wasteland. As more than forty dams were constructed along rivers in this century, old-timers witnessed hundreds of miles of riparian corridors dry up. Groundwater overdraft has also impoverished desert and riparian vegetation, as farms and cities pump millions more acre-feet out of the ground than rainfall in the region can naturally recharge. The roots of plants are left high and dry above the water table. Most of the Sonoran Desert was not at all naturally barren, but our misunderstandings have impoverished one of the richest arid landscapes on the planet. That is why the Desert Museum has endorsed a long-term Conservation Mission Statement which begins with these words from ecologist D.M. Bowman:

"So what is biodiversity?...the variety of life on this planet is like an extra-ordinarily complex, unfinished, and incomplete manuscript with a hugely varied alphabet, an ever-expanding lexicon, and a poorly understood grammar....Ripping the manuscript to pieces because we want to use the paper makes little sense, especially if the manuscript says that 'to survive you shall not destroy what you do not understand'. Our mission as ecologists must be to *interpret* the meaning of biodiversity. The urgent need for this mission, and our current ecological ignorance, must be forcefully communicated to the public."

Instead of seeing future inhabitants rip out any more pages essential to the desert's story, the conservation organizations of the region have begun to work together to ensure that the most important corridors and secluded refugia for desert flora and fauna are identified and protected or restored. These critical areas—essential to the flow of diversity from source to sink, from headwaters to river mouth, and from tropical wintergrounds to summer nesting areas—must be kept from further fraying if the fabric of the Sonoran Desert is to remain intact. Scientists can prioritize such areas in terms of their value to biodiversity, but they will be safeguarded for future generations only if a broad spectrum of society is involved in endorsing their protection.

References

Austin, Mary. *The Land of Little Rain.* Albuquerque: University of New Mexico Press, 1974.

Bowman, D.M.S.J. "So What Is Biodiversity?" *The Biodiversity Letter* 1(1): 1 (1995).

Broyles, B. and R.S. Felger, eds. *Dry Borders. Journal of the Southwest* 39 (3-4): 303-860 (1998).

Buchmann, Stephen L. and Gary Paul Nabhan. *Forgotten Pollinators.* Washington D.C.: Island Press, 1996.

Daily, Gretchen C., ed. *Nature's Services: Societal Dependence on Natural Ecosystems.* Washington, D.C.: Island Press, 1997.

Felger, R.S, P.L. Warren, L.S. Anderson and G.P. Nabhan. "Vascular Plants of a Desert Oasis: Flora and Ethnobotany of Quitobaquito, Organ Pipe Cactus National Monument, Arizona." *Proceedings of the San Diego Society of Natural History* 8: 1-39 (1992).

Nabhan, G.P. and A.R. Holdsworth. *State of the Desert Biome: Uniqueness, Biodiversity, Threats, and Adequacy of Protection in the Sonoran Region.* Tucson: The Wildlands Project, 1998.

Redford, Kent. "Science and the Nature Conservancy." *Nature Conservancy* 44 (1): 14-15 (1994).

Rondeau, R., T.R. Van Devender, C.D. Bertelson, P. Jenkins, R.K. Wilson and M. Dimmitt. "Annotated Flora and Vegetation of the Tucson Mountains, Pima County, Arizona." *Desert Plants* 12 (2): 1-47 (1996).

Villasenor, J.L. and T.S. Elias. "Analysis de Especies Endemicas para Identificar Areas de Proteccion en Baja California, Mexico." *Conservacion de Plantas en Peligro de Extincion.* E. Linares et al., eds. Mexico City: UNAM, 1995.

Wilson, Edward O. *The Diversity of Life.* Cambridge: Harvard University (Belnap) Press, 1992.

Plants

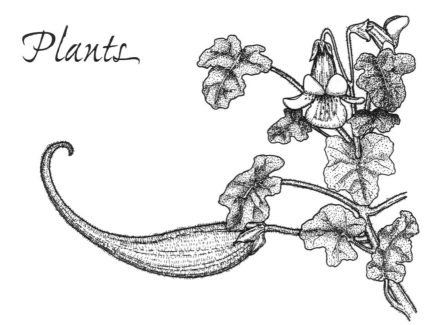

Plant Ecology of the Sonoran Desert Region

Mark A. Dimmitt

You could easily recognize a desert even if you were blindfolded. You would discover that you could walk fairly long distances without bumping into plants, and when you did the encounter would likely be painful. Even standing still you would have unmistakable clues about your location. You'd feel the arid atmosphere pulling moisture out of your body and experience a sensation of pressure on your skin from the intense sunlight. On really hot, dry days you could smell pungent, aromatic terpenes and oils exuded by the parched vegetation.

With the blindfold removed, you would see that most desert plants also look different from those in other habitats—they are often spiny, almost always tiny-leafed, and rarely "leaf green." Many have bold, sculptural growth forms characterized by swollen stems or starkly exposed stems unconcealed by foliage. At the other extreme is a unique desert growth form that

landscape architect Iain Robertson calls "diaphanous plants." The stems and foliage of these plants are so fine-textured and sparse that the eye tends to look right through them.

These tactile, olfactory, and visual experiences offer clues to desert plants' adaptations to their rigorous environment. Before exploring these special characteristics, it is important that you understand something about plant structures, functions, and classification.

BASIC PLANT ANATOMY AND CLASSIFICATION

Many people mistakenly identify ocotillos, agaves, African euphorbias, and numerous other plants as cacti because of their succulent or spiny stems, when in fact these plants are not related to each other or to cacti. Frequently plants (and animals) are similar to each other in outward appearance because their ancestors have adapted to the same environmental challenges by evolving similar forms or structures. This similarity in response to environment, despite lack of common recent ancestors, is known as *convergent evolution.* The very similar outward appearance of some New World cacti and Old World succulent euphorbias is an excellent example.

Unlike overall form or vegetative structures, the sexual parts of plants (flowers and fruits) are reliable indicators of interrelationships and means of identification. Flowers must function successfully if a plant is to reproduce. Therefore the floral structures tend to remain more consistent within a species than do vegetative parts. Floral structures also form complex patterns that are more readily traceable as plants evolve. The parts of flowers and fruits are also easier to identify and describe than the vegetative organs (leaves, stems, and roots). Moreover, qualitative vegetative characters are hard to describe precisely even when the overall appearance (gestalt) is distinctive. For example, nearly every hiker knows poison ivy on sight. But try to describe the foliage so precisely that someone who has never seen the plant can distinguish it from skunk bush (*Rhus trilobata*). It's quite difficult to delineate the leaves' different shades of green, degrees of hairiness, and the scalloping of the margins, especially if you lack the minutely-detailed vocabulary of the botanist. For example, the terms pubescent, puberulent, lanate, villous, hirsute, hirsutulous, ciliate, tomentose, strigose, pilose, and hispid are just some of those used to describe different kinds and degrees of hairiness. Vegetative parts are also more plastic; that is, they vary greatly—even in the same individual—under environmental

influences. The leaves of brittlebush grow much larger and greener in shade or during rainy periods than in sun or in drier conditions. (See photo on page 131; see also species account.)

The complex parts of flowers and fruits are arranged in distinctive patterns that can be characterized exactly. Petals, stamens, and other structures can be counted and their lengths and widths measured (and these are usually less variable than the dimensions of leaves). The point of attachment of the stamens to the petals (or other parts) can be described unambiguously. For example, a flower that has many (more than ten) petals and sepals that intergrade into one another, many stamens (usually hundreds), a two- to multi-lobed stigma, and an ovary enveloped in stem tissue may be unequivocally identified as belonging to a member of the cactus family. All 2000 species of cacti possess some variation of this basic pattern, and no other plant group does.

To recognize floral patterns you must be able to identify the parts of a flower. The drawing on page 131 shows the anatomy of a generalized flower.

Flower Anatomy, from Outside to Inside

In the game "Twenty Questions" players attempt to identify an unknown by asking the person who knows the answer a series of yes-or-no questions. If done well, twenty questions is sufficient to eliminate every other possibility in the world and leave the correct answer standing. Assume, for example, that the unknown thing is a dog. First question: "Is it a concept?"

The size and hairiness of the leaves of brittlebush (*Encelia farinosa*) vary with soil moisture, as illustrated by these stems cut from several plants. The leaves on the far left stem were produced under cool, moist conditions. The leaves on the two middle stems grew under moderately to very dry soil and increased heat. The stem on the far right was from a bush growing on rocky soil on a hot south slope; it has dried out and shed its leaves entirely.

(No; therefore it's an object.) Second question: "Is it alive?" (Yes.) Third question: "Is it a plant?" (No). Fourth question: "Is it a vertebrate? (Yes). Fifth question: "Is it an herbivore?" (No.) The enormous inventory of the universe has been narrowed to a very short list in only five questions.

Botanists identify plants (and zoologists, animals) unknown to them with a Twenty Questions-like procedure called a *dichotomous key* (or simply, *key*). A key is a nested series of dual choices that quickly narrows the possibilities to a single species. For example, the first pair of choices might ask you whether the flower has three petals versus four or five. Each of the two possible answers leads to another pair of choices, and so on, until you have identified

your quarry out of 300,000 species of flowering plants. But before you can use such a key effectively, or before you can describe your unknown to someone who will identify it for you, you must know the parts of the flower and plant you are examining.

The *sepals* collectively make up the *calyx*. They enclose all other flower parts in the bud, usually completely concealing the rest of the flower until it opens.

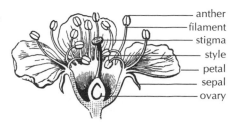

anther
filament
stigma
style
petal
sepal
ovary

The *petals* collectively make up the *corolla*. Petals are frequently the visual advertising banner that attracts pollinators. Petals and sepals look similar in many flowers, such as in lilies and agaves. By definition the sepals are the parts on the outside; petals are typically concealed in the bud stage. When sepals and petals can't be readily distinguished, they are called *tepals*.

The corolla and calyx make up the *perianth*. The perianth parts may be separate or fused together for part or all of their length. Often there is only one series of perianth parts. Of necessity these must be on the outside and therefore they are sepals, even if they are large and colorful.

The female part of a flower is the *pistil*, composed of *stigma, style* and *ovary* (also called *ovulary*). The ovary contains ovules, which develop into seeds when fertilized by the sperm in pollen. Seeds are plant embryos encased in a protective membrane, usually along with stored energy to fuel germination. If the ovary is visible beneath the calyx, it is said to be *inferior*. It is superior if you must look inside the flower to see it (that is to say, it is above the calyx).

The male part of a flower is the *stamen*, composed of the *anther* and the *filament*. Anthers produce pollen grains, which contain sperm cells.

Photosynthesis

Chlorophyll is one of the most consequential chemicals in the biosphere. Nearly all life on the planet depends on it. Living organisms seem to defy the law of entropy, the universal tendency toward increasing disorder in a closed system. By using energy acquired from outside they prevent themselves—temporarily—from dying and disintegrating into simple, dissociated molecules (becoming disordered). A small number of species derive their energy from metabolizing sulfur compounds. All others, including all the organisms that we encounter in everyday life, depend on solar energy (light) to maintain their orderly existence. Light, however, is unmanageable; it can't be concentrated and stored for later use. (Outside of science fiction there is no such thing as a photon battery). Enter photosynthesis. Green plants use light energy to combine low-energy molecules (carbon dioxide and water) into high-energy molecules (carbohydrates), which they accumulate and store as energy reserves. *Chlorophyll* (the green pigment in plants) is the only known substance in the universe that can capture volatile light energy and convert it into a stable form usable for biological processes (chemical energy).

*"See it with your eyes:
Earth reenergized
by the sun's rays every day."*

— Moody Blues

Almost without exception living organisms—plants, animals, and even fungi and bacteria—obtain energy for sustaining life from carbohydrates (sugars and starches) by the metabolic process of *aerobic respiration*. ("Respiration" is colloquially and medically used to mean breathing. The mechanical act of breathing, however, is only the first step in the physiological process of respiration.) Respiration is the chemical pathway through which carbohydrate is broken down (oxidized) into carbon dioxide and water, releasing the energy stored in the carbohydrate molecules. This is represented by the formula: Carbohydrate $+ O_2 \rightarrow\rightarrow\rightarrow\rightarrow H_2O + CO_2 +$ Energy. (The multiple arrows indicate many sequential chemical reactions.) Green plants manufacture carbohydrates by photosynthesis. Animals acquire their carbohydrates by eating plants or other animals.

Photosynthesis is somewhat the reverse of respiration: carbon dioxide and water are combined to form larger molecules of carbohydrate, with the addition of energy from sunlight: $H_2O + CO_2 +$ Energy $\rightarrow\rightarrow\rightarrow\rightarrow$ Carbohydrate $+ O_2$. Water is absorbed through the roots, and CO_2 diffuses into the leaves through the *stomates* (valved pores in leaf and stem surfaces). The plant joins several carbon dioxide molecules and adds hydrogen atoms split from water molecules to form molecules of sugar (simple carbohydrate). Surplus oxygen atoms from the water molecules are released through the stomates as oxygen gas (O_2).

When you see the word "carbohydrate," think "stored energy" and "calories." Plants store energy for long-term use in the form of starch, which is a complex carbohydrate consisting of long chains of sugar molecules. When a plant needs energy to grow new leaves or flowers, it does exactly what animals do—it respires carbohydrate to release the stored energy. The complex respiratory pathway of scores of individual chemical reactions is nearly identical in all life forms: bacteria, mushrooms, saguaros, coyotes, and even in the highest life forms such as toads.

In contrast to plants, animals use fat as their main energy store; it has twice the number of calories per gram as do carbohydrate and protein. When animals in need of energy run low on the small amount of carbohydrate stored in the liver or circulating in the blood, they convert fat (or protein, if they run out of fat) into carbohydrate and then respire it.

The most common form of photosynthesis creates a 3-carbon sugar as its first stable product, so it's called C_3 photosynthesis. Other sugars with more carbon atoms are later synthesized from this first one. More than 90 percent of all plant species use C_3 exclusively, but there are two specialized supplementary variations.

One variant is called C_4 photosynthesis because the first stable product is a 4-carbon sugar. Plants with C_4 metabolism actively transport carbon dioxide to localized bundles of photosynthetic tissue. This process offers improved efficiencies under hot, sunny conditions. C_4 plants use carbon dioxide more efficiently (by bypassing photorespiration) and lose less water through *transpiration* (water evaporated from inside plants) per unit of carbohydrate made. The overall result is that C_4 plants can grow much faster under high temperatures than most C_3 plants. The majority of summer-growing grasses in warm climates are C_4. So are many other summer-growing plants, especially weeds (invasive pioneer plants) that seem to spring up overnight, such as pigweed (*Amaranthus* spp.) and summer spurges (*Euphorbia hyssopifolia* and others), as well as devil's claw (*Proboscidea* spp.), and many saltbushes (*Atriplex* spp.). Only about 3 percent of all the Earth's plant species are known to use C_4, but a number of them are vital crops, such as corn, sorghum, and sugar cane. Another variant of photosynthesis, CAM, is discussed under the succulence section of this chapter on page 135.

COPING WITH DESERT CLIMATE

The impression that the desert environment is hostile is strictly an outsider's viewpoint. Adaptation enables indigenous organisms not merely to survive here, but to thrive. Furthermore, specialized adaptations often result in a requirement for the seasonal drought and heat. For example, the saguaro, well adapted to its subtropical desert habitat, cannot survive in a rain forest or in any other biome, not even a cold desert. In these other places it would rot, freeze, or be shaded out by faster growing plants.

Aridity is the major—and almost the only—environmental factor that creates a desert, and it is this functional water deficit that serves as the primary limitation to which desert organisms must adapt. Desert plants survive the long rainless periods with three main adaptive strategies: succulence, drought tolerance, and drought evasion. Each of these is a different but effective suite of adaptations for prospering under conditions that would kill plants from other regions.

Succulence

As a group succulents are the most picturesque desert plants. They capture our attention because they look nothing like the familiar plants of the temperate zone where most people live. Their vernacular names suggest how they command our attention: elephant tree, boojum, jumping cholla, creeping devil, and shindagger. Spanish names translate into equally colorful terms such as dragon's blood, child-killer, and old man's head. Even some scientific names are inspired by the plants' characteristics: *Ferocactus* (as in ferocious), *Opuntia molesta* (the molesting-spined cactus), *O. invicta* (the invincible point), and *Agave jaiboli* (as in a highball cocktail, because liquor is made from it).

Succulent plants store water in fleshy leaves, stems or roots in compounds or cells from which it is not easily lost (see photo on page 135). All cacti are succulents, as are such non-cactus desert dwellers as agaves, aloes, elephant trees, and many euphorbias. Several other adaptations are essential for the water-storing habit to be effective.

GETTING WATER Succulents must be able to absorb large quantities of water in short periods, and they must do so under unfavorable conditions. Because roots take up water by passive diffusion, succulents can absorb water only from soil that is wetter than their own moist interiors. Desert soils seldom get this wet and don't retain surplus moisture for long. Desert rains are often light and brief, barely wetting the top few inches (centimeters) of soil, which may dry out after just a day or two of

In this cross section of a barrel cactus the thin dark outer layer is epidermis and photosynthetic tissue. All of the lighter color is water storage tissue.

summer heat. To cope with these conditions, nearly all succulents have extensive, shallow root systems. A giant saguaro's root system is just beneath the soil surface and radiates as far as the plant is tall. The roots of a two-foot-tall cholla in an extremely arid site may be thirty feet (9 m) long. Most succulents, in fact, rarely have roots more than four inches (10 cm) below the surface and the water-absorbing feeder roots are mostly within the upper ½ inch (1.3 cm). Agaves are an exception in lacking extensive root systems; most of the roots don't extend much beyond the spread of the leaf rosette. Instead, the leaves of these plants channel rain to the plants' bases.

CONSERVING WATER A succulent must be able to guard its water hoard in a desiccating environment and use it as efficiently as possible. The stems and leaves of most species have waxy cuticles that render them nearly waterproof when the stomates are closed.

Water is further conserved by reduced surface areas; most succulents have few leaves (agaves), no leaves (most cacti), or leaves that are deciduous in dry seasons (elephant tree [*Bursera* spp.], boojums [*Fouquieria columnaris*]). The water is also bound in extracellular mucilages and inulins—compounds that hold tightly onto the water.

Many succulents possess a water-efficient variant of photosynthesis called CAM, an acronym for *Crassulacean Acid Metabolism*. The first word refers to the stonecrop family (Crassulaceae) in which the phenomenon was first discovered. (*Dudleya* is in this family, as are hen-and-chickens and jade plant.) CAM plants open their stomates for gas exchange at night and store carbon dioxide in the form of an organic acid. During the day the stomates are closed and the plants are nearly completely sealed against water loss; photosynthesis is conducted using the stored carbon dioxide. At night the temperatures are lower and humidity higher than during the day, so less water is lost through transpiration. Plants using CAM lose about one-tenth as much water per unit of carbohydrate synthesized as do those using standard C_3 photosynthesis. But there is a trade-off: the overall rate of photosynthesis is slower, so CAM plants grow more slowly than most C_3 plants. (An additional limitation is the reduced photosynthetic surface area of most succulents compared with "ordinary" plants.)

The equilibrium between gaseous carbon dioxide and the organic acid is dependent on temperature. Acid formation (carbon dioxide storage) is favored at cool temperatures; higher temperatures stimulate release of carbon dioxide from the acid. Thus CAM works most efficiently in climates that have a large daily temperature range, such as arid lands. Cool nights allow much carbon dioxide to be stored as acid, and the warm days cause most of the carbon dioxide to be released for photosynthesis. (A note of interest: A plant in CAM mode will store enough acid to impart a sour taste in early morning; the flavor becomes bland by afternoon when the acid is used up. But don't taste indiscriminately—many succulents are poisonous!)

Many succulents possess CAM, as do semisucculents such as some yuccas, *epiphytic* (growing on trees or rocks) orchids, and *xerophytic* (arid-adapted) bromeliads. Exceptions are stem succulents with deciduous, non-succulent leaves, such as elephant trees (*Bursera* spp.), limberbushes (*Jatropha* spp.), and desert roses (*Adenium* spp.). Succulents from hot, humid climates that lack substantial daily temperature fluctuations also usually do not use CAM. Some succulents, such as *Agave deserti*, can switch from CAM to C_3 photosynthesis when water is abundant, allowing faster growth. Over five percent of all plant species spread among thirty or more plant families are known to use CAM.

Another crucial attribute of CAM plants is their idling metabolism during droughts. When CAM plants become water-stressed, the stomates remain closed both day and night and the fine (water-permeable) roots are sloughed off. The plant's stored water is essentially sealed inside and gas exchange greatly decreases. However, a low level of respiration (oxidation of carbohydrate into water, carbon dioxide and energy) is

carried out within the still-moist tissues. The carbon dioxide released by respiration is recycled into the photosynthetic pathway to make more carbohydrate, and the oxygen released by photosynthesis is recycled for respiration. Thus the plant never goes completely dormant but is metabolizing slowly—idling. (This sounds like perpetual motion, but it isn't. The recycling isn't 100 percent efficient, so the plant will eventually exhaust its resources.) Just as an idling engine can rev up to full speed more quickly than a cold one, an idling CAM plant can resume full growth in twenty-four to forty-eight hours after a rain. Agaves can sprout visible new roots just five hours after a rain, whereas it may take a couple of weeks for a dormant nonsucculent shrub to resume full metabolic activity. Therefore, succulents can take rapid and maximum advantage of the soil moisture from a summer rain before it quickly evaporates. The combination of shallow roots and the CAM-idling which allows rapid response enables succulents to benefit from rain even in amounts less than ¼ inch (6 mm).

PROTECTION Stored water in an arid environment requires protection from thirsty animals. Most succulent plants are spiny, bitter, or toxic, and often all three. Some unarmed, nontoxic species are restricted to inaccessible locations. Smooth prickly pear (*Opuntia phaeacantha* var. *laevis*) and live-forever (*Dudleya* spp.) grow on vertical cliffs or within the canopies of armored plants. Still others rely on camouflage; Arizona night-blooming cereus (*Peniocereus greggii*) closely resembles the dry stems of the shrubs in which it grows.

These adaptations are all deterrents that are never completely effective. Evolution is a continuous process in which some animals develop new inheritable behaviors to avoid spines or new metabolic pathways to neutralize the toxins of certain species. In response the plants are continually improving their defenses. For example, packrats can handle even the spiniest chollas and rarely get stuck. They also eat prickly pear for water and manage to excrete the oxalates which could clog the kidneys of some other animals. Toxin-tolerant insects often incorporate their host plant's toxins into their own tissues for protection against their predators.

Drought Tolerance

Drought-tolerant plants often appear to be dead or dying during the dry seasons. They're just bundles of dry sticks with brown or absent foliage, reinforcing the myth that desert organisms are engaged in a perpetual struggle for survival. They're simply waiting for rain in their own way, and are usually not suffering or dying any more than a napping dog is near death (see photo on page 138).

Drought tolerance or drought dormancy refers to desert plants' ability to withstand desiccation. A tomato plant will wilt and die within days after its soil dries out. But many nonsucculent desert plants survive months or even years with no rain. During the dry season the stems of brittlebush and bursage are so dehydrated that they can be used as kindling wood, yet they are alive. Drought-tolerant plants often shed leaves during dry periods and enter a deep dormancy analogous to *torpor* (a drastic lowering of metabolism) in

animals. Dropping leaves reduces the surface area of the plant and thus reduces transpiration. Some plants that usually retain their leaves through droughts have resinous or waxy coatings that retard water loss (creosote bush, for example).

The roots of desert shrubs and trees are more extensive than are those of plants of the same size in wetter climates. They extend laterally two to three times the diameter of the canopy. Most also exploit the soil at greater depths than the roots of succulents. The large expanses of exposed ground between plants in deserts are usually not empty. Dig a hole almost anywhere except in active sand dunes or the most barren desert pavement and you are likely to find roots.

Rooting depth controls opportunities for growth cycles. In contrast to the succulents' shallow-rooted, rapid-response strategy, a substantial rain is required to wet the deeper root zone of shrubs and trees. A half-inch is the minimum for even the smaller shrubs —more for larger, deeper-rooted plants. It takes a couple of weeks for dormant shrubs such as brittlebush (*Encelia farinosa*) and creosote bush (*Larrea tridentata*) to produce new roots and leaves and resume full metabolic activity after a soaking rain. The trade-off between this strategy and that of succulents is that once the deeper soil is wetted, it stays moist much longer than the surface layer; the deeper moisture sustains growth of shrubs and trees for several weeks.

Mesquite trees (*Prosopis* spp.) are renowned for having extremely deep roots, the champion reaching nearly 200 feet. But these riparian specimens

The ocotillo and several leafless shrubs in this photo are not dead; they are just dormant and waiting for rain. Anza-Borrego Desert State Park, California.

are not drought-tolerating trees—their roots are in the water table. Most large floodplain mesquites die if the water table drops below forty feet, and mesquites growing away from waterways remain short and shrubby. No desert plant is known to use very deep roots as a primary strategy for survival. In fact, the root systems of most trees—including mesquites—are mostly confined to the upper three feet of soil. Few rains penetrate deeper than this, and at greater depths there is little oxygen to support root respiration.

In contrast to succulents that can take up water only from nearly saturated soil, drought tolerant plants can absorb water from much drier soil. A creosote bush can obtain water from soil that feels dust-dry to the touch. Similarly these plants can continue to photosynthesize with low leaf-moisture contents that would be fatal to most plants.

Some plants in this adaptive group are notoriously difficult to cultivate, especially in containers. It seems paradoxical that desert ferns and creosote bushes, among the most drought-tolerant of desert plants, can be kept alive in containers only if they are never allowed to dry out. The reason is that these plants can survive drought only if they dry out slowly and have time to make gradual physiological adjustments. If a potted plant misses a watering, the small soil volume dries out too rapidly to allow the plant to prepare for dormancy, so it dies. Researchers showed that some spike mosses (*Selaginella* spp.) must dehydrate over a five to seven day period. If they dry more rapidly they lack time to adjust, and if drying takes longer than a week they exhaust their energy reserves and starve to death. (*Selaginella lepidophylla* from the Chihuahuan Desert is widely sold as a novelty under the name "resurrection fern." Rehydration and resumption of active life takes only a few hours.)

Drought Evasion

Interstate 40 from Barstow to Needles, California traverses some of the emptiest land in the West. It dashes as straight as it can through 130 miles (200 km) of dry valleys that are almost devoid of human settlements. The vegetation is simple, mostly widely-scattered creosote bushes. It's difficult to tell if you're driving through the Mohave or Sonoran desert. The small, rocky mountain ranges interrupting the valleys beckon to true desert lovers, but the drive is just plain bleak to most folks. The exits on this freeway average ten miles (16 km) apart and connect to two-lane roads that shoot straight over the distant horizon with no visible destinations. You rarely see a vehicle on any of them.

Frequent travelers on this freeway become accustomed to its monotony until they think they know what to expect. The creosote bush may turn green if there's been a rain; ocotillo always flowers in April; most of the time it's just brown gravel and brown bushes. Then one spring travelers were astonished to discover the ground between the bushes literally carpeted with flowers. It happened in March 1998, when for three weeks the freeway bisected a nearly unbroken blanket of desert sunflowers forty miles long and ten miles wide. At every exit-to-nowhere several cars and trucks were pulled off and people wandered through the two-foot-deep sea of yellow. Those with long memories may have recalled that the same thing happened in 1978.

Perhaps they wondered where these flowers came from, and where they were during the intervening twenty years.

Those desert sunflowers (*Geraea canescens*) were annual wildflowers, plants that escape unfavorable conditions by "not existing" during such periods. Annuals complete their life cycles during brief wet seasons, then die after channeling all of their life energy into producing seeds. Seeds are dormant *propagules* with almost no metabolism and great resistance to environmental extremes. (A propagule is any part of a plant that can separate from the parent and grow into a new plant, for example, a seed, an agave aerial plantlet, a cholla joint.) Seeds wait out adverse environmental conditions, sometimes for decades, and will germinate and grow only when specific requirements are met.

Wildflower spectacles like the one described above are rare events. Mass germination and prolific growth depend on rains that are both earlier and more plentiful than normal. The dazzling displays featured in photographic journals and on postcards occur about once a decade in a given place. In the six decades between 1940 and 1998 there have been only four documented drop-everything-and-go-see-it displays in southern Arizona: 1941, 1978, 1979, and 1998. During that period only the displays of 1978 and 1998 were widespread throughout both the Sonoran and Mohave deserts.

Annuals in the Sonoran Desert can be divided into three groups, based on time of germination and flowering. Winter-spring species are by far the most numerous. The showy wildflowers that attract human attention will germinate only during a narrow window of opportunity in the fall or winter, after summer heat has waned and before winter cold arrives. In most of the Sonoran Desert this temperature window seems to occur between early October and early December for most species. During this window there must be a soaking rain of at least one inch (2.5 cm) to induce mass germination. This combination of requirements is survival insurance: an inch of rain in the mild weather of fall will provide enough soil moisture that the resulting seedlings will probably mature and produce seeds even if almost no more rain falls in that season. (Remember that one of the characteristics of deserts is low and *undependable* rainfall.) If the subsequent rainfall is sparse, the plants remain small and may produce only a single flower and a few seeds, but this is enough to ensure a future generation (see photo on page 141). There is still further insurance: even under the best conditions not all of the seeds in the soil will germinate; some remain dormant. For example, a percentage of any year's crop of desert lupine seeds will not germinate until they are ten years old. The mechanisms that regulate this delayed germination are not well understood.

The seedlings produce rosettes of leaves during the mild fall weather, grow more slowly through the winter (staying warm in the daytime by remaining flat against the ground), and bolt into flower in the spring. Since the plants are inconspicuous until they begin the spring bolt, many people mistakenly think that spring rains produce desert wildflower displays.

There is a smaller group of annual species that grow only in response to summer rains. Annual devil's claw (*Proboscidea parviflora*) and Arizona poppy (*Kallstroemia grandiflora*) are among the few showy ones.

A third group consists of a few opportunistic species which will germinate in response to rain at almost any season. Most of these lack showy flowers and are known only to botanists, but desert marigold (*Baileya multiradiata*) is a conspicuous exception; it is actually not an annual, but rather a short-lived perennial in most of its range. A few species of buckwheats (*Eriogonum*) germinate in fall or winter and flower the following summer.

The annual habit is a very successful strategy for warm-arid climates. There are no annual plants in the polar regions or the wet tropics. In the polar zones the growing season is too short to complete a life cycle. In both habitats the intense competition for suitable growing sites favors longevity. (Once you've got it, you should hang onto it.) Annuals become common only in communities that have dry seasons, where the perennials are widely spaced because they must command a large soil area to survive the drier years. In the occasional wetter years, both open space and moisture are available to be exploited by plants that can do so rapidly. The more arid the habitat, the greater the proportion of annual species in North America. (The percentage decreases in the extremely arid parts of the Saharan-Arabian region.) Half of the Sonoran Desert's flora is comprised of annual species. In the driest habitats, such as the sandy flats near Yuma, Arizona, up to ninety percent of the plants are annuals.

Sun-cups (*Camissonia brevipes*) is a typical desert annual. The plant on the left grew in a rather dry year; it had just enough soil moisture to produce a single flower and a few seeds. The plant on the right grew in a wet year and is much larger; it produced hundreds of flowers.

Winter annuals provide most of the color for our famous wildflower shows. Woody perennials and succulents can be individually beautiful, but their adaptive strategies require them to be widely-spaced, so they usually don't create masses of color. A couple of exceptions are brittlebush when it occurs in pure stands, and extensive woodlands of foothill palo verde (*Cercidium microphyllum*). The most common of the showy winter annuals that contribute to these displays in southern Arizona are Mexican gold poppy (*Eschscholtzia mexicana*), lupine (*Lupinus sparsiflorus*), and owl-clover (*Castilleja exserta*, formerly *Orthocarpus purpurascens*).

One of the contributing factors to the great number of annual species is *niche* separation. (A niche is an organism's ecological role; for example, sand verbena is a butterfly-pollinated winter annual of sandy soils.) Most species have definite preferences for particular soil textures, and perhaps soil chemistry as well. For example, in the Pinacate region of northwestern Sonora there are places where gravels of volcanic cinder are dissected by drainage channels or wind deposits of fine silt. In wet years *Nama demissum* (purple mat) grows abundantly on the gravel and the related *Nama hispidum* (sand bells) on the silt. I have seen the two species within inches of each other where these soil types meet, but not one plant of either species could be found on the other soil. There are specialists in loose sand such as dune evening primrose (*Oenothera deltoides*) and sand verbena (*Abronia villosa*), and others are restricted to rocky soils, such as most caterpillar weeds (*Phacelia* spp.). This phenomenon of occupying different physical locations is *spatial* niche separation.

Another diversity-promoting phenomenon is *temporal* niche separation: the mix of species at the same location changes from year to year. Seeds of the various species have different germination requirements. The time of the season (which influences temperature) and quantity of the first germination-triggering rain determines which species will dominate, or even be present at all in that year. Of the three most common annuals of southern Arizona listed above, any one may occur in a nearly pure stand on a given hillside in different years, and occasionally all three are nearly equally abundant. This interpretation of the cause of these year-to-year variations is a hypothesis based on decades of empirical observation. Much more research is needed to discover the ecological requirements of most species of desert annuals. And of course the Sonoran Desert's two rainy seasons provide two major temporal niches. Summer and winter annuals almost never overlap.

The dramatic wildflower shows are only a small part of the ecological story of desert annuals. For each conspicuous species there are dozens of others that either have less colorful flowers or don't grow in large numbers. Every time the desert has a wet fall or winter it will turn green with annuals, but it will not always be ablaze with other colors. One of the most common winter annuals is desert plantain (*Plantago insularis*). It usually grows only a few inches tall and bears spikes of tiny greenish flowers, but billions of plants cover many square miles in good years. The tiny seeds are covered with a soluble fiber which forms a sticky mucilage when wet by rain; this aids germination by retaining water around the seed and sticking it to the ground. A related species from India is the commercial source of psyllium fiber (Metamucil® for example). The buckwheat family (Polygonaceae) is well-represented. There are more than a score of skeleton weeds (*Eriogonum* spp.) and half as many spiny buckwheats (*Chorizanthe* spp.), most of which go unnoticed except by botanists (see species accounts). Fiddlenecks (*Amsinckia* spp., Boraginaceae) may grow in solid masses over many acres, but the tiny yellow flowers don't significantly modify the dominant green of the foliage. These more modest species produce more biomass than the showy wildflowers in most years, and thus form the foundation of a great food pyramid.

Some perennials also evade drought much as annuals do, by having underground parts that send up stems, leaves, and flowers only during wet years. Coyote gourd (*Cucurbita digitata*) and perennial devil's claw (*Proboscidea althaeifolia*) have fleshy roots that remain dormant in dry years. Desert larkspur (*Delphinium parryi*) is a perennial that has woody rootstocks but also sprouts only in wetter years. Desert mariposa (*Calochortus kennedyi*) and desert lily (*Hesperocallis undulata*) have bulbs that may remain dormant for several years until a deep soaking rain awakens them.

Our desert wildflower displays are in jeopardy from invasive exotic plants. Species such as Russian thistle (*Salsola tragus,* also called *S. kali*), mustards (especially *Brassica tournefortii*), filaree (*Erodium cicutarium*), and Lehmann's lovegrass (*Eragrostis lehmannii*) are more aggressive than most of the native annuals and are crowding them out in many areas where they have become established. Some are still increasing their geographic ranges with every wet winter. Disturbed sites such as sand dunes, washes (naturally disturbed by wind and water, respectively), roadsides, and livestock-grazed lands are particularly vulnerable to invasion by these aliens.

Combined Drought Adaptations

These three basic drought-coping strategies—succulence, drought tolerance, and drought avoidance—are not exclusive categories. Ocotillo behaves as if it were a CAM-succulent, drought deciduous shrub, but it is neither CAM nor succulent (see details in the species accounts). The genus *Portulaca* contains species that are succulent annuals. The seeds may wait for a wet spell to germinate, but the resulting plants can tolerate a moderate drought. The semisucculent yuccas have some water storage capacity, but rely on deep roots to obtain most of their water. Mesquite trees are often *phreatophytes* (plants with their roots in the water table), but some species can also grow as stunted shrubs on drier sites where ground water is beyond their reach.

Adaptations to Other Desert Conditions

Water scarcity is the most important—but not the only—environmental challenge to desert organisms. The aridity allows the sun to shine unfiltered through the clear atmosphere continuously from sunrise to sunset. This intense solar radiation produces very high summer temperatures which are lethal to nonadapted plants. At night much of the accumulated heat radiates through the same clear atmosphere and the temperature drops dramatically. Daily fluctuations of 40°F (22°C) are not uncommon when the humidity is very low.

Microphylly (the trait of having small leaves) is primarily an adaptation to avoid overheating; it also reduces water loss. A broader surface has a deeper boundary layer of stagnant air at its surface, which impedes convective heat exchange. A leaf up to ½ inch (10 mm) across can stay below the lethal tissue temperature of about 115°F (46°C) on a calm day with its stomates closed. A larger leaf requires transpiration through open stomates for evaporative cooling. Since the hottest time of year is also the driest, water is not available for transpiration. Non-succulent large-leafed plants in the desert environment would overheat and be killed. Desert gardeners know that tomatoes will burn in full desert sun even if well watered; their leaves are just too big to stay cool. Desert plants that do have large leaves produce them only during the cool or rainy season or else live in shaded microhabitats. There are a few mysterious exceptions, such as jimson weed (*Datura wrightii*) and desert milkweed (*Asclepias erosa*). Perhaps their large tuberous roots provide enough water for transpiration even when the soil is dry.

Leaf or stem color, orientation, and self-shading are still more ways to adapt to intense light and heat. Desert foliage comes in many shades, but rarely in typical leaf-green. More often leaves are gray-green, blue-green, gray, or even white. The light color is usually due to a dense covering of *trichomes* (hairlike scales), but is sometimes from a waxy secretion on the leaf or stem surface. Lighter colors reflect more light (= heat) and thus remain cooler than dark green leaves. Brittlebush and white bursage leaves show no green through their trichomes during the dry season, while desert agave (*Agave deserti*) is light gray due to its thick, waxy cuticle. Other plants have leaves or stems with vertical orientations; two common examples are jojoba and prickly pear cactus. This orientation results in the photosynthetic surface facing the sun most directly in morning and late afternoon. Photosynthesis is more efficient during these cooler times of day. Prickly pear pads will burn in summer if their flat surfaces face upward. Some cacti create their own shade with a dense armament of spines; teddy bear cholla (*Opuntia bigelovii*) is one of the most striking examples.

POLLINATION ECOLOGY AND SEED DISPERSAL OF DESERT PLANTS

Flowers are very useful for identifying plants and providing aesthetic pleasure for humans, but they have a more vital function—they are the sexual reproductive organs of plants. Many plants also have methods of asexual (vegetative) reproduction, which produces offspring that are genetically identical to the parent: root-sprouting (limberbush, palo verde, aspen), stolons and rhizomes (agaves, strawberries, many grasses), and aerial plantlets (some agaves, mother-of-millions, kalanchoe). All of the progeny

of asexual reproduction are clones of their parent plants. (A clone is a group of organisms that are genetically identical; in the case of flowering plants each clone originates from a single seed.) Horticulturists have developed additional methods of plant cloning that are valuable in perpetuating superior varieties of plants: cutting, grafting, and tissue culture. The 'Kadota' fig is a cultivar (contraction for cultivated variety) that has been propagated by cuttings for at least two millennia; it is described under a different name in the writings of Pliny the Younger.

In contrast, sexual reproduction combines half the genes from each of two parents, so sexually produced offspring are different from either of their parents and from one another. This variation is the raw material of natural selection which in turn results in evolution. A species that cannot reproduce sexually—there are quite a few among both plants and animals—is at greater risk of extinction if its environment changes, because it cannot adapt to new conditions.

Pollination is the transfer of pollen from an anther onto the stigma of a flower. The pollen then grows a tube that penetrates the style down to the ovary; sperm cells swim down the tube and fertilize the ova. Fertilized ova develop into seeds, which are the sexual propagules of flowering plants.

Outcrossing (pollination by pollen from another plant) is evolutionarily advantageous because the offspring are more variable than those from self-pollination. Variability increases plants' probability of surviving in an ever-changing environment. (But self-pollination is still sexual reproduction which results in different combinations of genes and therefore allows evolutionary change, as vegetative cloning does not.) Plants have many adaptations that increase the likelihood of outcrossing.

Because plants are rooted in the ground and can't get together to mate, they must employ an agent to transport pollen between plants. From this need widespread and complex kinds of *mutualism* (mutually beneficial interactions) have evolved between plants and animals. The pollen-transporting agent is frequently an insect or other flying animal. (Flying animals are more mobile than grounded species, and thus more likely to visit widely-separated plants.) In order to get pollinated, a flower must both make its presence known (advertise), and provide an incentive (a reward) for an animal to make repeated visits to flowers of the same species. The advertisements are fragrance and/or conspicuous color. Two kinds of food are the usual reward. Nectar is a sugar solution that provides energy for flight. Flying requires much more energy than terrestrial locomotion. Pollen, besides being the male gene-bearer of a flower, is also rich in proteins essential for maintaining animal tissues and for raising young. In place of nectar some flowers offer oil (fat), another energy food. Others provide fragrances that the pollinator gathers to use for its own reproductive advertisement, and a few fascinating species employ deceit and provide no reward (see the species account on pipevine for an example).

The sugar in nectar and the protein in pollen are expensive to produce, so there is selective pressure to use these resources efficiently. It is important that animals other than the pollinators do not eat (steal) the nectar and pollen, and that the pollinators transport pollen to other flowers of the same species and deposit it in the right place. Natural selection has produced specialization: most plants with animal-pollinated flowers attract only a few species of animals which have the right size and behavior to reach the reward and pick up pollen. The more than 100 million years of coevolution between flowering plants and their pollinators has greatly contributed to the huge number of species in both kingdoms (300,000 flowering plants, 350 hummingbirds, and 15,000 known bees in the world). It also explains why there are so many different shapes and colors of flowers.

Flowers can be classified into several pollination *syndromes* according to their pollinators. (A syndrome is a set of characteristics associated with a specific phenomenon.) This is not the same classification as systematic taxonomy and does not reflect the evolutionary relationships among plants. Species in the same family or even the same genus may attract different pollinators.

The hummingbird pollination syndrome is one of the most easily recognized. Hummingbirds are large compared to most insects, almost unique in their ability to feed while hovering, and daytime-active; they have no sense of smell, but have long narrow beaks and tongues that can probe deep narrow tubes, and excellent color vision. Hummingbird flowers tend to be long-tubular, non-fragrant, sideways- or downward-facing, day-blooming, and brightly colored. Bees and most other animals cannot easily land on a hanging flower, and even if they succeed they cannot reach the nectar at the base of the narrow tube.

There are common misconceptions that all hummingbird flowers are red and that hummingbirds can see only the warm colors of the spectrum. It is true that most hummingbird flowers in the temperate biomes are red, but in the tropics they come in many colors. The predominance of red in temperate hummingbird flowers may be a disincentive to bees. Bees are aggressive pollen collectors in temperate climates. But they cannot see red, so red flowers do not appear conspicuous to them.

Wind-pollinated plants make no investment in attracting animals; their flowers lack fragrance or showy parts. Many people would not recognize them as flowers at all. Prodigious quantities of pollen are released, an infinitesimal proportion of which lands on a receptive stigma of the same species. While this seems inefficient, it is obviously effective, judging from the successful groups of plants with this syndrome. Conifers, most riparian trees (such as willows and sycamores), oaks, and grasses are all wind-pollinated. Conifers and grasses are the dominant plants in the two biomes that bear their names. Grasses occur in most biomes and comprise the sixth largest family of plants with about 9000 species worldwide. Wind pollination is not always entirely passive (see the species account for jojoba).

Seed Dispersal

Seeds generally need to be transported some distance from the parent plant in order to find a suitable site for establish-

ment. Some plants have wind-dispersed seeds, which are occasionally blown many miles from their origins. This means of dispersal is common among *pioneer plants* (plants that are adapted to colonizing disturbed habitats). Because of their superior ability to invade newly-disturbed ground, pioneer plants comprise many of our agricultural and garden weeds. Moreover, most annual crops are domesticated pioneer plants. That's why we need to plow (disturb) fields in order to grow them.

Many plants use animals to disperse their seeds in another complex coevolutionary process. Small, brightly-colored fruits such as hackberry and boxthorn are offered as food for birds that swallow them whole. Other fruits such as those of hedgehog cacti are large and birds feed on them repeatedly. Some bird fruits are sticky, such as mistletoe berries; a few stick to the bird's bill until wiped off on a branch while others are successfully swallowed. The seeds of bird fruits are typically small and hard; they pass through birds' guts undamaged and may be deposited many miles from the parent plant.

Mammal-dispersed fruits tend to be larger, aromatic, not colorful (most non-primate mammals have poor color vision), and usually have larger seeds than bird fruits. The animal often transports the fruits a short distance (compared to the flying distances of many birds) to a safer place before eating the pulp and dropping at least some of the seeds. The seeds of coyote gourds (*Cucurbita* spp.) may be dispersed in this manner. Coyotes swallow the whole fruits of palm trees; they digest the thin pulp and excrete the hard seeds intact. Since seeds contain energy stores to nourish the germinating embryo, seeds themselves are also nutritious food for mammals and birds. Some plants offer their seeds without juicy pulp to attract mammals. Pocket mice and antelope squirrels gather the abundant seeds of foothill palo verdes and bury them as food caches for the dry season. The animals don't eat all that they bury, so some seeds remain in the ground and germinate when the rains come. (Birds that specialize in eating seeds, as opposed to fruits containing seeds, crush and digest the seeds and therefore do not disperse viable propagules.)

Even in the desert some seeds are water-dispersed. Blue palo verde (*Cercidium floridum*) grows mostly along washes. Flash floods disperse the very hard, waterproof seeds downstream, *scarifying* (abrading the surface of) them in the process. In the absence of scarification these seeds must weather in the ground for a few years before the seed coats become permeable and permit germination.

The timing of seed maturation is crucial for many plants. The less time seeds are present before they sprout, the greater is their chance of survival. The tropically derived plants in our region germinate with the summer rains. These species usually flower in spring and their fruits ripen shortly before the arrival of the summer rainy season. Palo verde and saguaro are examples. Other plants produce large quantities of seeds and rely on camouflage or burial in the soil to conceal some of them from hungry animals. Brittlebush, for example, flowers and seeds in spring, but the seeds germinate with fall rains. Annuals do the same.

FLOWERING SEASONS IN THE SONORAN DESERT

The Spring Flowering Season

The spring flowering season in the Arizona Upland subdivision spans from mid February to mid June with a peak from mid March to late April depending on rainfall and temperatures during the growing season. In the warmest areas of the Lower Colorado River Valley subdivision it is normally a couple of weeks earlier, though it sometimes starts as early as November. The different life forms which dominate at different times vary in their showyness and reliability. The early-blooming winter annuals can create an incredible display, but do so only rarely. Later-blooming species bloom more dependably, but mostly not in great masses of color. The progression of spring bloom described below is for average years near Tucson. It may be three weeks earlier or later depending on weather, elevation, and latitude.

WINTER ANNUALS such as poppy (*Eschscholtzia mexicana*), lupines (*Lupinus sparsiflorus* and others), and owl-clover (*Castilleja exserta*) create the vast carpets of color for which the Sonoran and Mohave deserts are so famous. This event may occur between late February and mid April, usually in mid March. Annuals are highly dependent on rainfall. Massive and widespread displays occur only about once a decade, when the winter rainy season is both earlier and wetter than normal. Good shows happen in localized areas every three or four years. A good bloom cannot be reliably predicted more than a week or two before it begins, and usually lasts at peak beauty for only two weeks. Seeing such a bloom requires being able to travel on short notice, and perhaps great distances. Death Valley may be spectacular in a year when Organ Pipe Cactus National Monument is poor. The high Mohave Desert may peak two or three weeks later than the lower-elevation and more southerly Sonoran Desert. A good bloom may occur in a remote area and remain undiscovered.

HERBACEOUS PERENNIALS AND SMALL SHRUBS such as penstemon (especially *Penstemon parryi*, shown here), brittlebush (*Encelia farinosa*), and fairy duster (*Calliandra eriophylla*) also require rain to bloom but are less sensitive to its timing. They are somewhat more dependable than the annuals, making a good show in about half of the years and peaking some time in March. These species usually grow as individuals or in small patches and do not create masses of color.

CACTI, because they store water, are fairly independent of rain. They bloom well nearly every year though wetter years produce more flowers. The greatest diversity of spring-blooming species can be seen in April. The cactus show continues as the abundant prickly pears bloom in early May, followed by saguaros from mid May to mid June.

TREES AND LARGE SHRUBS are fairly dependable bloomers, though flowers will be sparse in dry years. Creosote bush (*Larrea tridentata*) and whitethorn acacia (*Acacia constricta*, shown here) both bloom mainly in spring and sometimes again in summer. Blue palo verde (*Cercidium floridum*) turns bright yellow in late April, followed two weeks later by the much more abundant but paler yellow foothills palo verde (*C. microphyllum*). Desert ironwood trees (*Olneya tesota*) bloom heavily about every other year with masses of lavender flowers, usually in late May. The abundant ocotillo reliably produces spikes of red flowers throughout April. These species bloom about two weeks earlier in western Arizona.

SUMMARY
If you want to see the famous carpets of color, keep abreast of local news from Palm Springs to Tucson and from Death Valley to northern Mexico. Begin checking in late February and be ready to travel on short notice. You'll find masses of annuals somewhere in this area about once every three or four years. If you want dependability and will settle for less quantity, success is almost guaranteed in the middle half of April.

The Summer Flowering Season

This season begins a few weeks after the first summer rain and continues into late fall. Though there are many beautiful species to be seen, there are rarely massive displays of color in this season, because the summer rains are more sporadic and localized than the winter rains and the soil dries rapidly in the heat.

SUMMER ANNUALS such as summer poppy (*Kallstroemia grandiflora*) and devil's claw (*Proboscidea parviflora*) germinate within a few days after the first soaking summer rain and begin to flower as soon as three weeks later. Chinchweed (*Pectis papposa*) is the most widely-adapted summer annual; it ranges from New Mexico into the central Mohave Desert where it is the only summer annual (summer rains are uncommon in the Mohave). It can form showy carpets of yellow when rains are abundant.

HERBACEOUS PERENNIALS
AND SMALL SHRUBS
bloom opportunistically if they get enough rain. Trailing four-o'clock (*Allionia incarnata*) and desert marigold (*Baileya multiradiata*) are nonseasonal, flowering in response to rain in all but the coldest months. Fairy duster will also bloom again in wet summers, but not as profusely as in spring. Sacred datura (*Datura wrightii*) is mainly a summer perennial though it may begin flowering as early as April in warmer areas. There are several woody shrubs that bloom in late fall. Most are composites such as turpentine bush (*Isocoma tenuisecta*) and desert broom (*Baccharis sarothroides*). Desert senna (*Senna covesii*) and Coulter hibiscus (*Hibiscus coulteri*) flower in response to any warm rain and peak in summer when most such rain occurs. Desert zinnia (*Zinnia pumila*) is truly biseasonal, flowering well in both rainy seasons.

CACTI include several summer-flowering species. The pincushion cactus *Mammillaria grahamii* makes buds during its previous growing season, then goes dormant during the dry season. The buds burst into bloom five days after each of the first two or three summer rains. The fishhook barrel cactus, *Ferocactus wislizeni*, is much larger than the pincushion and less dependent on rain; it flowers throughout August and September.

TREES AND LARGE SHRUBS are nearly all spring bloomers, but a few bloom again in summer if rains are generous. Whitethorn acacia (*Acacia constricta*) and velvet mesquite (*Prosopis velutina*) flower heavily in spring and often again in summer. Desert willow (*Chilopsis linearis*) flowers from spring through fall if it has enough water.

SUMMARY
Though the Sonoran Desert has two flowering peaks, there is almost always something in bloom. The only exceptions are after a hard winter freeze or during severe droughts.

Predicting Wildflower Blooms

Desert annual wildflower blooms are nearly impossible to predict for two reasons. First, the necessary conditions are not precisely known. Second, many interacting variables affect the phenomenon. Here is what we do know:

➤ *Spring-blooming annuals must germinate in the autumn.* This is a crucial fact that most people don't know. The "critical window" is probably between late September and early December, but differs with different species. The controlling environmental factor is temperature.

➤ *A "triggering rain" of at least one inch must occur during this autumn window,* the earlier the better after summer heat has waned. Rains at other times will seldom trigger germination of the showy-flowered species.

➤ *The triggering rain must be followed by regular rains* totaling at least an inch per month through March, a season total of at least five inches—seven or more are better.

In short, a really good wildflower bloom requires both an unusually early and an unusually wet winter rainy season. The rains must also be well spaced. Spectacular, widespread shows occur about once in ten years in the Sonoran and Mohave deserts. Good or better displays occur in localized areas perhaps every three or four years; these may be in remote regions and go unnoticed.

However, even when all the above conditions are met, the bloom may be mediocre or poor. And occasionally a good bloom occurs when the above conditions appear not to have been met. The latter can happen when an unusually warm rain triggers germination in winter, but the short growing season usually precludes a really good show. Factors which are suspected of preventing a show include:

➤ a few weeks of warm, windy weather; the water stress triggers premature flowering;
➤ a cold winter that retards growth of the seedlings;
➤ high population levels of herbivores: rodents, rabbits, quail, or insects; and
➤ a wet preceding summer, resulting in thick growth of summer vegetation, which in turn prevents germination of winter annuals.

When it does happen, the peak typically lasts only two weeks at a given location, sometime between late February and mid April. Most often it happens in early to mid March.

Furthermore, only certain areas ever have mass displays; soil type and vegetation cover are important factors. The rocky and densely-vegetated Tucson Mountains rarely if ever have mass blooms, whereas Picacho Peak and the Tohono O'odham (Papago) Indian Reservation do fairly regularly.

The above information pertains only to annual wildflowers such as poppies, lupines, and owl-clover. Perennials are less fussy about the timing of rainfall. Thus a late but wet rainy season can still produce good blooms of penstemon, larkspur, brittlebush, and other perennials. Some plants such as palo verdes, ocotillo, and most cacti flower nearly every year regardless of rainfall. (See other page for details.) But it is the annuals alone that produce the desert's famous carpets of color.

The Invisible Larder

I conducted a wildlife survey in the Lower Colorado River Valley in the 1970s. The site had received almost no biologically effective rainfall for three years. Creosote bushes were almost the only plants present; they were widely-spaced and had shed most of their leaves. Yet in the kilometer (6/10 mile) long by fifty meter (150 foot) wide transect I trapped one pocket mouse overnight, and in the morning observed a whiptail lizard, a rock wren, and two black-throated sparrows. These are all resident species; not transitory migrants. What were they living on?

A persistent, large soil seed bank is an extremely important resource in arid habitats. It provides an unseen (by humans) food source for desert animals as well as survival insurance for plant species. The greater density of seed-eating animals and the abundance of decomposing microbes in the moist soils of wetter regions greatly shortens the viability of seeds. In deserts viable—and nutritious—seeds persist in large numbers through decades of drought. After a wet year there may be 200,000 seeds per square meter (square yard) of soil. Even after several dry years with little or no additional seed production there are still several thousand seeds per square meter, enough to sustain low populations of seed-eaters such as harvester ants, kangaroo rats, and sparrows. The whiptail was foraging for insects that fed on the seeds or plant *detritus* (partially decomposed organic matter) in the soil. As the statistician in the movie "Jurassic Park" said, "Life will find a way."

Additional Readings

Bowers, Janice E. *A Full Life in a Small Place and Other Essays from a Desert Garden*. Tucson, University of Arizona Press, 1993.

Buchmann, Steven L. and Gary P. Nabhan. *The Forgotten Pollinators*. Washington, D.C.: Island Press, 1996.

Dykinga, Jack W. and Charles Bowden. *The Sonoran Desert*. NY: Harry N. Abrams, 1992.

Hanson, Roseann Beggy and Jonathan Hanson. *Southern Arizona Nature Almanac*. Boulder: Pruett Publishing Co, 1996.

Hartmann, William K. *Desert Heart: Chronicles of the Sonoran Desert*. Tucson: Fisher Books, 1989.

Imes, Rick. 1990. *The Practical Botanist: An Essential Field Guide to Studying, Classifying, and Collecting Plants*. New York: Fireside Books/Simon and Schuster, 1990.

Larson, Gary. *There's a Hair in My Dirt! A Worm's Story*. New York: Harper-Collins, 1998.*

Nabhan, Gary P. *Gathering the Desert*. Tucson: University of Arizona Press, 1985.

———. *The Desert Smells like Rain*. San Francisco: North Point Press, 1982.

Seuss, Dr. *The Lorax*. New York: Random House, 1971.*

* Though these two books are found in children's literature, they convey the essence of ecology better than any scientific treatise I have encountered.

Section Contents

Flowering Plants of the Sonoran Desert

Mark A. Dimmitt

This section focuses on the most common, conspicuous, or interesting plants of the Sonoran Desert. Many are treated as groups based on their taxonomy (for example legume trees) or ecology (such as annual wildflowers).

About Plant Names

Most plants can be identified positively only by their scientific (Latin) names. We will also list regional English and Spanish vernacular (common) names when we know them. Scientific names are officially recognized worldwide and are validated by the regular reports of the International Code of Botanical Nomenclature, which also lists retained synonyms. On the other hand, there is really no such thing as a common name for the vast majority of the planet's 300,000 species of flowering plants. There are committees that establish common English names for birds and several other animal groups, but no organization has assumed the responsibility for standardizing common plant names worldwide (the U.S. Department of Agriculture has "officialized" some for the United States).

Moreover, a minority of plants have names that are common in the sense of being widely recognized, such as apple, rose, or carnation. The names of less well-known plants, and even widely-known plants, often vary geographically, and there may be numerous "common" names that are unknown outside of the given region. For example, *Antigonon leptopus* has more than a dozen vernacular names in the Southwest, including, queen's wreath, coral vine*, confederate vine, San Miguel, *coronillo*, Mexican creeper, love vine*, chain of love, mountain rose*, queen's jewels, and *bellissima*. The starred (*) names are shared by other, unrelated plants. Some vernacular names, such as spider lily, are given to at least 20 unrelated flowers. The majority of plants have no vernacular name at all because they are not well-known species. For this reason we use the term "vernacular name" in this book. (Because they are not official, plant vernacular names should not be capitalized except for proper names, for example, Parry penstemon.)

We use Thorne's system for family names, all of which end in "-aceae" and are named for the original named Latin genus in the family.

Agavaceae (agave family)
Nolinaceae (nolina family)

Mark A. Dimmitt

Behind the formidable appearance of agaves and the plants formerly grouped with them is a wealth of uses. They have long been and still are used extensively by indigenous peoples throughout North and Central America for food, fiber, and medicine. The roasted, sugar-rich agave hearts have been an important food for numerous Native American groups. Juice from the mature plants is consumed both fresh and fermented. Fermented liquid from the cooked heads is distilled into *mescal*. *Tequila*, the best-known variety of mescal, is distilled from one species, *Agave tequilana*. The tequila agave (*mezcal azul*) is a significant economic crop in southern Mexico; North Americans alone consume more than a million gallons of tequila a year.

Fiber from the leaves of *Agave sisalana* is the source of sisal rope, and *A. fourcroydes* yields henequen fiber. Sisal is a major economic product widely cultivated in Africa, Asia, as well as in the New World; it provides 70 percent of the world's hard, long fiber for ropes, rugs, and bags in recent decades. Numerous native American peoples weave baskets from the fibers of yuccas and nolinas.

The complex chemicals in this family have many uses. Compresses for wounds have been made from macerated agave pulp, and juices from leaves and roots were used in tonics. But beware—sap from many agaves can cause severe dermatitis. The juice of the more virulent agaves has been used as fish poison and arrow poison. Agaves and yuccas are used in Mexico to make soap. Yuccas were once used to provide the foam of root beer and are still used in livestock deodorant. More recently steroid drugs have been synthesized from extracts of several species in the family.

Today these plants are appreciated for their beauty and are widely used to add accents to landscape designs and mark property boundaries all over the world. Howard S. Gentry's book *Agaves of Continental North America* was so popular that it was reprinted in 1992, an unusual event for a botanical monograph.

As originally described by Gentry, Agavaceae consists of 18 genera and a little over 400 species, many of them native to western North America. This family is difficult to define; it has been revised by taxonomists several times in recent years.

Agaves, yuccas, and relatives were once included in the very large lily family Liliaceae. Gentry segregated agaves, yuccas, and other genera into their own family Agavaceae in the 1970s. More recently other botanists have split the old Liliaceae into many more families and removed some genera from Agavaceae, among them *Nolina, Dasylirion, Sansevieria*, and *Dracaena*. Whatever their taxonomic status, these are highly useful plants with dramatic forms.

Genus *Agave*

English names: agave, century plant
Spanish names: mezcal, mescal, maguey, amole (general names that vary by
region and use of the plant)

Agaves are among the most conspicuous plants of arid North America; their bold forms attract attention in any landscape whether natural or designed. All are characterized by succulent or semisucculent leaves that form rosettes from a few inches to several feet across, but there are many variations on this basic pattern. Most species are essentially stemless but a few grow trunks that creep along the ground. Some species have only one rosette; most multiply by underground suckers and may develop into large colonies. Agave leaves vary from green through bluish to silver-gray, and are often strikingly banded with different shades of color. Leaves range from long and narrow to short and broad, and from arrow-straight to gracefully recurved or haphazardly twisted. The leaf margins are typically lined with large, sharp spines (teeth) and each leaf is usually tipped with a hard, sharp spine. A few species have leathery, unarmed leaves. The leaves are so tightly compacted in the growing tip that the teeth leave imprints on both surfaces of adjacent leaves after they unfurl, overlaying their own complex patterns over the banding. These imposing features make agaves popular among succulent collectors and landscape designers.

Various agave species have also been and continue to be important sources of food, fences, rope, medicine, and liquor. The Mescalero Apaches were named for their dependence on this plant. Before them the Hohokam cultivated agaves as a major food crop (see *A. murpheyi*).

Agaves flower on tall, branched or unbranched stalks that grow from the center of the leaf rosette. As a plant approaches maturity at 10 to 30 years of age it accumulates a great quantity of sugar and starch in the heart tissue. These carbohydrates provide the energy that fuels the rapid development of the inflorescence (the flowering structure, including supporting stems), which is usually massive compared to the plant that produces it. In all but a few species the rosette dies after flowering and fruiting, having spent all of its life energy to produce a huge quantity of seeds—a *monocarpic* (once-fruiting) life cycle. The plants literally flower themselves to death.

Though the flowering rosette usually dies, many species produce vegetative offsets (*suckers* or *pups* in English, *hijos* or "sons" in Spanish) before or after flowering. In this way *clones* (multiple, genetically identical, individuals that originated from a single original seed) form colonies that may persist for centuries or longer.

RANGE
The genus ranges from Utah in western North America through Mexico (where the most species are found), with a few in northern South America and on Caribbean Islands. The majority of agave species occur in semiarid habitats above the desert, especially in desert grasslands and oak-pine woodlands. About 40 of

the 150 North American species occur in the Sonoran Desert region.

ECOLOGY

Agaves typically grow on well-drained, rocky slopes. Different species are adapted for pollination by insects, nectar-eating bats, and hummingbirds. Seeds are dispersed by wind, usually only a short distance from the parent.

The two major groups, or *subgenera,* of agaves are distinguished from each other by whether their inflorescences are obviously branched or not. Most species in the subgenus Agave (branched inflorescences) developed features that enable them to be pollinated by nectar-feeding bats, although other pollinators may currently be more important. They grow whitish to yellow flowers which produce copious nectar and pollen at night. Bats are attracted by the fragrant nectar, which usually smells unpleasant to humans—like ammonia or rotting fruit, depending on the species. Agave nectar and pollen are major food sources on the northernmost and southward legs of the bats' migratory routes. After wintering in the Mexican tropics, bats migrate northward through the desert following the south-to-north wave of spring-blooming columnar cacti. They raise their young in southern Arizona, then return south via the mountains, feeding on agaves. Hawkmoths are also common visitors to the night-flowering agaves and are probably effective pollinators; bees and other diurnal insects aid in pollination as well.

Some species in the subgenus Agave occur outside the range of the bats, or flower at a season when the bats are not present. These have colored diurnal flowers that are pollinated by humming-birds and bees. Species in the subgenus Littaea (which bear unbranched spikes) are pollinated mainly by insects and sometimes hummingbirds.

Of the 40 or more agaves in our region, several are ecologically or horticulturally important. *Agave americana,* a very large plant with rosettes up to 12 feet (3.7 m) broad, is by far the most commonly used in gardens in the southwest U.S. and worldwide, although most of its natural range is outside the Sonoran Desert region. In our species accounts, we focus on some of the more interesting local species.

ETHNOBOTANY

Several beverages are made from the sugar-rich juices of mature agaves. The extracted juice is drunk fresh as *aguamiel* (honey-water) or fermented into *pulque;* both are popular beverages south of the Sonoran Desert. Steamed heads or central stalks are mashed and allowed to ferment with added liquid. After several days, the resulting fluid is distilled into the potent liquor mescal. The most widely known local ("bootleg") variety of mescal is Bacanora, named after that Sonoran town and made from *Agave angustifolia.* Other varieties of bootleg mescal are made in nearly every Mexican village within agave habitat. Tequila, the most famous legal variety of mescal, is made from the single species *Agave tequilana,* grown near the town Tequila in Jalisco. Tequila is to mescal much as Chardonnay is to wine.

Mature agaves also provide food. The leaves are cut off near their bases, leaving a *cabeza* ("head") resembling a giant pineapple, weighing up to 70 pounds (32 kg). The cabezas are usually pit-roasted in large numbers and eaten during fiestas. (Note: The *raw* flesh of many agaves is caustic and can even blister skin.)

➤ *Agave deserti*

English name: desert agave
Spanish name: amul

DESCRIPTION
This is a small species with rosettes rarely over 1½ feet (0.5 m) in diameter. The leaves are nearly straight and light gray to bluish-gray, with marginal teeth. Most varieties sucker near the base to form dense colonies, but some offset by long underground rhizomes, and a couple of forms rarely offset at all. The branched inflorescence bears bright yellow flowers that are attractive to hummingbirds. Some populations have red buds, which contrast beautifully with the flowers.

RANGE
One of the most desert-adapted agaves, it is native to rocky or gravelly soils in the Lower Colorado River Valley subdivision of the Sonoran Desert. Its range extends barely into Arizona Upland and the Mohave Desert.

COMMENTS
Agave deserti is a very slow-growing species; even in cultivation with generous irrigation it takes at least 20 years to flower. As spent rosettes of wild plants die and decompose, new ones replace them on the outer margin, eventually forming ring-shaped colonies. Rings 20 feet (6 m) in diameter in California's Anza-Borrego Desert State Park may be more than a millennium old. One of the more edible agaves,

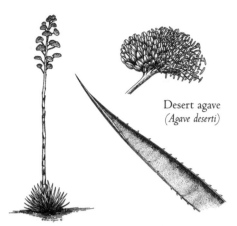

Desert agave
(*Agave deserti*)

Agave deserti has been extensively harvested by desert peoples.

In Baja California this species can be confused with the straight-leaved varieties of *Agave cerulata*, which also grows in extreme desert habitats. *Agave chrysantha* is larger with golden-yellow blossoms; it descends into the higher elevations of Arizona Upland in south-central Arizona. This is the most frequently-encountered agave on the northeastern boundary of the Sonoran Desert north and east of both Phoenix and Tucson.

➤ *Agave murpheyi*

English name: Hohokam agave, Murphey's agave
Spanish name: maguey (a general name for many agaves)

DESCRIPTION
This agave consists of freely suckering rosettes to 3 feet (1 m) across of narrow, straight, toothy light-green leaves. It rarely sets seeds, but produces many plantlets (also called *bulbils* and *semillas*, or "seeds") on the flowering stalk.

RANGE
Only a few populations are known in southern Arizona and northern Sonora.

Hohokam agave
(*Agave murpheyi*)

COMMENTS

The origin of this agave is unknown; all of the populations are associated with ancient Indian sites. It was extensively cultivated for food and fiber by the Hohokam Indians. They planted this species in the desert (and *A. delamateri* at higher elevations) along low rock check-dams built on bajadas to slow runoff and increase water penetration. Hundreds of thousands of acres of habitat were modified in this way before the Hohokam culture dispersed 800 years ago. The check-dams and associated roasting pits and agave harvesting tools are still abundant, and in a few sites colonies of the agave survive to this day. All of the populations from Caborca, Sonora, to New River, Arizona, are so similar that they may be one genetic clone. Proof of this would further substantiate the plant's cultural dispersal as one of the few domesticated north of Mesoamerica.

➢ *Agave schottii*

English name: shindagger
Spanish names: amole, maguey, amolillo

DESCRIPTION

The narrow, straight leaves—up to a foot (30 cm) long in small rosettes—have no marginal teeth, but are very sharp-tipped, hence the English vernacular name. Shindagger suckers profusely to form dense colonies that sometimes merge with others and cover many acres (hectares). The inflorescence is a narrow spike 6 to 8 feet (2-2.4 m) tall, bearing long-tubular, fragrant yellow flowers.

RANGE

It grows in southern Arizona and northern Sonora, mostly in desert grassland and oak habitats and often on steep, rocky slopes. It is occasionally found in Arizona Upland, e.g., near the summit of the Tucson Mountains.

COMMENTS

There are several other small agaves that may be confused with this species, especially *A. felgeri, A. toumeyana, A. polianthiflora*, and *A. parviflora*. The last two are only half as large as *A. schottii; A. polianthiflora's* flowers are pink to red, while the other 3 have shorter (less than 1½ inches, 30 mm) flowers.

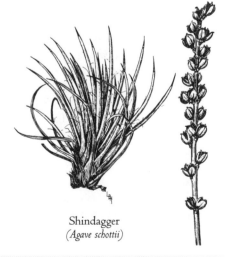

Shindagger
(*Agave schottii*)

> *Agave pelona*

English name: none
Spanish name: mescal pelón
   (bald agave)

> *Agave zebra*

English name: zebra agave
Spanish name: none

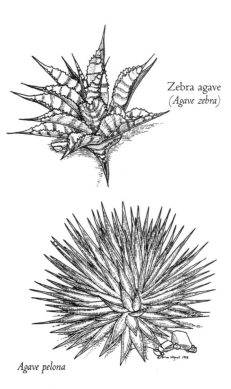

Zebra agave
(*Agave zebra*)

Agave pelona

DESCRIPTION

Agave pelona: The dense rosette of narrow, toothless leaves is dark blue-green or yellow-green, 2 to 4 feet (0.6-1.2 m) across. A long reddish spine tips each leaf. Since they do not cluster, the singular rosettes are very symmetrical. The flowers are an unusual deep reddish-purple; they're borne on a tall, unbranched spike in the spring.

Agave zebra: These rosettes—3 to 4 feet (1-1.2 m) across—do sucker. The light gray, strongly silver-banded and spine-imprinted leaves are stiff, channeled (folded into a "V"-shape in cross section), and usually strongly recurved. The grayish flowers are borne on a tall, narrow main stem (*panicle*).

RANGE

These two species are described together partly because they are both narrowly endemic to the Sierra del Viejo and Cerro Aquituni southwest of Caborca, Sonora. They grow intermingled on steep limestone slopes, where their different growth forms contrast dramatically.

COMMENTS

The other reason these two are grouped is because they share similar past and present uses. Apparently neither species was much harvested by local peoples for traditional ethnobotanical uses. But their rarity and beauty has given both of them value in the new ethnobotany of aesthetic horticulture, where their popularity increases among succulent collectors.

Genus *Yucca*

English names: yucca, Spanish bayonet
Spanish name: palmilla

Yuccas are usually easy to distinguish from agaves even when out of flower. Like all members of the family, they bear leaves in rosettes. But yucca foliage is only semisucculent or nonsucculent and the leaves are usually straight; many species grow trunks. When in bloom, all are easily recognized by their large, white, fleshy, bell-shaped flowers. Unlike most agaves, most yuccas are *polycarpic* (blooming more than once).

RANGE

Like agaves, most yucca species occur in semiarid habitats above the desert. Habitats range from the northern Great Plains through woodlands and the dry tropics of Mexico. One species occurs in the southeastern U.S. and the West Indies. About 10 species occur in the Sonoran Desert region.

ECOLOGY

Yucca pollination ecology is an example of a tight *symbiosis* called a *mutualism*. (Symbiosis refers to a close association between two species in that at least one benefits from the association. Mutualism is a symbiotic relationship in which each species depends on the other for survival.) With only one exception, yucca reproduction depends on moths (genera *Parategiticula*, *Tegiticula*) which deliberately cross-pollinate the flowers. (*Yucca aloifolia* of the southeastern U.S. is pollinated by bees.) The blossoms need pollen from a different plant in order to produce seed, and it must be packed into a deep receptacle on the stigma, an event that could not occur by chance visitation. The moth is equally dependent on the yucca. It lays eggs on each pollinated ovary, and the hatched larvae eat some of the developing seeds.

Biologists have only recently determined that almost every species of yucca has its own species of yucca moth; some yuccas have two moth species. Such a tight mutualism has risks for both partners. Emergence of adult moths must coincide with yucca flowering for the reproductive needs of both species to be met. However, the synchronization of moth emergence with flowering is frequently poor and seed set and moth reproduction in such years are low. Furthermore, yucca populations may flower sparsely or not at all in dry years. Yuccas don't have to set seed every year because they flower many times in their long lives. The yucca moths employ a survival strategy analagous to that of desert annual plants. The full-grown larvae exit the ripening yucca fruit and burrow into the ground, where they enter a deep dormancy (*diapause*). Like the seeds of many annuals, only some of the larvae will metamorphose and emerge as moths in the following flowering season. The rest remain in diapause for two or more years.

ETHNOBOTANY

Yucca flowers and fruits are edible fresh or dried. Chemicals in the roots of some species are used to make soap. (The Spanish name *amole* is applied to a number of unrelated plant species from which soap is made.) The roots of Mohave yucca (*Y. schidigera*) were used to provide the foaming agent in root beer, and the stems are still (over-) harvested to produce livestock deodorant.

➤ *Yucca arizonica* (formerly *Yucca baccata* spp. *arizonica*)

English names: blue yucca, Spanish bayonet, Thornber yucca
Spanish names: dátil (date), palma criolla (creole or native palm)

Blue yucca
(*Yucca arizonica*)

DESCRIPTION

The soaptree yucca has a simple or branched trunk up to 23 feet (7 m) tall. The numerous 2-foot (0.6 m) long, thin, flexible leaves are clustered at the ends of the stems, making the plant appear somewhat like a palm. Flowers are creamy white, borne in a great cloud on the upper half of a stalk up to 10 feet (3 m) tall, usually in May and June.

RANGE

Soaptrees grow mainly in desert grassland from central Arizona to west Texas and northern Mexico. The range extends into the upper margin of Arizona Upland.

COMMENTS

The leaves yield the major basketry fiber for the Tohono O'odham, who know how to harvest the tender new leaves in a way that promotes branching instead of killing the plant. The English common name refers to another of its uses.

DESCRIPTION

A variable semisucculent species, this yucca usually occurs as dense clusters of stems to 8 feet (2.5 m) tall, tipped with rigid bluish to yellowish leaves. The lower half of the wide inflorescence is typically concealed within the leaves.

RANGE

From the Sonoran Desert into oak-pine woodland, southern Arizona to central Sonora.

COMMENTS

It was formerly considered a subspecies of banana yucca (*Y. baccata*), a more widespread, usually stemless or recumbent species. Its roots are used for red fiber in Tohono O'odham baskets.

➤ *Yucca elata*

English names: soaptree yucca, soapweed
Spanish names: palmilla, palmito, soyate, cortadillo

Soaptree yucca
(*Yucca elata*)

> *Hesperoyucca whipplei* (formerly
  Yucca whipplei)

English names: Spanish bayonet, our
    Lord's candle, chaparral yucca
Spanish names: sotolillo,
    lechuguilla, quiote

DESCRIPTION

The bluish-green rosette 3 to 6 feet
(1-2 m) in diameter consists of about a
hundred long, narrow, and dangerously
rigid, sharp-tipped leaves. Rosettes are
single to multiple in different subspecies. The inflorescence rises well
above the leaves, usually about 8 feet
(2.5 m) tall in the desert forms, bearing
dense masses of creamy-white flowers
(tinged with purple in some populations). The flowering rosette dies, so the
nonclustering subspecies are monocarpic.

RANGE

This plant is primarily a chaparral
species of the Californias, with a few
desert populations in the Lower

Spanish bayonet
(Hesperoyucca whipplei)

Colorado River Valley and gulf side
of the Baja California peninsula. The
desert form is usually about 3 feet
(1 m) in diameter and nonclustering.

COMMENTS

This species was recently removed from
the genus Yucca because it was determined to be only distantly related. Its
pollinating moth is also distinct.

Nolinaceae (formerly Agavaceae), nolina family

> *Dasylirion wheeleri*

English name: desert spoon
Spanish names: sotol (sotole is
    a palm), saño

DESCRIPTION

This perennial evergreen consists of a
rosette of hundreds of long, narrow
leaves armed with small, sharp, marginal teeth. The rosettes are usually stemless and about 6 feet (2 m) across. Old
specimens may develop trunks to 6 feet
(2 m) tall, and these sometimes branch.
Sotols are *dioecious* (producing only male
or only female flowers on each plant).
The inflorescence emerges from the
center of the rosette in early summer
and grows to 12 feet (3.7 m) tall. Its
numerous, dense branches bear thousands of tiny, green- or violet-tinged
whitish flowers, followed by winged

fruits on female plants. Desert spoon does not die after flowering. The stem branches at the base of the inflorescence and continues growing.

RANGE
Desert spoon grows on rocky hillsides and slopes at 3000 to 6000 foot (900-1800 m) elevation in southeastern Arizona, southwestern New Mexico, northern and eastern Sonora, Mexico, and to west Texas. Despite its English name, it is primarily a grassland species that extends into the desert.

COMMENTS
The rosettes flower only once in several years. Blooming plants attract huge numbers of insects, including flies, bees, wasps, and butterflies.

Until a few decades ago, the Tohono O'odham wove beautiful sleeping mats by plaiting together sotol leaves after removing marginal teeth from the leaves.

6 foot (2 m) branched trunk. The tall, densely-branched inflorescence bears thousands of small whitish flowers. But the overall appearance is a greenish plume, drying to an equally attractive straw color.

COMMENTS
The Coahuila Indians ate *N. bigelovii* flowering stalks after roasting them in pits.

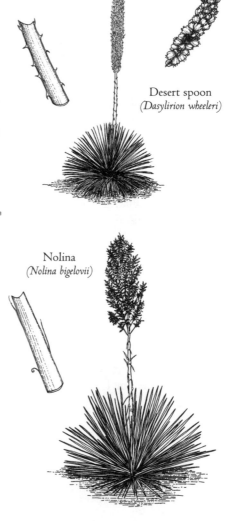

Desert spoon
(*Dasylirion wheeleri*)

Nolina
(*Nolina bigelovii*)

Genus *Nolina*

English names: nolina, beargrass
Spanish names: yuca, sotol, palmita

Nolina bigelovii is the common northern Sonoran Desert species of this genus; it has no vernacular name, save that borrowed from its botanical name— "Bigelow nolina." The similar *N. beldingii* occurs in Baja California; you can probably guess its vernacular name. Both resemble desert spoon (*Dasylirion*) with no marginal teeth on the leaves. The yellowish- to bluish-green rosette of nonsucculent foliage begins at ground level and in an old plant may top a

Aracaceae (Palmae) (palm family)

Mark A. Dimmitt

Palms are among the most distinctive plant life forms and thus scarcely need description. Virtually all of the more than 2600 species in the world are instantly recognizable as palms, and only the unrelated cycads and another small family of tropical plants might confuse a nonbotanist. Because of their high recognition they're widely used in all kinds of media—in movie sets to landscape designs and cartoons—to represent a tropical setting or, in stark contrast, a desert oasis.

Palms are indeed tropical plants. Almost all species are native to tropical forests and very few can tolerate frost or dry soil. The existence of palms in the Sonoran Desert is one of many indications of its tropical connection.

➤ *Washingtonia filifera*

English name: California fan palm, Washington palm
Spanish names: palma de Castilla, palma de abanico (fan palm)

➤ *Washingtonia robusta*

English names: Mexican fan palm, skyduster palm
Spanish names: palma colorada (red-brown palm), palma blanca (white palm), palma de Castilla

DESCRIPTIONS

Washingtonia filifera has a very thick trunk and grows slowly to about 45 feet (14 m) tall. *Washingtonia robusta* has a thin trunk, but grows rapidly to twice as tall. Both species have large, palmate leaves with spiny *petioles* clustered at the top of the trunk. (A petiole is the stalk that connects the leaf to the stem.) Dead leaves hang vertically and form a persistent skirt around the trunk. Inflorescences extend beyond the leaves and bear masses of tiny white to cream-colored flowers. During the fall months, large clusters of small, hard fruit hang from the tree. The palms may live 150 to 200 years.

RANGE

Washingtonia filifera occurs naturally in about 70 desert oases, most in southeastern California and northern Baja California. A few groves occur in Arizona, all but one of which are close to the Colorado River. *Washingtonia robusta* is found in similar habitat in northern Baja California, and there are a few populations on the coast of Sonora near Guaymas.

NOTES

Both Washington palms require a constant source of water and, because they lack deep roots, they can grow only where water is near the surface. Nonetheless these are true desert plants, because an oasis is a desert environment. (Water sources in other habitats are not properly called oases.) These palms will die if the temperature is below freezing for more than 24 hours, but they can survive a few hours near 15°F (-10°C). In the Coachella Valley of California, numerous palm oases occur in a nearly straight line along the San Andreas fault, which forces subterranean water to the surface.

The fruit is produced in huge bunches; each one has a thin, but sweet, layer of pulp covering the large seed. The fruits are eaten by coyotes and were also harvested by native Americans. Both species dispersed the seeds from canyon to canyon, establishing many of the genetically uniform populations throughout the species' ranges.

The giant palm borer (*Dinapate wrighti*) is a thumb-sized beetle (to 2 inches; 52 mm) long. It belongs to the family of branch and twig borers (Bostrichidae), and is by far its largest member. Adults fly during the hottest summer nights, mate, and lay eggs in a California fan palm crown. The grubs feed on the fibrous trunk tissue with their large mandibles. During their 3 to perhaps 10 year larval life they burrow down the trunk nearly to the ground, then part-way back up before pupating. Heavily-infested trees can snap in strong winds. Commercial transport of palms has spread the borer to the urban oases of Las Vegas and Phoenix where they now infest date palms (*Phoenix* spp.) as well.

The persistent skirts of dried leaves that clothe the trunks provide

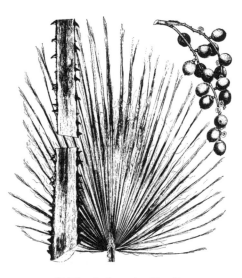

California fan palm (detail)

shelter for a myriad of wildlife species. Orioles have a particular affinity for nesting among the leaves. Paper wasps (Vespidae) and bees often nest there too. Multitudes of scorpions and spiders can be found living in the thatch; lizards prey on the invertebrates and hide from their predators under the skirts.

California fan palm
(*Washingtonia filifera*)

Many wild palms lack skirts because vandals set fire to them. The trees themselves are fire-resistant, but the destruction of the thatch ruins much of the palms' value as wildlife habitat. There may be a benefit, however, in the reduction of palm borer populations. Cultivated palms are usually trimmed of their dead leaves to "improve" their appearance (a matter of taste), or to eliminate habitat for insects and rodents, which are perceived as nuisances. The wasps, at least, are beneficial as predators on many garden pests and are rarely aggressive towards humans (except those trimming the palms).

Wherever palms grow, their leaves may be used for making thatched roofs. The large broad leaves can easily be made into a rainproof shelter that will last for several years. Palm-thatched houses are fairly common in southern Sonora outside the desert. Leaves from natural and cultivated populations are transported to resorts on the desert coast to cover *palapas* (beach shelters). The trunks are used to make corrals and fences.

These palms are uncommon in nature, but there are millions of them in landscapes all over the warm-temperate world. *Washingtonia robusta* is the palm planted so abundantly in Southern California that it has become a symbol for that region, even though it is not native there. Warren Jones, one of the founding fathers of desert landscape architecture, says they are the Sonoran Desert's greatest contribution to ornamental horticulture.

➢ *Brahea* spp.

English names: blue palm, hesper palm
Spanish names: palma azul (blue palm), palma ceniza (ashy palm)

DESCRIPTION
Several species of fan (palmate-leafed) palms occur in desert oases and wet tropical canyons in the Sonoran Desert region. They resemble the Washington palms except that they are smaller in stature and most have gray-green to blue-green foliage. *Brahea armata* in Baja California is the bluest; it creates an extraordinary sight in nature, where it is often found among white granite boulders with boojums and elephant trees (see Plate 27), or in designed landscapes. It is uncommon in cultivation because it's very slow growing.

RANGE
Brahea armata grows in desert oases in north-central Baja California. Other species occur in oases in Sonora and in tropical deciduous forest.

Aristolochiaceae (pipevine family)

Mark A. Dimmitt

This is a tropical family of 400 species in the world, 300 of which are in the genus *Aristolochia*. They are mostly vines with strangely-shaped flowers and foliage that exudes a distinctive, unpleasant odor. Only a single species occurs in the Sonoran Desert.

> *Aristolochia watsonii*

English names: Southwestern pipevine, snakeroot, birthwort
Spanish names: hierba del indio (Indian herb), guaco

DESCRIPTION
A trailing or climbing vine with stems up to 3 feet (0.9 m) long. The 1 inch (2.5 cm) long, arrow-shaped leaves are usually dark brownish-green when growing in full sun. The tubular-funnel form flowers are about 1½ inches (3.8 cm) long, green with brown spotting.

FLOWERING
Throughout the warm months.

RANGE
Aristolochia watsonii grows from Southern Arizona and adjacent Mexico to the tip of Baja California.

NOTES
Though this species is inconspicuous and rarely encountered unless actively sought, it is included here because of three fascinating ecological stories.

One story is its pollination. Most pipevines are pollinated by deceiving insects into visiting the flowers; no reward is available. Usually the flowers smell like carrion or dung to attract insects that are seeking a place to lay eggs. The story of our species is even more remarkable. It is pollinated by Ceratopogonid flies; these are the small, blood-sucking flies that pester humans and other mammals in the humid summer. The flowers of *Aristolochia watsonii* resemble a mouse's ear—translucent funnels with fur and veins—and give off a musty odor. The fly apparently expects to find a blood meal, and instead is trapped inside the flower tube overnight. During the night the flower releases pollen. The following morning the flower releases the pollen-covered fly. If the fly visits another flower it effects pollination.

All pipevines contain a variety of powerful toxins that humans have used for medicines. The name birthwort comes from its use during difficult births; it stimulates expulsion of the fetus and placenta. The plant has also been used to treat snakebite, paralysis, malaria, impotence, intestinal worms, and infections. Though pipevine seems to be effective for numerous ailments, the side effects are horrific.

Despite the virulent toxins, pipevines are the larval food plant of the pipevine swallowtail, a large, showy butterfly. The caterpillars are black or red with red tubercles and grow to about 3 inches (7.6 cm) long. This conspicuous coloration warns potential predators that the larvae are protected by the pipevine's toxins, which they store in their bodies without harm to themselves. Several caterpillars can often be found feeding on one small plant of *A. watsonii.* They will completely defoliate the plant (it later resprouts from its perennial root). Without this species of pipevine, the pipevine swallowtail would not occur in the Sonoran Desert.

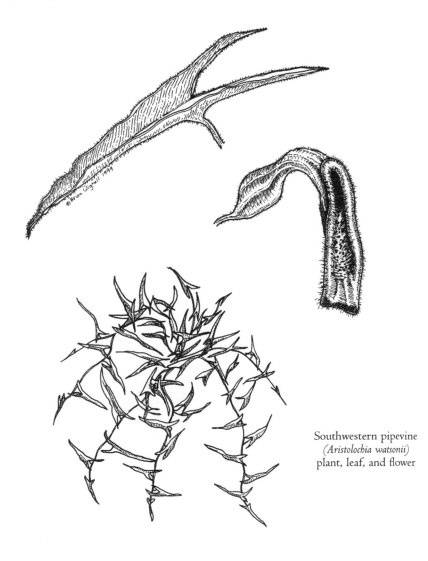

Southwestern pipevine
(*Aristolochia watsonii*)
plant, leaf, and flower

Asteraceae or Compositae (sunflower family)

Mark A. Dimmitt

The sunflower family is stunningly successful. It is the largest plant family, or the second largest after orchids, depending on the criteria used, with over 20,000 species occupying almost all of the world's habitats except underwater. The genus *Senecio* (groundsel) alone has 1000 species. So many variations within a single group make composites the bane of many botanists seeking to identify the species. Because of this and the great preponderance of yellow flowers, unidentified plants are often semi-affectionately dubbed "DYCs" (damn yellow composites). This term has been adopted by some Mexican botanists as *"PCAs"* (*pinchi compuestas amarillas*). The family is well represented in the Sonoran Desert, constituting, for example, 16% of the Tucson Mountains' flora (105 species and subspecies).

The flowers, also called *florets*, are nearly always clustered into heads, with each subtended by a whorl or whorls of modified leaves called *bracts* (the *involucre*). There are two general forms of flowers. A disk flower, in its most complete form, has 5 petals fused into a tube, with a tube of 5 fused anthers inside the petal tube, and a 2-lobed stigma exserted through the anther tube. A ray flower (a "petal" of a daisy) is similar, except that some of the fused petals extend on one side into a flat strap-like *ligule*. Flowers may be unisexual or sterile, lacking either or both "male" and "female" sexual parts. Each functionally "female" flower, whether ray or disk, has a single inferior ovary with a single ovule. If the ovule is fertilized, it will develop into a single seed in a special fruit called an *achene*. Each head may have only ray flowers or disk flowers, or both.

Genus *Ambrosia*

On the surface ambrosias don't seem to be very interesting. They are wind-pollinated, so they need no animal allies for reproduction. Virtually nothing eats them. And most species are so ordinary in appearance that they are rarely noticed despite their abundance throughout much of our desert. But on closer inspection, these plants reveal themselves to be essential, crucial species of the Sonoran Desert.

Each subdivision of the Sonoran Desert has at least one common species of bursage that contributes subliminally to the visual character of the landscape. The wide range of this group reflects the bursages' ability to adapt and move into numerous environmental niches. Triangleleaf bursage (*Ambrosia deltoidea*) grows abundantly among saguaro cacti and palo verde trees where some rain falls in both winter and summer. Canyon ragweed (*A. ambrosioides*) is a fairly tropical, big-leafed species of somewhat moist habitats that seems to require some summer rain, while white bursage (*A. dumosa*) rivals creosote bush as the most arid-adapted perennial in North America where rainfall is mostly in winter, and summers are brutally hot. *Ambrosia chenopodifolia* grows on the mild Pacific coast of Baja California under a winter rainfall regime. Hollyleaf bursage (*A. ilicifolia*), a bursage with especially attractive foliage, grows in the wetter microhabitats within the hottest, most arid parts of the Sonoran Desert.

Bursages are vitally important to the community as *nurse plants*. The seedlings of most desert plants cannot survive the extreme environmental conditions in exposed ground; they must start life in the shelter of another (nurse) plant. Bursages are among the few plants that can pioneer exposed sites. Over the following centuries they may be replaced by a series of other species that establish under one another's canopies. For one possible scenario, imagine a bursage growing today being replaced by another in mid-century, then by a barrel cactus in the 22nd century, a creosote bush in the 23rd, a foothill palo verde in the 26th, and a saguaro in the 27th. Visualize that saguaro producing its first flowers in the year 2670 and you begin to grasp the time-scale on which desert communities function. And here's hoping you'll never think of bursages as just drab little allergenic bushes again.

DESCRIPTION
What follows is a general description for the genus *Ambrosia*. Bursages and ragweeds are mostly inconspicuous shrubs and subshrubs (although some are herbaceous perennials), with tiny to fairly large, usually grayish, leaves. The plants bear inconspicuous *staminate* (male) and *pistillate* (female) flower heads on the same plant (*monoecious* flowering). After pollination by wind, some species (the bursages) develop spiny fruits (burs) which are dispersed when they cling to the fur or feathers of passing animals.

RANGE
Ambrosia is a mostly American genus of about 24 species, half of which occur in the Sonoran Desert region.

NOTES
Despite the abundance of bursages and ragweeds, few vertebrates browse them. Nor do they seem to host many insects; they definitely do not rely on insects for pollination. The copious wind-borne pollen is highly allergenic and a major cause of hay fever.

➤ *Ambrosia ambrosioides*

English name: canyon ragweed, bursage
Spanish name: chicura

DESCRIPTION
This is a woody shrub to about 6 feet (1.8 m) tall with sparsely-branched, wand-like stems bearing

large (to 7 inches, 18 cm long), broad, dark gray-green leaves. It flowers in late winter into spring and occasionally at other times. Though it's called a ragweed, the fruit is a bur.

Canyon ragweed
(*Ambrosia ambrosioides*)

RANGE
Canyon ragweed is a subtropical shrub common throughout Sonora, southwestern Arizona, and the southern half of Baja California. There is also a disjunct population in Durango, Mexico. In the drier parts of its range it occurs mainly in drainages and along roadsides where it receives extra water from runoff. In the northern edge of its range it is restricted to the warmer slopes above valley floors.

NOTES
O'odham used canyon ragweed in sweat baths to relieve arthritic pain. They spread hot coals on flat ground and covered them with a layer of ragweed leaves, then laid the afflicted person atop them and covered him or her with a blanket.

➤ *Ambrosia deltoidea*

English names: triangleleaf bursage, burrobush, rabbitbush
Spanish name: estafiate

DESCRIPTION
Triangleleaf bursage is a densely-branched subshrub to about 2 feet (0.6 m) tall with triangular, finely toothed, gray-green leaves. The leaves are lost in very dry periods. Though it can be confused with brittlebush when not in flower; this bursage is a smaller plant with smaller, duller gray leaves. It flowers from late winter into spring, with burs ripening in late spring.

Triangle-leaf bursage
(*Ambrosia deltoidea*)

RANGE
This species is characteristic of Arizona Upland desert, where it is often the dominant understory plant. It also occurs in the eastern portion of the Lower Colorado River Valley subdivision and in a few localities in Baja California.

The Fundamental Importance of Bursage

The Arizona-Sonora Desert Museum has always endeavored to restore the natural areas of its grounds disturbed by construction, our goal being to do it so well that no one could detect that humans had ever even walked on that spot, let alone cleared and trenched through it. Our first attempts were dismal. Although we had restored the land contours, replanted the cacti and trees, and painstakingly replaced the surface rocks right-side-up and partly buried, the site still looked devastated. Upon comparing our restoration to the adjacent undisturbed ground, we recognized that we had replaced less than a quarter of the original vegetation. In our particular type of Sonoran Desertscrub called Arizona Upland, about half the ground's surface is covered by small shrubs. At the Desert Museum the dominant small-shrub species is triangleleaf bursage. These and other small shrubs greatly contribute to the base color and texture of the terrain. Today we are proud of our undetectable desert restoration efforts, and commercial nurseries propagate tens of thousands of bursage plants for use in desert revegetation projects. Things that we normally overlook can be of fundamental importance.

NOTES

This and other bursages are called burrobush and rabbitbush in pollen reports. Don't confuse it with the plant more commonly called rabbitbrush, *Chrysothamnus nauseosus*, which flowers only in the fall and is much less allergenic. Bursage is the least ambiguous of the English vernacular names; the others are also applied to unrelated plants. Triangleleaf bursage lives as long as 50 years.

Among the plants sheltered by this bursage is the local pincushion cactus, *Mammillaria grahamii*. On some rocky bajadas nearly every bursage has several pincushions growing within its modest canopy.

➤ *Ambrosia dumosa*

English names: white bursage, burro-weed, burrobush
Spanish names: chicurilla, hierba del burro (burro herb), huizapol

DESCRIPTION

This subshrub to about 2 feet (0.6 m) tall has small, white, deeply-divided leaves. Plants are leafless in dry periods, which is most of the year in its habitat. It flowers in late winter to spring, depending on rainfall.

White bursage
(*Ambrosia dumosa*)

RANGE

White bursage is common in the driest areas of the Sonoran Desert and southern Mohave Desert. In the parched flats of the Lower Colorado River Valley white bursage and creosote bush are often the only perennial species. In Arizona Upland desert it grows in the fine soils of the valleys; in contrast, triangleleaf bursage grows on rocky or caliche soils.

NOTES

Recent research shows that the growth of bursage roots are inhibited by secretions of creosote bush roots. These are the two most common species in the arid valleys of the Lower Colorado River Valley and therefore the most likely competitors for scarce water resources. Perhaps even more interesting is the finding that the roots of different bursage plants seem able to detect one another and respond by growing in different soil volumes. It is likely that many other plants have similar interactions, but we still know very little about what is going on within the soil.

Other Representative Genera in the Composite Family

➤ *Baccharis sarothroides*

English names: desert broom, broom baccharis
Spanish names: romerillo (rosemary), escoba amarga (bitter broom), hierba del pasmo

DESCRIPTION

Desert broom is a vertical, evergreen, densely-branched shrub usually 3 to 6 feet tall (0.9-1.8 m), occasionally to 10 feet (3 m). The many fine twigs are green; the tiny, linear leaves are deciduous during dry periods. The plants are *dioecious* (that is, each individual plant bears only "male" or "female" flowers) and blooms in the fall. The wind-dispersed, white-tasseled seeds are produced by the female plants in such abundance that the plants and nearby ground appear to be snow-covered.

RANGE

Desert broom grows in desert, desert grassland, and chaparral from 1000 to 5000 feet (300-7500 m) elevation in Arizona, California, Sonora, and Baja California.

NOTES

Desert broom is a pioneer plant that is efficient at colonizing newly-disturbed soil. Therefore it is often abundant in washes, roadsides, and abandoned plowed fields. Like many pioneer plants, it is rather short-lived, lasting only perhaps a couple of decades. Desert gardeners may dislike desert broom for its aggressively invasive nature, but it can be useful. As the vernacular names suggest, the thin terminal twigs can be bundled together at the end of a pole to make a broom. The flowers attract hordes of butterflies and other insects. Because it flowers in autumn, it's a good plant for extending the season of a butterfly garden. Grow male plants (offered by some nurseries), if you wish to prevent unwanted volunteers. Unwanted plants

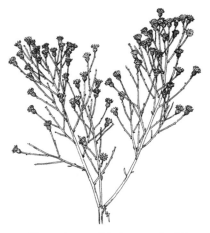

Desert broom (*Baccharis sarothroides*)
male plant

have another horticultural use: desert broom branches create ideal shade to protect newly planted cacti and other plants from sunburn. Several layers of tip branches provide 50-75% shade; the twigs slowly disintegrate over several months and the sheltered plant gradually becomes acclimated to full desert sun. Such redeeming value can be found in some of the worst weeds.

The resinous leaves and stems are rarely eaten, except by jackrabbits during droughts when little else is available.

There are several other species of Baccharis. Waterweed (*B. sergiloides, escoba amarga*) grows in permanently wet soils. Seep-willow (*B. salicifolia*, formerly *B. glutinosa*) commonly forms dense thickets along waterways. Other plants called "broom" are in the legume family (Fabaceae or Leguminosae).

➤ *Baileya multiradiata*

English name: desert marigold, wild marigold, desert baileya
Spanish name: hierba amarilla (yellow herb)

DESCRIPTION
Desert marigold is an annual or short-lived perennial with white-wooly leaves forming a dense mound about 6 to 8 inches (15-20 cm) in diameter. Flower heads perch well above the foliage on 1 foot long (30 cm) stems. The flat 1 inch (2.5 cm) daisies are bright yellow.

RANGE
Desert marigold grows mostly on bajadas and in valleys throughout the northern Sonoran, southern Mohave, and northern Chihuahuan deserts. It is one of the most common wildflowers, often abundant on disturbed soils such as road shoulders.

NOTES
This is perhaps our most consistent showy roadside wildflower; it blooms in any season after a moderate rain. On hard soils such as caliche, desert marigold is replaced by the somewhat similar *Bahia absinthifolia* (which has no vernacular name except its genus name). The foliage of bahia is more sparsely distributed; its stems arise from perennial underground rhizomes that form colonies. Bahia leaves are less woolly and the flowers slightly smaller than are those of desert marigold. Bahia's range extends into pine forests.

Desert marigold
(*Baileya multiradiata*)

➤ *Bebbia juncea*

English names: chuckwalla's delight, sweetbush
Spanish names: chuparosa (sucked, e.g., by insects for nectar), junco (generic name for rushes)

Chuckwalla's delight
(*Bebbia juncea*)

DESCRIPTION
Bebbia is an intricately-branched shrub to about 4 feet (1.2 m) tall and wide. The threadlike leaves are absent most of the year; the pencil-lead-thin stems are green and photosynthetic. The cylindrical, yellow flower heads are composed only of disk flowers. Flowering is almost year round, most heavily following rains.

RANGE
Bebbia is widespread in the arid and semiarid habitats of our region.

NOTES
The vernacular name "chuckwalla's delight" was invented by naturalist Edmund Jaeger, who observed that chuckwallas "feed greedily on the flowers." Many desert plants have yellow flowers, and chuckwallas seek out and eat most of them.

➤ *Encelia farinosa*

English name: brittlebush
Spanish names: rama blanca (white branch), incienso (incense), hierba de las animas (herb of the souls), hierba del bazo (spleen-herb), palo blanco (white stick), hierba ceniza (gray herb)

DESCRIPTION
Brittlebush is a somewhat woody shrub 3 to 5 feet (1-1.5 m) tall. The dense branching pattern tends to form a hemispherical mound, especially in very arid conditions. The leaves range from nearly hairless bright green through gray-green to white with a dense covering of soft matted hairs. Small yellow flowers are borne on multiply-branched stalks well above the leafy stems; they are usually produced in late winter to mid spring, with occasional bloom at other times.

RANGE
This distinctive plant is widespread and common in most of the Sonoran Desert and in warmer parts of the Mohave Desert.

NOTES
Brittlebush is very drought resistant and often forms nearly pure stands in extremely arid habitats. There is a widespread myth that creosote bush secretes toxins that kill all other plants growing nearby. In fact creosote bush doesn't inhibit many annual plants, but brittlebush does. Rain water dripping off the foliage dissolves chemicals that inhibit the seed germination of many species. For this reason only a few species of annual wildflowers can grow under brittlebushes.

The more arid the conditions of the growing season, the smaller and whiter are the leaves produced. During prolonged drought the leaves are completely lost. It is fairly frost-tender; in Arizona Upland it tends to be restricted to slopes above the cold valley floors. The stems exude a gum that can be chewed or used for incense (hence the Spanish name *incienso*). The gum was once exported to Europe by the mission priests and is still used by the Tohono O'odham. The English name was given to it by members of the 1907 Sykes expedition to the Pinacate region; they dubbed it "white brittlebush." (Read the fascinating tale of this pioneering adventure in *Camp-Fires on Desert and Lava*, by William Hornaday.)

Brittlebush
(*Encelia farinosa*)

> *Trixis californica*

English name: trixis
Spanish names: plumilla (little feather), arnica

DESCRIPTION
Trixis is an evergreen shrub 1 to 3 feet (30-90 cm) tall that grows compact and upright in sun, but sparse and floppy when in the shade. The 1- to 2-inch (2.5-5 cm) long, dull green leaves emit a rank odor when bruised. The small yellow flower heads have a spicy fragrance. They appear at any time after a rain, with a peak bloom in spring.

RANGE
Trixis grows from southern California to west Texas, and south to central Mexico.

NOTES
The Seri regard this plant as a panacea, claiming that there is nothing that this medicinal herb is not good for. Among its wide range of uses, trixis is smoked like tobacco and administered as an aid to childbirth.

Trixis is successful in many habitats from low desert on the Colorado River to woodlands at 5500 feet (1700 m) elevation. If enough soil moisture is available, it grows anywhere temperatures remain above freezing. Moreover, it belongs to a subgroup of composites (tribe Mutiseae) that is mostly South American. There are many plant and animal groups with similar distribution patterns—a large number of species in their "home" areas, a few of which become very successful, extending their ranges far beyond their places of origin. There is something special about the biology of trixis, but we don't know what it is.

Trixis
(*Trixis californica*)

Bignoniaceae (bignonia family)

Mark A. Dimmitt

The bignonia family has about 725 mostly tropical species worldwide, of which only four occur in our region. Well-known members include catalpa, jacaranda, and trumpet vine (*Campsis radicans*).

➤ Chilopsis linearis

English name: desert willow
Spanish names: mimbre (wicker),
    jano

DESCRIPTION
Desert willow is a deciduous small tree to 30 feet (9 m) tall that resembles a willow, with its long, narrow leaves and slightly drooping branches. Large flowers shaped like irregular trumpets range from nearly white in the western end of its range to deep purple with yellow *nectar guides* (visual signals enticing to bees) at the eastern end. Wild trees bloom from mid-spring through mid-summer with sufficient rainfall; selected cultivars may bloom into fall with irrigation.

RANGE
C. linearis occurs from west Texas to southern Nevada, Arizona, southern California, and northern Mexico. Washes from the low desert to grasslands are its primary habitats.

NOTES
Large bees are the main pollinators, though the flowers are also attractive to other insects and to hummingbirds.

This is one of the few trees in the north half of the Sonoran Desert that is not a legume. Another plant in this family, *Tecoma stans* (yellow trumpet bush), attains small-tree size in the tropics, but not in the desert. It is a shrub in the north because it periodically freezes to the ground. *Tecoma* is one of many examples of the Sonoran Desert's tropical legacy.

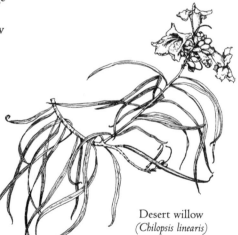

Desert willow
(*Chilopsis linearis*)

Brassicaceae (Cruciferae) (mustard family)

Mark A. Dimmitt

The mustards number about 3000 species worldwide, most of which are herbs. Some important vegetables such as radish, broccoli and cauliflower are in this family. Most Sonoran Desert species go unnoticed except by botanists, because they are inconspicuous even though sometimes abundant. Two attractive wildflowers are bladderpod (*Lesquerella gordoni*) and silver bells (*Streptanthus carinatus*).

The greater story of mustards in the desert is that some alien species are pernicious weeds. The most disturbing is probably Sahara or Moroccan mustard (*Brassica tournefortii*). First discovered in the United States only in the 1930s, it has by now invaded most of the dry sandy areas of the Sonoran Desert. Unlike most pioneer plants (weeds) it doesn't require disturbed soil to invade. In wet years it forms extensive, dense stands that crowd out the native annuals. Because most birds and mammals will not eat it, this mustard's displacement of native species presumably has detrimental effects all the way up the food web. Many potential wildflower displays have also been thwarted by this one weed. Another alien mustard, London rocket (*Sisymbrium irio*), is a major pest in disturbed soils and gardens.

Bladderpod
(Lesquerella gordonii)

Burseraceae (torchwood family)

Mark A. Dimmitt

The New World elephant trees are nearly unknown to the general public, but nearly everyone has heard of their Old World relatives. The aromatic sap of frankincense (*Boswellia sacra*) and myrrh (*Commiphora* spp.) was once worth its weight in gold. It's still quite valuable, and most wild frankincense trees are sacred, hereditary property and are zealously protected. Our native species are also valuable to those who know them. The tropical species south of the desert are used in products as diverse as incense, masks, and asthma treatment.

The torchwood family contains 550 species of shrubs and trees worldwide, many of which have succulent stems and aromatic foliage. There are about 12 species of *Bursera* in the Sonoran Desert region, about a third of which occur in the desert.

➤ *Bursera microphylla*

English name: elephant tree
Spanish names: torote blanco, copal,
palo colorado (red stick), torote
colorado

DESCRIPTION

This large shrub or small tree grows to 25 feet (8 m) tall; it usually has several contorted trunks and reddish branches. The trunks and main branches are swollen with water-storage tissue and covered with whitish sheets of thin, peeling bark. The thin bark transmits sunlight to chlorophyll-bearing tissue in the stems, which can thus conduct photosynthesis when the plant is leaf-less. The stems and the finely divided, shiny green leaves are highly aromatic. It may leaf out in any month in response to rain, but stem growth is mainly in summer. The flowers, which open around the time of the summer solstice (the hottest, driest time of the year, when almost nothing else is blooming), are tiny and inconspicuous.

RANGE

This is a true desert species, occurring in western Sonora and almost all of Baja California with a few marginal populations in south-central Arizona (reaching northern limits in some Phoenix-area mountain parks) and extreme southern California. The plants in some Arizona populations look quite different from Mexican plants and may be hybrids or a new species.

NOTES

While the aromatic sap of this and other burseras smells pleasant to humans, it tastes foul and functions as an herbivore deterrent. The sap is under pressure; when a leaf is broken, a thin stream of sap spurts out an inch

(2.5 cm) or so. This "squirt-in-the-eye" defense is apparently effective, since the elephant tree's foliage is nearly always intact.

The bitter fruit stimulates salivation and the Seri sometimes chew it to quench thirst. They used the soft wood to make boats. The sap is used to seal cracks in boats and pottery. The aromatic oils are used in a variety of medicines, including treatments for stingray wounds, lice, cuts, and gonorrhea.

The other desert species of *Bursera*, except for one in Baja California, are scarcely succulent and not easily confused with *B. microphylla*. But two unrelated plants look almost identical when not in leaf (which is most of the year). *Pachycormus discolor* (Baja elephant tree, torote blanco, copalquin) often grows side by side with *B. microphylla* in Baja California, but responds to rainfall in the winter, when burseras are not actively growing. *Pachycormus* is in the cashew family (Anacardiaceae). *Jatropha cordata* (sangregado, mata muchachos) grows next to burseras in Sonora. A member of the spurge family (Euphorbiaceae), *J. cordata* is more closely related to *Poinsettia* than to *Bursera*. These three look-alike species are prime examples of convergent evolution.

Elephant tree
(*Bursera microphylla*)
branch with fruit

Cactaceae (cactus family)

Mark A. Dimmitt

The enormous popularity of cacti among gardeners and plant collectors is surpassed only by that of roses and orchids. Their appeal extends far beyond their native habitat; there are legions of devotees in the eastern United States, Europe, and Japan. The desire to possess these strange yet beautiful plants supports hundreds of specialty nurseries; the largest shops grow and sell millions of plants annually. Cacti are one of the reasons tourists visit the American southwest.

DESCRIPTION OF FAMILY

Most people think they know a cactus when they see one, but they are often mistaken. All cacti are succulents, but not all succulents are cacti. Agaves, ocotillos, aloes, and the succulent euphorbias (such as African milk trees) are among the swollen or spiny plants often mistaken for cacti. However, the term *cactus* refers to a particular family of plants defined by a distinctive flower pattern. To be a cactus, the plant MUST produce flowers with the following characteristics: many *tepals* (combined sepals and petals) that intergrade with each other; many stamens (usually hundreds), and numerous stigma lobes (rarely only three). If a plant lacks such a flower, it cannot be a cactus.

RANGE

The cactus family is nearly endemic to the New World from southern Canada to southern South America. There is an exception—one of the 1800 species occurs naturally in Africa, Sri Lanka, and Madagascar. However, introduced cacti have gone wild and sometimes become pests in several regions of the Old World. Cacti are most common (in numbers of both plants and species) in semiarid habitats with low rainfall, yet with dependable rainy seasons. A few species occur in extremely arid deserts and wet tropical forests. About 300 species occur in the Sonoran Desert region.

NOTES

The majority of cactus species are pollinated by numerous species of bees, a number of which specialize in cacti. Cactus bees are all solitary, but in some species the females congregate by hundreds of thousands at nesting sites to dig their individual nest burrows, which are densely concentrated in an area of a few thousand square feet. Cactus pollen is packed into these burrows to feed the grubs, which the parents do not tend. Some cacti are pollinated by birds, moths, or bats.

Giant Columnar Cacti:
Saguaro, Organpipe, Senita, and Cardón

Two defining life forms of the Sonoran Desert's vegetation are giant columnar cacti and legume trees. Both are characteristic of arid tropical habitats (the cacti only in the "New World") and their presence in the Sonoran Desert reflects its affinity to the tropics. Arizona Upland (a subdivison of the Sonoran Desert) experiences frequent frosts, and only one columnar cactus (the saguaro) is sufficiently cold-hardy to be widespread in this subdivision. The Lower Colorado River Valley subdivision is too dry for columnar cacti and they are almost absent there. Several species of giant columnar cacti occur in each of the other four subdivisions.

In addition to the giants described below, there are several smaller columnar species in the Sonoran Desert, and even more small and large species in the adjacent tropical communities. *Myrtillocactus cochal* (*cochal*) occurs in Baja California. It branches profusely from a short trunk, forming a large candelabra-shaped mass of stems up to 13 feet (4 m) tall and wide. Repeat photography indicates that it lives only a few decades. *Bergerocactus emoryi* (golden torch cactus), also from Baja California, has very slender stems to about 7 feet (2 m) tall densely clothed with yellow spines. *Pachycereus pecten-aboriginum* (*etcho, hecho, cardón barbón*) resembles a skinny cardón. Its fruit is very bristly and used as a hair brush. It is mainly a thornscrub species that enters the southern edge of the Sonoran Desert in southern Sonora and southern Baja California.

➤ *Carnegiea gigantea*

English names: saguaro, giant cactus
Spanish names: saguaro

The huge green columns of saguaros have captivated the attention of nearly every tourist who has set eyes on one, let alone a whole "forest" of them. The United States has even devoted a national park to this plant. Saguaros are even more important to the O'odham peoples who have lived in their habitat for centuries. The high esteem O'odham have for saguaros is reflected in their many creation stories for this plant, which tend to share the common theme of people being turned into saguaros. These giant cacti are not plants to the Tohono O'odham; they're another form of humanity.

The saguaro is the most thoroughly studied plant in the Sonoran Desert. Therefore its ecology can be described in considerable detail. Nearly every other organism in its range (including humans) can be ecologically connected to it in some way.

DESCRIPTION
The saguaro is the largest cactus in the United States, commonly reaching 40 feet (12 m) tall; a few have attained 60 feet (18 m) and one was measured at 78 feet (23.8 m). The cylindrical stems are accordion-pleated; the ridges

(outer "ribs") are lined with clusters of hard spines along the lower 8 feet (2.4 m) and flexible bristles above this height. White flowers are about 3 inches (8 cm) in diameter; they bloom mainly in May and June and are followed a month later by juicy red fruit.

RANGE

The saguaro's range is almost completely restricted to southern Arizona and western Sonora. A few plants grow just across the political borders in California and Sinaloa. Saguaros reach their greatest abundance in Arizona Upland. Plants grow from sea level to about 4000 feet (1200 m). In the northern part of their range they are most numerous on warmer south-facing slopes.

ANATOMY

This anatomical description of the saguaro is generally applicable to other cacti as well, except that most smaller species have less well-developed woody skeletons. The *epidermis* ("skin") is covered with a thick, waxy cuticle that waterproofs the surface and restricts *transpiration* (loss of water vapor) almost exclusively to the *stomates* (pores for gas exchange). The outer surface is folded into pleats (commonly called "ribs," but not to be confused with the internal, woody ribs). These pleats enable the stem to expand in girth during water uptake without stretching and bursting. *Areoles*, the roundish pads from which the spines and usually the flowers are produced, are distributed

at 1-inch (2.5 cm) intervals along the ridges of the ribs. Each areole bears a cluster of about 30 spines, the longest up to 2 inches (5 cm) long. The spines are stout and sharp on young plants. Stems taller than about 8 feet (2.4 m) produce bristly spines. Spines serve multiple functions, primarily protection from herbivorous animals and from sun; their shade reduces heat load and consequent water loss. The lower trunks of old saguaros lose their spines and develop dark, corky bark.

Immediately beneath the epidermis is a thin layer of chlorophyll-containing cells where most photosynthesis takes place. The deeper interior—most of the bulk of the plant—consists of water storage tissue (*parenchyma*). Water comprises most of the weight of the saguaro. A fully hydrated large stem is more than 90 percent water and weighs 80 pounds per foot (120 kg per meter).

The great, mostly aqueous bulk of larger plants protects them from temperature extremes. Heat absorbed through the surface during the day is stored in the mass of interior tissue, resulting in a fairly small temperature rise that doesn't reach a lethal level. The heat is slowly radiated and conducted back into the air during the cooler night. The same thermal inertia usually keeps the tissues above freezing on cold winter nights.

Near the center of the stem is a cylinder of 13 to 20 woody ribs running the length of the main stem and branching into the arms. In the upper part of a stem the ribs are separate; as the stem ages the ribs continue to grow and fuse into a latticed cylinder.

A tap root extends downward to more than 2 feet (60 cm). The rest of the extensive root system is shallow, as is the case for most succulents. Roots are rarely more than 4 inches (10 cm) deep and radiate horizontally about as far from the plant as the plant is tall.

Occasionally abnormal growths occur, the best-known type called a "crest." This fan-topped form results when the *apical meristem* (the actively proliferating tissue at the growing tip) broadens from its normal point into a line of dividing cells. The cause in saguaros and other cacti is not known, other than that it usually seems to follow damage to the growing point. It does not harm the plant, which frequently continues to produce flowers and fruit. Crests are occasionally found in nearly all plant species; the phenomenon is especially noticeable in saguaros because of their size. (The garden cockscomb, found in many nurseries, is a genetically stable crested mutant of the Chinese woolflower, *Celosia argentea*.) Other saguaro anomalies occasionally encountered are ribs that undulate or spiral.

REPRODUCTIVE ECOLOGY—
FLOWERING TO GERMINATION
Saguaros flower mostly near the stem tips during the dry foresummer; peak production is from mid-May to mid-June. The sturdy white flowers open late at night and remain open until midafternoon of the next day. They are about 3 inches (8 cm) in diameter and emit an aroma like that of overripe melon. The flowers are self-sterile; cross-pollination is necessary for fruit set.

Characteristics of the flowers point almost unambiguously to pollination by bats. The nocturnal opening of buds, maturation of pollen, and production of nectar; their exposed position high above the ground, their heavy texture, the particular fragrance emitted at night, and the copious nectar and pollen, are all characteristic of bat flowers. Even the proportions of amino acids in the pollen protein matches that of bats' nutritional needs more closely than that of other animals. The only anomalies are that the flowers remain open well into the next day and produce more nectar after sunrise. For years the bat-saguaro *mutualism* (mutually beneficial relationship) went unquestioned, until a field biologist examined the relationship closely. Reality turns out to be more complicated, as is usual in nature.

Many bats feed on nectar and fruit rather than on insects. Two species, the lesser long-nosed bat (*Leptonycteris curasoae*) and the Mexican long-tongued bat (*Choeronycteris mexicana*), occur in saguaro habitat. After wintering in tropical Mexico, the lesser long-nosed bats migrate up the arid coast of Sinaloa and Sonora beginning in March, feeding mostly on columnar cactus flowers. The flowers supply their complete diet during migration. The nectar provides energy-rich carbohydrate for flight. The pollen that clings to their furry faces while they're lapping up nectar is swallowed as the bats groom themselves at roost. Pollen provides most of their dietary protein. Their feeding activity also effectively pollinates flowers.

Near Kino Bay, Sonora, where there is a major *Leptonycteris* bat roost, Ted Fleming and associates caged different saguaro flowers against animal visitors during either daytime or nighttime. To their surprise, the flowers that were exposed during the day had a much higher rate of pollination (resulting in fruit set) than did those exposed at night. During the day, flowers were visited by great numbers of bees (mostly introduced honeybees) and birds (mostly White-winged Doves). At night the bats were present, but spent most of their time feeding on cardón and organpipe cacti, which were also common at the site.

One cool spring when the bats arrived from the south before the cardón and organpipe cacti began to flower, they fed heavily on saguaros. But in most years bats seem to be minor players in saguaro pollination. The saguaro story is similar farther north. The range of cardón stops south of the U.S.-Mexico border, and organpipes barely extend into southern Arizona. Saguaros occur well north of Phoenix in eastern Arizona and almost as far north as Kingman in western Arizona, but the bats rarely venture north of Tucson. While bats depend heavily on saguaros at the northern limit of their summer migration in southern Arizona, they are apparently too few in number to be important saguaro pollinators there. The many saguaro populations north of the bats' range must depend entirely on other pollinators. Among them is the native bee *Diadasia opuntiae*, which despite its name prefers saguaros over prickly pears or chollas.

As is often the case in scientific research, the Fleming study raises more questions than it answers. If bees and doves accomplish most of the pollination, why do saguaro flowers still have mostly bat-attracting characteristics? Have we encountered the early stage of an evolutionary shift? In 100,000 years will saguaros have diurnal yellow flowers? Or do the bats still have sufficient, yet-unknown influence on saguaro evolution to maintain the status quo?

In June and early July the pollinated flowers mature into 3-inch (8 cm) long fruits containing up to 2000 tiny seeds embedded in juicy, red pulp. The rind splits into 3 or 4 sections and peels back to expose the pulp and a red inner lining. These open fruits are often mistaken for red flowers.

The fruit ripens in tremendous abundance during the peak of the foresummer drought, and is nearly the only moist food available during this hottest, driest time of the year. It becomes a staple for many birds, mammals, and insects during late June and early July. The primary effective seed dispersers are several species of fruit-eating birds, such as White-winged Doves, Gila Woodpeckers, and House Finches. The birds digest the pulp, and the seeds pass through their guts intact. Birds also tend to defecate while perched in trees, thus depositing the seeds in favorable environments for establishment. Long-nosed

Saguaro flower

Saguaro fruit

bats also consume cactus fruits, but they defecate while flying or roosting in caves, so most seeds land where they can't grow. (See the story of the bats' southward migration in the section on Agavaceae.)

Fruit ripening occurs just before the summer rainy season arrives in the eastern Sonoran Desert. Seeds germinate in about 5 days after a rain. The rest of the plants' life cycle is spent at a much slower pace.

GROWTH FROM SEEDLING TO MATURITY

Seedlings grow very slowly during their first few years and are extremely vulnerable. They are tiny, heat- and frost-tender, soft-spined canteens. Rodents, rabbits, and birds eat the seedlings they can find, and many others succumb to desiccation or winter freezes. Nearly all survivors are located beneath the canopies of *nurse plants*, where they are sheltered from weather extremes and concealed from herbivores. Most desert plant species—not only saguaros—must begin life under nurse plants. Creosote bush, bursage, and desert zinnia are among the few perennials that can successfully establish in fully exposed sites.

Even in the shelter of nurse plants, most saguaro seedlings perish from drought, frost, or predation. Significant establishment of seedlings requires several consecutive years of milder and wetter-than-average weather. Such conditions occur only a few times a century in Arizona Upland, and even less frequently in the Lower Colorado River Valley community. This results in saguaro populations with only a few size and age classes. For example, 60 years after a favorable establishment period, there will be a large group (called a *cohort*) of recently-matured plants averaging 10 feet (3 m) tall. The same area may have only 2 or 3 other size-age cohorts; one several decades younger, another several decades older, with few plants in between. Southern Arizona experienced wet summers and mild winters during the first half of the 1980s. The cohort of seedlings that probably established during this period will begin protruding above their palo verdes (*Cercidium* spp., the most common saguaro nurse plants in Arizona Upland) at about year 2030.

Growth rate is controlled mainly by the amount of rainfall, plant size, and soil type. A tiny seedling has very little water-storage tissue and a relatively large surface area through which water is lost. Soon after a rain it exhausts its meager supply, stops growing, and goes into CAM idling mode. (See the discussion of CAM in the chapter "Plant Ecology of the Sonoran Desert Region.") Larger plants contain more water relative to the enclosing surface area, and can continue to grow for several weeks after rain. Therefore saguaro growth rates increase as the plants get larger.

In the Tucson Mountains, which averages 14 inches (355 mm) annual rainfall, a saguaro takes about 10 years to attain 1½ inches (3.8 cm) in height and 30 years to reach 2 feet (61 cm). Saguaros begin to flower at about 8 feet tall (2.4 m), which takes an average of 55 years. Compare this with 40 years to first flowering in the wetter eastern unit of Saguaro National Park (16 inches, 406 mm, average annual rainfall) and 75 years in the drier Organ Pipe Cactus National Monument (9 inches, 230 mm). No one has studied the populations near

Yuma, which receive about 3½ inches (90 mm) average annual rainfall.

Saguaros may begin to grow arms when the plant is between 50 and 100 years of age (in the Tucson Mountains), usually just above the stem's maximum girth at about 7 to 9 feet (2.1 to 2.7 m) above-ground. The number of arms and overall size of a plant seem to be correlated with soil and rainfall. Saguaros on bajadas with finer, more water-retentive soils tend to grow larger and produce more arms than do those on steep, rocky slopes. A few saguaros have been observed with as many as 50 arms; many never grow arms. Saguaro arms always grow upwards. The drooping arms seen on many old saguaros is a result of wilting after frost damage. The growing tips will turn upwards in time. There is a myth that arms are produced so as to balance the plants, but research shows arm-sprouting to be random. Many saguaros can be found with several arms all on the same side of the main stem.

ECOLOGY OF MATURE SAGUAROS

Old saguaros often stand alone on open ground. They not only outlive their nurse plants, they may hasten their deaths. Since a saguaro's root system is shallower than that of palo verdes or other nonsucculents, in dry years the saguaro intercepts most of the meager rainfall. Trees and shrubs within the root zone of a saguaro are thus more likely to succumb to drought than those that lack such competition.

Animal associations with saguaros are many and varied. Numerous species of vertebrates and invertebrates use this plant for food, shelter, and perching sites. Some of these associations are described throughout this book.

Saguaros make excellent nesting places for many birds. The primary nesting associates are Gila Woodpeckers and Gilded Flickers, both of which excavate nest holes in the fleshy stems. The plant seals the wound with scar tissue called *callus*, which quickly becomes very hard and impervious to microbial infection. Callus tissue decomposes more slowly than most of the rest of the saguaro and can be found on the ground among the debris of dead plants. Because of their shapes the callus remains of nesting holes are called "saguaro boots." The Seri used both saguaro and cardón cactus boots to carry and store food.

Gila Woodpeckers typically excavate nest holes in the thick, older sections of stems between the epidermis and the cylinder of woody ribs. These holes seem to cause no serious damage to the saguaro unless they're very numerous. In contrast, flickers usually excavate larger holes higher on the stem and penetrate the interior rib cylinder. This interrupts water and nutrient transport, and the stem's further growth is greatly impeded. Stems weakened by flicker holes are also more easily broken by storm winds. Broken stems may themselves branch to produce one to several arms; this increases the potential seed production of the plant. But research also suggests that flicker holes may do sufficient damage to cause death of the plant.

Woodpeckers usually excavate new nesting holes each year. Several other hole-nesting but non-excavating birds occupy abandoned woodpecker nests, including Elf Owls, House Finches, Ash-throated Flycatchers, and Purple Martins. Invertebrates also inhabit them.

Larger birds such as Red-tailed Hawks build nests in the angles between the main stems and the arms. Tall saguaros also make good hunting and resting perches for many birds. Cactus Wrens sometimes nest among the arms or, less often, in a woodpecker cavity.

MORTALITY

Saguaros die from different causes at different sizes. The vast majority of seedlings die in the first year from drought or frost. Some birds, such as Curve-billed Thrashers, dig them up in search of insects. Larger seedlings up to about a foot (30 cm) tall, or 2 decades of age, suffer high mortality from being eaten by rodents and rabbits. The few that survive to about 3 feet (1 m) tall at about 40 years of age are much more resistent to weather extremes and animal damage, and mortality is low from then into old age.

Some mature saguaros are killed by lightning strikes, windthrow, and perhaps from flicker damage. The chief agent of mortality of mature saguaros in the Arizona Upland is freezing temperatures. The saguaro is a tropical cactus with limited frost tolerance, and it reaches the northern, coldest limit of its range in Arizona Upland. Small to medium-sized plants are protected by the canopy of their nurse plants. Larger plants in the open are protected by their thermal mass and small *surface to volume ratio* (that is, a great volume for storing heat and relatively small surface area for losing it). But saguaro stems begin growing thinner above about 12 feet (3.7 m), while continuing to add outer ribs (pleats). This increases the surface to volume ratio and makes older plants more vulnerable to freezing.

It is difficult to determine the lethal temperature for a saguaro or other plant. The seasonal timing and duration of freezing temperatures are at least as important as the minimum temperature. Healthy middle-aged saguaros have survived 10°F (-12°C) for a few hours in mid-winter, while 12 hours of 20°F (-7°C) in late fall have caused widespread damage and death.

Several times each century strong winter storms push arctic air masses deep into the Sonoran Desert. These hard freezes damage or kill the smallest and largest saguaros, as well as other tropical elements of the flora. Most often these freezes kill a portion of the outer layer of saguaro tissue, which forms a brown scab tissue. Since saguaros have a relatively small photosynthetic area to begin with, a significant reduction of this area results in slow starvation. A frost-damaged saguaro may survive for another decade or even longer, but eventually it weakens until it can no longer resist infection. Bacterial rot caused by *Erwinia cacticida* turns the flesh of weakened plants into an odoriferous black liquid. *Erwinia* is at most a weak pathogen; healthy cacti normally wall off small infections and continue to thrive. Several species of fruit flies and other insects are specialized to feed on rotting saguaro tissue.

The *Erwinia* bacterium is common in the environment. One *vector* (an agent of transmission) is the blue cactus borer, the larva of the moth *Cactobrosis fernaldialis*. The maggot-like caterpillar burrows into the flesh of saguaros and other cacti, feeding on the bacterial rot it introduces. Living saguaros typically have many round,

½-inch (1.3 cm) scabs on their surfaces. Sifting through a rotted carcass will reveal that these scabs are the outer ends of contorted cylinders of callus tissue; they are the healed tunnels left behind by cactus borers. There are other shapes and sizes of calluses in between these "worm holes" and woodpecker boots; their causes remain a mystery.

Fallen saguaros become homes for many small animals. Snakes, rodents, lizards, and invertebrates find shelter beneath and within them, until the community of reducers and decomposers completely recycle the remains.

A MODERN MYTH:
THE IMPENDING DOOM
OF THE GIANT CACTUS
The imminent demise of the saguaros is a recurring rumor dating back several decades. Its most recent incarnation began in the early 1990s and refuses to die, despite having been soundly refuted. It is an interesting story involving flawed science compounded by bad journalism. Briefly, here is what happened:

Visiting biologists who had not previously worked in the Sonoran Desert noticed that many of the large saguaros had brown, scabby skin and were evidently in declining health. They published a paper labeling the problem "brown decline" and postulated several possible causes, including air pollution and ultraviolet radiation through the hole in the ozone layer.

The authors misinterpreted the situation they observed because they were not familiar with saguaro ecology. The misinterpretation might be analagous to a situation in which extraterrestrials briefly visit earth and collect a sample of people from a nursing home. Their specimens would be mostly old and in poor health, and the aliens would have concluded that the human species was on the brink of extinction. Should they sample an elementary school they would reach a very different but equally wrong conclusion.

The large, conspicuous saguaros that dominate the Sonoran Desert landscape are analogous to the nursing home population. They were indeed old and many were declining. These comprised the few members of their age group that had barely survived the catastrophic freeze of December 1978. The scientists, however, failed to notice the large population of young, healthy saguaros in the "elementary schools" that were concealed under nurse trees.

The alarm sounded by this article would not have become widespread except for the unfortunate tendency of many journalists to seek sensational stories. The disappearance of our giant cacti was a major story, but their health and continued existence are barely newsworthy. The media picked up the scientific paper and published its gloomy speculation. Numerous local biologists were interviewed by journalists all over the world. All of the local biologists refuted the idea that saguaro populations were declining. Some media decided not to carry the story upon hearing the contrary evidence. But others published or broadcast dramatic stories forecasting impending doom for the giant saguaro, and only briefly mentioned, or omitted altogether, the contradictory opinions. This erroneous story recirculates when writers quote these past inaccurate publications.

Deeper History

The saguaro doom story first surfaced in the 1940s; at that time little was known about saguaro ecology. Saguaro National Monument was established in 1933 east of Tucson. That bottom-land area was chosen because it had a tremendous population of giant old saguaros. (It had few young or middle-aged saguaros, due to the effects of livestock grazing and the cutting of potential nurse trees since the late 1800s.) But there was a catastrophic freeze in 1937, and during the next decade the giant forest was suffering massive mortality from bacterial necrosis. The Park Service bulldozed and buried thousands of rotting cacti in the hope of stopping what it mistook as a new, virulent disease. These efforts failed, and the alarm over the presumed fate of the saguaros became a factor in the establishment of the West unit of Saguaro National Monument on the other side of Tucson in 1961. This location in the Tucson Mountains did not have a cohort of giants so the impact of the freeze of 1937 was less evident, or according to the view of that time, the bacterial necrosis disease had not infected this population. (The western unit had mature stands of giants by the mid 1970s; this cohort was devastated by the freeze of 1978.) This first misinterpretation of the ecology of saguaros had a positive outcome—it engendered the preservation of another tract of splendid desert. Both units of the National Monument were designated Saguaro National Park in 1994.

So are saguaros declining? The answer is, yes, most of the time. So are most species of desert plants and animals—that's the nature of this ecosystem. In most years there is slightly higher mortality than *recruitment* (the successful establishment of new individuals), so populations decrease. In years of severe droughts or freezes the mortality can be dramatic. In the occasional wet years mass recruitment reverses the trend of decline with a reproductive boom. In the case of saguaros these episodes of net recruitment seem to occur less than a half-dozen times per century in Saguaro National Park (west), and less often in the drier regions.

Hard data are available for establishment frequency of another plant on the western edge of the desert. Desert agaves (*Agave deserti*) successfully established seedlings in the foothills near Palm Springs, California only once in 17 years of study. Therefore this agave population was either maintaining the status quo or losing ground during 16 of those 17 years. Judging from the huge number of plants growing there, it's obvious that they're getting along just fine with these population dynamics.

Ethnobotany

Saguaros are good examples of the temporary, ever-changing nature of plant distributions. Its current geographic range is quite recent; saguaros have been expanding from their Ice Age refugia for the last 10,000 years. They arrived in the Tucson vicinity only 8000 years ago, according to the fossil record. Humans have been living in southern Arizona longer than have saguaros. (See "Deep History of the

Sonoran Desert" chapter.)

The saguaro is a focal point in the culture of the Tohono O'odham Indians. Some O'odham calendar months are named for seasonal changes of the saguaro. The woody ribs are used for fence-making and combined with ocotillo, grasses and mud in building dome-shaped houses.

Saguaro fruit is an important seasonal food, available immediately after the season called the "Painful Moon" or "Hunger Hurting Moon." There is a scarcity of natural food during the foresummer drought preceding the saguaro fruit ripening. The beginning of the Tohono O'odham year is the "Saguaro Harvest Moon." In June and early July, O'odham live in temporary camps in the saguaro forests and conduct the saguaro fruit harvest (*basañ bahidaj*). Women use poles made from saguaro ribs to knock off the saguaro fruit. The pulp is boiled down into syrup. Some of the syrup is used to make a wine that is consumed during the rainmaking ceremony. The seeds are dried and used in the winter as a snack, or ground into flour and made into a gruel. Saguaro fruit is nutritious, containing 10 percent protein and 70 percent carbohydrates; the seeds are 30 percent fat.

CULTIVATION

Saguaros are popular for landscaping. Large plants are collected from the wild, which is legal with permits from the state of Arizona and the landowner. However, the success rate of transplanting large saguaros is disappointing and the practice should be discouraged. The five-year survival rate of saguaros over 15 feet (4.6 m) tall is less than 10 percent, despite the best care the

Desert Museum can provide. The 5-year criterion is necessary because a mortally damaged saguaro can take several years to die. (The plants occasionally survive for a decade or more in an emaciated state, continuing to flower yearly. A transplant is not counted successful unless the cactus regains its former girth and resumes normal growth.) Observations of large landscaping and salvage jobs throughout southern Arizona indicate that similarly low success rates are the norm.

Saguaros up to 10 to 12 feet (3-3.7 m) tall nearly always transplant successfully *if* they are treated properly. At least a foot (30 cm) of the 2 to 4 major lateral roots must be excavated without damage. The plant must be replanted no deeper than it was growing originally; covering any of the trunk with soil is frequently fatal just as it is to almost all trees. The hole must be filled with soil with excellent drainage. If the cactus is not positioned as it was originally oriented, then shade must be provided to prevent sunburn on its side to the sunny south. Lastly, the plant needs occasional irrigation for the first year.

Saguaros are not necessarily slow growers in cultivation. Nursery-grown seedlings 1 to 2 feet (30-60 cm) tall are widely available. If planted in a suitable climate and watered and fed regularly, these cacti can reach flowering size in as little as 15 years. Saguaros are some of the few cacti that do not root readily from cuttings. Though cuttings do occasionally root, they do not regain their juvenile growth form (tapered trunk with stout spines), and so always look like saguaro arms stuck in the ground.

➤ *Stenocereus thurberi*

English name: organpipe cactus
Spanish names: mehuelé, pitahaya dulce
(sweet pitahaya), órgano, marismeña

DESCRIPTION

The many, usually unbranched stems
that arise from ground level readily
distinguish an organpipe cactus from
a saguaro. The stems also are thinner,
and have solid woody cores. Plants are
usually 9 to 11 feet (2.7-3.4 m) in
height, but occasionally exceed 20 feet
(6.1 m). The pinkish-white flowers,
produced from April through August,
open after dark and close shortly after
sunrise. The spines on the fruit loosen
and fall at ripeness. The juicy, sweet,
red pulp contains many tiny seeds.

RANGE

In Arizona the organpipe cactus is
found mostly in Organ Pipe Cactus
National Monument and the adjacent
Tohono O'odham Indian Reservation
in the south-central part of the state.
In Mexico it occurs almost throughout
Sonora, the southern half of Baja
California, and barely extends into
Chihuahua and Sinaloa. The desert is
marginal habitat for this species; it is
more abundant in thornscrub and
tropical deciduous forest.

NOTES

This tropical cactus is more frost sensitive than the saguaro, so it is restricted to the warmest microhabitats at its
northern limit in the United States. It
is most often found on south-facing
rocky slopes below 3300 feet (1000 m)
in elevation.

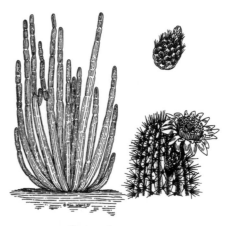

Organpipe cactus
(*Stenocereus thurberi*)

Nectar-feeding bats are the primary
pollinators and some of the major seed
dispersers. Because the flowers close at
daybreak, diurnal animals are not significant pollinators as they are of saguaros.

The fruits of organpipe cactus are
widely regarded as the second-best-
tasting fruit of all cacti (after those of
Stenocereus gummosus). Commercial harvest is feasible in some large populations, and fruits are sold in markets in
Sonora and Baja California.

Organpipe is the dominant plant in
a strip of thornscrub several miles wide
along the coast of southern Sonora;
they grow so densely that visibility is
seldom more than a few yards (meters),
and with the other thorny vegetation
they form nearly impenetrable thickets.

The ribs are used in housing. The
Seri made boat sealant from dried
organpipe flesh and animal fat. Slabs
of heated organpipe flesh were used
as compresses for aches.

➤ *Lophocereus schottii*
   [Pachycereus schottii]

   English names: senita (borrowed
   Spanish name; see Notes below)
   Spanish names: sina, sinita, garambullo,
     cabeza de viejo (old man's head), tuna
     (or pitahaya) barbona, hombre viejo
     (old man), musaro, senita (see note)

DESCRIPTION

Senita generally grows 10 to 13 feet
(3 to 4 m) tall and has the same general
form as organpipe cactus. It differs in
having stems with only 5 to 7 (rarely up
to10) ribs and very short spines on the
juvenile stems, giving them a sharply
angular aspect. Mature (flower-produc-
ing) stems are quite different; they're
densely covered with long, bristly, gray
spines. Pink, nocturnal flowers about
an inch (2.5 cm) in diameter emerge
through the bristles from April through
August; they emit an unpleasant odor.
They are followed by marble-sized red
fruits with juicy red pulp.

RANGE

Senita is found in desert and thornscrub in Sonora and Baja California.
Several small populations occur in
Arizona in Organ Pipe Cactus
National Monument and on the
Tohono O'odham Reservation. There
is also a historically transplanted
population on the Gila River Indian
Reservation in Sacaton, Arizona (south
of Phoenix). The plant grows in very
arid habitats, usually on fine-textured
valley soils. It often grows in close
association with desert ironwood trees.
Senita is more frost tolerant than

organpipe cactus, but because it grows
on the colder valley bottoms its range
doesn't extend as far north.

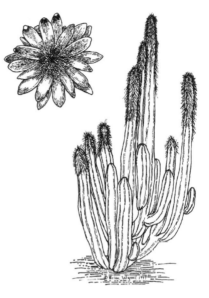

Senita
(*Lophocereus schottii*)

NOTES

Senita has a recently discovered mutu-
alistic relationship with a moth that
deliberately pollinates the flowers and
uses the developing fruit as food for its
larvae. The relationship is very similar
to that of the yucca and its pollinating
moth, and only the third such example
of a tight pollination-related mutualism known in the world. Senitas are
very long-lived. When sites in Baja
California photographed in 1905 were
revisited in the 1990s, nearly every
senita was still present.

The word "senita" seems to be
derived from the Spanish root meaning
"old," in which case it roughly translates as "little old woman." This is
rather descriptive of the gray-bristled

mature stems. But "senita" is not a legitimate Spanish word. The Mexican name of the cactus is borrowed from the original Indian name, sina. Some feel that the Spanish vernacular name should therefore be spelled "sinita."

> *Pachycereus pringlei*

English name: elephant cactus
Spanish names: cardón, sagüera, sagueso, sahuaso

DESCRIPTION

The cardón resembles the saguaro in growth form but it is much more massive. It develops a very thick trunk and the branches are closer to the ground and often more numerous than those of a typical saguaro. In sheltered locations plants may exceed 60 feet (18 m) tall. Young stems are armed with stout spines; mature stems are nearly spineless and have bluish epidermis between the rows of closely-spaced felty areoles on the external ribs. The flowers are similar to those of a saguaro, but with more and narrower tepals. The ovoid fruits are densely covered with felty areoles; on different plants they range from spineless to very long-spiny. The juicy pulp of ripe fruits ranges from white to red and contains large, hard seeds—very different from the tiny seeds of the saguaro.

RANGE

The cardón occurs in most of Baja California, on the coast of Sonora, and on the islands in the Gulf of California. Its northern limit is determined by frost, to which it is very intolerant. The northernmost plants in Sonora are near Caborca.

NOTES

Though the flowers may open before dark and remain open past dawn, nectar-feeding bats perform most of the pollination of this species. Cardón was recently found to have a *trioecious* breeding system; that is, plants have 1 of 3 variations of gender. Some have only male flowers that produce pollen but no seeds, some have only female flowers, and some have perfect flowers (each flower produces pollen and has fertile ovaries). There are also a few plants that have neuter flowers, producing neither pollen nor seeds.

The Seri eat the pulp and seeds of cardón, which is more abundant in their land than saguaro. Historically, house walls and walking sticks were made out of the ribs.

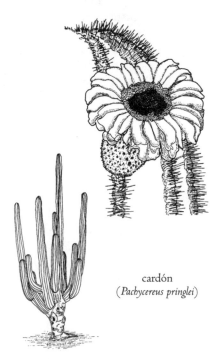

cardón
(*Pachycereus pringlei*)

Genus *Peniocereus*

➤ *Peniocereus greggii*

English names: desert night-blooming
cereus, Arizona queen-of-the-night
Spanish names: saramatraca,
reina-de-la-noche

DESCRIPTION

The stems of this cactus are usually few
in number, thin and barely succulent, to
about 3 feet (1 m) tall. The plant grows
from a large, starchy underground tuber
that can occasionally weigh more than
40 pounds (18 kg). The nocturnal
white flowers are about 3 inches (8 cm)
across with very long floral tubes, and
are strongly scented. (Some people can't
smell them.) They close soon after the
morning sun reaches them.

RANGE

This cactus can be found in Southern
Arizona to southern Texas and adja-
cent northern Mexico, as well as in
Baja California.

NOTES

The plants grow under desert ironwood,
creosote bush, and other desert shrubs;
the stems are extremely difficult to distinguish from those of the shrubs that
provide shade, physical support, and
concealment. When the stems are eaten
by herbivores such as white-throated
woodrats (packrats) and cactus borers
(*Cactobrosis fernaldialis*), new stems soon
sprout from the tuberous root.

This is one of the Sonoran
Desert's most famous yet least encountered plants. It is virtually invisible
most of the year, but on a few nights it
becomes stunningly conspicuous. Some

experienced field botanists spent
months scouring a small tract of land
to locate and tag all of the plants for
an ecological study. Then on a night
when the plants flowered they found
enough new plants to double their
known population size.

Plants in each population bloom
in synchrony; large ones can produce
a score of flowers at once. The legend

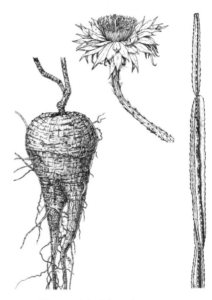

Desert night-blooming cereus
(*Peniocereus greggii*)

is that they all bloom on a single night
per year. Reality is almost as intriguing:
each plant produces only 3 to 5 flushes
of flowers between late May and early
July. During each flush most of the
flowers open on one night, with a few
stragglers the night before or after the
big bang.

In ideal cultivated situations where
the plants are protected from predators, these cacti can grow hundreds of
times larger than they do in nature.

Archaeologist Julian Hayden had a plant in his Tucson yard that was over 8 feet (2.4 m) tall and perhaps twice as wide. Its great tangle of stems produced 200 flowers on one night and another 100 on the following night.

Desert night-blooming cereus plants usually occur as widely-separated individuals, and the flowers are not self-fertile. The flowers are cross-pollinated by hawk moths (*Sphingidae*) which fly hundreds of yards between plants in their search for the nectar reward. The cactus fruit turns red when ripe, attracting birds that eat the pulp and disperse the seeds in their droppings. The root is used medicinally to treat diabetes and other maladies.

This cactus is a desert version of the canary in the coal mine, an early warning that something is wrong in the ecosystem. Where pesticides are heavily used in agricultural areas adjacent to natural habitat, the hawkmoth populations are devastated and most of the flowers fail to fruit. This is an example of *chemical habitat fragmentation*; the habitat appears to be intact, but some of its ecological processes have been destroyed or degraded.

Genus *Stenocereus*

➤ *Stenocereus eruca*

English name: creeping devil
Spanish name: chirinola, casa de ratas
(rat house)

DESCRIPTION
This bizarre cactus migrates across the desert during its lifetime. The living sections of the very spiny stems are usually 5 to 10 feet (1.5-3 m) long; they lie prostrate with only the terminal few inches raised above the surface. Stems root near the growing tips and older stem portions die and disintegrate, so the plants literally creep across the landscape over time. In some areas they occur as widely scattered individual stems; in favorable localities they form impenetrable patches of branching stems several yards (meters) across. Large, nocturnal white flowers are produced sparingly, probably in response to rain.

RANGE
The creeping devil is found only on sandy soils on the central Pacific coast of Baja California Sur. (See Plate 22.)

NOTES
In addition to its unusual growth habit, creeping devil is elusive. To find some populations one must venture onto 4-wheel-drive trails that shift locations with the dunes; the maps always seem to be wrong. On a recent expedition we headed westward from a tiny community that our map indicated was 5 miles from the coast. An hour later our odometer and compass said we should have driven into the Pacific 5 miles before, but the ocean was beyond several more ridges of dunes. But we did find the creeping devils. Hundreds of them were scattered in the clearings, aimed in all directions. They looked so alien that we joked about whether it would be safe to camp among them. In Baja California where there are so many strange-looking plants, such thoughts are mundane.

Creeping devil grows rapidly in

cool maritime climates like its home. In the mild climate at the Huntington Botanic Gardens, stems grow in excess of 2 feet (50 cm) a year; at the hot, arid Desert Museum, the plant grows only about 2 feet a decade.

> *Stenocereus gummosus*

English name: none
Spanish name: pitahaya agria
(sour pitahaya)

DESCRIPTION
This is a sprawling, shrubby plant with many sparsely-branched stems from the ground to about 10 feet (3 m) or more tall. Taller stems bend from their weight and lean on one another; those that touch the ground take root and branch, eventually forming large thickets. Big white nocturnal flowers are followed by fruits the size of small oranges with bright red skin and pulp. Flowering is mainly in summer, but also at other times in response to rain.

RANGE
This cactus is widespread in most of Baja California except in the high mountains and the arid northeast coast. Disjunct populations occur on the midriff islands and coastal Sonora opposite Tiburón Island.

NOTES
The fruit is widely regarded as the tastiest in the whole cactus family; it is sweet yet slightly tart, with a crisp texture. The fruits are produced over a long season, but in such small numbers that commercial harvest is not viable. Plants are long-lived and extremely slow-growing. Most specimens identi-

fied in 1905 photos were still present when resurveyed in 1996; many had not increased appreciably in size. Because of these traits pitahaya agria will always be a rare, mouth-watering treat to be sought on treks to Baja California.

Genus *Echinocereus*

English name: hedgehog cactus,
strawberry cactus (many species)
Spanish names: See individual species

DESCRIPTION
The Sonoran Desert species of this genus have upright stems, usually not more than a foot (30 cm) tall, and typically in clusters. A few of the 44 currently recognized species in the genus have single stems. The ribbed stems help distinguish hedgehogs from the mammillarias, with which they are sometimes confused. Spines are straight or slightly curved, never hooked as in many mammillarias. The flowers are funnel-shaped or partly funnel-shaped, but abruptly flaring outward about 2½ inches (6 cm) across in most Sonoran Desert species. Some are long, narrow funnels conducive to hummingbird or moth pollination. Flowers come in many colors, including white, yellow, pink, purple, and red; some are bicolored. The stigmas are usually green. Some species of *Echinocereus* are readily distinguished from other cacti by the flower buds that rupture through the skin of the stem, rather than emerging from the areoles. Nearly all species flower in the spring, though a Baja California species flowers in August or September. One species

within the Sonoran Desert and a few beyond it have thin, trailing stems.

RANGE
Species occur in nearly all habitats in the arid southwestern United States and southward well into Mexico.

NOTES
The fruits of several species are large, juicy, and tasty. They were eaten by indigenous peoples when they could get them before the birds and squirrels did. The genus is popular among cactus collectors because most species are of moderate size and have beautiful flowers.

➤ *Echinocereus brandegeei*

English names: none known
Spanish names: pitayita (little pitaya), casa de rata (rat house)

DESCRIPTION
This is a variable cactus. The stems grow in clusters, which vary from a few stems to great mounds more than 6 feet (2 m) across. Individual stems on different plants may be short and erect or long and trailing. The spines are also extremely variable in size and color, with color ranging from brownish to clear white or bright yellow. The flower tubes flare abruptly outward to about 3 inches (8 cm) across and are purplish lavender to pale pink. This species is unusual in the genus—it flowers in late summer instead of spring.

RANGE
This cactus is widespread in the southern half of Baja California, where it typically grows on rocky hillsides.

NOTES
In some parts of its range *E. brandegeei* grows among devil's club cholla (*Opuntia invicta*). In these locations plants of the former species tend to have short, stout stems and very broad spines; they look remarkably like the cholla. The adaptive value of this convergence, if any, is unknown.

➤ *Echinocereus engelmannii*

English names: strawberry hedgehog, calico cactus
Spanish names: sinita barbona (bearded little sina), cacto fresa (strawberry cactus), pitayita

DESCRIPTION
This hedgehog grows in clusters of up to 60 cylindrical stems, each up to 1 foot (30 cm) tall (rarely 3 times as tall). Sprouting from the closely-spaced areoles are 2 to 6 (usually 4) long central spines and 6 to 14 radial spines. Brilliant, deep purplish-red to lavender flowers open in April. The fruit has a juicy, tasty, red pulp.

RANGE
This plant grows throughout much of the Sonoran and Mohave deserts and southern Great Basin Desert, mostly in rocky or gravelly soils.

NOTES
This species has several varieties, most of which are difficult to distinguish from *E. fasciculatus*. The latter species typically has 1 long central spine and 1 to 3 shorter central spines. There are two Chihuahuan Desert species with the same vernacular name, *E. enneacanthus* and *E. dasyacanthus*. (See also the next species.)

➤ *Echinocereus fasciculatus*

English names: strawberry hedgehog, robust hedgehog
Spanish name: pitahayita

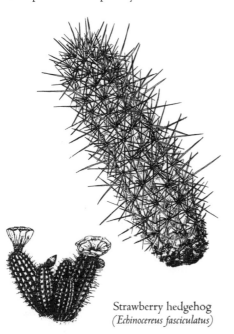

Strawberry hedgehog
(*Echinocereus fasciculatus*)

DESCRIPTION
This plant looks very similar to *E. engelmannii* described above (they may be a single species). On close inspection one can see that the areoles each have only 1 principal central spine and about 12 radial spines. The brilliant magenta to reddish-purple flowers bloom in the spring, usually in the first half of April in Tucson. The fruit—pale red outside when ripe—has bright red, fleshy pulp.

RANGE
This species is found in desert habitats from central Arizona to northern Sonora.

NOTES
Some taxonomists favor elevating the varieties of this species—*E. fasciculatus* var. *fasciculatus, E. fasciculatus* var. *boyce-thompsonii* and *E. fasciculatus* var. *bonkerae*—to full specific rank. *Echinocereus fasciculatus* var. *boyce-thompsonii* (Boyce Thompson hedgehog) differs from the typical variety in having much longer, downward-pointing spines. The flowers are usually very deep purple. *Echinocereus fasciculatus* var. *bonkerae* (Bonker hedgehog) has fewer (5 to 15) stems and very short white or pale gray spines. Its flowers are also deep purple. Most of the plants in this group have 4 sets of chromosomes, but at least one of the varieties includes individuals with 2 sets.

➤ *Echinocereus nicholii*

English name: Golden hedgehog
Spanish name: none known

DESCRIPTION
This is a large hedgehog with clusters of as many as 30 stems up to 2 feet (60 cm) tall, thickly clothed with long yellow spines. The light- to medium-pink flowers bloom in April. It was formerly classified as a variety of *E. engelmanii.*

RANGE
The golden hedgehog grows in south-central Arizona, mostly from Tucson to Organ Pipe Cactus National Monument, and also in adjacent Sonora.

NOTES
When backlit by the sun, the spines of this cactus create a dazzling golden-yellow halo.

Genera *Ferocactus* and *Echinocactus*

English name: Most species are called barrel cactus

Spanish name: biznaga (a general name for barrel-shaped cacti)

DESCRIPTION

The following is a general description for both genera of barrel cacti: The stems are globular to columnar, usually unbranched, and pleated, ranging from less than a foot (30 cm) tall in the smallest species to 6 to 12 feet (1.8-3.7 m). The central spines are the larger of two types and arise from the center of the areole; the principle central spine often has a different shape from that of all the other spines. Radial spines are smaller and arise from the margins of the areoles. The short, funnel-shaped flowers are very stiff and usually don't extend beyond the spiny armor. The thick rind of the fruit is moist in *Ferocactus*, dry at maturity in most *Echinocactus*. The seeds are packed in a dry interior, not embedded in pulp. *Echinocactus* differs from *Ferocactus* in having sharp-pointed scales on the flower tubes and woolly ovaries.

RANGE

There are 25 species of *Ferocactus* and 6 of *Echinocactus* in the world. In the Sonoran Desert there are 9 and 2 species, respectively.

NOTES

The majority of *Ferocactus* species flower in summer and are pollinated by bees in the genus *Lithurge*. The fruit and seeds are eaten by rodents, birds, mule deer, bighorn sheep, and javelina. The plant itself is eaten by cactus beetles (*Moneilema gigas* and other species), jackrabbits, packrats, and javelina.

The surface area of barrels and other more or less globose plants is small compared to the volume, so evaporative losses are small relative to the large volume of water stored. Repeat photography of sites in Baja California in 1905 and the 1990s indicate that the life spans of barrel cacti are typically less than a century.

One of the great fables of desert survival is that barrels and other cacti are reservoirs of water that can be easily tapped and drunk. It is true that indigenous peoples and a few other desert residents know how to obtain emergency water from cacti. But most city dwellers, including most aspiring survivalists, could not get water from any cactus if their lives depended on it (pun intended).

The first problem is getting to the pulp inside the very tough and spiny epidermis. A pocket knife is inadequate, and tools that are typically carried in a car, such as tire irons, aren't very effective either. The labor of cutting into a barrel on a hot day is likely to cause loss of more water from sweating (and perhaps bleeding) than one would gain from the cactus.

Secondly, the water in cactus pulp is tightly bound in a gooey mucilage. Most of the year the pulp is more like a damp sponge than a watermelon—you can't squeeze much liquid out of it. Furthermore, the raw pulp of many cacti is inedible. Some species have potentially toxic levels of oxalic acid (prickly pears), bitter and sometimes toxic alkaloids (senita and many other cacti), or other substances that cause diarrhea (some barrels) or vomiting. The best cactus for emergency water is *Ferocactus wislizeni*. See more details under that species. For a historical account of extracting water from a barrel cactus (in cool weather and when it was legal), see *Camp-Fires on Desert and Lava* by William T. Hornaday, pages 216-219.

Water can be obtained from cacti using a machete and solar still. But anyone with the foresight to pack these tools is smart enough to carry plenty of water and inform friends of the itinerary and expected return date!

Caution: native wild cacti are protected by state laws; it is illegal to take or destroy them. In addition, most barrels cannot branch, so cutting out the top is lethal.

➤ *Echinocactus polycephalus*

English names: many-headed barrel, cottontop cactus
Spanish name: biznaga de chilitos (little chile barrel)

DESCRIPTION

This *Echinocactus* is unique as the only barrel in our region that branches under normal conditions. The 8-inch (20 cm) diameter heads occur in clusters of up to 200, forming mounds to 3 feet (1 m) across and somewhat less high. The dense, stout spines obscure the plant bodies and restrict the small yellow flowers from opening fully; they appear in July. The brown spines appear bright red when wet from rain.

RANGE

This cactus occurs mostly on rocky and gravelly slopes in the driest parts of the Sonoran and Mohave deserts. It is rarely found where rainfall exceeds 5 inches (130 mm). (*E. polycephalus* var. *xeranthemoides*, occurs in the northern Mohave and Great Basin deserts, where there is more rain.)

NOTES

Many-headed barrel cacti are slow-growing and probably very long-lived. Plants grown from seed at the Desert Museum are just beginning to branch at nearly 20 years of age. This species is also geographically stable; its range has not changed for at least the past 30,000 years despite the dramatic climatic swing from ice age to a warm interglacial period. Though the fruits seem to be imprisoned within the spiny armor, birds and packrats can get to them and disperse the seeds.

Turk's head or eagle's claw cactus (*E. horizonthalonius* var. *nicholii*) is in the same genus but has a different growth habit; it nearly always has a single stem. One variety of Turk's head is common

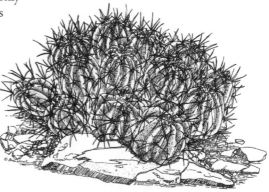

Many-headed barrel
(*Echinocactus polycephalus*)

and widespread in the Chihuahuan Desert. Our variety is endangered and restricted to 3 small populations in Arizona and Sonora on limestone. The flowers are pollinated by bees. The fruits remain buried among the spines; they eventually disintegrate and the seeds simply fall to the base of the mother plant. Bighorn sheep and javelina eat the whole plants and probably function as occasional long-distance seed dispersers. Javelina only recently migrated into the Sonoran Desert, and their added predation may be exterminating this cactus.

> *Ferocactus cylindraceus*
> *[F. acanthodes]*

English names: spiny barrel, California fire barrel, compass barrel
Spanish name: biznaga

DESCRIPTION
This barrel cactus is narrow-columnar rather than barrel-shaped. It grows up to 10 feet (3 m) tall, but is usually less than half that. Stems are nearly always single; like most other barrels this cactus branches only when the tip is damaged. The spines are usually reddish (bright red when wet), but yellow or brownish in some forms. The central spines are flat and curved at the tips, but not hooked. The many bristly radial spines obscure the body of the plant. Crowns of flowers are crowded among the dense spines at the stem tips and are almost always yellow, though some plants in the Pinacate region have orange to red flowers. Flowering is usually in March and April; some populations near Phoenix flower in summer.

RANGE
The spiny barrel occurs in northern Baja California and Sonora, southern California, and Arizona. It usually grows on steep rocky slopes, rarely occurring side-by-side with the fish-hook barrel, which generally is found on gravelly bajada soils.

NOTES
Because of this barrel's cylindrical shape, it could be confused with a young saguaro if one fails to notice the hooked spines. The Seri name for this species means "thinks it's a saguaro." In Baja California *F. cylindraceus* can be confused with *F. gracilis*; the latter species has red flowers while those of the former are almost always yellow.

> *Ferocactus emoryi*
> *[F. covillei]*

English name: Coville barrel, Emory's barrel
Spanish name: biznaga

DESCRIPTION
The description that follows applies to the desert populations of this barrel. Coville barrel is easily recognized by

Coville barrel
(*Ferocactus emoryi*)

the relatively few spines per areole, all of which are stout (not bristly). The main central spine in each areole, usually red, is strongly flattened and hooked. Young plants are first globular, then barrel-shaped, but more slender than fishhook barrels. They are typically 1 to 4 feet (30-120 cm) tall. The bright red flowers open in July to mid-August before *F. wislizeni* blooms. Juvenile plants are very distinctive in bearing their spines at the ends of large tubercles rather than along ribs; they look like giant mammillarias.

RANGE
This desert form occurs from south-central Arizona to central Sonora.

NOTES
The Seri name for this species translates "barrel that kills." Eating its flesh or juice causes nausea, diarrhea, and temporary paralysis. The Seri use the pulp as a pain-relieving poultice, further evidence that there are some active chemicals in this plant. However, the flowers and fruit are edible (no cactus fruit is poisonous, though some are inedible).

The desert populations were originally described as *F. covillei*. The populations in the thornscrub of southern Sonora (the original *F. emoryi*) are massive barrel-shaped plants up to 9 feet (2.7 m) tall and almost 3 feet (0.9 m) in diameter, with shorter curved but not hooked yellow spines and yellow flowers. Between Hermosillo and Guaymas, Sonora, the 2 forms intergrade, which is why they were combined into a single species. It's exciting to see the variability in such intergrading populations. We found one plant

that had both red and yellow flowers; they opened yellow and changed color with age.

➤ *Ferocactus emoryi* var. *rectispinus*

English name: straight-spined barrel
Spanish name: biznaga

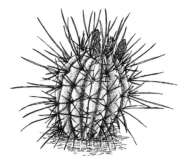

DESCRIPTION
Plants are barrel-shaped when young and rarely exceed 3 feet (1 m) tall. They become more cylindrical in age and occasionally reach 9 feet (2.7 m). This barrel cactus is easily recognized by its 6- to 11-inch (15-28 cm) long, straight central spines. When backlit by the sun on a rocky slope, it is surrounded by a glowing reddish aura that is much larger than the plant's body. It has a few stout radials, but lacks bristly spines. The yellow flowers bloom in August and September.

RANGE
This is a rare cactus that occurs in its pure form only near the gulf coast of central Baja California. A few miles inland it hybridizes freely with *F. peninsulae*, forming hybrid swarms with variable spine lengths and curvature, and red-tinted flowers.

Straight-spined barrel is one of the
many outstanding botanical marvels
of Baja California. Many cactophiles
have ventured off the main highway
to seek it and pay homage to it.
Because populations are very sparse
and the terrain extremely rugged, it is
a difficult quest, and finding even a
single specimen is a thrill.

➤ *Ferocactus wislizeni*

> English names: fishhook barrel,
>   Arizona barrel, compass barrel
> Spanish name: biznaga de agua
>   (water barrel)

DESCRIPTION

The thick, barrel-shaped body of this
cactus is usually 2 to 4 feet (0.6-1.2 m)
tall, occasionally reaching over 10 feet
(3 m). The ribs bear broad, flat, strong-
ly-hooked central spines as well as several bristly radial spines, but not enough
of these to obscure the stem. The flowers are not strongly crowded by the
spines and open wide. Flower color is
extremely variable; on most plants they
are some shade of orange, often with
a stripe of darker shade on each petal.
About 10percent of the plants have yellow or red flowers. They are produced
over a 2-month-long season, August
and September, much longer than the
blooming period of other barrels.

RANGE

Populations of this barrel are concentrated in south-central Arizona and
adjacent northern Sonora. There are
also some populations in southern
New Mexico and western Texas. The
distinct variety on Tiburón Island may
well be a distinct species. The fishhook
barrel is most abundant on gravelly
bajadas; it is less common on rocky
slopes or silty valley floors.

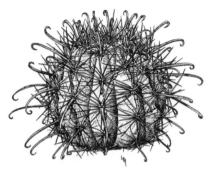

Fishhook barrel
(Ferocactus wislizeni)

NOTES

This is the barrel that the Seri and others prefer to use for emergency water;
Seris have survived on it for as long as a
month. The juice is flat-tasting but not
completely harmless. Drinking it on an
empty stomach often causes diarrhea,
and some Seri experience pain in their
bones if they walk a long distance after
drinking the juice. The seeds are not
enclosed in a juicy pulp, but are edible,
as is the sour rind. The O'odham eat
the persistent fruit much of the year as
an emergency food. The Seri also eat
the flowers and buds.

Cactus bees (*Lithurge* spp.) pollinate
the flowers. The fruit persists until it
is removed by animals and may remain
on the plant for more than a year.
Birds, squirrels, and large mammals
such as deer and javelina are the main
consumers of the fruit. Plants grow
fairly slowly, and large specimens are
more than a century old. In cultivation,
with supplemental water and fertilizer,

they reach flowering size of 10 inches (25 cm) diameter and the same in height in about 12 years.

Ferocactus wislizeni and *F. cylindraceus* have the vernacular name "compass barrel" because large plants of these species tend to lean toward the southwest. It has been hypothesized that the intense heat from the afternoon sun retards growth on the southwest side of the plants, and the greater growth on the cooler side pushes the plants over. Large leaning plants may topple when the soil is softened by heavy rains.

Ferocactus herrerae (twisted barrel) is included in *F. wislizeni* by some botanists. As the common name suggests, its ribs are often spiraled. It is found in southern Sonora and into Sinaloa. Its flower color (always yellow with brown tips) and its distribution (there is a 100 kilometer gap in Sonora where neither barrel occurs) give rise to some doubt that it belongs to the same species as fishhook barrel.

Genus *Mammillaria*

English names: pincushion cactus, mammillaria, nipple cactus

Small, attractive, diverse, and generally easy to cultivate, mammillarias enjoy enormous popularity with collectors. Perhaps because of this, many minor geographical variants have been described as species—resulting in more than 400 named species. Thorough field work and genetic analysis often reveal that populations described as separate species actually belong to a single, variable species.

Description of Genus

Even after duplicate names are sorted out, *Mammillaria* is the largest of several genera of diminutive cacti with about 175 species worldwide. The Sonoran Desert species are less than 6 inches (15 cm) tall, with closely-spaced areoles bearing many spines that obscure the body of the plant. The areoles are at the tips of long tubercles that are arranged in 2 spirals (one clockwise, the other counterclockwise). These tuberculate stems contrast with the ribbed stems of the genus *Echinocereus*. In addition, many mammillarias have hooked central spines, whereas no hedgehog does. Plants range from single stems to large clusters, depending on species. Flowers, usually less than an inch (25 mm) in diameter and either pink or white, are produced in a ring near the tip of the stem.

Range

Various species grow from the lowest desert habitats to cold conifer forests and fairly wet tropical forests. The genus ranges from southern California to central Texas and southward through Central America. About 25 described species occur in the Sonoran Desert; some of these may be combined in the future.

Notes

The fruits of mammillarias are edible, though their small size makes them difficult to gather in nutritionally significant quantities. Some are pleasantly tart, others bland. The O'odham call them "coyote's paws" to contrast them with the larger, juicy fruits of hedgehogs. (See *Cucurbita digitata*, coyote melon, for an explanation of "coyote" plants.)

Mammillarias often go through "boom-and-bust" population cycles. *Mammillaria thornberi*, for example, is so rare that it was considered for listing under the Endangered Species Act in the 1980s. But in the 1930s it occurred in countless millions in the Avra Valley west of Tucson. This vast population all but disappeared after the catastrophic freeze of 1937. The population has never recovered; no one knows why.

Mammillarias can be confused with plants of several other genera of small cacti such as *Coryphantha, Epithelantha,* and *Neolloydia.* Consult a field guide or taxonomic monograph to sort them out.

➤ *Mammillaria grahamii* [*M. microcarpa*]

English names: fishhook pincushion, Arizona fishhook
Spanish name: cabeza de viejo (old man's head)

DESCRIPTION
Single or clustered cylindrical stems grow to about 6 inches (15 cm) tall. The plant body is nearly concealed by dense, white, straight radial spines, in contrast to 1 to 3 central spines per areole, one of which is dark-colored and hooked. Crowns of bright pink flowers almost an inch (25 mm) across are borne just below the stem tips in early summer. The bright red, slightly fleshy fruits are presumably eaten by birds.

RANGE
Fishhook pincushion occurs in Arizona south of the Mogollon Rim south into Sonora in desert and woodland habitats.

NOTES
This is the most common mammillaria in Arizona Upland. It is especially abundant among debris under jumping cholla and partially concealed beneath desert shrubs such as bursage (*Ambrosia deltoidea*). Like many small desert plants, it cannot tolerate full sun.

The flowers can be used as a monsoon season rain gauge. The buds are produced during the preceding summer growing season and usually remain dormant through the winter, spring, and foresummer. The buds burst into bloom 5 days after the first rain of the summer and last about a week. Plants produce a second and sometimes a third flush of flowers after subsequent rains.

Fishhook pincushion
(*Mammillaria grahamii*)

Genus *Opuntia* (incl. *Cylindropuntia, Grusonia,* and *Corynopuntia*)

English names: prickly pear, cholla, opuntia, choya, cane cactus
Spanish names: nopal, tuna, cholla

The genus *Opuntia* is quite large, yet it is still replete with hidden diversity. The more closely botanists study these plants the more species they discover. People who live with and use these plants recognize even more differences between them than do botanists. Juanita Ahil was a Tohono O'odham who lived in the desert near Sells, Arizona. Ecologist Tony Burgess could recognize two species of opuntia growing in Juanita's yard. Juanita was able to distinguish five different kinds from the appearance of the prickly pear pads alone. Later, the characteristics of the fruits these plants produced confirmed that she was correct; the fruits differed accordingly in color, taste, and keeping qualities.

DESCRIPTION

Opuntia has been the main genus in the Opuntioid subfamily of Cactaceae. (A new classification system for this group is underway; see below. Until this revision, all of the other cacti in the Sonoran Desert region are in the Cereoid subfamily.) The Opuntioid are distinguished from other cacti by four characteristics. First, the stems grow in distinctly jointed segments. The elongation of joints is permanently terminated by the onset of the dry season; subsequent growth of the plant occurs by the initiation of new joints by branching from the areoles. (Other cacti have *indeterminate* growth. A saguaro stem, for example, grows ever longer each growing season until the plant dies or the stem

tip is damaged.) Second, whether or not they have regular spines, Opuntioid areoles bear *glochids* (usually small to minute, barbed spines that are very sharp and brittle, and very easily detached). Third, rudimentary leaves are present on new joints. Fourth, the seeds have a pale covering called an *aril;* most other cacti have shiny black seeds. The largest genus, *Opuntia,* has at least 300 species of shrubby or arborescent plants worldwide. The vernacular names are based on growth form. The chollas (most of our species are in the subgenus Cylindropuntia) have cylindrical stem segments, while those of prickly pears (subgenus Platyopuntia) are flattened.

RANGE

The genus ranges from southern Canada to southern South America, in habitats ranging from arid desert to tropical semiarid woodlands and high mountains.

NOTES

In general, chollas are more drought tolerant than prickly pears and extend into drier deserts. Less arid habitats tend to have more prickly pear species. All desert species are pollinated by a few species of bees that specialize in cacti; they collect the pollen to feed developing larvae. Some opuntia flowers emit a fragrance of damp earth; perhaps the smell resembles "home" to the ground-nesting bees. Mature fruit ranges from tan or green and dry to bright red or purple with juicy pulp; all contain large, very hard seeds. Many birds, mammals, and insects feed on the fleshy fruits; so may desert tortoises and spiny iguanas. Though all species flower, some rarely, if ever, produce viable seeds; these species reproduce almost entirely by vegetative means.

Both prickly pears and chollas provide shelter for numerous animals. White-throated wood rats (packrats) build nests of sticks and other debris within the shrubby prickly pears. (The shrubby ones are those with thickets of pads on the ground, as opposed to the trunked species, such as santa-rita prickly pear). Packrats cover their houses with a layer of cholla joints, if available. In more exposed locations the nest may consist almost entirely of piled-up cholla joints. These nests provide good protection from coyotes, but not from snakes, which can enter the nests along the trails used by packrats. (Rattlesnakes commonly take up residence in packrat nests, usually after eating or evicting the packrat.) Cactus Wrens prefer to build their nests among the stems of chollas. Curve-billed Thrashers, Mourning Doves, Roadrunners, and other birds also commonly use both chollas and prickly pears as fairly secure nesting and roosting sites.

The flesh of prickly pears and some chollas is eaten by jackrabbits, packrats, and javelina, all of which can deal with concentrations of oxalic acid that can be toxic to many animals and can clog the kidneys of species with highly concentrated urine. Several insects have also coevolved with cacti as their sole food source. The most conspicuous are the giant cactus beetle (*Moneilema gigas*), cactus weevils (*Metamasmius* spp.), cochineal (*Dactylopius* spp.), and a moth (*Copidryas cosyra*) with very colorful but rarely-seen larvae that feed on fresh cholla stems. Read about the blue cactus borer (*Cactobrosis*) in the saguaro section.

Opuntias are extensively used for food and other purposes by humans. The fleshy fruit (called *tuna* in Spanish) of some species (*O. engelmannii* in Arizona Upland) is edible and tasty. It can be eaten fresh, if care is taken to avoid the glochids on the rind. More often the brilliant red-purple and distinctly-flavored juice is expressed to make drinks, syrup, and jelly. Some prickly pear species are commercially cultivated for fruit production; numerous superior cultivated varieties have been selected.

The formidable flower buds of some chollas are eaten by O'odham and other desert dwelling peoples. The buds are rolled on the ground or another hard surface with sticks to remove the spines and glochids. The buds are pit-roasted for a day, and either eaten immediately or dried and pickled for later consumption. Cholla buds contain significant protein, but they are probably more important for their high calcium content and soluble fiber. The O'odham harvested and ate them early in the dry season after the last year's crops had been consumed, and before saguaro fruits ripened. Cholla buds are still part of the traditional O'odham diet. The Hohokam, predecessors to the O'odham, also ate cholla buds, and there is evidence that they cultivated chollas around their homes.

Millions of people cook and eat the tender young pads of several species of prickly pear. Besides being more tender, immature pads have less oxalic acid, which could be toxic in large amounts. *Nopales* (the edible species of prickly pear and the harvested whole pads of the same) are very nutritious. *Nopalitos* (small pads that are cut into bite-size pieces) are mucilaginous like okra, and good for thickening broths. The mucilage also helps control blood-sugar levels associated with adult-onset diabetes. Diabetes is a common affliction among native Americans who adopt Western high-fat, low-fiber diets. There is also clinical evidence that nopales reduce blood cholesterol. Widely ignored by Anglos, who often regard them as worthless nuisances, opuntias are abundant and healthy foods for those who know how to use them.

Prickly pears are a historically important reason that the Spaniards continued their conquest of the New World. They quickly looted the precious metals they were after, but they also discovered *cochineal*. Cochineal is a scale insect that feeds on prickly pears. Its body fluids contain a bright crimson, foul-tasting substance that protects it from predators. Ground up cochineal insects were used by native peoples to dye their textiles rich red or purple, depending on the processing. In Europe this color of dye was so rare that only royalty could afford it. In some kingdoms the colors "royal purple" (derived from a sea cucumber) and, after discovery of the New World, royal crimson from cochineal, were reserved for the king by law. The cultivation and export of cochineal dye became a major economic activity, and its source was kept secret for many years. The commercial cochineal was harvested and later cultivated from prickly pears in southern Mexico. Our Sonoran Desert species contain the same dye.

When competitors finally discovered the source of the coveted dye, they attempted to establish populations of prickly pears and cochineal in other countries such as India, Ceylon, South Africa, and Australia. In many cases the cochineal died out, but the prickly pears escaped into the wild and became serious pests.

The cochineal industry thrived until the late 1800s, when cheaper aniline dyes became available. However, there is still significant commercial cochineal production in several countries, including Mexico. Cochineal is one of a very few red dyes approved by the U.S. Food and Drug Administration. (Earlier red food colorings were found to be carcinogens and were banned.) Today red candies, beverages, and lipstick are often colored with bug blood raised on prickly pears. Despite its evolutionary origin as an anti-predator toxin, the quantities used to color foods are nontoxic to humans, except for occasional allergic reactions (which can be caused by almost any substance).

The juice expressed from prickly pears has been used for centuries to strengthen adobe mortar. It was recently used in the restoration of the San Xavier Mission in Tucson. It is being tested for similar applications, such as stabilizing dirt footpaths and erosion-prone slopes.

Recent genetic studies reveal that the different subgenera of this genus may be distinct enough to warrant their elevation to full genera. In our region there would be 3 new ones: *Opuntia* would include all Sonoran region prickly pears.

Most chollas will be in *Cylindropuntia*, and the few club chollas will be placed in *Grusonia* or *Corynopuntia*.

Numerous species of cholla and some prickly pears hybridize with one another. Hybrid populations are fairly common. Some of these hybrids may reproduce sexually; others are sexually sterile but can reproduce vegetatively. There are several such "clonal microspecies" in the Tucson area alone, some of which are restricted to a patch of just a few acres (hectares). Most of the descriptions of these species are published only in scientific monographs and have not yet appeared in general plant lists and keys. So if you are frustrated with being unable to identify a cholla or prickly pear from a field guide, be assured that you're not alone. No field guide can cover all the possible hybrids and "microspecies." (See also the note under *Opuntia acanthocarpa*.)

Opuntias have captured the imagination of botanists as reflected in some of their names. My two favorites are from Baja California. *Opuntia (Grusonia) invicta* roughly translates "invincible points"; its English vernacular name is devil's club cholla. This cactus forms low mounds of ovoid joints covered with broad, 3-inch (7.5 cm) long spines like little daggers. Even better is *Opuntia (Cylindropuntia) molesta*—no translation needed. It's a shrub sometimes over 8 feet (2.4 m) tall, and its spines are 2 inches (5 cm) long and incredibly sharp. I have mused that the botanist who named it might have discovered it in the same manner as I did. There were hundreds of 2 foot (60 cm) tall plants on a gravelly wash bed. It was a wet spring and the grass was 3 feet tall. . . .

➤ *Opuntia [Cylindropuntia] acanthocarpa*

English name: buckhorn cholla
Spanish names: cholla, tasajo

➤ *Opuntia [Cylindropuntia] versicolor*

English name: staghorn cholla
Spanish names: tasajo, cholla morada (purple cholla)

Buckhorn cholla
(*O. acanthocarpa*)

Staghorn cholla
(*O. versicolor*)

DESCRIPTION
These 2 species are rather difficult to differentiate where they occur together. Both have many thin branches arising from the ground or a short trunk. Buckhorn cholla is usually about 3 feet (0.9 m) tall, but can reach 13 feet (4 m). Staghorn cholla is usually 3 to 8 feet (0.9-2.4 m) tall, but occasionally reaches 15 feet (4.6 m). The stems of both species may be tinged with red or purple, especially in winter; staghorn cholla more often exhibits this trait. Both species have variable flower colors, ranging from red, yellow, orange, pink, purple, and greenish or brownish, often

all in the same local population. The anther-bearing filaments of buckhorn are dark red; staghorn filaments are yellowish green. The flowers of buckhorn cholla tend to be larger than those of staghorn; both bloom in spring.

The fruits provide the best way to distinguish the 2 species when they are not in bloom. At maturity the fruit of buckhorn cholla is dry, deeply tuberculate, and covered with long, barbed spines. Fruits fall from the plant after several months. Staghorn cholla fruit is fleshy at maturity, turgid, usually spineless, and persists on the plant for more than a year. Thus staghorn cholla is always in fruit when mature.

RANGE
Buckhorn cholla is widespread in the northern Sonoran and Mohave deserts to about 4000 feet (1220 m). Staghorn cholla is restricted to Pinal, Santa Cruz, and eastern Pima counties in Arizona and northern Sonora, Mexico, at 2000 to 3000 feet (600-900 m) elevation.

NOTES
Because they are less spiny than O. bigelovii and O. fulgida, which usually grow with them, buckhorn and staghorn chollas are less favored nesting sites for birds. Buckhorn, staghorn, and pencil chollas are the preferred sources of cholla buds eaten by the Tohono O'odham and other Native Americans.

Besides the variability of flower color, buckhorn cholla is extremely variable in vegetative characters, such as plant height and branching angle, color, number, and size of spines and size and shape of tubercles. Taxonomists have distinguished 4 varieties, and almost every mountain range and valley has a different-looking population. Perhaps they and many other opuntias are in the process of evolving into new species.

> *Opuntia arbuscula*
[Cylindropuntia arbuscula]

English name: pencil cholla
Spanish names: tasajo, cholla, chumbera, chollita

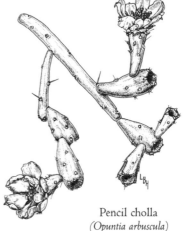

Pencil cholla
(*Opuntia arbuscula*)

DESCRIPTION
This cholla, densely-branched, but often with a distinct trunk, grows up to 9 feet (3 m) in height, but is usually less tall. The slender joints are 2 to 4 inches (5-10 cm) long and ½ inch (15 mm) in diameter. Usually joints are sparsely-spined, with each areole bearing 1 long yellow central spine and 1 to 3 short radial spines. Green, yellow, or brownish-red flowers in spring are followed by fleshy, greenish-purple fruits that remain on the plant for at least a year.

RANGE
This cactus occurs from Central Arizona to northern Sonora, at 1000 to 3000 feet (300-900 m) elevation.

➤ *Opuntia basilaris*

English name: beavertail cactus
Spanish name: nopal

DESCRIPTON
New pads of this species grow mostly from bases of older ones, resulting in sprawling plants seldom more than 2 pads tall. Clumps grow up to 6 feet (1.8 m) in diameter. The pads are spineless, but have many hair-like glochids that make the areoles look like dots of felt. The incandescent-pink flowers appear from late February at the lowest elevations to May at the highest. The dry fruits contain very large seeds, even for an opuntia.

Beavertail cactus
(*Opuntia basilaris*)

RANGE
Beavertail grows from near sea level to 6000 feet (1800 m) in the Mohave and northwestern Sonoran deserts to southern Utah; it barely enters Sonora.

NOTES
Nearly everyone who has had a close encounter with this or other "spineless" opuntias would rather have dealt with spines. Glochids are often too small to see, and they cause prickling pain and intense itching as the barbs work deeper into the skin with every movement. Removing hundreds or thousands of them after falling into such a plant is an exhausting and tormenting task. Some people shave them off at skin level, which somewhat reduces the irritation, even though this leaves the tips beneath the skin. A better remedy is to gently draw very sticky tape across the afflicted skin. Another effective treatment is to cover the area with a layer of white glue, then peel it off after it dries.

➤ *Opuntia bigelovii* [*Cylindropuntia bigelovii*]

English names: teddy bear cholla, jumping cactus, teddy bear cactus
Spanish names: cholla guera (blond cholla), cholla del oso (bear cholla), ciribe, cumbera

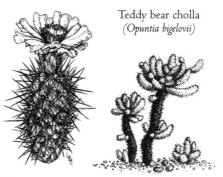

Teddy bear cholla
(*Opuntia bigelovii*)

DESCRIPTION
This distinctive cholla has a vertical trunk 3 to 5 feet (1-1.5 m) tall with densely-packed horizontal side branches on the upper foot (30 cm) or so. Older, lower side branches die and fall off. The joints are very densely spined; very little of the living surface can be seen through its armor. The spines are especially sharp and strongly-barbed. Young spines are yellow and become black with age. Yellow-green flowers in spring are followed by spineless fruits that usually contain no fertile seeds.

RANGE

Teddy bear cholla is widespread and abundant in the warmest parts of the Mohave Desert and the hotter, drier parts of the mainland Sonoran Desert. A few widely-scattered populations occur in Baja California. South-facing rocky slopes are preferred.

NOTES

The detached joints will readily generate new plants by rooting and branching. During the cooler months the terminal joints are detached by a slight touch by a passing animal, or even strong winds. The joints that attach to animals may be transported considerable distances before being dislodged. Since the fruits rarely contain viable seeds, this species reproduces almost entirely by this asexual process. Many plants have 3 sets of chromosomes instead of the ordinary 2; these are usually sterile. In some localities they form nearly impenetrable stands that occupy as much as 2 square miles of land almost to the exclusion of other plants. These giant, hillside-engulfing cholla forests may be a single (clonal) plant. See the next species for a remedy for close encounters with this plant.

➤ *Opuntia fulgida*
[Cylindropuntia fulgitda]

English names: jumping cholla, chain-fruit cholla

Spanish names: velas de coyote (coyote's candle), cholla, brincadora (jumper)

DESCRIPTION

Another tree cholla that somewhat resembles the preceding species, *Opuntia fulgida* differs from *O. bigelovii* in being less spiny, larger, and often having several trunks. It averages about 8 feet (2.5 m) in height; in good soils it may exceed 12 feet (3.7 m). Small, about 1 inch (2.5 cm) pink flowers open in the afternoons during the summer. Flowers grow from previous years' persistent fruits to form branching, pendant chains up to 2 feet (60 cm) long.

RANGE

Chain-fruit cholla is common in south-central Arizona and most of Sonora.

NOTES

The ecology of jumping cholla is much the same as that of *Opuntia bigelovii*. One difference is that *O. fulgida* fruits often contain viable seeds, though they rarely sprout. More commonly whole fruits fall to the ground or are transported by animals that eat them and,

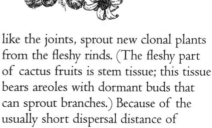

Jumping cholla
(*Opuntia fulgida*)

like the joints, sprout new clonal plants from the fleshy rinds. (The fleshy part of cactus fruits is stem tissue; this tissue bears areoles with dormant buds that can sprout branches.) Because of the usually short dispersal distance of detached joints and fruits, jumping cholla tends to form dense clonal colonies. In contrast to teddy bear cholla's preference for rocky habitat, jumping cholla grows better on the finer soils of lower bajadas and valleys. Extensive forests may consist of only a few clonal individuals.

Teddy bear and jumping chollas are surrounded by tall tales. One myth is that the joints are attracted to the moisture in animal flesh. Many people believe that they really do jump, and some even claim to have caught them in the act. The truth is that the very sharp spines are so well-barbed that even if one barely penetrates skin or clothing, its grip is stronger than the connection between joints. If you pass a jumping cholla and turn to look when you feel a tug on your clothing, you may see the joint detaching and flying through the air as the elastic recoil of the cloth snaps it into your flesh. The double surprise of seeing a plant moving faster than you and the sharp pain of impalement leaves a lasting impression.

The easiest way to remove a cholla joint is to place a comb between it and your (or your dog's) skin and quickly jerk it away. Because of the barbs it will take considerable force (and teeth-gritting) to dislodge it, and the joint may fly several feet. Make sure a hapless companion is not in the line of fire!

➤ *Opuntia leptocaulis*

English names: Christmas cholla, desert Christmas cactus
Spanish names: tasajillo, agujilla (little needle), tasajo

DESCRIPTION
This short, sparsely-branched shrub is composed of very thin, ⅛ inch (3 mm) stems to 2 feet (60 cm) tall, sometimes much larger when supported by other shrubs. There is one 2-inch (50 mm) spine per areole and many glochids. The ½-inch (13 mm) flowers are typically pale yellow; they open in late afternoon in May and June and close at dark. Mature fruit is bright red and persists on the plant through the winter, hence the English names.

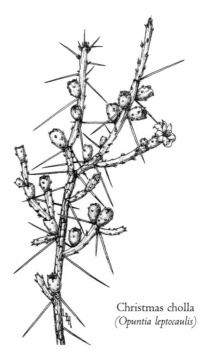

Christmas cholla
(*Opuntia leptocaulis*)

RANGE
This plant is widely distributed from Arizona to Oklahoma, Texas, and northern Mexico.

NOTES
Christmas cholla often clambers among the branches of shrubs, so it tends to be inconspicuous except when it has ripe fruit. These cacti, growing within apparently harmless bursages and

creosote bushes, have ambushed many hikers. This plant is rather unusual in having flowers that are open for only about 3 hours a day. One would expect that it has a specific relationship with a pollinator active in the late afternoon. But no special insects have been seen visiting it, just hummingbirds, honeybees, and one of the cactus bees. It does not seem to be attractive to moths despite its pale yellow color. Pencil cholla (*O. arbuscula*) looks similar, but is a larger plant with a distinct trunk and stems ½ inch in diameter.

> *Opuntia engelmannii*
  [Opuntia phaeacantha discata]

English name: Engelmann prickly pear
Spanish names: nopal, abrojo (spiny
  plant), vela de coyote (coyote candle)

> *Opuntia phaeacantha* var. *major*

English name: sprawling prickly pear
Spanish names: same as above

DESCRIPTION

Engelmann prickly pear is a shrubby cactus forming hemispherical mounds up to 5 feet (1.5 m) high and 2 or 3 times as wide. Pad size varies with individual plants; the largest are over a foot (30 cm) long. (These giants may be hybrids with other species.) The spines are also variable. The O'odham recognize this variability in applying 4 or 5 names to different forms. The flowers are yellow, about 3 inches (8 cm) in diameter, and bloom in May near the end of the spring flowering season. They last a day each, and those of some plants age to orange by afternoon. The

juicy fruit ripens to varying shades (from plant to plant) of rich purple to red. The specific status of this plant is still in dispute; it has been shuttled between *Opuntia phaeacantha*, *O. discata*, and *O. engelmannii* by different taxonomists during the past few decades.

Sprawling prickly pear in its pure form grows horizontally and is rarely over a foot (30 cm) high. Plants can grow to over 15 feet (4.5 m) across. The flowers and fruit are very similar to Engelmann prickly pear.

These 2 plants hybridize freely

Englemann prickly pear (*Opuntia engelmannii*)

with each other and sometimes with other species, so it is often difficult to identify a particular plant.

RANGE

These 2 species are the most common prickly pears over much of the arid Southwest from inland Southern California to central Texas and south into northern Baja California and central Sonora.

Engelmann and several other species of common, large prickly pears present an ecological mystery. The juicy, palatable fruits ripen in tremendous numbers in July and August. Most saguaro fruits, produced in similar abundance earlier in the summer, are devoured the same day they ripen. But Engelmann prickly pear fruits persist for several months. Though they are eaten by a wide variety of animals including rabbits, packrats, javelina, deer, squirrels, numerous birds, desert tortoises, and cactus beetles, there are far too many fruits for them to consume. Moreover, the fruits in the centers of large plants are out of reach of several of the wildlife species that would eat them; many of these fruits are still present in November fermenting and shriveling. Why would a plant seemingly waste energy in such overproduction? Could there be a vacant niche, a missing seed-disperser?

Sprawling prickly pear
(*Opuntia phaeacantha* var. *major*)

The large, very hard seeds offer another possible clue. It has been suggested that some prickly pears coevolved with the now-extinct giant mammals, such as mammoths and ground sloths. It's an intriguing theory and, if proven true, further illustrates the already-established fact that natural systems are anything but static.

Chenopodiaceae (goosefoot family)

Mark A. Dimmitt

The approximately 1300 species of chenopods worldwide range from annual herbs to trees. Many species have C$_4$ photosynthesis. The flowers are tiny and inconspicuous, but some species bear showy masses of fruits. Chenopods are common in deserts and especially in saline or alkaline soils. Spinach, beets, sugar beets, chard, and epazote are members with economic value. The most common chenopods in our region are several species of *Atriplex* (saltbush) and *Chenopodium* (goosefoot, pigweed); the latter genus contains both native and exotic herbs.

➤ *Atriplex canescens*

English name: fourwing saltbush
Spanish names: cenizo (ash-gray), chamizo (thatch, brush), costilla de vaca (cow's rib), saladillo (salted)

DESCRIPTION
Fourwing saltbush is a densely-branched, evergreen shrub 3 to 6 feet (1-2 m) tall with gray foliage. Female plants bear large masses of fruits, each with 4 large winglike membranes.

RANGE
This shrub is widespread in the desert Southwest, centered in the Great Basin and extending into Canada, Mexico, and the western Great Plains. It is often the dominant plant in valleys with saline or alkaline soils.

NOTES
There is something of a mystery concerning this saltbush's ecological needs. Despite the significant climate changes from the glacial to the present interglacial period, the plant has stayed put—its geographic distribution in the southwest U.S.

has not changed. On the other hand, its elevational range changes dramatically with soil type. It grows 1600 feet (500 m) higher on shale soils than on soils derived from igneous rocks.

The leaves of saltbush species appear grayish because they cope with saline soils by secreting excess salt into tiny hairs on

Fourwing saltbush
(*Atriplex canescens*)

the leaf surfaces. The hairs die from high salt concentration, leaving a deposit of salt crystals on the surface that reflects some of the intense light that would otherwise overload the photosynthetic system.

Cattlemen regard saltbush as drought insurance. Cattle rely on these bushes when grasses fail to produce a good crop due to insufficient rains.

Other species of *Atriplex* in our region include *A. hymenelytra* (desert holly) and *A. lentiformis* (quailbush). The former grows in mounds to 4 feet (1.2 m) tall and wide with white holly-shaped foliage; the latter gets much bigger, to 8 by 12 feet (2.4 by 3.7 m). Both species grow in the lowest desert elevations, as does another species with tiny leaves, *A. polycarpa*, desert saltbush, a shrub considered by far the most important browse plant in its range before agriculture displaced much of it. If water is available, these plants can photosynthesize on the hottest days, when most other plants are stressed and forced to shut down; they exemplify the adaptive value of C_4 photosynthesis. *A. lentiformis* seeds were gathered by the O'odham as emergency food, and soap was made from the leaves.

> *Salsola tragus*
> *[Salsola iberica, S. kali]*

English names: Russian thistle, tumbleweed, wind witch, leap the field
Spanish names: chamizo volador (flying bush), maromero

DESCRIPTION
Russian thistle is a nearly spherical annual herb, usually about 2 feet (60 cm) tall, but more than twice that in favorable conditions. The threadlike leaves are spine-tipped. The dried plant breaks from its root and blows across the ground, dispersing seeds as it tumbles.

RANGE
The plant is native to Eurasia, but now occurs all over western North America in disturbed soils.

NOTES
Though it's an integral part of Western lore, tumbleweed is an exotic, very harmful weed. (The familiar song "Tumbling Tumbleweeds" was written in Tucson and first published as a poem in a University of Arizona literary quarterly. The Sons of the Pioneers, a singing group of the 1940s, made the song popular, and retired to Tucson 40 years later.)

Tumbleweed was first noticed in South Dakota in the 1880s, brought in as a contaminant in crop seeds from Europe. By the turn of the century it had spread to the Pacific coast and the Mexican border. It has become a troublesome pest in disturbed soils such as agricultural fields and graded road shoulders. It is rare or absent in undisturbed habitat. In some naturally unstable habitats, such as sand dunes, Russian thistle has become the dominant plant and is crowding out native species.

Russian thistle
(Salsola tragus)

➤ *Salicornia bigelovii*
   English names: pickleweed, glasswort
   Spanish name: none known

DESCRIPTION
This pickleweed is an annual *halophyte*
(plant that tolerates salty soil). It consists
of succulent jointed stems that resemble
strings of little pickles; plants grow from
less than a foot (30 cm) to 3 feet (0.9 m)
tall in different populations. The inconspicuous flowers are wind-pollinated.

RANGE
It grows only in the intertidal zone of
estuarine salt marshes on both coasts of
North America. On the west coast it
occurs from Alaska to Nayarit, Mexico.

NOTES
Although it often grows in pure stands,
few people ever notice this plant, since
most folks avoid the wet ground where
it grows. But plantsman Jon Weeks, who
studied it for more than a decade, enthu-
siastically promotes its merits. Almost all
other true *halophytes* (plants that grow in
salty soil) are perennial, probably because
the metabolic rigors of dealing with
saline environments makes yearly seed
production difficult. But an annual must
make seeds every year, and *Salicornia
bigelovii* manages to devote up to 12% of
its biomass to seeds. It is also one of the
most extreme halophytes known—it can
survive in water that has 5 to 6 times the
salt concentration of sea water. It also
requires salt water to thrive and reproduce; it wilts when it gets rained on.

   Estuarine salt marshes are theaters of
evolution since each marsh is an isolated
island of saturated saline soil surrounded
by dry land and sea. Along the desert
coast of Sonora and elsewhere, each estu-

ary has a unique *ecotype* (distinct genetic
race) of pickleweed; these differ in such
characters as plant height, seed size, and
season of germination. Weeks can look
at a plant in the lab and tell which estuary it was collected from.

Pickleweed
(Salicornia bigelovii)

   This diversity will be valuable in
domesticating pickleweed as a crop plant.
The plant itself is of minimal utility for
forage because up to half its dry weight
is salt, though it might be used as fuel or
diluted with other feed for forage. But
the seeds have great potential. They are
about ⅓ oil (fat) of very high quality.
Different fractions of this oil are suitable
for such varied uses as diesel fuel, lubri-
cating oil, and margarine. The oil is more
than half linoleic acid; an essential dietary
fat for humans that does not contribute to
heart disease. For these reasons pickleweed
may become an economically viable crop,
one that can be irrigated with sea water.
If this comes to pass it will be both a
blessing and a curse. Though we could
increase food production by using
thus-far-unfarmable coastal deserts,
such agriculture would contribute to
the destruction of coastal deserts all
over the world.

Cucurbitaceae (cucumber family)

Mark A. Dimmitt

The cucurbits, as they are called, number 750 species worldwide; most species are vines. The flowers vary from barely noticeable to large and conspicuous. The fruits range from small and dry to large and tasty; the latter include cucumbers, squash, and melons. Gourds are hard-shelled squash that are used after they dry.

Two genera of bees (*Peponapis* and *Xenoglossa*) are tightly associated with this plant family. Squash or gourd bees are more effective pollinators of cucurbits than most other bees. They seem to be holding their own under competition from introduced honeybees (which collect the pollen but are inefficient pollinators of these flowers), probably because squash bees are active earlier in the morning.

➤ *Cucurbita digitata*

English names: coyote gourd, finger-leafed gourd
Spanish names: calabacilla (little squash), chichicayote, meloncillo (little melon), melón de coyote (coyote melon), calabaza amarga (bitter squash)

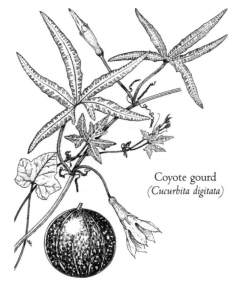

Coyote gourd
(*Cucurbita digitata*)

DESCRIPTION
The large, underground tuberous root produces several to many herbaceous stems as long as several yards (meters) in the summer. The stems typically lie flat on the open ground; occasionally they climb into surrounding vegetation with their tendrils. The palmately-divided leaves with very narrow lobes are widely spaced on the stems. (*Cucurbita palmata* is similar but has broader lobes; the 2 species intergrade in the western Sonoran Desert.) Vase-shaped, bright yellow flowers about 2 inches (5.1 cm) across open before dawn, and wilt by late morning. After pollination, female flowers devel-op 3-inch (7.5 cm) spheroid gourds, green with yellowish stripes when young, and maturing to straw-colored.

RANGE
This species occurs from the lowest and driest desert elevations into desert grasslands, from southeastern California to southern New Mexico and into northwestern Sonora.

The fruit pulp contains toxic and extremely bitter chemicals that humans use to make soap; clothing laundered in it reportedly repels body lice. Humans eat the nutritious seeds which contain up to 35% protein and 50% fat. Coyotes, porcupines, and some other animals can eat the seeds even when tainted by the pulp. Javelina dig up and eat the bitter tuberous roots, which they can sniff out even when there is no vine above ground. People have used the gourds as containers since prehistoric times.

The modifier "coyote" is in the name of a number of Southwestern plants. Most often it identifies wild relatives of domesticated plants. In the mythology of the O'odham and other native cultures, Coyote (a spirit who often appears in the form of the animal of the same name) is, among his other attributes, a trickster and all-around rascal who makes a great deal of mischief. One of the things he does is ruin useful objects by defecating on them. That's how the world got coyote gourds, coyote tobacco, and coyote passion flower, among others. The actual animal coyote marks its territory by defecating on conspicuous landmarks.

Cucurbita foetidissima (buffalo gourd) is a related species from higher elevations. It has a similar growth habit, but the leaves are large gray-furry triangles and have a rank odor when bruised. The starches from the very large root are edible after processing, and it is being developed as a potential feed crop. The seeds are a potential commercial source of oil.

Coyote gourd has a cool-season counterpart in *Marah gilensis*. It sprouts in early spring after winter rains and dies back to the huge root by the onset of the dry foresummer. During its brief season it produces a few very large seeds in each of its small, dry prickly fruits. Little else is known about the ecology of this plant.

Tumamoc globe-berry (*Tumamoca macdougalii*) provided an opportunity for the U.S. Endangered Species Act to work as it was intended. This diminutive plant, with a fist-sized underground tuber and wispy vines that grow and bear small red fruits during the summer rainy season, was thought to be very rare.

When a population was discovered in the path of the Central Arizona Project canal during the environmental impact study, the species

Buffalo gourd
(*Cucurbita foetidissima*)

was quickly listed as Endangered in 1986. The Endangered Species Act funded surveys and ecological studies on the plant over the next several years. The studies revealed the Tumamoc globe-berry to be both common and widespread; it was rarely encountered simply because it's difficult to see among the shrubs it climbs in, and few people had previously looked for it. The species was delisted in 1993.

Ephedraceae (ephedra family)

Mark A. Dimmitt

> *Ephedra* spp.

English names: ephedra, joint fir,
  Mormon tea
Spanish name: canutillo
  (and several spelling variants)

Ephedra
(*Ephedra trifurca*)

DESCRIPTION
These woody shrubs grow 2 to 5 feet
(0.6-1.5 m) tall and wide. The terminal
stems are thin, green, and essentially
leafless. These are conifers more primitive than pine trees; they bear papery
cones. (Their closest relative is the
bizarre *Welwitschia* spp. of the Namib
Desert, which looks like a beached
green octopus but has the same cone
structure as ephedra.) The various
species are similar in general appearance; distinguishing among them
requires close inspection.

RANGE
About 40 species occur in arid habitats
in the northern hemisphere and South
America. About half a dozen occur in
the Sonoran Desert region.

NOTES
The stems contain caffeine and
ephedrine (a drug that acts like adrena-
lin/epinephrin). The closely related
pseudoephedrine is now synthesized
commercially and is an ingredient in
commercial asthma and cold remedies,
e.g., Sudafed®. Pseudoephedrine is also a
precursor in the production of the dangerous illegal drug methamphetamine
("speed"). A tea with stimulant properties is made by steeping dried stems. It
has been used medicinally to treat a
variety of ailments including syphilis,
diabetes, and pneumonia. A Chinese
species is the source of *ma huang*, a tea
so potent that it has caused deaths from
overstimulation of the heart.

In the Gran Desierto and Algodones
Dunes ephedras grow much larger than
usual; bushes 15 feet (4.6 m) across can
be found. Occasionally a dune migrates
away from one of these plants, exposing
many feet of stems that had been slowly
buried as the dune marched over it and
the bush managed to keep a couple of
feet of living stems in the sunlight.
Observations of such blowouts suggest
that these plants are very old, though
they have not been documented to live
more than about 50 years. We don't
yet know how long ephedras live; they
have not been thoroughly studied by
biologists.

Euphorbiaceae (spurge family)

Mark A. Dimmitt

The 8000 worldwide species of spurges are as ecologically diverse as the composites. They occupy most habitats and exhibit nearly every growth form used by plants. In fact, annuals, perennials, trees, succulents, C_3, C_4, and CAM species can all be found in the single genus *Euphorbia*. Until recently this genus contained the little weedy spurges in our gardens, the spectacular Christmas poinsettia (which originated in tropical Mexico), and African succulents—some of which are nearly perfect mimics of the New World cacti. Now some taxonomists raise many of our species previously in a subgenus Chamaesyce to the genus *Chamaesyce*.

Recognizing such a diverse family by vegetative characteristics is virtually impossible. In most species the inflorescence is a *cyathium*, a single female flower with its distinctive 3-lobed ovary surrounded by a number of male flowers consisting of a single stamen each. The male cyathia may have colorful bracts that resemble petals (e.g., crown of thorns— *Euphorbia milii*), and some species have colored leaves that serve the same function

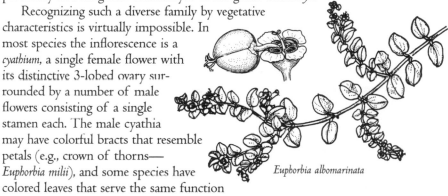

Euphorbia albomarinata

(e.g., poinsettia, *Euphorbia pulcherrima*). In some species male and female flowers are borne on separate plants. A few genera produce more typical flowers with true sepals and petals. The poinsettia is one of the largest ornamental nursery crops; tens of millions are produced annually for the Christmas season. Other economically important examples include manihot (the source of cassava and tapioca) and castor bean (the source of the once-popular health remedy, castor oil).

➤ *Jatropha cardiophylla*

English names: limberbush, dragon's blood (both applied to several other species)
Spanish names: sangregado, sangrengado (contractions for dragon's blood), torote, sangre de cristo (de drago) (Christ's or dragon's blood)

DESCRIPTION
Limberbush is a shrubby succulent to 3 feet (1 m) tall consisting of many usually unbranched stems arising from fleshy underground rhizomes. The branches are semisucculent and extremely pliable, and have smooth reddish bark. Bright green, heart-shaped leaves are present only during the brief

summer rainy season. The flowers are tiny whitish bells that appear during the rainy season. (This genus has "normal" flowers with petals.)

RANGE
South-central Arizona and most of Sonora below 4000 feet (1200 m) elevation.

NOTES
This is the northernmost North American *Jatropha*, a worldwide genus of 170 mostly frost-sensitive tropical species. The above-ground stems are succulent, but much of its biomass is in the massive underground roots, which can produce new stems quickly after being killed to the ground by periodic severe freezes. Coral bean (*Erythrina flabelliformis*) has a similar growth form in the north end of its range where it is occasionally killed to the ground by both frost and fires. Limberbush shares with ocotillo the trait of having both long shoots that produce stems, and short shoots at each original long-shoot leaf; the latter produce leaves periodically without sprouting new branches. The pollinator is a rare tiny fly.

Limberbush heralds the arrival of the summer monsoon. The increased humidity associated with this seasonal wind shift induces leaf production within a week or so. The leaves remain small until the first rain provides the moisture for their full expansion. The leaves turn yellow and fall soon after the air dries out at the end of the monsoon.

The O'odham use limberbush extensively in basketry. It is a wide-spread toothache treatment.

Jatropha cuneata grows in the driest of desert subdivisions. Its many stems arise from a very short trunk.

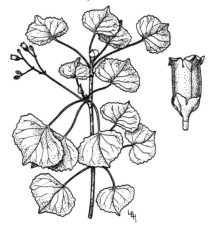

Limberbush
(*Jatropha cardiophylla*)

Fabaceae (legume family)

Mark A. Dimmitt

Legumes are a very large family of 16,000 species in nearly all of the world's habitats. Champion drought tolerators, they are most abundant in the arid tropics. Their prevalence in the Sonoran Desert flora (for example, there are 53 legume species in the Tucson Mountains, 8% of its plants) reflects this desert's tropical origin. North of the Mexican border most of the common Sonoran Desert trees are legumes.

DESCRIPTION

The family was named Leguminosae for its fruit, which in most species is a *legume* (the technical term for bean pod, a single-chambered capsule enclosing what appears to be a single row of seeds that is actually 2 rows—alternate seeds are attached to opposite halves of the pod). There are 3 subfamilies with flowers that look very different from one another at first glance, but arose from a common pattern:

CAESALPINIA SUBFAMILY

Probably the basic pattern from which the other 2 subfamilies evolved. The flowers have 5 separate, conspicuous petals, 1 of which (the *banner*) is always a different size, shape, or color from the other 4. There are 10 stamens, as in nearly all legumes; in this subfamily they are separate. All species are woody. Examples include royal poinciana or arbol de fuego (*Delonix regia*), palo verde (*Cercidium* and *Parkinsonia*), bird-of-paradise (*Caesalpinia*), and cassias (*Cassia*).

MIMOSA SUBFAMILY

The petals are fused in this group, but they're so tiny that they are not noticeable. What one sees is a powder puff of stamens. It's easy to visualize the derivation of flowers of this group from the above subfamily. Start with a caesalpinoid flower such as a palo verde blossom. Reduce the petals until they nearly disappear, greatly elongate the filaments of the stamens, and combine several to many flowers into a tight cluster. The visual result is a ball or cylinder of stamens (powder-puffs or *catkins*, respectively). All species are woody. Examples include acacias (*Acacia*), mesquite (*Prosopis*), fairy duster (*Calliandra*), and mimosa (*Albizia*).

PAPILIONOID (PEA) SUBFAMILY

This group is characterized by 3 upper petals and 2 lower fused petals, and 9 of the 10 filaments are fused with the 10th being separate. This subfamily's flowers diverged from Caesalpinoids in another direction from the mimosoids. Again, visualize a palo verde flower. Enlarge the banner (top) petal until it's the largest part of the flower. Fold the two adjacent petals forward until they touch at their tips; these are called the *wing petals*. Finally, reduce the remaining lower two petals and fuse them along their bottom edges to form a boat-shaped structure called the keel. Conceal most of the keel between the wings, and hide the stamens and pistil inside the *keel*. The result is a sweet pea-shaped flower. This subfamily includes many herbaceous as well as woody species. Examples are desert ironwood (*Olneya tesota*), all cultivated beans (*Phaseolus*, etc.), lupines (*Lupinus*), and of course, sweet peas (*Lathyrus*). In a few species, such as New World coral beans (*Erythrina*), the banner is reduced rather than enlarged.

NOTES

Plants require large quantities of three minerals: nitrogen, phosphorus, and potassium. The latter two elements are present in soil, but nitrogen is an atmospheric gas that plants cannot use directly. Some soil bacteria and *cyanobacteria* (blue-green algae) can *fix* nitrogen (convert it into nitrate or other compound) into a form which plants can use. Another major source of nitrogen is the decomposition of dead plants and animals. In arid soils especially, where decomposition of organic material is slow, plant growth is often limited by the available amount of soil nitrogen. Many legumes harbor colonies of nitrogen-fixing bacteria in their roots. The plant provides favorable habitat and carbon for the bacteria, and the bacteria in turn provide surplus nitrate to the plants. Nitrogen-fixing legumes have higher concentrations of nitrogen compounds in their tissues than non-fixing plants. When legume leaves decompose they release the nitrogen and enrich the soil. Nitrogen is an essential element in proteins, so nitrogen-fixing plants can make large crops of seeds with high protein contents (more than 50 percent in some species).

The typically large, nutritious, and abundant seeds of legumes are an important food source for many wildlife species, including insects such as bruchid beetles. Adult bruchids are flower beetles, while the larvae of most species are seed predators. Bruchids are not restricted to legumes, but there is a myriad of species that specialize on legume seeds. Some species are very host-specific, while others feed on a wide range of seeds. Decades of intensive study of the bruchid-seed relationship would likely not reveal all aspects of this tiny part of the ecological web.

> ## Acacia constricta

English name: whitethorn acacia
Spanish names: huizache, vinorama, chaparro prieto (squat and dark), vara prieta (dark stick), gigantillo (little giant), largoncillo

DESCRIPTION

This acacia is a large, deciduous shrub 6 to 10 feet (2-3 m) tall with twice-compound, gray-green leaves. It can become a small tree in deep soil. Inch-long (2.5 cm), white spines are prominent on young plants; mature shrubs are often thornless. The flowers are bright yellow balls of stamens; they bloom in late spring and again after summer rains.

Whitethorn acacia
(*Acacia constricta*)

RANGE
This plant is common in the southern half of Arizona south to central Mexico and west to the gulf coast of Mexico and Texas. There are disjunct populations in northern and southern Baja California.

NOTES
A. constricta's life span is about 70 years. It is odd that the flowers advertise strongly with both bright yellow color and strong fragrance, but they offer little reward—no nectar and sparse pollen. Pollinators do not visit in large numbers, but whitethorn manages to set good seed crops. The Seri used the leaves and seeds in a medicinal tea for stomach problems.

SIMILAR SPECIES
Acacia neovernicosa has nearly identical foliage and flowers; it is distinguished by its sparsely-branched, erect-but-crooked stems. It's a Chihuahuan Desert shrub.

➢ *Acacia greggii*

English names: catclaw acacia, wait-a-minute bush
Spanish names: uña de gato (cat's claw), tésota, gatuño, palo chino (Chinese stick), tepame, algarroba

DESCRIPTION
Catclaw is a deciduous shrub or small tree, sometimes 20 feet (6 m) tall, but usually less than half that. The stems bear curved, very sharp prickles. Gray-green, twice-compound leaves with small leaflets are seldom dense enough to block the view through the plant. The pale yellow flowers are borne in dense, short catkins in late spring.

Catclaw acacia
(*Acacia greggii*)

RANGE
It is common from southern Nevada south to central Baja California and southern Sonora and west to Texas and Chihuahua. Among the woody

legumes, this species penetrates the farthest north and the farthest into arid habitat.

NOTES
The O'odham drink a tea from the roots for both stomach and kidney problems. The Seri and Yaqui use the wood in bows.

The fragrant flowers, unlike those of *A. constricta*, produce nectar and attract many insects including bees, flies, and butterflies. Tree ring counting shows that this acacia can live at least 130 years.

Tree catclaw, *A. occidentalis*, is vegetatively very similar to catclaw acacia, but grows to 40 feet (12 m) tall. The inflorescences are balls instead of catkins and are extremely fragrant. Tree catclaw is nearly endemic to Sonora and occasionally found in cultivation.

Acacia willardiana (palo blanco, "white stick") is more typical of thornscrub than desert, but needs to be mentioned because it is distinctive for several reasons. It is an upright, sparsely-branched tree to almost 30 feet (9 m) tall. Its bark is smooth and a striking white color in the finest specimens; it peels off in papery sheets. The foliage is so sparse that it casts a scarcely detectable shade. Finally, the leaves are unique among New World acacias in being *phyllodes*. A phyllode is a flat, expanded petiole that resembles a leaf; the real leaflets fall off soon after they're produced. (Most of the one thousand Australian acacia species have phyllodes rather than typical leaves.) This palo blanco is endemic to Sonora, but another white-trunked legume tree in Baja California, *Lysiloma candidum*, is also called palo blanco.

> *Calliandra eriophylla*

English name: fairy duster
Spanish names: cosahui, huajillo, cabeza (pelo) de ángel (angel's head, angel's hair)

DESCRIPTION
This is a spreading shrub, usually about 2 feet (60 cm) tall and twice as wide with sparse semi-evergreen foliage of small, twice-compound, leaves. One-inch (2.5 cm) mimosoid flower heads are pale to deep pink. The plants flower profusely in late winter and sporadically at almost any season in response to good rains.

Fairy duster
(*Calliandra eriophylla*)

RANGE
Fairy duster grows from southern Arizona and southeastern California south to central Mexico and northern Baja California.

NOTES
Fairy duster has several pollinators, including bees, flies, and butterflies. It is among the first perennials to bloom in Arizona Upland, typically in February.

➤ *Cercidium floridum [Parkinsonia torreyana, Cercidium peninsulare]*

English name: blue palo verde
Spanish names: palo verde,
  retama (bloom)

DESCRIPTION

Blue palo verde is a multi-trunked deciduous tree to 40 feet (12 m) tall. Terminal twigs are dense and branches droop to the ground, creating a hemispherical mound. (Floods may break off lower limbs, exposing the trunks and creating more normal-looking trees.) The bark of twigs and young branches is bluish-green. Blue palo verde has twice-pinnate leaves, each segment having only 2 to 4 pairs of relatively large leaflets. A short, straight spine is hidden beneath each leaf (often absent in mature trees). Trees flower for 2 weeks in mid-spring, usually April in Arizona Upland. The bright yellow flowers are produced in such profusion that they completely conceal the branches. The banner petal is slightly larger than the other 4 and has small orange dots. Pods contain 1 or 2 flattened, extremely hard seeds the size of small lima beans.

RANGE

This tree occurs from southern Arizona and southeastern California to northern Sinaloa. It also occurs in the southern half of Baja California, but not in the north, presumably due to the north's extreme aridity. In most of its range this is a dry-riparian species, restricted to washes because it needs more water than is provided by rainfall. Just after the last Ice Age the climate was wetter than today, and blue palo verdes grew beyond the washes as foothill palo verdes do today.

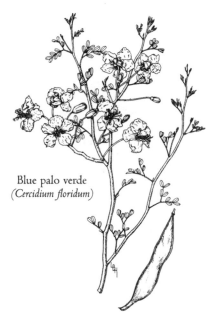

Blue palo verde
(*Cercidium floridum*)

NOTES

The green bark color of palo verdes is due to chlorophyll-bearing tissue. As do many other species, these trees conserve water by dropping their leaves during dry seasons, but palo verdes can still carry on some photosynthesis. Blue palo verde also drops its leaves in response to cool autumn temperatures.

The flowers are pollinated by numerous species of solitary bees that gather pollen and cache it in burrows for their grubs to feed on. The main palo verde bee is *Centris pallida*, but there are 6 other species of *Centris* that visit palo verdes, in addition to sweat bees, several leaf-cutter bees, bumblebees, and carpenter bees. Bee specialist Bob Schmalzel casually identified about 20 different species of bees foraging in a single palo verde tree at the same time.

Mature pods fall and remain under the trees for several weeks, where bruchid beetle larvae consume many of the seeds. Javelina and rock squirrels are among the few other animals that commonly eat the very hard seeds. A band of javelina feeding on these beans sounds like a rock crusher; the noise carries a considerable distance. Most indigenous peoples did not eat the seeds, but they are a fairly important food source for the Seri, who grind them into flour. Freshly-matured seeds with hardened seed coats are impermeable to water; they cannot germinate until the seed coat is either scarified by a flash flood or weathered for a few years in the soil.

Blue palo verde is fast-growing and fairly short-lived. Plants in cultivation seem to become senescent after about 30 years, and probably don't live as long as a century in the wild. The soft wood makes poor lumber. It is also poor for firewood, burning rapidly while emitting a strong, unpleasant odor, and making no coals.

➤ *Cercidium microphyllum*
   [Parkinsonia microphyllum]

   English names: foothill palo verde,
      littleleaf palo verde, yellow palo verde
   Spanish names: palo verde, lebón
      retama, palo brea (tar stick)

DESCRIPTION
Foothill palo verde is a multitrunked, deciduous, large shrub or small tree to about 15 feet (4.6 m) tall, rarely to 30 feet (9 m) in deep soils. The yellowish-green branches are stiff and strongly upright, not drooping as in blue palo verde. Leaves are twice-pinnate, each fork bearing 3 to 5 pairs of tiny leaflets. Foliage is sparse even at the peak of the rainy season, and absent during the dry season. There are no spines beneath the leaves as in blue

Foothill palo verde
(*Cercidium microphyllum*)

palo verde, but each twig terminates in a thorn. The tree flowers profusely for 2 weeks in late spring, beginning just as blue palo verdes in the same area are finishing. Flowers are less bright than those of blue palo verde; the petals are pale yellow and the banner is white. The pods contain 2 to 4 or more navy-bean sized seeds with thinner, softer shells than those of blue palo verde.

RANGE
This plant occurs almost throughout the Sonoran Desert except in the driest regions at the head of the Gulf of California and in the state of California where there is insufficient summer rainfall.

NOTES

See blue palo verde for information about photosynthesis and pollination. Foothill palo verde is more drought resistant than blue palo verde; it is not restricted to washes and is common on rocky hillsides. These very slow-growing trees are more than a century old at maturity, and desert ecologist Forest Shreve thought they could live up to 400 years. (Palo verdes do not make growth rings, so age-dating by counting rings is not possible.) The thin-shelled seeds do not require scarification to germinate. After maturing and falling they are quickly gathered and consumed by a variety of rodents and other seed eaters. Antelope ground squirrels and pocket mice are major seed dispersers. They bury seeds for later consumption; burial terminates reinfestation by bruchid beetles and preserves the viability of uninfested seeds. Seeds that have not been retrieved and eaten germinate when the summer rains come.

Humans have also relied on the seeds for food; crops are abundant in most years. The O'odham preferred to eat the green seeds or pods; young seeds are tender and taste much like fresh peas. The Seri ate the fresh green seeds and also toasted, ground, and ate the mature seeds in a gruel. They usually raided pack rat nests to obtain the mature pods, and then winnowed the seeds.

Four species of palo verde grow in the Sonoran Desert and sometimes naturally hybridize. Foothill palo verde hybridizes fairly commonly with Mexican palo verde (*Parkinsonia*) where both species grow in proximity. There are a handful of foothill/blue palo verde hybrids in the Tucson area and probably elsewhere. A single hybrid of blue and Mexican palo verdes is documented (blooming seasons rarely overlap). Hybrids with desirable traits are being propagated, notably the complex hybrid 'Desert Museum' palo verde, which is erect, fast-growing, thornless, long-blooming, and relatively tidy.

➤ *Lupinus* spp.

English name: lupine
Spanish names: lupino, altramuz

DESCRIPTION

Lupines number about 200 species of annual and perennial herbs and semi-woody shrubs worldwide. Our desert species are all annuals. *Lupinus sparsiflorus* (now including *L. arizonicus*) has palmate leaves and flowering spikes about a foot (30 cm) tall. The blue to purple pea-shaped flowers are about a ½-inch (13 mm) long, usually bloom in March in Arizona Upland.

Arizona lupine
(*Lupinus sparsiflorus*)

RANGE

One or more species of lupine can be found in nearly every habitat in the Sonoran Desert region. *Lupinus sparsiflorus* is widespread in the Mohave and northern Sonoran deserts.

NOTES

Lupines are among the 3 most common showy spring annuals of Arizona Upland. They cover many square miles of desert in wet years.

In some parts of the world lupine seeds are gathered and eaten. But some species have toxic seeds, and *Lupinus sparsiflorus* is especially poisonous. Neither the Seri nor O'odham have a practical use for lupine, but both cultures know it well enough to have named it for its sun-tracking habit. The hand-shaped leaves move to face the sun all day, then fold up at night. The Seri name means "sun-watcher" and the O'odham name means "sun-hand."

➤ *Olneya tesota*

English name: desert ironwood
Spanish names: palo fierro
   (iron stick), tésota, palo de
   hierro (iron stick), comitín

DESCRIPTION

Desert ironwood is a tree up to 35 feet (11 m) tall with twice-compound leaves and a pair of sharp, curved spines at each node. The nearly evergreen foliage is dense and deep green in wet years, sparser and gray-green during drought. In the extremely arid islands in the central Gulf of California the trees become completely leafless at times. Mature trees also shed their leaves a few weeks before flowering in May, then re-leaf when the summer rains come; only limbs that will flower drop foliage. Bloom is heavy only about 2 out of 5 years. The pea-shaped flowers are usually pale lavender, but occasionally rich purple-violet. The flowering period is brief, lasting 10-18 days.

RANGE

Desert ironwood ranges throughout the Sonoran Desert and is almost completely confined to it. Cold air drainage restricts it to the warmer slopes above valley floors in Arizona Upland. Seedlings cannot establish where temperatures regularly fall below 20°F (-6°C). In the most arid subdivisions, such as Lower Colorado River Valley, it is restricted to washes.

Desert ironwood
(*Olneya tesota*)

NOTES

The seeds are eaten by many animals; they are fairly soft-shelled, taste somewhat like peanuts (another famous legume), and are high in protein. They are mildly toxic and probably should not be eaten in quantity without proper preparation. The Seri cook them in 2 changes of water.

Desert ironwood replaces palo verde and mesquite as the major nurse tree in the Central Gulf Coast and parts of the Lower Colorado River Valley subdivisions. In those habitats 165 species have been documented having ironwood as nurse trees, and some of them require this species.

Desert ironwood grows slowly and lives long. Unlike many other trees, desert ironwood's rings are often incomplete or missing, so reliable counting to determine exact age is not feasible. On the basis of long-term observations, some botanists estimate that the trees can live for at least 300 years, and perhaps twice that. One ancient-looking tree at the Desert Museum was determined by carbon-dating to be about 180 years old. In more arid habitats some trees are very likely to be much older.

Decomposition after death is extremely slow. The dark-colored heartwood is rich in toxic chemicals and essentially non-biodegradable; it physically weathers away over many centuries. Dead trees can remain standing for a millennium; small firewood-sized chunks have been carbon-dated at 1600 years.

The wood is extremely dense; it will not float in water. (There are several such trees in the world that are called ironwood, so the full name desert ironwood should be used to avoid confusion.) It's the favorite wood used by the Seri Indians to make their famous wood carvings. This craft developed in the early 1960s for tourist trade purposes; more recently, non-Seri neighbors of the Seris in Mexico have mass-produced similar (but often cruder) figures.

Desert ironwood makes excellent firewood; it burns long and hot and makes good coals. Harvest for wood-carvings and charcoal has nearly extirpated large trees in most of Sonora, and campers and illegal woodcutters are depleting accessible populations in the United States. Because of this and their slow growth rate, it is no longer ethical to burn desert ironwood. It is also illegal; this tree is protected in both Sonora and Arizona.

> *Parkinsonia aculeata*

English names: Mexican palo verde, Jerusalem thorn, retama
Spanish names: bagote, junco marino, cacaporo, guacóporo, retama, palo verde, mezquite verde, espinillo

DESCRIPTION
This is a tree with strongly ascending branches to 40 feet (12 m) tall with sparse foliage that casts little shade. The leaves are very large, each consisting of up to 4 narrow *pinnae* up to a

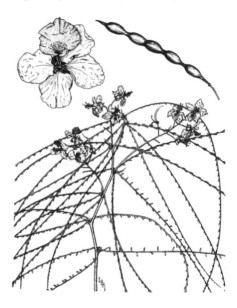

Mexican palo verde
(*Parkinsonia aculeata*)

foot (30 cm) long. Each pinna (plural pinnae) of this tree is a narrow leaf axis bearing 2 rows of leaflets, somewhat resembling its namesake—a feather. The 10 to 40 pairs of tiny leaflets fall off at the first sign of water stress. The pinnae are shed during more severe droughts. There are 2 long, sharp spines at the base of each leaf; these are sometimes forked on vigorous growth. The bright yellow flowers are borne profusely for a month or more in late spring, and sporadically during the other warm months. The banner of older flowers is orange. Pods usually contain 3 or more seeds.

RANGE

This tree is native to the New World tropics and has been introduced worldwide. It was probably introduced to Arizona, where it is primarily associated with human-altered sites such as vacant lots and roadsides.

NOTES

Though it can grow 8 feet (2.4 m) a year, Mexican palo verde has numerous drawbacks as a residential tree. The vicious spines make pruning an unpleasant chore. The long, stringy leaves are too sparse to cast useful shade and are shed in great quantities that create a continuous mess beneath the tree, especially when they become lodged in other landscape plants. Lastly, seedlings volunteer aggressively, and trees don't live longer than about 30 years.

Mesquites

➤ *Prosopis velutina*
[Prosopis juliflora var. *velutina]*

English name: velvet mesquite
Spanish names: mezquite, algarroba, chachaca, péchita (pods only)

➤ *Prosopis glandulosa* var. *torreyana*
[Prosopis juliflora var. *torreyana]*

English name: honey mesquite
Spanish names: mezquite, algarroba, chachaca

Following Lt. Edward Beale's exploration of the Chihuahuan Desert in the 1850s, his report touted mesquite as the key to settling the West. It was superb for lumber, firewood, and food and shelter for both man and livestock. He was correct, but he was ahead of his time. It's ironic that many ranchers today consider mesquite a rangeland pest, especially in Texas.

DESCRIPTION

Both of these mesquites are deciduous large shrubs or trees. Honey mesquite grows 10 to 30 feet (3-9 m) tall; velvet mesquite may attain 55 feet (17 m).

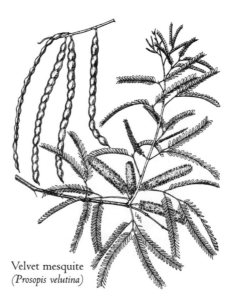

Velvet mesquite
(*Prosopis velutina*)

Honey mesquite usually has 2 pinnae with smooth or hairy, bright green leaflets. Velvet mesquite usually has 4 pinnae with fuzzy, dull green leaflets. The inflorescences are clusters of pale yellow catkins about 3 inches (7 cm) long. They bloom in spring and sometimes again in midsummer. Flattened, non-splitting pods are about 8 inches (20 cm) long, consisting of 3 distinct parts. Each of several lentil-shaped seeds is enclosed in a very durable woody capsule, the *endocarp*. Between the thin skin (*exocarp*) and endocarp is the *mesocarp*, a mealy pulp. Mesquite pods are thus not technically legumes (bean pods), but the trees are still in the legume family because of the flower structure.

RANGE
Both species occur mainly below 5000 feet (1500 m) elevation. Honey mesquite is widespread in Arizona, Sonora, and Baja California, and the desert habitats of California. Velvet mesquite is found only in Arizona and Sonora. The 2 species hybridize where they occur together. (Texas honey mesquite, *P. glandulosa* var. *glandulosa*, occurs from New Mexico to Trans Pecos Texas. It has recently colonized Arizona along I-10; the seeds are being dispersed by trucks transporting cattle and their seed-laden cowpies from Texas to California.)

NOTES
The flowers are pollinated mainly by bees, and the resulting pods are produced in huge quantities in good years. Fallen pods are quickly infested with bruchid beetle larvae, and eaten by a variety of larger animals. Germination

Honey mesquite
(*Prosopis glandulosa*
var. *torreyana*)

is enhanced by passage through the guts of large animals; otherwise a few years of weathering is needed to release the seeds from the endocarp. The foliage is also an important browse for numerous animals. Mesquite is long-lived, probably a couple of centuries in favorable sites.

The mesquite's root system is the deepest documented; a live root was discovered in a copper mine over 160 feet (50 m) below the surface. Like all known trees, however, 90% of mesquite roots are in the upper 3 feet of soil. This is where most of the water and oxygen are. The deep roots presumably enable a mesquite to survive severe droughts, but they are not its main life support.

Dense mesquite stands are called bosques (pronounced BOSE case). Once abundant on floodplains in the southwest U.S., most have been cut down or killed by rapid lowering of water tables. Only scattered remnants still exist. In the low desert, velvet mesquite is restricted to flood plains and large washes. At higher elevations it also occurs on dry hillsides. On the rocky slopes of Arizona Upland it is sparse and dwarfed to shrub size.

Understanding the natural habitat and ecology of mesquite in southern Arizona requires a deep time perspective. Mesquite coevolved with large herbivores, such as mastodons and ground sloths, which ate the pods and then dispersed them widely in their feces. The bruchid seed parasites were killed by the gut juices, greatly increasing seed viability. When this Pleistocene megafauna became extinct about 10,000 years ago, mesquite's range largely contracted to flood plains and washes, where seeds were scarified by floods or weathered in wet soil. Climate was also influential, since mesquite persisted on the slopes of the Waterman Mountains west of Tucson until 3800 years ago.

The introduction of cattle to North America refilled the vacant ecological niche of the extinct native herbivores, and mesquite began recolonizing its former hillside habitats. Many ranchers viewed this range expansion as an undesirable new invasion and attempted to eradicate the plant. A low density of mesquite trees in grassland is actually beneficial to livestock and wildlife; grass growing beneath these trees is more nutritious and remains green longer into the dry season. The shade is also beneficial for large animals. On the other hand, overgrazing causes a reduction in fire that controls mesquite populations in grassland; this allows the trees to become so numerous that they exclude grasses. Because dense mesquite outcompetes grass for water and light and because mesquite groves don't support fire, this conversion is permanent (on a human time scale) without physical intervention.

Velvet mesquite has been a major food source for indigenous peoples. The mesocarp is sweet, containing 20 percent to 30 percent sugars in the best trees. (The pods of honey mesquite are bitter.) The seeds contain about 35 percent protein, more than soybeans, though it's difficult to separate them efficiently from the inedible endocarp. Mesquite meal (the mesocarp) is tasty as well as sweet, and its popularity in modern cuisine is increasing. Pod production is sufficient to make mesquite commercially viable as a cultivated crop; research and development toward this end are ongoing. The Tohono O'odham appear to be on the verge of commercial success with this crop. Mesquite flour also has major conservation potential, in that it can be made into "bread" without baking. Most of the world's people cook over wood fires, and demand for fuel-wood is a major cause of deforestation.

The wood is hard, attractive, and in high demand for quality furniture. The sapwood is yellow, the heartwood rich reddish-brown. It is evermore expensive now that most of the great bosques have been lost to habitat destruction.

Mesquite has recently surpassed hickory as the most popular smoke flavoring for food. Because of the overharvesting, its wood should not be used for this purpose; burning dried pods imparts the same flavor.

The inner bark furnished both Indians and early settlers with material for basketry, coarse fabrics, and medicine to treat a variety of disorders. Gum exuded from the stem is used for manufacture of candy (gumdrops), mucilage for mending pottery, and black dye.

> *Senna covesii* and *S. bauhinioides*
> *[Cassia spp.]*

English name: desert senna
Spanish names: dais, hoja sen,
    rosamaría (rosemary)

DESCRIPTION
Both species are small subshrubs with
fuzzy gray-green leaves. They are most
easily distinguished by the number of
leaflets; *S. bauhinioides* has 2, *S. covesii*
has 4 or 6. Foliage is shed during dry
seasons. Rains trigger short spikes of
yellow caesalpinoid flowers, mostly in
spring and late summer.

RANGE
Both species are widespread in the
Sonoran Desert and into the arid trop-
ics of Mexico. Their ranges overlap in
Arizona.

NOTES
These and other sennas are *buzz-pollinated*
by large carpenter bees and bumblebees.
The anthers of buzz-pollinated flowers
don't split open lengthwise to expose
their pollen as in most flowers. Instead
each anther has a small pore at one end
that is too small for even the smallest
insects to enter. The flower is so oriented that the bee lands on it hanging
upside down with the anther pores
facing downward. The bee then vibrates
its wing muscles, making an audible
buzzing sound different from that of
the bee in flight. The pollen is thus
shaken out of the anthers onto the

Desert senna
(Senna covesii)

hanging bee. The pollen grains are tiny
and nonsticky to prevent them from
clogging the anther pores. Carpenter
bees have especially fine hairs on their
bodies for catching this pollen. As the
bees visit other flowers, some of the
pollen on their bodies gets stuck on the
stigmas and effects pollination.

In addition to the sennas, a number
of plants in the nightshade family
(Solanaceae) are buzz-pollinated,
including tomatoes. Though they are
self-fertile, tomato flowers must be
buzzed—ordinarily by wild bees—
to dislodge pollen so that it falls onto
the stigmas. As recently as the 1980s,
greenhouse growers hired workers to
vibrate the flowers with electric toothbrushes to ensure fruit set. Today they
purchase commercially-produced
bumblebee hives. The bees are not
only more efficient pollinators than
humans with toothbrushes, they are
also less expensive.

Fouquieriaceae (ocotillo family)

Mark A. Dimmitt

The ocotillo family is a small one of only 13 species restricted to the warm-arid section of North America. Members of this family are odd-looking plants, some even bizarre. They are characterized by spiny stems with bundles of seasonal leaves at each spine. A few species are stem succulents, the rest barely semisucculent. The fouquierias have a curious parallel with the Didiereaceae. The few species of this exclusively Madagascan family closely resemble some of the ocotillos in growth habit, differing from them in growing much larger and having succulent leaves. The didiereas are distantly related to the cacti and not at all to the ocotillos, so this is an example of convergent evolution.

➤ *Fouquieria columnaris*
  [Idria columnaris]

> English name: boojum
> Spanish name: cirio (a tall,
>   tapered candle)

DESCRIPTION
Boojum is a stem succulent with a single or branched, tapered primary stem to more than 60 feet (18 m) tall. Unbranched specimens resemble upside-down carrots. The succulent stems produce hundreds of horizontal, non-succulent, secondary branches armed with spines. The succulent stems elongate only in late winter and spring in direct relation to the amount of moisture received. Leaves and secondary branches may sprout any time moisture is available, but the plants are usually leafless and dormant in summer. Clusters of white, fragrant flowers are borne at the tops of primary stems from July to September.

RANGE
This plant is nearly endemic to central Baja California, where the desert climate is moderated by the cool, moist winds off the Pacific Ocean. It is most common in

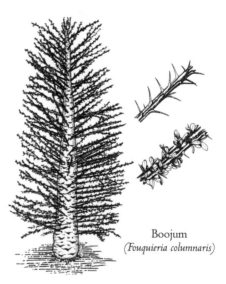

Boojum
(*Fouquieria columnaris*)

the Vizcaino subdivision. There is a small disjunct population on the Gulf coast of Sonora, where an upwelling of cold water just offshore simulates the Pacific's influence on the peninsula.

NOTES

Baja California is so well-endowed with strange plants that its landscapes scarcely seem to belong to this planet. But even here, the boojum clearly stands out as the strangest of the strange. It is surely among the top ten most bizarre plants in the world, rivaling the *Welwitschia* of the Namib Desert, the baobabs of Africa and Australia, and the titan arum of Sumatra. A strange plant of a remote area it is, but not rare. It grows by the millions in the central third of the peninsula. Most plants are erect and unbranched, but some bend over and even form loops (see plate 21); others branch in odd patterns. On the windy Pacific coast they're scarcely a foot tall, but grow many feet laterally, hugging the ground.

The English vernacular name comes from Lewis Carroll's *The Hunting of the Snark*, a fictitious account of exploration of far-away places. The book contains a mythical creature called the "boojum" which inhabited distant shores. When explorer Godfrey Sykes encountered the plants growing on the desolate Sonoran coast in 1922, he was reminded of Carroll's story and dubbed them boojums.

The boojum is essentially a succulent ocotillo (*Fouquieria splendens*). It differs from that close relative in being a winter grower. Earlier work by Robert Humphrey of the University of Arizona suggested that boojums grow only a few inches a year and that the tallest ones were up to 700 years old. But a more recent study indicates that their life spans may typically be a century or so. Of the boojums identified in photos taken in Baja California in 1905, not one was still present when the sites were revisited in the 1990s. And of those in photos taken in the 1950s, very few were still present only 40 years later. Every few decades a given boojum population experiences a direct hit from a hurricane. Boojum, cardón, and senita are especially vulnerable to high winds and suffer significant losses of large individuals from these events. The tallest known boojum was discovered by Robert Humphrey in Montevideo Canyon near Bahia de Los Angeles in the 1970s; at the time it was 81 feet (24.6 m) tall. It grew several more feet during the next 20 years. In 1998 it and the 60 foot plus (more than 18 m) cardón next to it were gone, probably casualties of Hurricane Nora which crossed that part of the peninsula in September 1997. We could not find any boojums over 50 feet (15 m) tall in that canyon in 1998.

At the time of the first draft of this manuscript, we knew only that boojums are pollinated by a variety of insects. Then Steven Buchmann and colleagues at the Carl Hayden Bee Research Center completed a detailed pollination study of this species and reported a fascinating story. While this plant is indeed pollinated by a large number of insects, in each of the 20 years of study, there was a very different array of species collected in the same boojum populations in Baja California and Sonora. Many species were not seen again for several years. These pollinators may be another example of temporal niche separation (see the section on drought evasion in the Plant Ecology chapter). The boojums flower every year, but different insect pollinators emerge in different years in response to as yet unknown environmental cues.

➤ *Fouquieria splendens*

English names: ocotillo, coachwhip
Spanish names: ocotillo, albarda, ocotillo del corral, barda

DESCRIPTION

Ocotillo is a woody shrub 10 to 20 feet (3-6 m) tall. Long, thin, and nearly unbranched, spiny stems arise from a very short trunk or caudex. The stems range from nearly vertical to widely-spreading in different individuals. Dense spikes of tubular, red to red-orange flowers sprout from the stem tips in spring. The flowering season begins as early as February at the lowest elevations in the lowest-

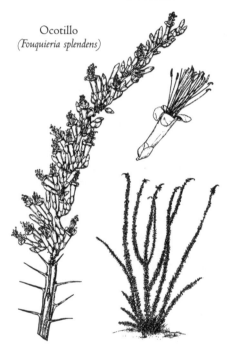

Ocotillo
(*Fouquieria splendens*)

elevation and lowest-latitude desert habitats, and as late as May in the grassland and woodland habitats. At the higher elevations some plants bloom again in late summer or fall. Each population blooms for a month.

RANGE

Ocotillo is the most widespread species in the family. It is common in most of the Sonoran and Chihuahuan deserts and extends north to the southern Mohave Desert, south to central Mexico, and east to central Texas.

NOTES

All species in this genus have primary and secondary leaves and stems. The primary stems elongate and produce leaves with stout *petioles* (leaf stalks). When the leaf blade falls off most of the petiole and the base of the leaf midrib become woody and form a spine. At the base of each spine is a growing point that produces only secondary leaves which lack petioles. These axillary growing points are the secondary "branches," properly termed short shoots; they grow only a couple of millimeters long during many decades of producing leaves.

Though ocotillos look like (albeit weird) woody shrubs, they behave as if they were CAM succulents. They have very shallow roots, as do succulents. Full-grown leaves are produced as quickly as within three days of a summer rain, and fall off after a couple of weeks of dry weather. This rapid leaf production suggests that the thin layer of moist, green tissue beneath the brown cuticle would have CAM-idling metabolism (see "Plant Ecology of the Sonoran Desert Region"). But it doesn't; it's C_3, as are the leaves. It photosynthesizes during the day in the dry season when the stems are leafless. The idling metabolism of ocotillo remains a scientific mystery.

Ocotillos exhibit an interesting correlation with elevation and geology—specifically, soil types. They occur at their highest elevations (6000 feet, 1800 m) on limestone and at their lowest (sea level) on granite. Desert vegetation in general extends about 1000 feet (300 m) higher in elevation on limestone; the reason is twofold. First, limestone weathers slowly without cracking and makes a very thin soil that favors drought- and sun-tolerant plants. Second, limestone has a high specific heat; it stores more heat than most other rocks and thus offers more frost protection on winter nights. At the lowest elevation extreme, ocotillos are limited by water rather than frost. Granite rock disintegrates to form a gravelly soil that creates a moisture-retaining mulch. At the limit of their ecological tolerance for aridity, ocotillos and some other plants can grow on granite, but not on other, drier soils.

The floral characteristics seem to point clearly to hummingbirds as the primary pollinator, but, as with the saguaro, the story is more complicated. Ocotillos are important to hummingbirds, but the reverse is less true. Several species of hummingbirds migrate northward through the desert in large numbers in spring on their way to breeding grounds as far north as Alaska. Ocotillo flowers are a crucial energy source for this migration, since it is the only desert hummingbird flower that blooms abundantly even in dry years. (Like the nectar-feeding bats, hummingbirds return south along the mountains in late summer, feeding on higher-elevation flowers.)

Many other animals will consume nectar if they can get to it, but their behavior and body forms usually make them ineffective pollinators compared to each plant's primary flower visitors that have coevolved a *symbiosis* (mutually beneficial relationship) . The long tubes of ocotillo flowers tend to prevent animals with mouthparts shorter than hummingbirds from reaching the nectar at the bottom of the tubes. But some animals have learned to cheat—they steal the nectar (that is, they take the reward without performing the pollination service). Verdins and carpenter bees can often be seen visiting blooming ocotillos, slitting the flower tubes at their bases and sucking out the nectar. It was assumed that this behavior failed to transfer pollen among flowers, a failure demonstrated for other nectar thieves.

But there's a twist in this story. Peter Scott and colleagues from Southern Indiana University discovered that hummingbirds rarely visit ocotillos in the Big Bend area of Texas. But carpenter bees do, and they transfer pollen effectively while crawling around on the inflorescences biting through the flower tubes. So at least in this population, what appears to be a nectar thief is in fact the major pollinator.

Antelope ground squirrels climb into the branches and feed on the flowers and seeds.

Ocotillo is used for fencing, house walls, and ramada roofs by Indians and ranchers. The cut, buried stems often root, creating a living fence. The flowers are soaked in cold water to make a refreshing beverage.

Krameraceae (krameria family)

Mark A. Dimmitt

➤ *Krameria* spp.

English name: ratany
Spanish names: mamelique, chacaté,
  cósahui, guisapol colorado,
  mezquitillo (little mesquite)

DESCRIPTION
This is a small family of 15 species. Ours
are finely-branched woody shrubs to about
three feet (1 m) tall and wide. The tiny
leaves are less noticeable than the dense
stems. Small, irregularly-shaped, deep
reddish-purple flowers are borne in
profusion in late spring. The two common species in our region, *K. grayi* and
K. erecta, are extremely similar in general
appearance.

RANGE
Warm deserts from southern Nevada to
northern Mexico, California to Texas.

NOTES
Ratanys are partial parasites. They have
chlorophyll and photosynthesize, but
their roots invade those of other plants
to usurp nutrients. Nearly every fruit is
infested with a tiny moth larva that eats
the developing seed, and fruits with viable
seeds are not common. Nothing is known
about the moth's life cycle; it may even be
an unnamed species.

Ratany is among the small number
of plants with flowers that produce oil
instead of nectar as a food reward. Bees
in the genus *Centris* pollinate it. Oil bees
have special squeegee-like hind legs specially adapted for scraping up the oil.
These bees also feed on nectar at other
flowers; they provision their nests with the
oil as larval food. In our region flowers of
plants in the family Malpighiaceae (e.g.,
Janusia, Mascagnia) also produce oil.

Cósahui is sold all over Mexico as a
major medicinal plant. The O'odham use
the roots for a red dye in basketmaking.

White ratany
(*Krameria grayi*)

Liliaceae (lily family)

Mark A. Dimmitt

At about 5000 species worldwide, the lily family is fairly large, even though several other large groups were split off into their own families in recent years. This more narrowly-defined family still includes our region's ajo lily (desert lily), mariposa, and the true lilies of wetter climates. Former lilies that have been placed in their own families include agaves, yuccas, onions, asparagus, and the aloes.

➤ Calochortus kennedyi

English name: desert mariposa
Spanish name: mariposa (butterfly)

DESCRIPTION
This is a perennial herb with only 1 to 3 small leaves that sprout from a deeply buried bulb (technically, a *corm*). The 1 to several flowers are about 2 inches (5 cm) across and open 1 or 2 at a time. The flowers have 3 broad red-orange or yellow petals that form an open cup. Each population blooms for about 2 weeks in late spring at desert elevations, given sufficient rainfall.

RANGE
It grows in heavy or rocky soils from 2000 to 6500 feet (600-2000 m) elevation, from southeastern California and southern Nevada to northern Sonora.

NOTES
Desert mariposa produces the most brilliant flowers in the Sonoran Desert; a single incandescent vermillion chalice can be seen from a ½-mile (almost a km) away. The yellow-flowered form is most common at elevations above the desert. Despite their brilliance mari-

posas are not commonly encountered. Though their geographic range is great, individual populations are widely scattered, and they don't bloom every year.

In our region this species grows only in heavy clay soils where the small bulbs are buried deeply, a necessity resulting from their Mediterranean climate ancestry. The genus *Calochortus* evolved in the winter-rain habitats of the Pacific coast of North America, and most of its 60

Desert mariposa
(*Calochortus kennedyi*)

species are extremely vulnerable to rot if they get wet during summer. The Sonoran Desert's summer rains are too brief in duration to wet clay-rich soil to the depth of the bulbs, but the slow, soaking winter rains do reach them in wet years. The bulbs behave as do the seeds of annual drought-evading species, remaining dormant for several successive years if rain is sparse.

Many other lily species also have bulbs that are deeply buried or grow in very hard soils. Protection from birds and rodents may be another advantage of this habit. We once sowed over 100,000 desert hyacinth (*Dichelostemma pulchella*) seeds in sandy soil in the Desert Museum's cactus garden. They germinated like grass under irrigation, but within a few years almost none remained. I've had the same experience at home with a variety of small bulbs in friable garden soil. One year I planted 1000 crocus bulbs, only 3 of which survived to flower. The rest were destroyed as soon as they sprouted, mostly by Curve-billed Thrashers that excavated the bulbs in search of insects. Once the bulbs were exposed, other birds and rodents ate them.

Knowing these facts about their biology is not sufficient to enable us to cultivate desert mariposas. Even though wild ones are abundant in King Canyon, only a few hundred yards (½ km) from the Desert Museum, we have never been able to grow them on the grounds. Germination of seeds is easy, but the resulting first-year bulbs never sprout again, though they may survive several years. We don't know how to break their dormancy.

Bulbs of all *Calochortus* species are edible, though digging them is labor-intensive because they are typically small and deeply buried in heavy soils. A minute percentage of ground bulbs added to wheat flour imparts a more desirable texture to bread; experimental field cultivation of some species is under way.

➤ *Hesperocallis undulata*

English names: Desert lily, ajo lily
Spanish name: ajo (garlic)

DESCRIPTION
A rosette of strap-shaped, undulate leaves sprouts from a deeply-buried bulb. Spikes bearing 2-inch (5 cm) long trumpet-shaped white flowers rise above the foliage late in the spring flowering season. The spikes are usually less than a foot (30 cm) tall, but occasionally exceed 6 feet (2 m) in wet years.

Desert lily
(*Hesperocallis undulata*)

RANGE
Desert lily is most common in loose sandy soils from the eastern Mohave Desert south through Arizona to northern Sonora and central Baja California.

NOTES
This species is widespread and conspicuous when in flower. Like desert mariposa, the bulbs behave as drought-evading annual seeds, flowering only in wet years. They are also very difficult to cultivate. In sand the bulbs may be 2 feet below the surface. This deep location probably protects them from both hungry animals and the summer rains, which might otherwise rot them.

Malvaceae (mallow family)

Mark A. Dimmitt

The mallows number about 1500 species worldwide; most are characterized by alternate, simple, palmate leaves with *stellate* (branched, star-shaped) hairs. Mallow flowers are easily recognized by the filaments, which are united into a tube surrounding the style. Familiar examples include hibiscus, cotton, okra, and hollyhock.

➤ Hibiscus denudatus

English names: rock hibiscus,
  rock rose mallow
Spanish name: none known

DESCRIPTION
Rock hibiscus is a subshrub to a foot (30 cm) tall. The thin stems bear sparse foliage of triangular, gray leaves. One-inch (2.5 cm) flowers resemble those of the familiar tropical hibiscus; color ranges from white in the western part of its geographic range to deep purple-pink in the eastern end. Plants flower in response to rain during the warm season. Flowers can be found in almost any month below 1000 feet (300 m) elevation, but only in summer at 4000 feet (1200 m).

Rock hibiscus
(Hibiscus denudatus)

NOTES
It grows throughout the Sonoran Desert to western Texas and central Mexico.

➤ Hibiscus coulteri

English names: desert hibiscus,
  desert rose mallow
Spanish names: tulipán (tulip), hibisco

DESCRIPTION
Desert hibiscus is a weak-stemmed, sparsely-branched shrub to 3 feet (1 m) tall. The upper leaves are 3-lobed. Showy flowers almost 2 inches (5 cm) wide are light yellow with a purple spot on each of the 5 petals. They bloom nearly year round in response to rain. Plants are difficult to find when not in flower

Desert hibiscus
(Hibiscus coulteri)

because they nearly always grow among other shrubs for support.

RANGE
It occurs from near Tucson to western Texas and northern Mexico.

➤ *Sphaeralcea ambigua*

English names: desert globemallow, sore-eye poppy
Spanish names: mal de ojo (sore eye), malvia, plantas muy malas (very bad plants)

DESCRIPTION
Desert globemallow is a short-lived subshrub; many stems with slightly woody bases arise from near ground level grow up to 3 feet (I m) long. The stems grow in all directions from erect to horizontal forming a hemispherical mound when growing in the open. The triangular leaves are gray-hairy. Goblet-shaped flowers are about ¾ of an inch (20 mm) across. Throughout most of its range the flowers are apricot-colored to bright orange. Scattered populations with pink, lavender, red, or white flowers comprise the variety *rosacea* (Parish mallow). Plants flower profusely in spring and sparsely at other times following rains.

RANGE
Desert globemallow is common in the deserts of California and Arizona and from Utah into Mexico.

NOTES
There are several species of *Sphaeralcea* in the Sonoran Desert, and only experts on the genus can distinguish them with confidence. *Sphaeralcea laxa*, "caliche globemallow," is most often confused with *S. ambigua*. These 2 perennial species can be distinguished only by microscopic examination of the seeds. *Sphaeralcea emoryi* and *coulteri* are winter annuals that are usually shorter and less branched than are the 2 perennial species.

Like the Spanish name, the O'odham name also means "sore eyes"; the name should be taken seriously. The very wet winter of 1998 produced huge fields of globemallows 5 feet (1.5 m) tall in many desert areas. On a Desert Museum expedition to Baja California we drove through a patch on a warm day with the windows down. The road was so narrow that the side mirrors slapped the mallows on both sides of the vehicle. Our resultant misery reminded us that stellate hairs probably evolved to discourage herbivores.

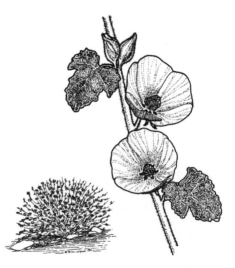

Desert globemallow (*Sphaeralcea ambigua*)

Pedaliaceae (sesame family)

(the family Martyniaceae is included here)

Mark A. Dimmitt

Within a small family of less than 100 species; the genus *Proboscidea* (also called *Martynia*) has but 9 species in tropical America and 2 in the Sonoran Desert. Many species have woody, hooked fruits that are transported on the fur or feet of large animals. Our local devil's claws are adapted to catch onto the hooves of deer or other ungulates. The two long, sharp hooks grasp a hoof in such a way that the thick, woody seed-bearing capsule of the fruit is positioned beneath the hoof. As the animal walks, the capsule is ground away and the 20 or so seeds are gradually released over a considerable distance. The fruit hooks hikers' ankles in the same manner; the sensation of being "grabbed" is startling!

➢ *Proboscidea altheaefolia*

English names: devil's claw, unicorn plant, elephant tusks
Spanish names: cuernos (uña) del diablo (devil's horns[claw]), espuela del diablo (devil's spur), gato (cat), campanita (little bell)

DESCRIPTION
This devil's claw is a prostrate perennial that grows from a large tuberous root. The stems emerge during warm weather in response to rain and produce a mat of sticky foliage. Two-inch (5 cm) bright yellow flowers with deeper yellow streaks are borne on stems above the foliage; they're sweetly fragrant. The fruit is typical for the genus (see illustration).

RANGE
Perennial devil's claw is widespread in much of the Sonoran Desert, including southeastern California, where it sprouts only every few years, west to Texas and south to Sinaloa; it is also found in Peru. It grows mostly in sandy soils, especially in disturbed places such as washes and road shoulders.

NOTES
There are several spellings of the specific epithet in literature, including *althaeifolia* and *altheaiefolia.*

Pollination is accomplished by a number of bees, particularly bumblebees and carpenter bees. These large bees barely fit into the flower tube and cannot help but brush against the anthers and stigma on the upper surface of the tube. The small bee *Perdita hurdi* also pollinates these flowers, but in an odd manner. The pollen ripens before the flower opens. Female *Perdita* cut a hole in the buds to enter and steal the pollen to provision their nests. They also visit open flowers to get nectar; however, this bee is too small to contact the stigma in the roof of the large floral tube. But male bees are sometimes waiting for a potential mate to show up. While the pair grapples to copulate, the female may touch the stigma and leave some pollen stuck to it. As is the case with ocotillos, what seems like an illegitimate relationship may accomplish pollination after all.

> *Proboscidea parviflora*

English names: devil's claw, unicorn plant
Spanish names: aguaro, cuernero (horned)

DESCRIPTION
This devil's claw is a summer annual to 2 feet (60 cm) tall and 2 to 3 times as wide with large, ill-smelling, sticky foliage. The 1-inch (2.5 cm) flowers, purplish with yellow spots, are nearly hidden beneath the foliage. The distinctive fruit is typical of the genus.

RANGE
It is distributed in semiarid habitats from Arizona to Texas, south to central Mexico. Found mostly in disturbed soils, it's a common weed in agricultural areas.

NOTES
The variety *hohokamiana* is a cultivar developed by the O'odham. It differs from the wild type in 2 important ways. The cultivar has claws up to a foot (30 cm) long with softer fibers. The black fibers in the claws are used in basket-making, especially by the Tohono O'odham. The longer, softer fiber in the domesticated claws are easier to work with. Secondly, the seeds of the cultivar are white instead of black, and lack germination inhibitors. While seeds of the wild type must lie in the ground for a couple of years before they will germinate, the white seeds sprout as soon as

they get wet in hot weather and are thus easier to cultivate. This is one of the few plants domesticated north of Mexico, and this seems to have been accomplished only late in the last century. There is a theory that the introduction of cattle was the catalyst. Cattle will eat devil's claw plants, and O'odham women may have been forced to save seeds and grow them in more protected areas than previously. Among the saved seeds was a variant with longer claws and white seeds. The cultivar is now grown by more than 25 native cultures, some of whom live far beyond the natural range of the wild devil's claw.

The same *Perdita* bee that visits *P. altheaefolia* also visits the cultivar, but not the wild type of annual devil's claw. This is probably a result of the cultivar having paler flowers, which might be mistaken for the yellow flowers of the perennial species. The seeds and young fruits are edible.

In the tropical deciduous forest south of the Sonoran Desert is another species, *Proboscidea louisianica fragrans* (syn. *P. sinaloensis*) that looks the same vegetatively. But its larger, more colorful, and fragrant flowers are borne above the leaves. It ranges into the southeastern U.S.

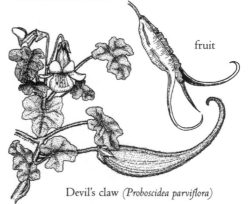

fruit

Devil's claw (*Proboscidea parviflora*)

Polygonaceae (buckwheat family)

Mark A. Dimmitt

The buckwheat family has 1200 species of trees, shrubs, herbs, and vines world-wide and is well represented in the Sonoran Desert. Buckwheat, rhubarb, and sorrel are edible members.

➤ *Eriogonum* spp.

English names: skeleton weed (many species), wild buckwheat
Spanish names: (vary with species)

DESCRIPTION

Species in this genus vary in growth form from herbaceous annuals and perennials to woody shrubs. Most of the approximately 100 species in the Sonoran Desert region can be readily recognized by their general appearance. The herbaceous species are called skeleton weeds. Their basal rosettes of leaves are rather inconspicuous, but their inflorescences are distinctive. One to several of them arise from the basal rosette and branch profusely, often trifurcately, from a few inches to 2 feet (60 cm) tall. The flowering stems are leafless or nearly so, and bear tiny flowers at each node. Then they dry out and persist as skeletons for a year or more. Each of the 20 or so desert species has a distinct skeletal form, several of which are very attractive and are used in dried arrangements.

RANGE

The genus is distributed mainly in western North America. California is the center of distribution with more than 100 species.

NOTES

The buckwheats have a wide variety of life histories. The most common of the skeleton weeds in our region is *Eriogonum deflexum* (flat-topped buckwheat); the skeletons are inverted, flat-topped cones that dry to a rusty-brown color. The seeds germinate in the fall with other winter

Skeleton weed
(Eriogonum deflexum)

annuals, but they don't flower until summer. The most conspicuous species is *E. inflatum* (desert trumpet, *guinagua*). The flowering stems of this perennial grow to more than 3 feet (1 m) tall and are strongly swollen below each node. The swelling is caused by irritation from a moth larva that lives inside the hollow stem; the few stems not infested remain uninflated. The winter rains induce leaf production, and summer rain triggers flowering. The Seri drank a tea of desert trumpet as a cold remedy.

The commonest shrubby buckwheat is *Eriogonum fasciculatum* (California buckwheat, *maderista, valeriana*). It is a densely branched, rounded shrub to about 2 feet (60 cm) tall and usually twice as wide, clothed with short, leathery, linear leaves. Many I-inch (2.5 cm) clusters of tiny white flowers are supported above the foliage; they dry to a rusty brown and persist for several months. *Eriogonum deserticola* (desert buckwheat) is a long-lived woody shrub endemic to the sand sea of the Gran Desierto and the contiguous Algodones Dunes.

The few dozen species in the genus *Chorizanthe* are commonly called "spineflowers." These herbs persist as dry skeletons like the skeleton weeds. They are smaller than most *Eriogonum* species, and most are prickly.

Skeleton weed
(*Eriogonum inflatum*)

Sapindaceae (soapberry family)

Mark A. Dimmitt

The soapberry family has over 1300 species worldwide, many of which are toxic. The only edible product that is likely to be familiar is lychee; other species are lumber trees or oilseed crops.

➤ *Dodonaea viscosa*
  [D. angustifolia and *D. viscosa*
  var. *angustifolia]*

  English name: hopbush
  Spanish names: jarilla (little jar),
    tarachi, ocotillo, alamillo (little
    poplar), guayabillo (little guava),
    granadina, varal, munditos,
    chapuliztle

Hopbush
(*Dodonaea viscosa*)
female plant

DESCRIPTION

Hopbush is a multi-trunked, evergreen shrub typically 5 to 7 feet (1.5-2.1 m) tall, occasionally to 10 feet (3 m), with bright green foliage. Inconspicuous clusters of small, greenish flowers are followed by showy winged fruits resembling hops on female plants. Flowering and fruiting are variable but mostly in summer and fall.

RANGE

In the Sonoran Desert region, hopbush occurs from central Arizona to Sonora and Baja California, from 2000 to 5000 feet (600-1500 m) elevation, always at the upper margin of the desert and often in acidic soils. The same species occurs in warm regions worldwide, even Australia. Exceedingly few plant species have such

a wide distribution; tanglehead grass (*Heteropogon contortus*) is another.

NOTES

The fruit has been used as a substitute for hops in the brewing of beer. True hops, *Humulus lupulus,* is in an unrelated family. The local form of hopbush has narrow leaves. In higher and wetter areas of Sonora and elsewhere the leaves are much wider, up to an inch (2.5 cm) across. Some botanists distinguish the wide-leafed form as a different species. Though hopbush is not succulent it shows evidence of CAM photosynthesis. The Seri use it as an external remedy for aches. It is used extensively in desert landscapes because there are very few fast-growing, evergreen, native shrubs.

Scrophulariaceae (snapdragon family)

Mark A. Dimmitt

This family numbers 4500 species worldwide and includes snapdragons, penstemons, monkeyflowers, owl clover, and Texas ranger. They are a northern (temperate) component of the Sonoran Desert flora; most species respond to the winter rains.

➤ *Castilleja exserta*
  [Orthocarpus purpurascens]

English name: purple owl-clover
Spanish name: escobita
  (whisk-broom)

DESCRIPTION
Owl-clover is a winter annual 6 to 8 inches (15-20 cm) tall. The small, snapdragon-like flowers are nearly hidden among the large, bright pink bracts that tip the stems.

RANGE
It grows in grassland and deserts from central California to northwestern Mexico.

NOTES
All members of this genus are partial parasites, usually on the roots of composite shrubs. Owl-clover is one of our showiest spring wildflowers, occurring in huge, dense stands after wet winters. There are no known ethnobotanical uses, but the O'odham admire it for its beauty.

➤ *Penstemon parryi*

English name: Parry penstemon
Spanish name: jarritos (little jars)

DESCRIPTION
Parry penstemon is an annual or short-lived perennial with a rosette of basal leaves and flowering stems to 4 feet (1.2 m) tall. Snapdragon-like flowers

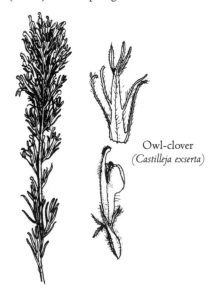

Owl-clover
(*Castilleja exserta*)

(but with open throats) are bright pink to purple-pink.

RANGE
It occurs mostly in the desert from central Arizona to southern Sonora.

NOTES
This is one of the earliest of the showy wildflowers; it's usually in bloom by March 1 in Arizona Upland. Seeds germinate in the fall and flower the following spring. If summers are not too dry, the plants may survive for 3 or 4 years. This is the only desert penstemon that reseeds reliably in desert gardens.

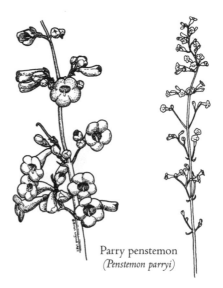

Parry penstemon
(*Penstemon parryi*)

Simmondsiaceae (jojoba family)

Mark A. Dimmitt

➤ *Simmondsia chinensis*

English names: jojoba, goat nut, deer nut, pignut, wild hazel, quinine nut, coffeeberry, gray box bush
Spanish name: jojoba

Jojoba is currently the Sonoran Desert's second most economically valuable native plant, overshadowed only by the Washington palms used in ornamental horticulture. Most people have at least heard of it though they may not know its proper pronunciation (hoe-HOE-buh).

DESCRIPTION

This is a woody, evergreen shrub, averaging 2 to 5 feet (0.6-1.5 m) tall and wide, sometimes to 10 feet (3 m), with leathery, grayish-green leaves. The pale green female flowers of this dioecious species are borne singly at each leaf node. The yellowish-green male flowers are borne in clusters. Plants bloom in winter and female plants ripen their acorn-shaped and -sized seeds in summer.

RANGE

Jojoba occurs throughout the Sonoran Desert where annual average rainfall exceeds 5 inches (125 mm). It extends beyond the desert into the coastal mountain ranges of southern California.

NOTES

The O'odham, who named the plant "jojoba," use a paste of the nut as an antioxidant salve on burns. It was taken back to Spain and the Vatican as early as 1716 as a cure for baldness. (There is no evidence that it is effective in this regard.)

This is the sole species in the family Simmondsiaceae; it is sometimes placed in the box family, Buxaceae. Plants produce flower buds on the new summer growth. The plants must experience a certain number of chilling hours, after which the buds will mature and open in response to a late winter rain.

Though the flowers are wind-pollinated, the process is not completely passive. As breezes pass through the plants the leaves create vortexes. Pollen grains spiral around the female flowers hanging beneath the leaves and thus have several opportunities to contact the stigmas. An average-size bush produces more than 2 pounds (a kilogram) of pollen. The ratio of male to female plants is 4 to 1 in Arizona. Fortunately, few people are allergic to the pollen.

Though it's usually an evergreen, jojoba will shed its leaves during severe droughts. The vertical orientation of jojoba leaves is an adaptation to the extreme desert heat. During midday in summer, when high leaf temperatures make photosynthesis less efficient, the sun is shining on the edges of the leaves. In morning and late afternoon, light reaches the flat leaf surfaces more directly, at a time when temperatures are more favorable for energy capture.

Jojoba "beans" contain more than 40

percent "oil," which chemically is actually a liquid wax. The wax is highly resistant to oxidation (going rancid) and is stable at high temperatures. These properties make it a very high quality lubricant, equal to sperm whale oil. Only sperm whale or jojoba oil is acceptable for some industrial applications. The wax is also used in cosmetics. For these reasons and because sperm whales are endangered, jojoba is being developed as a commercial crop in several countries. Selective breeding is developing plants that produce more beans with higher wax content, as well as other characteristics that will facilitate harvesting.

The foliage provides year-round browse for many animals, especially deer, javelina, bighorn sheep, and livestock. Nuts are eaten by many animals, including squirrels and other rodents, rabbits and larger birds. However, only Bailey's pocket mouse is known to be able to digest the wax. In large quantities, the seed meal is toxic to many mammals, and the indigestible wax acts as a laxative in humans. The Seri, who utilize nearly every edible plant in their territory, don't regard the beans as real food and in the past ate it only in emergencies.

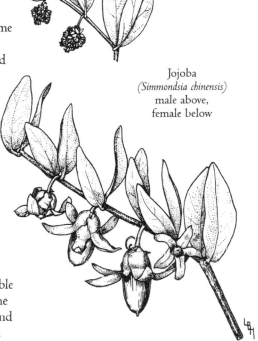

Jojoba
(*Simmondsia chinensis*)
male above,
female below

Solanaceae (nightshade family)

Mark A. Dimmitt

The nightshade family has about 2600 species worldwide and includes herbs, shrubs, trees, and vines. While many are highly poisonous (deadly nightshade, Jimson weed) others are major food crops such as potatoes, tomatoes, and chile peppers. Tobacco is an extremely toxic plant that is grown in huge quantities for two uses: as an addictive drug essential in maintaining cigarette sales and as an important agricultural insecticide. In both products the active ingredient is nicotine. Several useful pharmaceutical drugs are derived from various species of nightshades.

➤ *Datura wrightii*
  [D. meteloides]

  English names: sacred datura, jimson
    weed, angel's trumpet, devil's weed
  Spanish names: toloache grande,
    tecuyaui, belladona

Sacred datura
(*Datura wrightii*)

DESCRIPTION
Jimson weed (its most widespread English name) is a perennial herb 2 to 5 feet (0.60-1.5 m) tall and up to several feet (a couple of meters) wide from a large tuberous root. The foliage is dark green and sticky with an unpleasant odor when crushed. Big funnelform, white flowers are 6 inches long and 3 inches wide (15 x 8 cm); they open by dusk and emit a strong, sweet fragrance. The bloom is sporadic throughout the warm months if soil moisture is present, but it flowers most heavily in late summer. The spiny, golf-ball-sized fruit contains numerous disk-shaped seeds.

RANGE
Jimson weed occurs from central California to Texas and Mexico and into northern South America.

NOTES
All parts of these plants contain numerous toxic alkaloids. One of them is scopolamine, a common ingredient in cold and nausea remedies. Shamans in various cultures have ingested datura to induce visions. This is one of the most dangerous plants used for this purpose, because not only do individual plants vary in potency, but humans also differ in their tolerance to the toxins. Despite widely-published

warnings, every year a few people suffer life-threatening poisoning from eating this plant; some of them don't survive.

This beautiful plant is a useful ornamental if there is sufficient space for its large size and one is willing to put up with its winter disappearance below ground. Hawkmoths pollinate the flowers and lay eggs on the foliage. The caterpillars (called "hornworms" in this family) incorporate the plant's toxins into their own tissues and become toxic to their potential predators.

It would seem that a desert plant should not have such large leaves. But it grows them during the hottest weather and even when there seems to be little soil moisture available. See Plant Ecology chapter for a discussion.

Datura wrightii is our only perennial datura. *Datura discolor* (desert thorn apple, toloache) and *D. stramonium* (jimson weed, hierba del diablo—"herb of the devil") are annuals with smaller dimensions.

➢ *Lycium* spp.

English names: wolfberry, boxthorn
Spanish names: frutilla (generic);
   L. berlandieri: barchata, josó, hosó
   cilindrillo, tomatillo

DESCRIPTION

The several species of *Lycium* in our region are all densely-branched and usually spiny shrubs that are leafless during dry seasons. They range from about 2 feet (60 cm) to over 8 feet (2.4 m) tall and much wider, depending on species and water availability. Leaves are most commonly present during the cool season. Masses of small, greenish to purplish flowers are followed by pea-sized red berries resembling tiny tomatoes.

RANGE

The 100 species of this cosmopolitan genus occur in warm-temperate and sub-tropical habitats. About 15 species occur in the Sonoran Desert.

Wolfberry
(*Lycium pallidum*)

NOTES

Nearly every part of the Sonoran Desert has at least one species of *Lycium*. The structure of the flowers suggests bee pollination, and bees do visit them profusely, but butterflies and hummingbirds also visit in great numbers. Birds relish the fruits. People also highly value the small, tasty berries as a snack. The squeamish should be forewarned that the berries often have insect larvae inside them. This doesn't bother the Seri, one of whom said "Those aren't maggots, they're just live things."

Lac insects (*Tachardiella* spp.) can be found on wolfberries as well as on creosote bushes. A related species, the Indian lac insect, is harvested in huge quantities to make shellac and varnish.

Viscaceae (Loranthaceae) (mistletoe family)

Mark A. Dimmitt

Most of the 1000 worldwide mistletoe species are partial parasites. They have chlorophyll and thus photosynthesize, but take water and nutrients from their host plants.

➤ *Phoradendron californicum*

English name: desert mistletoe
Spanish names: toji, toje, chile de espino (spine chile), guhoja

DESCRIPTION
Desert mistletoe is an essentially leafless plant with dense clusters of brittle, jointed stems. The winter flowers are inconspicuous but strongly fragrant. Female plants produce small red berries.

RANGE
This species occurs in the desert from southern Nevada and California south to central Baja California and southern Sonora. The main host is mesquite; it is also found on other woody legumes and occasionally on *Condalia* and creosote bush.

NOTES
Mistletoe berries are the main winter food of the Phainopepla (Silky Flycatcher). The seeds are extremely sticky and are deposited on other host plants when birds wipe their bills on branches or deposit droppings. A heavy infestation of mistletoe can damage or kill the host plant, but this is uncommon. Both the Seri and O'odham eat the berries, which are sweet when growing on most legumes, and rather bitter when growing on palo verde or non-legume hosts. The Seri made a medicinal tea from the stems.

Big-leaf mistletoe, *Phoradendron macrophyllum* (formerly *P. flavescens* in part), has true leaves and white berries. It parasitizes broadleaf deciduous trees, particularly cottonwoods, willows, and sycamore. Its white berries are bitter and poisonous.

Desert mistletoe
(*Phoradendron californicum*)

Zygophyllaceae (caltrop family)

Mark A. Dimmitt

The caltrop family is a small one of about 250 species. The best-known (and most-hated) species is goat-head or puncture vine (*Tribulus terrestris*), a weed introduced from Europe. Lignum vitae (*Guaiacum* spp.) is perhaps the hardest wood in the world. Though there are only a few caltrop species in our region, one of them is among the most abundant and widespread desert plants and another is a showy summer wildflower.

➤ *Kallstroemia grandiflora*

English names: Summer poppy, Arizona poppy, orange caltrop
Spanish names: vaivurín, mal de ojo (sore eye)

DESCRIPTION
This is a summer annual that grows nearly prostrate to 3 feet (0.9 m) across with sparse divided foliage. Flowers are over an inch across (about 3 cm) with 5 petals. In early morning they open a bright orange with darker striations, then fade to pale orange with the afternoon heat. In the southern part of the range, pink-flowered plants appear as well.

RANGE
Summer poppy grows from northern Arizona to Texas and south to Colima, Mexico; it is also found in southern Baja California. It occurs in desert and tropical habitats but is most abundant in desert grassland, where it creates mass displays of color reminiscent of the desert blooms of spring annuals.

In the desert it occurs as widely scattered individuals or occasional small patches.

NOTES
Summer poppy seeds have a long dormancy. They will usually not

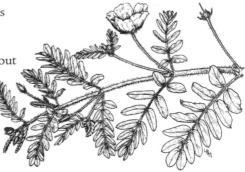

Summer poppy
(*Kallstroemia grandiflora*)

germinate until they are several years old, and presumably remain viable in the soil for many decades.

Kallstroemia parviflora and *K. californica* look vegetatively similar but have much smaller orange flowers. Seedling summer poppies may also be confused with the related puncture vine or goat-head (*Tribulus terrestris*). Puncture vine has tiny yellow flowers and bur-like seeds with strong, sharp spines.

The seeds of puncture vine, a European weed, are well adapted to dispersal by rubber tires, and the plant now grows along the runways of every temperate airport in the world (not to mention most bicycle paths). Once a serious pest in North America, its numbers have been effectively reduced by introducing a weevil that feeds specifically on puncture vine seeds. There is disturbing evidence, however, that the beetle is now attacking summer poppy. Biological pest control (integrated pest management) is much less damaging to the environment than the use of broad-spectrum pesticides, but it too can go awry and must be employed very carefully.

> *Larrea tridentata*
> *[L. divaricata]*

English names: creosote bush,
greasewood, covillea
Spanish names: gobernadora
(governor), hediondilla
(little stinker), guamis

Alphabetically creosote bush comes at the very end of our list of Sonoran Desert plants. If the list were arranged in order of importance—either ecological or cultural—this species would be near the top. Most visitors and Anglo residents notice it only because of its abundance, but those who live intimately with the desert have a very different outlook.

DESCRIPTION
This is an evergreen shrub, 3 to 6 feet (0.9 to 1.8 m) tall, sometimes taller. Many stems rise at an angle from the ground. The small, dark green, resinous leaves are borne mostly at the tips of twigs and the rest of the branches are bare. Small yellow flowers appear throughout the year following rains, most abundantly in spring. The fruit is a small, woolly ball which separates into 5 segments when mature.

RANGE
Creosote bush is the most common and widespread shrub in 3 of the 4 North American deserts. Cold winters exclude it from the Great Basin Desert. It also occurs in Argentina as the same or closely related species *L. divaricata.*

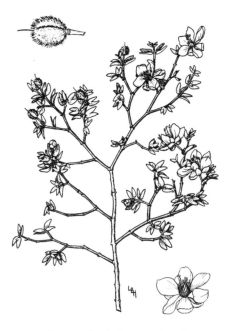

Creosote bush *(Larrea tridentata)*

NOTES

The foliage emits a strong aroma, especially after rains. This fragrance is identified as "the smell of rain" wherever creosote bush grows. Though pungent, it is not repulsive like the wood preservative called creosote oil, a petroleum product unrelated to the creosote bush.

This is the most drought-tolerant perennial plant in North America. Traveling west from Tucson to the mouth of the Colorado River the rainfall decreases from 12 inches (305 mm) to 3 inches (76 mm). Along this gradient the number of perennial plant species steadily decreases from more than 300 to fewer than 12. Along the lower river and the head of the Gulf of California there are valleys with only creosote bush and white bursage on the flats between drainages, and in some places even the bursage drops out, leaving pure stands of creosote. It can live for at least 2 years with no rain, losing its leaves and even shedding branches to reserve its last water and nutrients for the crown.

Although creosote bush roots have been shown to inhibit the growth of bursage roots and the roots of other creosote bushes in their vicinity, the shrub is an important nurse plant. It accumulates enriched soil beneath it and its canopy casts a little shade under which many plants thrive. Most of these are annual species so they are present only in wet years. Creosote bush also commonly shelters cacti such as *Echinocereus, Mammillaria,* and *Peniocereus.* Creosote roots also secrete a germination inhibitor that affects its own seeds only. The average distance between creosote plants depends upon the amount of rainfall and soil moisture-holding capacity for the given area.

Jackrabbits are almost the only mammal that eats creosote bush leaves, and only during droughts when little else is available. The soil beneath these shrubs is a favorite place for numerous rodents to dig their burrows, which are in turn used as shelter by even more species of reptiles and invertebrates. More than 60 species of insects are associated with this plant, including 22 species of bees that feed only on its flowers. Many are specific to it, such as the creosote katydid (*Insara covillei*) and creosote grasshopper (*Bootettix argentatus*), which are so camouflaged that they are very difficult to find. Lac insects (*Tachardiella larreae,* a scale insect) can occasionally be found on its stems. Desert peoples used its sticky secretions as a multipurpose sealant and glue. Ball-shaped leafy galls are common on stems. They are produced by the creosote gall midge (*Asphondylia*); larvae of these small flies live in the protective mass of tissue. The Seri smoked the galls like tobacco.

Creosote bush is among the longest-lived plants. Though each stem lives only a couple of centuries, new ones are continually produced from the outer edge of the root crown. As the dead stems in the center decay, an open ring of stems forms. With passing centuries the ring very slowly expands and breaks into separate bushes, with outliers all of one clone, because they descended from a single original seed. One such ring in the Mohave Desert in California is 26 feet (7.8 m) in diameter and is several thousand years old. Based on current growth rates its age could exceed 11,000 years, but the climate has not always been as arid as today and the plants probably grew somewhat faster in earlier millennia.

Creosote bush is very important to native peoples. The O'odham say it was the first plant created. It is the single most widely-used and frequently-employed medicinal herb in the Sonoran Desert. One of its medicinal names is chaparral tea, though it does not grow in chaparral. The Food and Drug Administration has considered banning its sale based on a couple of deaths attributed to drinking it. But innumerable native peoples and some knowledgeable ethnobotanists drink large quantities of it for a wide variety of ailments with no detectable ill effect (other than gagging from its awful taste). Its antioxidant properties were used in foods and paints through the 1950s and are now being evaluated as anti-cancer agents.

This plant is known as "grease-wood" among the O'odham and many ranchers, but to most other people greasewood is *Sarcobatus*, a Mohave and Great Basin shrub. This is another example of the problem of so-called common names. The Spanish name *gobernadora* is a political commentary. This fairly new name was invented in northern Mexico and meant to be associated with the already established name of *hediondilla* ("little stinker").

Creosote bush originated in South America where there are 5 species of *Larrea*. It is a mystery how it got to North America; today there is no suitable habitat in the intervening thousands of miles. It arrived here long enough ago to have evolved different chromosome races. In the Chihuahuan Desert the plants have the normal *diploid* chromosome number (2 of each chromosome, 26 in this case), as do the ones in Argentina. The Sonoran Desert plants are *tetraploid,* (having 4 sets of chromosomes) and those in the Mohave Desert are *hexaploid* (6 sets). Plants near their ecological limits often have greater chromosome numbers; the extra genetic material seems to confer more adaptability. In the case of creosote, the limiting factor may be cold temperatures; the Mohave is the coldest of these 3 deserts and its creosotes have the most chromosomes.

Desert Grasses

Thomas R. Van Devender and Mark A. Dimmitt

The grass family Poaceae (also known as Gramineae) has some 10,000 species and 650 to 900 genera. Only the sunflower (Asteraceae, or Compositae), legume (Fabaceae, or Leguminosae), and orchid (Orchidaceae) families are larger. The grass family has more individual plants and a wider environmental range than does any other family, reaching the limits of vegetation in polar regions and on mountaintops, enduring extremes of cold, heat, and drought, and dominating various landscapes worldwide. Grasses are the most successful *monocots* (seed-bearing plants with single seed leaves), and the most beneficial plants for humankind, providing highly nutritional grains and livestock forage, and preventing soil erosion. Of the 5 crops that provide almost two-thirds of the food we eat, including corn, barley, potatoes, rice, and wheat, only potatoes are not a grain.

Grasses are highly specialized monocots distinguished by certain characteristics of the stems, leaves, and inflores-cences. The jointed stems (*culms*) are round or flattened (never triangular); they are usually hollow except at the nodes (points on the stem from which leaves arise), where they are solid. The leaves generally consist of 2 main parts: a tubular *sheath* around the plant's stem and a *blade* (the broader, expanded part). Leaves are usually single at each node, but are 2-ranked, that is, spaced alternatively on opposite sides of the stems. Very small, individual,

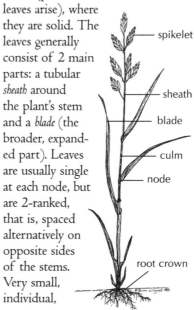

The structure of California brome

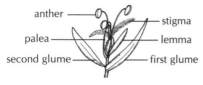

anther — stigma
palea — lemma
second glume — first glume

California brome floret

simple flowers (*florets*) are grouped in inflorescences called *spikelets*, which are subtended by small leaf-like bracts, the *glumes*. The florets are enclosed by other specialized bracts (the *lemma* and *palea*).

There are grasses that use both C_3 and C_4 photosynthetic pathways to fix carbon into tissue (see pages 132-134). Primitive grasses such as the bamboos and modern grasses that live at higher latitudes or elevations where temperatures in the growing season are relatively cool, mostly have the more typical C_3 photosynthesis predominating. However, in the intense sunlight of warmer habitats at lower latitudes and elevations, grasses with the more specialized C_4 metabolism are better able to fix carbon. C_4 grasses are important in tropical forests, semiarid grasslands, and warm deserts.

The earliest grasses lived in shady tropical forests. The evolution and spread of grasses undoubtedly resulted from their ability to adapt to seasonally dry habitats created as tropical deciduous forests developed in the Eocene (58 to 34 mya, million years ago). Considering their importance and taxonomic diversity, grasses have a relatively poor fossil record. While the earliest potential fossil grass pollen was described from late Cretaceous sediments, the oldest reliable megafossil grass fossils were spikelets and inflorescences from the latest Paleocene (about 58 mya). These were primitive proto-

bamboos with broad leaves, quite unlike the narrowleaf modern grasses of desert grasslands and deserts. Although the early fossils could not be assigned to living grasses, their morphological details reflect the early evolution of wind pollination in a seasonally dry tropical environment. In the late Oligocene (about 30 to 24 mya) fossils of central North America, more diverse grass fossils were found, including both archaic forms and quite a few living genera. By the Miocene (24 mya), many

A primitive broad-leaved, C_3 grass (*Pharus cornutus*) found in Coclé, Panama

more modern genera appeared in the fossil record. A great deal of the history of grasses was clearly not captured in the fossil record, notably the evolutionary radiation from primitive proto-bamboos to modern grasses in the Eocene and Oligocene. By the early Miocene, however, grasses in all our modern subfamilies were present, indicating that our modern taxonomic and physiologic diversity had been well established by that time.

DESERT GRASSES

The plants that are most conspicuous in Sonoran desertscrub communities are long-lived trees, such as foothills palo verde (*Cercidium microphyllum*), and cacti, such as the saguaro (*Carnegiea gigantea*), with grasses usually not so noticeable. Yet grasses are surprisingly numerous and play important ecological roles in desert communities.

Growth Forms

Many plants lose parts in times of stress. In temperate and tropical forests, deciduous trees drop their leaves once a year in response to cold or drought. Ocotillo (*Fouquieria splendens*) loses its leaves with drought, but leafs out several times a year depending on rainfall. Desert trees and shrubs such as creosote bush (*Larrea tridentata*), foothills palo verde, triangleleaf bursage (*Ambrosia deltoidea*), and many others, actually lose branches during drought or frost, becoming smaller plants; they could be considered stem- or branch-deciduous. Perhaps the ultimate type of deciduousness is that of the *herbaceous perennials*, including the bunch grasses of semiarid grasslands and deserts. (Perennials are plants that grow and reproduce for many years; herbaceous plants, unlike shrubs and trees, do not become woody.) With drought, frost, fire, and grazing, these plants die

Bush muhly
(*Muhlenbergia porteri*)

back to perennial crowns or roots—they are almost "whole plant deciduous."

When suitable moisture and temperature conditions return, bunch grasses must grow stems and photosynthetic leaf blades before flowering, and their new stems and leaves are pushed up from basal *meristems*. (Meristems are growing points that persist from year to year; these are basal—that is, at or below ground level—in many grasses.) This is in sharp contrast to nearly all other plants, in which the meristems are at the upper or outer stem tips of the plant and leaf-bearing stems are laid down beneath them. The advantage of this basal meristematic growth form is that when the aboveground plant parts are eaten by grazing animals or burned, the stems and leaves can be replaced rapidly, using energy stored in the crown and roots. Perennial desert grasses range in size from low turf grasses like curly mesquite grass (*Hilaria belangeri*) and false grama (*Cathestecum brevifolium*) to the nearly 3-foot (0.9 m) tall Arizona cottontop (*Digitaria californica*) and tanglehead (*Heteropogon contortus*).

There are two distinct alternative growth forms in desert grasses that differ from that of bunch grasses. A few larger grasses, notably big galleta (*Pleuraphis rigida*, formerly *Hilaria rigida*), bush muhly (*Muhlenbergia porteri*), and the introduced buffelgrass (*Pennisetum ciliare*), are functionally more like shrubs than they are like bunch grasses. During normal dry and cool periods, their leaves die back to nodes along the stems. This gives shrubby grasses an advantage: water can move from the soil to the living nodes through the intact stems after rains; new leaves and then flowers form rapidly. With fire or severe frost, the plants die back to the crown, essentially reverting to a bunch-grass life form; massive roots allow for rapid regrowth. Bush muhly is a widespread perennial that apparently has declined dramatically with cattle grazing. Repeat photography in the Grand Canyon has shown that big galleta, bush muhly, and Reverchon threeawn (*Aristida purpurea* var. *nealleyi*) can live for more than a century.

The other common growth form in desert grasses is the *annual*. (Annuals are plants that complete their life cycles—from seed through reproduction and death—in one year or less.) A common trend in grasses and many other desert plants is to evolve toward smaller size and shorter life spans. The best-known adaptations for surviving the heat and drought of fluctuating desert climates are water storage in the succulent stems of cacti, and periodic physiological dormancy in shrubs like creosote bush and brittlebush (*Encelia farinosa*). However, in avoiding environmental extremes by existing most of the time as

Sixweeks fescue
(*Vulpia octoflora*)

Needle grama
(*Bouteloua aristidoides*)

seeds in the soil, annual plants have an equally effective adaptation. Often the entire life cycle of desert annuals is compressed into 6 weeks or less; such short-lived plants are aptly called *ephemerals*.

Desert annuals mostly fall into two groups—winter-spring annuals, related most closely to plants from more temperate zones, and summer annuals, derived from plants of the tropics. Native spring annual grasses include Bigelow bluegrass (*Poa bigelovii*), little barley (*Hordeum pusillum*), and sixweeks fescue (*Vulpia octoflora*). Summer annuals include needle and sixweeks gramas (*Bouteloua aristidoides* and *B. barbata*), and panic grasses (*Brachiaria arizonica, Panicum hirticaule*). Gulf sandbur (*Cenchrus palmeri*), littleseed muhly (*Muhlenbergia microsperma*), and sixweeks threeawn (*Aristida adscensionis*) are opportunistic species that flower in summer or spring. It is interesting that the spring annuals are C$_3$ species derived from temperate relatives, while the summer species, as well as those not restricted to a particular season, are C$_4$'s related to grasses of the New World tropics. Sixweeks fescue is widespread in North America from southern Canada to northern Mexico. It reaches its southern limit in tropical deciduous forest in southern Sonora, where winter rainfall from Pacific frontal storms diminishes.

Several desert grasses are at evolutionary stages between perennial and annual. California brome (*Bromus carinatus*) is a tufted C_3 perennial, mostly 15 to 30 inches (40-80 cm) tall, that occurs from Arizona to California and Washington. In Arizona, this perennial was thought to be restricted to elevations of 5500 to 9000 feet (1675 to 2750 m) in the mountains. In 1944, smaller annual plants from the Sonoran Desert near Tucson were described as *Bromus arizonicus*. Today both plants are considered to be a single species that is perennial in the mountains and annual in the deserts.

Fluffgrass (*Erioneuron pulchellum*) is a widespread C_4 grass in dry rocky slopes up to about 5500 feet (1675 m) from western Texas to Utah and Nevada south into northern Mexico. In summer

Fluffgrass (*Erioneuron pulchellum*)

rainfall climates from the Chihuahuan Desert in Texas west to the northeastern Sonoran Desert near Tucson, it is a tufted perennial no more than 6 inches (15 cm) tall, that mostly flowers in summer or fall, but also sometimes flowers in the spring. On closer examination, individual plants in these populations can be seen to function as

either perennials or annuals. Large, 4 to 8 inch (10-20 cm) diameter individuals live in safe sites and last from season to season (perennials). Seedlings colonize open areas in between. These small plants usually do not survive the season, behaving as annuals, but augment seed production and thus chances of recruitment of new individuals into the population. In the Mohave Desert of southern Nevada, fluffgrass is strictly a spring annual.

Sixweeks grama is a widespread summer C_4 annual found from Texas and southern Colorado to Arizona and northern Mexico. However, some plants near Tucson are intermediates between sixweeks grama and Rothrock grama (*Bouteloua rothrockii*), a perennial species endemic to the Sonoran Desert. Both species are part of a closely related group of gramas with distinctive comb-like flowering spikes, the forms of which range from desert grassland bunch grasses (*B. gracilis*, *B. hirsuta*) to desert annuals (*B. barbata*). Interestingly, the Mayo grama (*B. barbata* var. *sonorae*) in southern Sonora is a perennial variety restricted to sandy soils in coastal thornscrub. It proliferates by *stolons* (above-ground stems that produce new plants, sometimes difficult to differentiate from *rhizomes*, which are below-ground stems). In this case, the perennial form (as opposed to the desert annual) occurs in more tropical habitats rather than in higher-elevation woodlands and forests. In contrast, dwarf hairy grama (*B. quieriegoensis*), another southern Sonora thornscrub *endemic* (it grows nowhere else), is derived from hairy grama (*B. hirsuta*), a desert grassland species.

Grasses in the Desert

The terms "desert grassland," "grassland in the desert," and "grasses in the desert" are sometimes confused.

Desert Grassland

Desert grassland is a distinctive vegetation type in semiarid climates that is widespread at 3500 to 5500 feet (1060 to 1675 m) and is transitional between the short-grass prairie of the Great Plains and desertscrub. The desert grasslands of the Sonoran Desert region are the western edge of the vast grasslands of the middle of the North American continent that extend from the central Canadian provinces south through the Great Plains onto the Mexican Plateau in Chihuahua and Durango, and west across the Continental Divide into New Mexico, southeastern Arizona, and northeastern Sonora. Desert grassland may be dominated by bunch grasses or, especially when disturbed, by various shrubs. (See plate 4.)

Grasslands in the Desert

Grasslands are found within the Sonoran Desert in very special and interesting ecological situations. In local areas on bajadas and in valley bottoms where clay-rich soils prevent water from draining, grass-dominated communities can develop. At Ventana Ranch between Ajo and Sells on the Tohono O'odham Reservation in Arizona, tobosa (*Pleuraphis mutica*) covers a periodically flooded valley at 2100 feet (635 m) elevation. This is an isolated, low elevation, western example of the tobosa swales of the Chihuahuan Desert.

Within the Sonoran Desert, from southern Arizona to the broad valleys of central Sonora, the vegetation types change, reflecting milder winter temperatures and modest increases in summer rainfall and relative humidity. The desert changes from the palo verde-saguaro desertscrub of the Arizona Upland (a subdivision of the Sonoran Desert) to the Plains of Sonora (another subdivision), where trees such as desert ironwood (*Olneya tesota*), palo brea (*Cercidium praecox*), and tree ocotillo (*Fouquieria macdougalii*) are common, although columnar cacti are not. At times and in places, grasses are common and important in the vegetation. With the monsoonal rains of July and August, two summer annuals, needle grama and sixweeks grama, flourish in the open spaces between trees and shrubs, especially in the area from Hermosillo to Guaymas. In September and October, their dried stems provide a subtle light yellow fall color to the desert slopes. Although the gramas are visually conspicuous, their slender stems rarely provide enough fine fuel to support a fire, unlike some non-native species (see "Exotic Species and Ecological Threats" on page 274). In northern Sinaloa, the Mayo Indians collect the grama spikelets from ant mounds and mix them with the clay to make their pots less brittle. Their name, *saitilla*, applies to needle grama and also to the unrelated Spanish needles (*Bidens* spp.—an annual composite), both with sharp fruits that readily travel in fur or socks.

On the basis of dense grasses evident in old photographs and the historical presence of the Masked Bobwhite (*Colinus virginianus ridgwayi*), portions of central Sonora near Benjamin Hill and the Altar Valley of southern Arizona were combined in a new vegetation type called "Sonoran savanna grassland." The combination of these diverse areas under the new term is confusing and misleading. "Savanna" is best restricted to seasonally-flooded tropical grasslands with scattered trees or shrubs. In Africa, "savanna" has been loosely used to describe any vegetation that has both woody plants and grasses, even if the grasses are subordinate or annuals. In Mexico, tropical savannas mostly occur along the Pacific coast from Sinaloa southward. However, near Benjamin Hill in the Plains of Sonora, long-lived trees and shrubs, rather than short-lived grasses, are generally dominant. The lush grass in old photos was most likely Rothrock grama, a short-lived perennial that may be thick and luxurious in a series of wet years, only to virtually disappear during drought. In contrast, true desert grassland with a rich complement of perennial bunch grasses reaches its western limit in the Buenos Aires National Wildlife Refuge in the Altar Valley of Arizona. Other than the bobwhite and such legumes as velvet mesquite (*Prosopis velutina*) and white-ball acacia (*Acacia angustissima*), the Arizona and Sonoran areas in "Sonoran savanna grassland" share few similarities in either community composition or structure.

The Benjamin Hill area does have local areas dominated by false grama—essentially an endemic type of "desert grassland." False grama is a dwarf, tufted perennial grass that spreads prolifically by stolons and forms surprising turfs in open areas. In wetter summers it becomes thick and 6 inches (15 cm) tall. In drought it survives as minute root crowns at the soil surface. The Spanish common name *zacate borreguero* ("sheep grass") reflects its ability to withstand heavy grazing by cattle and sheep. False grama is widespread and common in Sonora but has only been found in Arizona in a single population near Raggedtop in the Avra Valley west of Tucson. With selection for greater cold tolerance, it has potential as a low-water use grass in urban landscapes in Arizona. In the hyperarid lowlands of the Lower Colorado River Valley subdivision of the Sonoran Desert in southwestern Arizona, southeastern California, and the Gran Desierto of northwestern Sonora, the vegetation is a simple desertscrub dominated by creosote bush, white bursage (*Ambrosia dumosa*), and big galleta. In some of the windblown valleys in the Pinacate Region of Sonora, sand piles up on the downwind sides of barren volcanic mountains and big galleta forms pure stands in a starkly beautiful desert landscape.

Rothrock grama
(*Bouteloua rothrockii*)

Grasses in the Desert

In the Sonoran Desert, the grasses are very diverse; over a dozen of them reach the western limits of their ranges in Sonoran desertscrub in the Tucson Mountains. Examples include bull grass (*Muhlenbergia emersleyi*), cane beardgrass (*Bothriochloa barbinodis*, formerly called *Andropogon barbinodis*), curly mesquite grass, gramas (*Bouteloua chondrosioides, B. eriopoda, B. gracilis, B. hirsuta*), plains bristlegrass (*Setaria macrostachya*), plains lovegrass (*Eragrostis intermedia*), and squirrel tail (*Elymus elymoides*, formerly called *Sitanion hystrix*). Studies

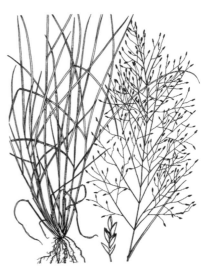

Plains lovegrass (*Eragrostis intermedia*)

Curly mesquite grass (*Hilaria belangeri*) plant and spikelet cluster

of black grama (*B. eriopoda*) in the Jornada del Muerto in the northern Chihuahuan Desert in New Mexico suggest that successful reproduction has only occurred during a few periods of unusually wet years in the last century; favorable conditions for reproduction most likely occur even less often on its western range limits.

In addition, a suite of arid-adapted grasses live in the desert communities in Arizona and Sonora, including Arizona cottontop, bush muhly, fluffgrass, tanglehead, and four different threeawns: *Aristida californica, A. parishii, A. purpurea, A. ternipes*. Grasses are the second most common family (15 percent, 92 taxa) in the Tucson Mountains (610 species total), a desert range usually dominated by palo verde-saguaro desertscrub.

A number of grasses are endemic to the Sonoran Desert. Some of them are widespread from southern Arizona and southeastern California southward into Sonora and Baja California, for example, the threeawns *Aristida californica, A. glabrata*, and *A. parishii*, and Rothrock grama. *Chloris brandegeei* is an annual found in Baja California and on several islands in the Gulf of California. *Aristida peninsularis* and *Bouteloua annua* only occur in Baja California. The Gulf sandbur is an annual endemic to sandy habitats in Baja California and

Sonora on both sides of the Gulf of California. Sandburs (the genus *Cenchrus*) have bristles on the fruit that are fused at their bases and modified into the spiny burs that allow the seed to be very effectively transported from place to place by mammals—including

Arizona cottontop *(Digitaria californica)*

humans. Sandburs are closely related to grasses in the genus *Pennisetum*, which do not have their bristles fused. (Buffelgrass, an exotic species discussed below, is confusing in this regard, since it has weakly-fused bristles; it is placed in either genus by different specialists.) Chromosome evidence indicates that sandburs evolved from *Pennisetum*.

Desert needlegrass *(Stipa speciosa)* is one of the few C_3 spring-flowering perennial grasses in the Sonoran Desert. It occurs from Colorado to Arizona and southern California. In the Mohave Desert of western Arizona, it flowers in the spring, and is occasionally encoun-

tered in desert mountains, where it reaches its southern distributional limit. Fossils in ancient packrat *(Neotoma* sp.) middens demonstrate that the range of desert needlegrass expanded southward in the winter-rainfall dominated climates of the late Wisconsin glacial period, and especially the early Holocene, 11,000 to 9,000 years ago (see page 63). This grass also occurs in southern South America, as do a number of other species with *amphitropical discontinuous distributions* (that is, occurring on both sides of the tropics, but not in them—for example, in both North and South America). Other amphitropical species are creosote bush, little barley *(Hordeum pusillum)*, allthorn *(Koeberlinia spinosa)*, and *Atamisquea emarginata.*

Tanglehead *(Heteropogon contortus)* is widespread in tropics of the world, a distribution pattern shared by hopbush *(Dodonaea viscosa)*. Tanglehead is an example of a tropical species reaching its northern limit in desert grassland in southwestern United States at about 5500 feet (1675 m) elevation. It is "wedged out" by cold temperatures at higher elevations and by drought in the desert below. Other tropical species with similar distributions are coral bean *(Erythrina flabelliformis)*, kidneywood or palo dulce *(Eysenhardtia polystachya)*, brown vine snake *(Oxybelis aeneus)*, and green ratsnake *(Elaphe triaspis)*.

Threeawns, genus *Aristida*, are a taxonomically difficult group containing about 150 species in temperate and subtropical areas in North America and about 40 species in the United States. The fruit usually has 3 awns (stiff hairs, or bristles), thus the common name. The threeawns in the Sonoran Desert present several interesting patterns. Sixweeks threeawn is found from Missouri and Texas west to California and southward to Argentina, and also in warmer parts of the Old World. It is a highly adaptable and opportunistic

Purple threeawn *(Aristida purpurea)*

annual that grows in both spring and summer and can be from 2 inches to 2 feet (50-600 mm) tall, depending on rainfall. Purple threeawn *(Aristida purpurea)* is a common perennial in Arizona Upland desertscrub in Arizona. Typical purple threeawn *(A. p.* var. *purpurea)* with striking, nodding red infloresences commonly occurs on roadsides. (The edges of roads and highways are pseudo-riparian habitats with disturbed soils and concentrated runoff water—easy dispersal corridors for many native and introduced grasses.) If trimmed at the end of the spring and summer growing seasons, purple threeawn is an attractive ornamental in xeriscapes.

In contrast, Reverchon threeawn *(A. p.* var. *nealleyi)* only occurs in rocky desertscrub habitats and rarely on roadsides. Superficially it is very similar to *A. parishii*, which lives in the same upland habitats, mostly differing in details of the fruit. In this threeawn and in many other desert plants, including borages *(Amsinckia* spp., *Cryptantha* spp., *Plagiobothrys* spp.), spiderlings *(Boerhavia* spp.), and spurges *(Euphorbia* spp.), there has been far more evolution of the seeds and fruits than of the vegetative plant parts, likely reflecting selection for differences in dispersal methods or microhabitats. In *Aristida ternipes, A. t.* var. *gentilis* has 3 awns (bristles) and mostly occurs in desert grassland, whereas spider grass *(A. t.* var. *ternipes)* is a single-awned "threeawn" that is widespread in desert and tropical lowlands. Spider grass is very responsive to rainfall and, like many perennial grasses, can flower in its first year. A seedling encountered in early September in the Tucson Mountains was flowering only 5 weeks after germination in response to a 1.8 inch (46 mm) rainfall. In the desert, spider grass is a low, rounded bunch grass 1 or 2 feet (30-60 cm) tall. In a tropical deciduous forest near Alamos in southern Sonora, with annual rainfall of 20-25 inches (500-630 mm), roughly twice that of the Tucson area, spider grass grows to be an erect, elegant 4 to 5 foot (120-150 cm) tall plant.

Exotic Species and Ecological Threats

There are introduced exotic plants throughout the Sonoran Desert region in Arizona and Sonora. With the notable exception of riparian habitats in Arizona and northern Sonora, introduced species usually account for relatively low percentages of local species present, and are mostly innocuous with few serious impacts on the vegetation. Introduced species are most diverse and abundant in riparian habitats (river bottoms, arroyos, washes) and pseudoriparian habitats (road edges) because they are disturbed, unstable dispersal corridors that harvest water, nutrients, and seeds from large areas. Successful invaders are often short-lived, fast-growing plants, which have high reproductive effort. Longer-lived exotics are usually "mortality resistant"—survivors not easily killed by environmental stresses (floods, fire, drought, freezes, heavy grazing).

Unfortunately a few exotic species have the potential to cause ecological havoc in the Sonoran Desert region and threaten irrevocable landscape changes; most of the worst of these are grasses. Competition with native species is typically intense and can alter the composition of the flora and vegetation. When a new ecological process is introduced into an ecosystem as a result of an introduction of a new species, the impact on the vegetation is more severe—the entire vegetation type can be converted to a different assemblage.

In subtropical desertscrub, tropical thornscrub, and deciduous forest, fire fueled by exotic grasses can be devastating, as most of the community dominants have little or no adaptation to fire.

Some annual grasses native to the Mediterranean areas of Europe are especially troublesome. They are preadapted to the winter rainfall climate and fire regimes of the chaparral vegetation in California. As they have moved eastward into the Mohave Desert, with its winter rainfall, they

have directly competed with the native spring flora and introduced fuel and structure conducive to fire.

As they moved further eastward and southward into the biseasonal climatic regimes of the Sonoran Desert, their ecological interactions have been more complex. In desert grassland, competing subshrubs (winter-spring active) and perennial grasses (summer active) have differing water-use strategies. In desertscrub, a similar seasonal competition for soil nutrients occurs between spring and summer annuals. In the spring, introduced annuals compete directly with native spring herbs for water, space, and nutrients. The roots of the introduced annual grasses, including mouse barley (*Hordeum murinum*), red brome (*Bromus rubens*), and wild oats (*Avena fatua*), are active at relatively cool soil temperatures, accelerating their growth compared to native annuals. Often they are so prolific that few nutrients remain for summer ephemerals. However, the alterations of community structure and competition due to fires are much more serious.

Red brome is a weedy Mediterranean annual that was established in California by 1848, and is presently common through much of western United States. It is seasonally abundant and widespread in the Sonoran Desert region. In lower, more arid areas, it is mostly found in disturbed habitats, while at higher elevations throughout the Arizona Upland it also occurs in undisturbed habitats. Observations in a Lower Colorado River Valley creosote bush desertscrub at 1560 feet (475 m) elevation in the Eagletail Mountains west of Phoenix illustrate the variability of these annuals. In the spring of 1992, red brome a foot

(30 cm) tall covered the slopes in response to excellent winter-spring rains. The spring of 1996 was very different. In response to a light rain in late February, tiny, approximately 2 inch (50 mm) tall sixweeks threeawn were widespread, but red brome did not germinate, suggesting that native species have some advantages when rainfall is

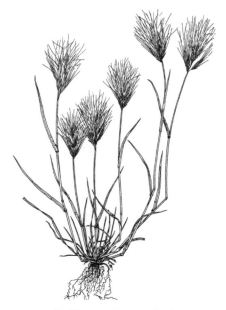

Red brome *(Bromus rubens)*

light and soils are relatively warm. Nonetheless, in the last decade or so, early summer wildfires in Sonoran desertscrub, mostly fueled by dried red brome, have increased 10- to 20-fold.

Arabian grass *(Schismus arabicus)* and Mediterranean grass *(S. barbatus)* are closely-related winter ephemerals that are geographically segregated in the Old World: Arabian grass from southwest Africa to the western and northern Sahara and the western Mediterranean region, and Mediterranean grass from Kashmir and southern Russia west to

Greece. The differences between them are subtle and these plants are difficult to tell apart.

During years of favorable winter rains, these grasses can be abundant across much of the northern part of the Sonoran Desert, forming extensive, dense carpets on relatively flat, sandy terrains. The first stems and leaves often spread out close to the ground, effectively excluding or preventing other ephemerals from sprouting. The earliest record for Arabian grass in North America seems to be from 1933. Mediterranean grass does not seem to be as common in our region as Arabian grass, and apparently does not extend much south of the Arizona border. Interestingly, the earliest North American records for this grass are from 1926 in southern Arizona and from 1935 in Fresno County, California—the reverse of the typical dispersal pattern of Mediterranean weeds, which usually first settle in California and follow the highways eastward. Today

Mediterranean grass *(Schismus barbatus)*

this exotic grass is a preferred food for desert tortoises *(Gopherus agassizii)* in the Mohave Desert, instead of native annuals such as lupines *(Lupinus* spp.), and deer vetch *(Lotus* spp.). This may cause physiological problems, because tortoises in the Mohave Desert may need the higher water and protein content of the native annuals to survive the dry summers. In the Tucson Mountains, the introduced red brome joins native curly mesquite grass and fluffgrass as a preferred food of the desert tortoise.

Buffelgrass is a shrubby savannah grass native to the warmer parts of Africa, Madagascar, and India, that is widely introduced and established in hot, semiarid regions of the world for forage and fodder. A strain from the Turkana area of Kenya was officially released for planting by the U.S. Soil Conservation Service in San Antonio, Texas, in 1946. The shrubby habit of buffelgrass allows rapid growth of leaves from the nodes and copious seed production in response to rain. It is very adaptable, likely due to its great genetic variability and largely *apomictic* reproduction. (Apomixis is a form of asexual reproduction where seeds are produced without fertilization or any sexual union.) The plant burns readily (even when green) and recovers quickly from fires.

Since the 1960s buffelgrass has been extensively introduced for livestock forage in Arizona and Sonora. In Arizona the expansion was initially slow in areas where buffelgrass was not actively seeded into cleared areas. However, in recent years, buffelgrass has begun to spread rapidly along highway shoulders both in lower and higher elevations. It also is invading desertscrub communities on rocky slopes away from roads. In 1992 when a floristic survey of the Tucson Mountains was completed, buffelgrass was only occasionally encountered, and only noticeably invasive on Tucson's

Tumamoc Hill and "A" Mountain. (On "A" Mountain, fires accidentally set during the annual Fourth of July fireworks stimulated buffelgrass expansion.) In the past 6 years, dense patches of buffelgrass have become established in several new areas. On Radio Towers Peak, a dirt road provided a dispersal corridor and entry into palo verde-

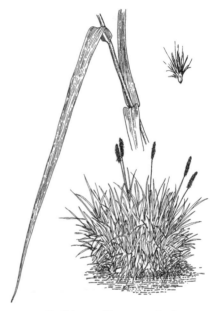

Buffelgrass (*Pennisetum ciliare*)

saguaro desertscrub, while on Panther Peak buffelgrass established itself away from roads. In metropolitan Phoenix, buffelgrass is well established on North Mountain and on many other mountains, especially on upper south-facing slopes, where it apparently outcompetes and eliminates native cacti and trees. Such upper slopes appear sandy-colored in winter when buffelgrass is dormant, and are readily distinguishable to the informed eye.

In central Sonora, more than a million hectares (470 thousand acres) of desertscrub and thornscrub have been cleared to plant buffelgrass, often as part of government-subsidized programs to support the ranching industry. Fire cycles in buffelgrass allow the grass to expand rapidly into large areas of uncleared vegetation, very much like cancer spreads in a human body. Along many highways, especially in central Sonora, dense monocultures of buffelgrass have replaced other weedy roadside species, especially brittlebush. Recurrent fires maintain the buffelgrass. Ranchers have been forced to replace wooden posts with metal ones, and the bases of power poles are shrouded with metal or cement. Fires fueled by buffelgrass are now a serious urban problem in Hermosillo, Sonora.

In southern Sonora, buffelgrass has been planted in clearings in tropical deciduous forest but apparently is not able to invade undisturbed shady forests. With management using fire, buffelgrass *praderas* (prairies) can be maintained indefinitely. However, ranchers are reluctant to use fire lest they lose the last tuft of forage. In many areas, native species including boat-thorn acacia (*Acacia cochliacantha*), Indian mallow or pintapan (*Abutilon abutiloides*), and Sonoran bursage (*Ambrosia cordata*) invade buffelgrass pastures. Without fire, boat-thorn acacia grows tall enough to shade buffelgrass, and succession backwards to tropical deciduous forest may well occur.

This "grasslandification" of desertscrub and thornscrub now occurring is the opposite of the well-known "desertification" of the southwestern United States—the expansion of shrubs into desert grassland beginning in the late-19th century. Ironically, because buffelgrass is such a hardy, drought-

resistant grass and excellent forage, great effort has been expended to introduce it and to select hardier varieties. The ecological result of buffelgrass introduction into fire-intolerant subtropical and tropical communities is a permanent (on our time scale) conversion to a savanna, much like those in Africa, with drastically reduced plant cover and overall diversity.

Fountain grass (*Pennisetum setaceum*) is another robust, often 3 foot (1 m) tall perennial clump grass from Africa. It is a common landscape ornamental in southern Arizona, where it is slowly spreading into natural habitats, especially in desert riparian canyons. It has been cultivated in Tucson since 1940 and

Fountain grass (*Pennisetum setaceum*)

was established in the nearby Santa Catalina Mountains by 1946. The larger, purplish cultivar 'Cupreum' is a better choice for landscape plantings, since it does not volunteer, that is, invade areas where it is not planted.

Natal grass (*Rhynchelytrum repens*) is another African grass that is common on roadsides in many areas in Sonora. This attractive grass with fluffy, rose-colored inflorscences is invasive in desert grassland in Sonora south of Nogales and near Maycoba in the Sierra Madre Occidental, as well as eastward into Chihuahua. In the mountains north of Guaymas on the Gulf of California, it has invaded steep slopes in undisturbed palm canyons. High humidity likely helps it survive in the arid Central Gulf Coast desertscrub. In these conditions, it could potentially fuel fires in vulnerable desertscrub vegetation.

The introduction of the South African Lehmann lovegrass (*Eragrostis lehmanniana*) has begun a transformation of desert grasslands in southeastern Arizona. It produces 2 to 4 times the annual biomass of native grasses and responds favorably to grazing and fire. It is replacing native grasses in vast areas, dramatically altering community composition without conversion to desertscrub. It has followed major highways well into the Sonoran Desert west of Tucson.

Grasses are one of the largest, most diverse families of plants, found worldwide in virtually all climatic regimes and vegetation types. They are the most important plants economically, providing edible grains for people and forage for livestock. Superbly adapted to live in variable desert environments, they avoid extremes in drought and heat through life history strategies. Well-hydrated plants grow rapidly in response to spring or summer rainfall, dying back to stem nodes (shrubby grasses), root crowns (bunch grasses), or seeds (annuals) during dry periods. The same adaptations help grasses to survive and recover rapidly from fires, grazing, and other severe disturbances.

Some of the many exotic species reaching North America in today's intercontinental travel and trade are grasses. Successful species from other arid and semiarid lands, especially Africa and the Mediterranean, are rapidly expanding in the Sonoran Desert region, competing with native species. When new ecological processes such as fire are introduced, the impacts are devastating.

Fire-intolerant desertscrub and thornscrub can be permanently converted to Africanized "savannas" or weedy Mediterranean annual "grasslands." Increases in fires and in the total area of disturbed communities in the last few decades in Arizona and Sonora should be a call to arms to battle these invasive Old World grasses.

Lehmann lovegrass (*Eragrostis lehmanniana*)

Additional Readings

Gould, Frank W. *Grasses of Southwestern United States.* Tucson: University of Arizona Press, 1977.

McClaran, Mitchel, and Thomas R. Van Devender, eds. *The Desert Grassland.* Tucson: University of Arizona Press, 1995.

Webb, Robert H. *Grand Canyon: A Century of Change. Rephotography of the 1889–1890 Stanton Expedition.* Tucson: University of Arizona Press, 1996. (Included here is a table of plants recorded to live more than a hundred years, including grasses.)

Invertebrates

Section Contents

A Vertebrate Looks at Arthropods

Barbara Terkanian

At close range even a familiar arthropod like a grasshopper may seem appallingly strange. The creature's expressionless eyes give no indication that it can think or feel. Its colored armor plating seems more suited to a machine than to an animal. There are too many legs jointed in too many places. The abdomen pulses, the antennae twitch, and tiny oral appendages shift food to the mouth. Lean a little closer and the thing leaps, almost into one's face, to veer off on crackling wings.

Such an experience may strengthen the conclusion that arthropods are so different from us that they are essen-tially alien life forms. But such an extreme first impression is not really justified. Arthropods are classified as animals, exactly as we are, and so they must have characteristics in common with us and with other members of the animal kingdom. Certainly they do. Arthropods share numerous physical, biochemical, and developmental charac-teristics with the other creatures classi-fied into the 30-odd phyla composing the Kingdom Animalia.

We can concede that arthropods qualify as animals, but how does the grasshopper in the garden fare in comparison to vertebrate animals like ourselves? In fact, arthropods provide

Kingdom: Animalia
Phylum: Arthropoda
   Animals with jointed external skeletons, such as lobsters, spiders, centipedes, and insects.
Phylum: Chordata
   Animals with dorsal nerve cords (vertebrates and others). The vertebrates—fishes, amphibians, reptiles, birds, and mammals—have bone or cartilage skeletons. Non-vertebrate members of this phylum lack skeletons.

an interesting counterpoint to vertebrates. Curious parallels and strong contrasts exist between the two groups. Similarities in skeletal function are associated with historical parallels in the colonization of land and the evolution of flight. The giant size of vertebrates contrasts strongly with the much smaller size of arthropods, while the extraordinary diversity of arthropods eclipses the modest radiation of vertebrates.

The Design of Arthropod and Vertebrate Skeletons

Both arthropods and vertebrates have *articulated* skeletons. Components of the skeleton meet (articulate) at joints, which allows one part of the body to move in relation to another. Muscles spanning joints and anchored to different parts of the skeleton provide the power for movement. Articulated skeletons serve two functions. First, they allow their owners to retain a characteristic physical form. Second, they support an organism's weight and resist the stresses of locomotion. Other kinds of animals, like snails, have hard shells into which they may retreat for protection, but these structures do not define the animal's form or support its weight.

Arthropod and vertebrate skeletons are quite distinct from each other. Basically, the vertebrate skeleton is internal (an *endo*skeleton) while the arthropod skeleton is external (an *exo*skeleton). Here, both kinds will be referred to as *skeletons*. The vertebrate skeleton is buried under skin and muscle. Within the body, skull, and vertebral bones encase the brain and spinal cord, while ribs protect the heart and other organs. Long bones form the internal core of

the legs. With arthropods, virtually all external structures as well as a few internal ones are covered by exoskeletal material, including eyes, mouthparts, antennae, body, legs, the fore and hind sections of the digestive tract, and some respiratory surfaces. Regions of flexible, unhardened exoskeleton serve as joints between neighboring segments.

That articulated skeletons arose only in arthropods and vertebrates is simply an evolutionary coincidence. The two groups are not closely related to each other. But functional advantages common to skeletons are related to two novel parallels in arthropod and vertebrate evolution: first, vertebrates and arthropods are perhaps the most successful of all terrestrial animals; and second, they are the only organisms ever to evolve genuine powers of flight.

Life on Land and in the Air

A way to appreciate the advantages of having a skeleton in a terrestrial environment is by comparing animals with skeletons to "soft-bodied" animals. The latter category includes snails, slugs, and a great variety of worms. Some of these animals are ancient residents of land environments. But compared to animals with skeletons, soft-bodied terrestrial animals display a limited variety of form and diversity of locomotion.

The external forms of soft-bodied terrestrial animals resemble each other despite substantial differences in each species' internal anatomy. Nearly all have fleshy, cylindrical bodies without legs, and most lie full length upon the ground. These creatures are limited to slug-like forms because they depend on muscle to retain form and support their weight. Their aquatic relatives (octopus, squid, and various worms) can evolve complex forms because they live in water, which is denser than air and provides more support for attenuated appendages like arms. Even among vertebrates, purely muscular "limbs," such as an elephant's trunk or a chameleon's tongue, are not employed to support their owner's weight.

Animals with skeletons can use the passive strength of bone or hardened exoskeleton to support their bodies in air. Consequently, terrestrial vertebrates and arthropods have more diverse forms, unlike soft-bodied land animals. Turtles, flamingos, ballet dancers, and rhinos look nothing like slugs. Praying mantids, stalk-eyed flies, and scorpions cannot be said to resemble worms. Toes, legs, necks, antennae, tails, and wings are great departures from a basic cylindrical form. And all of these departures are of functional value to their owners. Long, jointed limbs would hardly seem to be an evolutionary innovation, but they and other jointed appendages account for the great freedom of movement enjoyed by terrestrial arthropods and vertebrates.

Without skeletons, most soft-bodied animals creep or writhe over surfaces by contracting muscles of the body wall. Terrestrial animals with skeletons use their limbs to walk, trot, run, leap, brachiate, and fly. They can do so because they are constructed differently from soft-bodied animals. Typically, arthropods and vertebrates stand with their bellies clear of the ground. This means that as one of these animals moves, its body passes through air, instead of over the ground, where friction has a greater effect. The foot of a running arthropod or vertebrate touches the ground just long enough to propel the animal forward with each stride. This enables it to run swiftly, compared to ground-bound, soft-bodied animals. Some arthropods and vertebrates are swifter yet, because they have evolved the ability to fly.

Flight is exclusive to animals with skeletons. Birds, bats, and a few species of dinosaurs have evolved flight. Among arthropods, only insects can fly, but since insects amount to 70 percent of all animal species, arthropods account for the vast majority of flying animals. All flying animals must be able to satisfy stringent flight requirements. They must have extensive respiratory systems to supply quantities of oxygen to flight muscles, sophisticated nervous systems to coordinate rapid flight responses, and of course, they must have functional pairs of wings.

Wings have arisen by different pathways in arthropods and vertebrates. In vertebrates the foreleg is modified into a wing; as a consequence the limb becomes useless or compromised for quadrupedal locomotion. Birds walk only on their hind legs. Bats use their wings as legs, while keeping their extraordinarily long fingers and extensive wing membranes folded out of the way. Arthropods have sacrificed nothing for flight; their wings evolved as completely new structures rather than as modifications of existing limbs. If one counts, insects retain three pairs of walking legs and two pairs of wings. All three pairs of legs may be used for locomotion or adapted for specialized tasks such as prey capture. Sometimes one pair of wings is modified into yet another arthropod tool while the remaining pair is used for flying. For example, beetle forewings have changed into sturdy protective coverings for the membranous hind wings. The hind wings of flies have become balancing devices necessary for their dizzying aerial maneuvers.

Interestingly enough, there appear to be limits to the sizes flying animals can attain. The largest flying birds weigh about 30 pounds (13.5 kg). For tiny arthropods, the very density of air becomes an important factor. Insects weighing less than $3/10{,}000$ of an ounce (a milligram) can drift wingless through air just as plankton does through water. Size is an important matter to a flying animal. It is also an important distinction between vertebrates and arthropods.

Vertebrates Are Giants; Arthropods Are Small

Vertebrates are the biggest animals ever to evolve on Earth. We know of giant dinosaurs, huge amphibians, massive sloths, enormous moas, and the colossal blue whale. In comparison, the very biggest terrestrial arthropods can be held in the palm of the hand. *Pterygotus*, a long-extinct aquatic arthropod, did qualify as a giant—it approached 10 feet (three meters) in length. Ancient exceptions aside, there is little overlap in the size ranges for species belonging to each of the two groups. Small vertebrates are still larger than the great majority of arthropods, the tiniest of which are best viewed under a microscope. So why does gigantism arise frequently in vertebrates but not in arthropods?

An increase in size has physical consequences for any creature. As an animal doubles in size, its weight increases eight-fold, but the weight bearing capacity of its skeleton is only quadrupled and the strength of its muscles is merely doubled. Because of their great weight, large vertebrates have skeletons which are disproportionately heavy and robust compared to those of small vertebrates. Presumably, terrestrial arthropods could reach horror-movie size simply by developing big, sturdy skeletons. But they have not done so during hundreds of millions of years on Earth. It seems that the costs associated with large size affect arthropods more strongly than vertebrates.

A heavy, cumbersome skeleton, risk of injury, and complications during molting all become more serious problems at large size. The arthropod skeleton accounts for a greater proportion of its owner's total weight than does the skeleton of a vertebrate. As an arthropod gets larger, the proportion of weight attributed to the skeleton will increase faster than it does for a vertebrate. At some point, the advantages of increased size will not compensate for the difficulties associated with a heavy skeleton. When that happens, natural selection will favor the smaller individuals in a population.

One important difficulty for large arthropods is the risk of injury. As the outermost part of an arthropod's body, it is the rigid skeleton that comes in contact with the environment. Without the cushioning effect of soft tissues, it is more vulnerable to abrasion and impact damage than the internal skeleton of vertebrates. Running becomes hazardous because all of the weight of a heavy arthropod would come down on the relatively small area of the foot. Without the shock absorption provided by the hooves, paw pads, cartilage, and ligaments found in vertebrate extremities, an external skeleton might be expected to fracture under the force of impact. A simple fall might be even more damaging.

Finally, molting, necessary for growth, causes other problems at large sizes. Just after molting an arthropod is essentially a soft-bodied invertebrate. The skeleton is still soft, and does not provide good support. Worse, an arthropod cannot rely on muscles to define form the way soft-bodied animals do, because arthropod muscles are designed to exert force against a rigid skeleton, and until the skeleton hardens, many muscles are useless. Instead, an arthropod gulps air or water in order to hold its form until the skeleton hardens. The larger an arthropod's size, the more difficult this process becomes. Each time an arthropod molts it must undergo this risk. Vertebrates do not molt their skeletons as part of growth so they escape these risks completely.

Arthropods may be unable to attain the impressive sizes of vertebrates, but their small size is related to a big distinction—their extraordinary diversification.

More than a Million Arthropod Species

Arthropods are the most diverse of all animals, comprising over 85 percent of all living animal species. Estimates for the number of species in one class of arthropods, the insects, range from 1 to 10 million. In contrast, the entire Phylum Chordata—vertebrates and their close relatives—amounts to fewer than 5 percent of all living animals. The remaining 10 percent are account-ed for by other invertebrate phyla, such as molluscs. Why is there such an overwhelming number of arthropod species compared to all other kinds of animals? Why are there relatively few vertebrate species, despite their sophisticated internal skeletons and access to terrestrial environments? The answers have to do with arthropods' relatively small size, their capacity for rapid change, and their long tenure on Earth.

Small animals can exploit habitats more fully than large ones. A single plant may be a meal to a vertebrate, but to arthropods it can be a universe. One species might complete larval development in a flower bud, while another species spends its entire life feeding on the woody stems. A large plant like the saguaro can support an entire community of arthropods throughout its life and after its death. Other habitats, such as the surface of water and the bodies of other animals, are used by arthropods, but are inaccessible even to the smallest vertebrates. Winged insects or ballooning spiders can travel great distances, colonizing new habitats quickly. As they invade new habitats arthropods undergo selection which favors individuals best equipped to survive in the new conditions. Over time, the better-equipped individuals may come to differ so much from their ancestors that they become distinct species.

Arthropod populations can undergo rapid change. Agricultural pests are well-known for swiftly evolving tolerance to previously devastating pesticides. Short generations, multiple generations per year, and large populations are conducive to the prompt emergence of new forms, and under the right conditions, new species. Vertebrates are also capable of change and speciation, but because of their longer generation intervals these processes tend to require more time. While some arthropod species—fruit flies for instance—change and diversify rapidly, others, such as scorpions and horeshoe crabs, settled into a useful design early on and have remained unchanged for millions of years.

Finally, arthropods have been around for a long time. Trilobites, an early and now extinct group of marine arthropods, lived 500 million years ago. Early terrestrial forms, like scorpions, are known from 400-million-year-old fossils. Reptiles, the first entirely terrestrial vertebrates, did not arise until the Carboniferous period, approximately 100 million years later. Insects evolved flight 150 million years before birds, dinosaurs, and mammals. There has been plenty of time for diversification and evolution of many exquisite, bizarre, and intriguing arthropod species.

Similar, Different, and Interesting

Remarkable parallels and contrasts can be developed when arthropods and vertebrates are compared. But there are reasons to focus on arthropods alone, without considering them in relation to other animals. In the name of survival, arthropods have evolved forms ranging from familiar to outrageous to beautiful. They evade enemies, feed, and reproduce by methods that are sometimes ruthless, sometimes subtle, and frequently ingenious. As is their habit, arthropods in the Sonoran Desert have diversified, giving this region a rich inventory of fascinating species.

You are invited to form a closer acquaintance with native Sonoran Desert arthropods by reading about them in the following chapters. The chapters are organized on the basis of arthropod phylogeny, which reflects the evolutionary relationships between species. As you proceed through this section, you can use the phylogenetic listing following this introduction to keep track of arthropod groups.

These intriguing and largely harmless creatures are among the most visible and approachable of all Sonoran Desert animals. Keep an eye out on your next desert walk, and you may see an arthropod or two that you have previously encountered in the pages of this book.

Additional Readings

Alcock, John. *In a Desert Garden: Love and Death Among the Insects.* NY: W.W. Norton, 1997.
Conniff, Richard. *Spineless Wonders: Strange Tales from the Invertebrate World.* NY: Henry Holt, 1997.
Evans, Arthur V. *An Inordinate Fondness for Beetles.* New York: Henry Holt & Co., 1996.
Friederici, Peter. *Strangers in Our Midst: The Startling World of Sonoran Desert Arthropods.* Tucson: Arizona-Sonora Desert Museum Press, 1997.
Imes, Rick. *The Practical Entomologist.* New York: Fireside, 1992.
Smith, Robert L. *Venomous Animals of Arizona.* Tucson: University of Arizona, 1982.
Werner, Floyd G. and Carl Olson. *Learning about & Living with Insects of the Southwest: How to Identify Helpful, Harmful and Venomous Insects.* Tucson: Fisher Books, 1994.

A Phylogeny for Arthropods Mentioned in this Book

In essence, phylogenies are family trees. Researchers group living things into increasingly specific categories based on reconstructions of the organisms' evolutionary histories. Since most of the branching in the arthropod family tree took place before humans came into existence, the tree structure is deduced from the study of fossils, and the examination of molecular, developmental, anatomical, and behavioral characteristics of contemporary species.

Many characters important in determining relationships between species are not observable in the field. The informal notes in this phylogeny are intended to help readers develop a feel for arthropod classification by using visual characteristics alone.

Kingdom Animalia

    Phylum Arthropoda

Animals with an exoskeleton, a segmented body, and jointed legs. The earliest arthropods probably had one pair of appendages per body segment, but there have been many divergences from the ancestral arrangement. Segments may be fused or grouped into body regions and appendages may be exaggerated, modified, or lost.

        Subphylum Chelicerata

Members of this subphylum have two major body divisions, the cephalothorax (the head and mid section combined) and the abdomen. The first pair of appendages on the cephalothorax is modified into jaw-like structures. Chelicerates have simple eyes resembling unfaceted beads. All species in this category lack antennae and wings.

            Class Arachnida

The arachnid cephalothorax appears to be unsegmented. In some orders, the abdomen is clearly segmented (scorpions); in others (most spiders) no segments are apparent. Arachnids have four pairs of legs; however, some arachnid mouthparts (such as scorpion claws) have evolved into structures that could be mistaken for limbs. Almost all arachnids are terrestrial.

                Order Scorpiones (scorpions)
                Order Uropygi (vinegaroons)
                Order Araneae (spiders)
                Order Amblypygi (tailless whipscorpions)
                Order Pseudoscorpiones (pseudoscorpions)
                Order Solifugae (sun spiders)
                Order Opiliones (daddy-long-legs)

Subphylum Crustacea

Members of this group have two pairs of antennae and branched (biramous) appendages. Five pairs of appendages are associated with the head, including a pair of jointed mandibles. Many species have compound eyes, which are made up of simple eyes grouped together to form faceted spheres. Most species are marine.

Class Malacostraca

This class contains some of the most familiar crustaceans. Members have eight trunk segments and six abdominal segments. All abdominal segments bear appendages. Only a few species belonging to this group are found in the Southwest.

Order Decapoda (crayfish)

Subphylum Uniramia

Species in this subphylum are distinguished from ones in subphylum Crustacea by having a single pair of antennae, unbranched (uniramous) appendages, and mandibles that are usually unjointed.

Class Chilopoda

These animals have a long, flattened body with numerous segments. One pair of legs arises per trunk segment and 15 or more pairs of legs are present.

Order Scolopendromorpha (centipedes)

Class Diplopoda

Members of this class have long, cylindrical bodies. Every other body segment is fused to the one ahead of it, so the animals appear to have two pairs of legs per body segment. They have many pairs of small legs, hence the name millipede or "thousand feet."

Order Spirostreptida (millipedes)

Class Insecta

Insects have three distinct body regions: the head, thorax, and abdomen. Three pairs of legs, and often two pairs of wings, arise from the thorax. Many insects have well developed compound eyes.

Order Ephemeroptera (mayflies)
Order Odonata (dragonflies, damselflies)
Order Isoptera (termites)
Order Plecoptera (stoneflies)
Order Orthoptera (grasshoppers, crickets, katydids)
Order Phasmatodea (walkingsticks)
Order Hemiptera (true bugs)
Order Homoptera (cicadas, leafhoppers, aphids)
Order Megaloptera (dobsonflies)
Order Coleoptera (beetles)
Order Diptera (flies and mosquitoes)
Order Lepidoptera (butterflies and moths)
Order Trichoptera (caddisflies)
Order Hymenoptera (bees, wasps, ants)

Scorpions

Steven J. Prchal

Scorpions have changed little in the 350 to 400 million years since they first climbed from the primal seas and took their place among earth's first terrestrial arthropods.

The long, segmented body of the scorpion is divided into two obvious sections: the elliptically shaped body and the trade-mark "tail." The body of the scorpion is divided into two parts, the *cephalothorax* and the *mesosoma* or pre-abdomen. The cephalothorax contains all of the sensory, locomotion, and feeding appendages. Two pairs of *chelicerae*, positioned on either side of the mouth, allow the scorpion to rip and tear its prey while feeding. Combining the sensitivity of antennae with the grasping ability of a hand, the *pedipalps* (pincers) are used for sensing as well as holding prey while envenomating or eating. Male and female scorpions also use the pedipalps to clasp their mates during elaborate mating dances. Like all arachnids, scorpions have four pairs of jointed legs. Sensory hairs on the legs can detect the vibrations of prey up to 1 foot (30 cm) away. Dorsally the cephalothorax is covered by the *carapace*. A pair of median eyes atop the carapace, as well as several lateral eyes arranged into two groups along its front edge, give the animal its limited vision.

The *mesosoma* contains the paired genital openings, the *spiracle slits* that open into the tracheal system allowing the scorpion to breath, and the *pectines*, a pair of comb-like appendages on the ventral surface that sweep the ground possibly as contact pheromone detectors. The pectines may also help the male find a suitable place to deposit his spermatophore during mating.

Class: Arachnida
Order: Scorpiones
Families: Buthidae, Diplocentridae, Iuridae, Superstitionidae, Vaejovidae
Arizona Upland genera: *Hadrurus, Vaejovis, Centruruoides, Superstitionia*
Spanish name: alacran

The *metasoma* (tail) of the scorpion is actually an extension of the abdomen. It consists of five segments, each one longer than the last; at the tip is the *telson* (stinger), which is not considered a true segment.

A scorpion uses its venomous sting primarily to subdue insect prey. It also uses the sting defensively, readily stinging a predator or the mistakenly placed bare foot.

Three species of scorpions are commonly found in the Arizona Upland subdivision of the Sonoran Desert. They are the bark scorpion, *Centruroides exilicauda*, the striped tail or devil's scorpion, *Vaejovis spinigerus*, and the giant or desert hairy scorpion, *Hadrurus arizonensis*.

Although more than 30 species of scorpions are found in Arizona, only the sting of the bark scorpion is considered to be truly life threatening. Its slender shape, and its long, delicate pincers and tail distinguish it from other more stoutly-built species in the state. Of the three most commonly seen species, the bark scorpion is the only one that prefers to climb, and it may be found many feet above the ground on trees and rock faces. Because bark scorpions display *negative geotaxis*, that is they orient themselves upside down, people are often stung by them as they pick up an object and press against a scorpion clinging to the underside. Defensive stinging is usually a series of quick jabs, after which the scorpion makes a hasty retreat.

Its preference for climbing and natural attraction to cool moist areas and air flows makes *C. exilicauda* a frequent urban guest. Inside the house, scorpions may be seen trapped in sinks and bathtubs or hiding in dark areas of the closet or storage room. They may also be found climbing walls or clinging to the ceiling. Outside, *Centruroides exilicauda* frequently lives in lumber or brick piles. The only species tolerant of others, bark scorpions may be found in large aggregations, especially during their winter hibernation. Bark scorpions live in Baja California del Norte, northern Sonora, southeastern California, extreme southeastern Utah, Arizona and southwestern New Mexico.

Bark scorpion (above)
Stripe tailed scorpion (below)

The stripe tailed scorpion is Arizona's most common species of scorpion. This species occurs in a variety of habitats from near sea level to 7000 feet (2100 m) in Texas, New Mexico, Arizona, southern California, Sonora, and northeastern Baja California del Norte. These sturdy, medium-sized scorpions are usually under rocks during the day. Like all scorpions, they are nocturnal and venture from their shelters at night to forage for prey. A stout tail with darkly-marked ridges running lengthwise along the underside and a total body length of about 2 inches (5 cm) identify this most common desert ground dweller.

The giant hairy scorpion is one of the least common of Arizona's desert scorpions and the largest scorpion in the United States (up to 6 inches [15 cm] long). It can be found at lower elevations in southern Utah, southern Nevada, southeastern California, Arizona, and northern Sonora. Its large size allows it to feed readily on other scorpions and a variety of other prey, including lizards. Burrowing deep in the desert soil, *H. arizonensis* often follows the moisture line. As summer progresses and the moisture level in the soil recedes, the scorpion follows it, creating burrows as deep as 8 feet (2.5 m) below the surface. A giant hairy scorpion frequently assumes a strong defensive posture when threatened, curling its body and tail high overhead and spreading its pincers. The stinging action is swift and well directed, but the sting is mild, causing only local pain and swelling.

ECOLOGY

While scorpions are well equipped for survival, they are not without their natural enemies. Scorpions not only feed upon each other but are prey to other animals as well. Elf Owls have been photographed bringing scorpions with their telsons removed to their young. Lizards and small fossorial snakes in the genera *Chilomeniscus, Chionactis,* and *Sonora* also find them suitable food. Grasshopper mice and desert shrews are known to feed on scorpions, as are pallid bats.

Giant hairy scorpion

Scorpions give birth to live young through the summer months, frequently having retained sperm from mating the previous year before going into hibernation. Scorpions are not fully developed when they are born, and will continue to develop until the first molt of their exoskeleton in 7 to 21 days, depending on the species. As the babies are born, they quickly crawl up their mother's pincers and legs to take a position on her back, where they will safely ride until they molt. Should they fall off, they can become prey—not only for a variety of arthropod predators, but also even for their mother.

One of the most fascinating things about scorpions is that they fluoresce under ultraviolet light, probably due to the complex substance in the epidermis that makes it impermeable. To truly appreciate the lives of scorpions, take a black light out to the desert on a warm, moonless night. You'll be amazed at how common scorpions are in the undisturbed Arizona Upland habitat. You can also use this technique to observe these ancient nocturnal arachnids as they detect and capture prey, court and mate, and dig burrows. These behaviors are never seen when scorpions are encountered by turning over rocks and other materials where they spend the daylight hours. Not only can you find scorpions in nature using ultraviolet light, you can use this same technique to look for scorpions in and around your home.

Spiders

Renée Lizotte

S piders and spiderlike animals belong to the class Arachnida. Arachnids differ from most other arthropods in having no antennae. All adult arachnids have four pairs of legs and have no wings. They usually lack mandibles, having instead fang-like mouthparts to pierce and break up prey.

Spiders are soft-bodied arachnids with two body parts: the fused head and thorax, called the *cephalothorax*, and the abdomen, also known as the *opisthosoma*. Spiders have four pairs of walking legs and a fifth pair of appendages, located just behind their mouthparts, known as *pedipalps*. Pedipalps are not used for locomotion but often assist in touching and maneuvering prey. Each male pedipalp is equipped with a special structure that is used to transfer sperm to the female. This apparatus makes the male palp enlarged; it is often described as resembling a boxing glove. The mouthparts, called *chelicerae*, each end with a fang. The fang is connected to the venom gland, which enables the spider to inject venom into its prey.

VENOM

All but a few species of spiders are venomous. (A notable non-venomous example in the Southwest is the feather-legged spider.) While most spiders are venomous, few are dangerous to humans. In Arizona, only the widows (black widow) and the brown spiders (brown recluse types) have venom dangerous to people. Other species may bite and cause some local swelling, but unless one is allergic to the venom, no medical attention is necessary.

Venom in spiders has two functions: prey capture and defense. In its most common use, spiders bite their prey and inject venom, which immobilizes the prey and starts the process of digestion. Spiders have no teeth and rely on the venom to liquefy their prey in order that their stomachs, known as sucking stomachs, can draw in the meal. Some of the larger spiders, such as tarantulas and wolf spiders, have projections on the inside of their chelicerae that help break the prey into smaller pieces to aid digestion. These hunting spiders typically leave a small pellet of crushed exoskeleton after eating, whereas the smaller web-building spiders leave an empty shell of the former prey, neatly cocooned in silk.

The second function of venom is for defense. A spider that is threatened by restraint or by touching may bite. The spider controls the amount of venom injected and may inject none at all. A black widow could bite when its hiding place under a rock or a log is exposed. People are sometimes bitten by brown spiders that have taken up residence in clothes that have been left piled on the floor or in a tent overnight.

SENSES

Spiders typically have eight eyes, although some species have a reduced number. The brown spiders have only six, while some cave-dwelling spiders have no eyes at all. Although most spiders have eyes, only one group, the jumping spiders, rely on their eyes to hunt. Other spiders use their eyes to orient when wandering or to detect motion in the initial stages of prey capture. Spiders are equipped with special hairs, mostly on their legs, through which they feel, taste and hear. Even at a distance, spiders can "feel" where their prey is by the displacement of air around these hairs.

SILK

The diverse use of silk is the most characteristic feature of spider, which have at least four, but more typically six, *spinnerets* (appendages that spin silk). A spider may have up to six types of silk gland, each producing a different kind of silk. It is actually a complex strand of proteins that is produced as a liquid, and solidifies under tension. Silk is used to build webs, catch food, line burrows,

protect eggs, detect prey (as trip lines), and even to aid in dispersal. This dispersal phenomenon is known as *ballooning*. Spiderlings (baby spiders) climb to the top of an object or plant and let out silk. The spiderlings are so light that, as the air currents take up the silk, the spider is transported, sometimes miles away. Some of the larger spiders, such as tarantulas, are too heavy to balloon, but most smaller species have this ability.

ECOLOGY AND LIFE HISTORY

Spiders are predators. They survive by eating other animals such as insects, crustaceans, or even other spiders. Thus when spiders get together it can be potentially life-threatening. But spiders have developed special ways of communicating with each other to avoid cannibalism. Male spiders warn female spiders of their presence by plucking their webs, exchanging chemical signals, tapping the surfaces females are resting on, or producing audible drumming or scrapings. Each species has a specific code, and its use prevents most cannibalism.

Among insects, the primary predators on spiders are spider wasps in the family Pompilidae. Different species of wasps have different strategies, but the result is the same: the wasp lays an egg on the immobilized spider, and when the wasp larva hatches, it eats the spider. Most wasp species do not hunt particular spider species, though they may prefer one over another. Interestingly, the spiders rarely defend themselves when attacked by wasps.

Male spiders, which reach maturity at an earlier age than females, must search for a mate. With the exception of tarantulas, spiders typically mate one time during their lives. The average spider lives for approximately one year, but some of the larger wolf spiders live about two years; and tarantulas may live up to twenty years. The male web-building spider sometimes guards the web of an immature female from other males, mating with her immediately after the molt in which she becomes an adult. Males of other species must go through an elaborate courtship of song and dance before the females will accept them. Once the female has mated, she can store sperm for several months up to a year. She may produce 2 to 3 egg cases during that time. Depending on the species, a few to a few hundred eggs may be laid per egg case, but usually each subsequent egg case is smaller with fewer viable spiderlings.

ADAPTATION TO DESERT LIFE
Spiders are fairly tolerant of environmental extremes. They can adjust their body temperatures to be higher or lower than the ambient temperature by, for example, sunning to warm up, or by escaping into the shade or a burrow to cool down. High body temperatures threaten the loss of water as vapor through transpiration; a loss amounting to over 20 percent of the spider's body weight is lethal. Spiders can drink, even from moist soil. Many spiders are nocturnal, which means that they are active only during the cooler, more humid parts of the day. Spiders can survive cold winter temperatures by moving to relatively warm microclimates such as burrows or leaf litter; they curl up to reduce exposure, become rigid and reduce their metabolic rate.

➤ Arizona blond tarantula
Aphonopelma chalcodes

Family: Theraphosidae
Other common names: desert
  tarantula, western tarantula
Spanish name: tarántula

DESCRIPTION
This 3 to 4 inch (70 to 100 mm) large bodied, burrowing spider is commonly seen during the summer rainy season in southwestern deserts. The female is usually a uniform tan color. The male has black legs, a copper-colored cephalothorax and a reddish abdomen. Their burrows can be as large as 1 to 2 inches (25 to 50 mm) in diameter, with some strands of silk across the opening.

DISTRIBUTION AND HABITAT
The Arizona blond tarantula is typically found in saguaro-dominated plant communities. There are many similar species throughout the desert southwest, but they are difficult to differentiate.

ECOLOGY
Tarantulas are nocturnal predators that never venture far from their burrows unless it is mating season. In winter they plug their burrows with soil, rocks, and silk and survive in a relatively inactive state. During this time the animals live off stored fat reserves.

Tarantulas have an interesting defensive capability in addition to venom. Some of the hairs on the top of the abdomen are specialized for defense. These *urticating hairs*, as they are called, are tipped with backward pointing barbs. If a tarantula is threatened in any way, it brushes these hairs into the face, paw or other body part of its attacker. Once these hairs are embedded, they are irritating and very difficult to remove because of the barbs.

Arizona blond tarantula

LIFE HISTORY
Male tarantulas mature when they are 10 to 12 years of age, at which time they leave their burrows in search of females. Upon finding the burrow of a mature female—she's usually at least 10 years old—the male will announce himself by stroking the silk at the top of the burrow and tapping particular sequences that the female responds to. During mating, the male must reach under the female to insert his pedipalp into her gonopore to deposit sperm. He is particularly vulnerable to predation by the female when mating. The male's first pair of legs has a "spur" located behind the knee which

he uses to hold the female above him during copulation. After copulation the male makes a hasty retreat. The female lays her eggs in a burrow, sometimes staying with them. The young remain in the burrow until they disperse.

Each time a female tarantula molts, typically once a year, she also molts the lining of her *epigynum* (the female reproductive structure) where the sperm are stored, so she must mate again before she can produce fertile eggs. The many tarantulas seen on the roads in Arizona during the summer rains (July, August, September) are usually males searching for mates. The male tarantula does not survive long after his summer mating. Sometimes the female makes a meal of the male, or another predator kills him. Sometimes he dies of exposure to heat and cold. Even in captivity, out of harm's way, males only survive a few months after mating.

➤ wolf spider
Hogna carolinensis

Family: Lycosidae
Spanish names: mordelena, carga hijas, buena madre

DESCRIPTION
Wolf spiders are large, with a 1 inch (25 mm) body length; like tarantulas, they live in burrows. Wolf spider burrows can be differentiated from tarantula burrows by the turret of silk and twigs that extends vertically from the wolf spider's hole. The wolf spider can be from gray to dark brown with distinctive peach or orange coloration on the front of the chelicerae.

Distribution and Habitat

This particular species of wolf spider is found throughout the United States in habitats ranging from desertscrub to woodlands.

Ecology

These spiders are most often found in Arizona Upland habitat, where their burrows are quite conspicuous. They are typically active from March through October, when their green eye shine can be easily seen as they cross roads at night. The eye shine is caused by a *tapetum* in the eye which reflects light rays back through the eye retina and probably enhances the spider's night vision. Wolf spiders are primarily nocturnal predators and are rarely seen during the day.

Wolf spider

Life History

Wolf spiders are expert and vigorous hunters. Adult males can be found wandering throughout the summer rainy season, presumably searching for mates. The male must give the female appropriate signals when he finds her, to avoid being perceived as a threat. He does this by tapping his legs in a particular fashion. He also drums with his palps, and in a procedure called *stridulation*, he produces sounds by scraping the palp against itself. After the female lays eggs, she carries the egg case with her wherever she goes, attached to her spinnerets. Sometimes she suns the egg case, sticking her rump, with egg case attached, outside the burrow entrance. The spiderlings hatch after about a month and climb onto the mother's back, holding onto specialized hairs. After another month, they disperse, sometimes by ballooning. A female wolf spider may live up to 2 years.

➤ labyrinth spider
Metepeira arizonica

Family: Araneidae

Description

The web of this spider is more easily recognizable than its body. The labyrinth spider most commonly constructs its web between the pads of prickly pear cactus. It builds an orb web, behind which is a tangled, disordered web made with some debris woven into the silk. Below the debris is a silken retreat in which the spider hides. The spider is small, about ½-inch (12 mm) long, with a bulbous abdomen and thin delicate legs. The carapace is brown or grey; the abdomen is dark with a distinct white pattern.

Distribution

This spider is found in Arizona and California.

Ecology

Like many other spiders, the labyrinth spider is a predator that uses the orb portion of its web to ensnare prey. It is a passive hunter that lets the web do the work. Small insects that visit the prickly pear cactus to drink nectar or eat the fruits sometimes inadvertently fly into the labyrinth spider's web.

LIFE HISTORY

The labyrinth spider is active, with its webs visible in prickly pears, from March through October. During the rainy season the female mates and lays eggs. She builds an egg case around the eggs and hangs it in her web near her retreat where it is camouflaged by other debris in her web. Once the young emerge they are self-sufficient; they disperse by ballooning.

➤ funnel-web spider
Calilena arizonica, Hololena hola, Novalena lutzi

Family: Agelenidae
Spanish name: zacatera

DESCRIPTION

The funnel-web spider is similar in appearance to the wolf spider, but it is smaller and more delicate, with a body length of about ½ inch (12 mm). It builds a sheet-like web with a distinct funnel shape leading to a retreat. Because these webs are often built in grasses, a common name for these arachnids is "grass spider."

DISTRIBUTION AND HABITAT

Funnel-web spiders live world-wide; these three species are common in southern Arizona. They build webs in grass or leaf litter, on stones, or in the corners of buildings.

ECOLOGY

These spiders use their webs to catch prey. The sheet of the web acts as a catch basin for insects that blunder onto it, becoming stuck in the sticky silk. The spider, sensing the vibrations in the web, goes out to retrieve its meal. If the prey item is small enough, the spider will cut it out of the web and bring it down into its retreat to feed on.

LIFE HISTORY

Funnel-web spiders are active from March through October. A male spider must communicate to the female via stridulations and web stroking. Once the female accepts him onto her web, mating takes place. The female lays 100 to 200 eggs in an egg case. She weaves the egg case into her web, near or within her funnel retreat. After about a month the eggs hatch and the young disperse, often by ballooning.

➤ giant crab spider
Olios giganteus

Family: Hetropodidae (Sparassidae)
Other common names: huntsman
Spanish name: cazadora del desierto

DESCRIPTION

One of the largest in this area, this spider has a leg span of 2 to 2½ inches (50 to 64 mm). It is medium to light brown. It often extends its legs at right angles to its body. It can move sideways rapidly, hence the name "crab" spider. Despite its large size, it is capable of climbing fairly smooth vertical surfaces and is often seen high on walls or even ceilings of dwellings. This is one easy way to distinguish it from the wolf spider, a non-climber.

DISTRIBUTION AND HABITAT

Though it belongs to a group of spiders which is mostly tropical, the giant crab spider is found throughout Arizona and Sonora, in a variety of habitats, such as in dead saguaros, under rocks, and in dwellings.

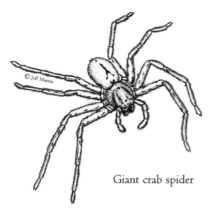

Giant crab spider

ECOLOGY

This is a hunting spider that wanders in search of insect prey, then relies on speed to catch it. During the day it hides, its flattened body perfectly designed for fitting into narrow cracks or fissures. At night it comes out to hunt. Reportedly, its bite is painful, though it is not dangerous to humans. These spiders generally settle into one place only at egg-laying time. Females produce large egg bags that they hide in and guard.

➤ green lynx spider
Peucetia viridans

Family: Oxyopidae
Spanish name: araña verde

DESCRIPTION

This is a very bright green spider, about ¾ inch (19 mm) in length, with long, spiny legs and an oblong to oval abdomen.

DISTRIBUTION AND HABITAT

This spider, which is a member of a spider family that is mostly tropical, is found in southern United States from coast to coast, and also in Mexico and Central America. It often lives in clumps of prickly pear cactus.

ECOLOGY

Lynx spiders are hunters specialized for living on plants. This species does not use a web to capture its prey. In our region, it often lies in wait for insects in the blooms or on the pads of prickly pear, for which its bright green color offers ideal camouflage. It pounces on its prey in a cat-like manner, which is the reason for the name "lynx." It is active during the day.

LIFE HISTORY

The inseminated female lays her eggs in a sac that she hangs in a web. She hangs above it, hugging it with her legs. The female guards her eggs and the newly hatched young until their first molt.

Green lynx spider

➤ brown spider
Loxosceles spp.

Family: Sicariidae (Loxoscelidae)
Other common names: violin
  spider, Arizona brown spider,
  fiddle spider, necrotizing
  spider, brown recluse
Spanish name: uvari

DESCRIPTION
This is a small, inconspicuous brown
spider with slightly darker brown
markings on the cephalothorax. These
markings vaguely resemble the shape
of a violin, hence the common names
"violin" or "fiddle" spider. The species
native to this area are closely related to
the infamous brown recluse of the mid-
western United States, but the markings
are less obvious.

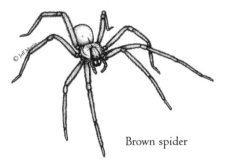

Brown spider

DISTRIBUTION AND HABITAT
These spiders occur in the southern
United States through Mexico and
Central America. Not all species live
in the desert, but several species of
Loxosceles are found in Arizona. These
secretive spiders are found in debris
in the desert or around dwellings in
outlying areas.

ECOLOGY
A trapping spider, the brown spider
catches insects and other invertebrates
in sticky, irregular webs spun beneath
rocks and debris. The bite of this
spider is potentially dangerous to
humans: reportedly some have suffered
amputation and even death as the
result of bites. Although sometimes
the bite causes little harm, the most
common reaction is a spreading sore at
the site of the bite, which, if untreated,
may result in permanent tissue damage.
Those who suspect a brown spider bite
should see a physician.

LIFE HISTORY
The inseminated female lays eggs in
cases. Once hatched, the young may
live 2 to 3 years.

➤ black widow spider
Latrodectus hesperus

Family: Theridiidae
Other common names: western
  black widow
Spanish name: viuda negra

DESCRIPTION
The female of this sexually dimorphic
species is usually ½ to ¾ inch (12 to
19 mm) in body length, shiny black or
very dark brown, with a large rounded
or oval abdomen which is characterized
by a bright red-orange hourglass shape
on the underside. The male is less than
half the size of the female, medium
brown with cream-colored markings on
legs and abdomen. The young of both
sexes resemble the male.

DISTRIBUTION AND HABITAT
Different species of *Latrodectus* are found

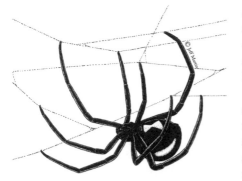

Black widow spider

colored egg sac made of her silk. One spider produces several sacs within its one- to two-year lifespan, but only one sac at a time. The spiderlings disperse by ballooning.

The black widow is one of two species within our region that is potentially dangerous to humans (the brown spider is the other). The bite can kill a human, but this is rare. More often, the bite is painful and causes serious reactions, including nausea, dizziness and abdominal cramps.

throughout most of North America, more commonly in warmer climates. Black widows are common around man-made structures such as garages, lawn furniture, and woodpiles. They also live in a variety of natural habitats.

ECOLOGY
The black widow preys mainly upon insects that it traps in the web. The web is irregular and strong to the touch in comparison to other webs. Some species of spider wasps prey upon black widows. Black widows are shy, sedentary, and largely nocturnal. They are not aggressive, but will bite in self-defense.

LIFE HISTORY
The female mates only once in her life-time, retaining sperm for future egg-laying. The smaller male is sometimes eaten by the female following mating, hence the name "widow." This charac-teristic, however, is not limited to black widows, but can occur after mating in many arachnids, most of which are highly predatory. The female lays approximately 300 eggs at one time and encases them in a round, cream-

> trapdoor spider
Ummidia spp.

Family: Ctenizidae
Spanish name: araña terafosa

DESCRIPTION
Trapdoor spiders are close relatives of tarantulas, and their general appear-ance is similar, but they can be distinguished by their small size, less hairy abdomens, and legs that shine almost as if polished.

DISTRIBUTION AND HABITAT
These spiders range from Virginia south to Florida and west to California. Trapdoor spider tubes are usually found in the sides of banks in disturbed areas.

ECOLOGY
Trapdoor spiders prey on large terrestrial arthropods, and even occasionally on small lizards. They themselves are preyed on extensively in some areas by parasitic wasps of the family Pompilidae.

Perhaps the most interesting aspect of these spiders is their architecture. They build tube-like tunnels in the sides of banks in disturbed areas, along natural insect walkways. The tunnel is capped with an ingenious trapdoor. Trapdoor spiders are well-adapted for the strenuous activity of tunnel-building. Their chelicerae are equipped with digging rakes (*rastella*) that are used to loosen earth during the digging process,

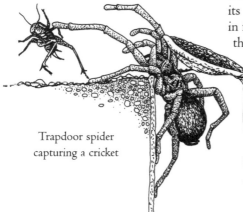

Trapdoor spider capturing a cricket

and then to roll it into a ball which is thrown from the developing burrow with strong, spined hind legs. Once the initial tunnel has been constructed, it is reinforced with a coating of a mixture of earth and saliva. Next a layer of silk is added, this being spun in one piece. The last step, the addition of the door, is the part of the process that differs from species to species. Two types of doors may be constructed. The most well-known is the "cork"-type door, which is very thick and beveled to fit the opening exactly. The other is the "wafer"-type door, which is a simply-constructed sheet of silk and dirt. The species also differ as to whether the tunnels are simple, or branching, with

multiple doors. In all cases, however, the doors are equipped with silk hinges for easy opening and closing.

The tunnel is used by the trapdoor spider as shelter from the elements and predators, as a nursery, and as a trapping device. The top of the door is usually camouflaged with bits of debris, such as twigs and rock, making its discovery very difficult. This results in fooling prey as well as predators, thereby making it a very effective shelter and trap.

When the spider is using the trap to capture prey, its chelicerae hold the lid shut on the end of the door farthest from the hinge. It awaits the vibrations of passing prey conducted by the silk, quickly throws open the door, grabs the prey and returns with it down the tube. Although the lid stays shut easily on its own, attacks by predators can be discouraged by the spider holding the lid closed with its chelicerae, and, at the same time, bracing its legs against the wall of the tunnel. The only predators that are not dissuaded by this seem to be parasitic wasps, which simply chew right through the door.

The tunnel is also used by the female as a nursery. She lays her eggs in the tube and immediately covers them in a sac which is attached to the tunnel wall. She remains with them until hatching and beyond, allowing them to remain unharmed in the burrow until they are as much as eight months old.

Tailless Whipscorpions & Sun Spiders

Renée Lizotte

ailless whipscorpions look at first glance like spiders. The first appendages (*pedipalps*) are modified for grasping prey, with hook-like projections. The first true pair of legs is modified to serve as "feelers," and are long, delicate, and whip-like, with many fine hairs.

© Jeff Martin

DISTRIBUTION
Amblypygids are found in tropical regions throughout the world. In Arizona, these animals live in abandoned rodent burrows and along dry river washes in the Arizona Upland foothills.

Tailless whipscorpion

Order: Amblypygi
Family: Tarantulidae
Sonoran Desert Genera: tailless whipscorpion (*Paraphrynus* spp.)
Other common names: amblypygid, whipspider

Order: Solifugae
Family: Eremobatidae
Sonoran Desert Genera: sun spider (*Eremobates* spp.)
Other common names: solpugid, windscorpion
Spanish name: matavenados

ECOLOGY

Tailless whipscorpions are reclusive predators of insects. They hunt nocturnally, using their long, delicate first pair of legs to find their food. The spined pedipalps impale and crush the prey and then transfer it to the *chelicerae* (jaws). Tailless whipscorpions can only pinch their prey; they lack venom glands.

LIFE HISTORY

After stroking females with the whiplike front legs, males deposit a spermatophore on the ground. The females then pick up the sperm masses with their gonopores. After females lay the eggs, they will carry them until they hatch, and then carry the young for 4 to 6 days.

➤ sun spiders

Sun spiders are 1 to 3 inches (25 to 75 mm) long and are yellow or tan in color. They have eight walking legs, long club-like pedipalps and large, muscular chelicerae. The tips of the chelicerae are equipped with pairs of pincers that are quite formidable. When moving, sunspiders often hold their pedipalps in the air.

DISTRIBUTION

Sun spiders are found throughout the world in mostly tropical and subtropical areas. They are also at home in the hottest, driest deserts of the world.

ECOLOGY

Sun spiders are good predators, able to run down their prey and catch it with great speed. Sun spiders feed upon insects and arachnids, and even small lizards. They are also good diggers and probably spend most of their time underground. They are most active in the desert Southwest during the warm months of May and June, and they remain active throughout the rainy season during July, August, and September.

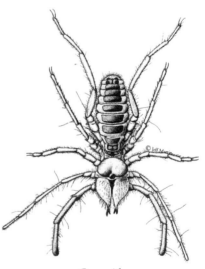

Sun spider

LIFE HISTORY

The mating process begins with a male encountering a receptive female. Some "dancing" and stroking ensues, with the male finally flipping the female onto her back, and with his chelicerae, inserting into her gonopore a sperm droplet. The female stores the sperm for later fertilization, after which she digs a burrow and deposits about 100 eggs. The juveniles are completely independent upon hatching.

Centipedes & Millipedes

Renée Lizotte

Centipedes are arthropods that have elongated bodies with one pair of legs per segment. They range in size from less than an inch to several inches. The giant desert centipede is usually 6 to 8 inches (15 to 20 cm) long, while the common desert centipede is 4 to 5 inches (10 to 13 cm) long. The larger giant desert centipede is orange with a black head and tail. This warning coloration advertises the centipede as dangerous. The smaller, brown and tan, common desert centipede is less so.

Giant desert centipede

While painful, neither bite is especially dangerous to humans.

DISTRIBUTION

Centipedes are found world-wide, in temperate and, more abundantly,

Order: Scolopendromorpha
Family: Scolopendridae
Sonoran Desert species: giant desert centipede *(Scolopendra heros)*, common desert centipede *(Scolopendra polymorpha)*
Spanish name: cien pies

Order: Spirostreptida
Sonoran Desert species: desert millipede *(Orthoperus ornatus)*
Other common names: rainworm
Spanish name: mil pies

in tropical areas. These 2 desert species are found throughout the southern United States and into Mexico.

ECOLOGY
Centipedes use structures called *gnathosomes* or *gnathopods* to inject venom into their prey. These are paired pincer-like appendages in front of the legs. The "bite" is actually a pinch. Centipedes are fast-moving predators that feed on any small creatures they can catch—mostly insects, but occasionally other arthropods, lizards, and even small rodents. Centipedes in the desert are strictly nocturnal and spend their days underground or concealed from the sun. They lack the waxy layer in their cuticle that other arthropods have, and are therefore more prone to desiccation than are other terrestrial arthropods.

LIFE HISTORY
Centipede mothers care for their eggs, coiling around them and grooming them. This grooming is thought to protect against mold and bacteria. Once the young hatch, the mother tends them as she did the eggs, until they disperse a few days later.

➤ millipedes

Millipedes have long, cylindrical bodies with 2 pairs of legs on each segment. New segments and pairs of legs are added each time the millipede sheds. Since it continues to grow and shed throughout its lifetime, it's impossible to say how many legs a millipede has without counting. The common millipede in southern Arizona is a dark reddish brown, but millipedes in other areas may be tan to golden brown. Most desert millipedes are 4 to 5 inches (10 to 13 cm) long.

DISTRIBUTION
Millipedes are found world-wide except in polar regions, but are more abundant in tropical climates.

ECOLOGY
Millipedes are *detritivores*, foraging for decaying organic material (in the desert, generally in sandy washes). They are nocturnal and prefer humid environments, often appearing on roads after soaking summer thunderstorms. They are good burrowers and spend most of their time underground. If disturbed, the millipede rolls into a coil. If further threatened, it exudes foul-tasting chemicals from openings along the sides of its body. These noxious substances are the millipede's only defense, since it doesn't bite.

LIFE HISTORY
Millipedes are egg layers that do not care for their eggs or young. The eggs are laid underground or in some other concealed area. Millipedes can live 10 years or more.

© Jeff Martin

Desert millipede

Grasshoppers

Goggy Davidowitz

Observant hikers in the Sonoran Desert soon notice grasshoppers jumping near their feet or flitting between bushes and cacti. These are among the most frequently encountered and easily observed of the Sonoran Desert insects in the late summer.

Grasshoppers have a fairly simple body design. The rounded head capsule contains the compound eyes, chewing mouth parts, and the short thread-like antennae, which are always shorter than the body (hence the name "short-horned" grasshoppers, in contrast to another suborder, the *katydids* or "long-horned" grasshoppers). The middle *thoracic* segments and part of the abdomen are covered by a shield-like *pronotum* that extends from the first thoracic segment. The forewings are leathery and not used for flight. Instead they protect the delicate hind wings, which are folded accordion-like beneath the forewings until they are unfolded for flight. However, all immature stages and the adults of many species lack wings altogether and cannot fly.

The most noticeable feature of grasshoppers is their long, jumping hind legs, which enable them to leap well over 20 times their body length (imagine a 6-foot tall person jumping 120 feet!). However, while the powerful jumping muscles of the hind legs provide the force necessary for leaping, they cannot propel the grasshopper in these impressive leaps unaided. Most of the kinetic energy to do this comes not from the muscles, but from the *semilunar crescent* located in the knee of the hind leg. This crescent-shaped organ is made of elastic fibers that store energy in preparation for a jump; they release this energy explosively, propelling the grasshopper forward many times its body length.

Grasshoppers develop through *incomplete metamorphosis*. The nymphs appear similar to the adults except that they lack wings and have incomplete reproductive organs. The number of *instars* (larval stages between molts) through which a grasshopper develops before reaching adulthood is fixed in some species (typically 4 to 6). In others, it depends on growing conditions:

the better the conditions, the fewer the immature stages.

LIFE HISTORY AND ECOLOGY

Most Sonoran Desert short-horned grasshoppers spend the winter months in the soil as eggs. These are laid in clutches from just a few to well over a hundred eggs, depending on the species. The eggs are enclosed in an egg pod made of a frothy material that protects them from parasites, desiccation and mechanical hazards. Grasshoppers hatch during 2 separate seasons in the Sonoran Desert. The smaller winter *cohort* (all members of a population hatching at about the same time) emerges from the egg stage after the winter rains, maturing to adults in April to May. The much larger summer cohort emerges after the monsoon rains, leading to a peak in adult abundance from late August to early October. The number of generations per year for many species depends largely on the duration and quantity of the summer monsoons. For example, the pallid-winged grasshopper (*Trimerotropis pallidipennis*) can produce two or three generations a year in a good rainy season, but only one or none at all in dry years.

Male grasshoppers attract females both visually and acoustically. Males of banded-winged grasshoppers (subfamily Oedipodinae) can typically be seen in the summer taking short flights, flashing their brightly-colored

---

Order: Orthoptera
Suborder: Caelifera (short-horned grasshoppers)
Family: Acrididae
Subfamilies and species discussed:
    Acridinae (Slant-faced grasshoppers)
        Achurum spp.
        Mermiria spp.
    Cyrtacanthacridinae (spur-throated grasshoppers)
        Schistocerca nitens (gray bird grasshopper)
        Schistocerca shoshone (green bird grasshopper)
    Gomphocerinae (tooth-legged grasshoppers)
        Bootettix argentatus (creosote bush grasshopper)
        Ligurotettix coquilletti (desert clicker)
    Melanoplinae
        Dactylotum variegatum (harlequin or rainbow grasshopper)
    Oedipodinae (banded-winged grasshoppers)
        Trimerotropis pallidipennis (pallid-winged grasshopper)
    Romaleinae (lubbers)
        Taeniopoda eques (horse lubber)
        Brachystola magna (plains lubber)
Suborder: Ensifera (long-horned grasshoppers)
Family: Tettigoniidae
        Insara covilleae (creosote bush katydid)

---

wings, snapping them together, or both, producing a distinct sound. These short flight noises are called *crepitation*; the sounds are usually species-specific. Males also attract females by *stridulation* (scraping the hind femora against the forewing); this too produces species-specific mating sounds. Not all sounds produced by grasshoppers function solely to attract females, however. While hiking in the Sonoran Desert, one is likely to hear short bursts of clicks emanating from a creosote bush. This incessant clicking comes from a male desert clicker (*Ligurotettix coquilletti*); it spends much of its life on a single creosote bush claimed as a territory, stridulating to warn away other males as well as to attract females.

FEEDING

About 70 percent of herbivorous insects eat only one or a few species or genera of plants. In contrast, grasshoppers are generalist feeders, eating plants from an extremely broad range of families. Grasshoppers tend to grow better and produce more offspring when their diet consists of a mixture of plants. But different species go about this in very different ways. The gray bird grasshopper (*Schistocerca nitens*) and the green bird grasshopper (*Schistocerca shoshone*) are cryptically colored, avoiding predators by spending most of the day on a single host plant. (Bird grasshoppers are so-named because they are among the largest of the Sonoran grasshoppers, with the females reaching 6.5 cm [2½ inches] in length.) They obtain the necessary

dietary mixtures after feeding for a long period on one host species, by eventually shifting to another. In contrast, the horse lubber (*Taeniopoda eques*) and the harlequin or rainbow grasshopper (*Dactylotum variegatum*) also eat a wide variety of plants, but switch frequently, taking a nibble here and a nibble there.

Not all grasshoppers are generalist plant-eaters, however. The creosote bush grasshopper (*Bootettix argentatus*) is the only one among the more than 8000 species of grasshoppers worldwide that eats a single species of plant—the creosote bush. Hiking in the desert scrub one is likely to come upon another exception, a large (8 cm [3¼ inches]), heavy-bodied, flightless grasshopper, the plains lubber (*Brachystola magna*). This short-winged grasshopper is a generalist herbivore, yet it has recently been shown to be predacious as well, pouncing on and eating other grasshoppers and insects.

PREDATOR AVOIDANCE

Grasshoppers employ a wide range of mechanisms to keep from being eaten. The foremost of these is *crypsis*, matching the background in color or texture. This is most evident in the banded-winged grasshoppers. These ground-dwelling grasshoppers superbly match the color of the soil they live on. In some species, such as the pallid-winged grasshopper, populations living on red soil have a predominantly reddish color, those living on white soil are white, and those on dark or black soil are dark brown or black. These

© Jeff Martin

Banded-winged grasshopper

Katydid or Grasshopper?

"It's green. It must be a Katydid!"

Although it's true that katydids are often green, there are some other ways to tell them apart from grasshoppers. For one thing, their appearance is different. Katydids have long antennae and sword-like ovipositors; grasshoppers have short antennae and blunt ovipositors. (The ovipositor is the egg-laying structure at the hind end of the abdomen of the female.) Another difference is in their egg-laying behavior. Katydids lay their eggs in plants whereas grasshoppers lay theirs in the ground.

differences can be seen over distances of only a few hundred meters. Species of the genera *Achurum* and *Mermiria* in the subfamily Acridinae (commonly called the slant-faced grasshoppers because their faces are positioned obliquely to the rest of the body) live and feed on grasses. They are long and slender and typically have bands running the length of the body, mimicking grass stalks. The casual observer will be hard-pressed to notice one of these insects clinging to a stalk of grass.

The creosote bush grasshopper is another excellent example of crypsis. This species eats only the leaves of the creosote bush and spends all its time among them. It is olive green, with shiny, pearly spots mimicking the green leaves with their shiny, oily secretions. The creosote bush katydid (*Insara covilleae*), in a separate suborder of long-horned grasshoppers, also lives solely on creosote bushes; it too is olive green with pearly, shiny patches—a fine example of convergent evolution.

Banded-winged grasshoppers take a different approach to escaping predators. Their hind wings are often brightly colored with red, orange, yellow, or white bands, in sharp contrast to the often drab brown of the forewings. When startled by a predator (or hiker), they take to the air. The predator focuses attention on the brightly colored and flashy hind wing, only to have the grasshopper disappear from sight when it folds its wings, lands, and again cryptically blends into the background.

Rather than hiding, some grasshoppers actually advertise their presence to predators. A hiker in the Sonoran Desert is likely to come upon the horse lubber (*Taeniopoda eques*), a large (6.5 cm [2½ inches]), black heavy-bodied grasshopper with yellow or orange stripes and antennae, greenish veins on the forewings, and pinkish-red hind wings. The harlequin or rainbow grasshopper (*Dactylotum variegatum*) is smaller (3.5 cm [1¼ inches]), short-winged, and black with bright blue, red, yellow, and white markings. These species sequester toxins from the plants they eat, making them unpalatable for most predators. Their *aposematic* (warning) coloration informs potential predators that they are poisonous—stay away!

Walkingsticks

Robert L. Smith

*I*n biology, success is measured by the number of genes an individual passes on to the next generation. In order to reproduce, an individual must live long enough to obtain from the environment sufficient raw materials with which to replicate its genes and give them a new start in its offspring. The survival of herbivores, animals whose diet consists of plant materials, is constantly threatened by an omnipresent community of predators, which obtain their energy from eating animal protein.

Many vertebrate predators, particularly birds, like to eat insects, so natural selection has favored various schemes that help insects avoid being devoured. Stick and leaf insects belong to the family Phasmatidae, a group of predominantly tropical plant-eating insects closely related to cockroaches, grasshoppers, crickets, and mantids, and they are survivalists extraordinaire. Usually long and slender—some species grow up to 30 centimeters (12 inches) in length—walkingsticks bear remarkable resemblance in both structure and color to twigs and leaves of the woody plants they eat. Many of the stick mimics are wingless, but some have added "leaves" to their twig disguises in the form of shortened wings and elaborate legs that look like foliage.

The external skeletons of a number of these arthropods have spines that resemble the thorns of their host plants, and body segments frequently duplicate the plant's *internodal distance* (the space between leaves). The *cuticle*, or outer cov-

© Jeff Martin

Order: Phasmatodea
Family: Phasmatidae

ering, may even be structured and colored to approximate nodes and scars. Some species, notably *Carausius morosus*, are able to change color, like chameleons, to blend into the background.

Such tactics are called *crypsis*, a group of behaviors which includes camouflage—blending to escape detection—and defensive *mimicry*—looking like something unpalatable or non-nutritous. Walkingsticks do both and survive quite well. (Very few animals eat sticks.)

If stick insects moved quickly or abruptly, they would betray their almost perfect disguises. So, to enhance their cryptic appearance, walkingsticks move very slowly, if at all, during the day. Most species wisely restrict their activities to nighttime.

Yet, a walkingstick that remained still on a shaking plant would be much more conspicuous than one that moved in concert with the plant. So when a stick insect is disturbed, perhaps by a bird alighting nearby or a slight breeze causing the plant to tremble, it flexes its legs randomly, making its body quiver. This subtle behavior, called *quaking*, produces small, irregular movements not likely to be noticed by birds and other predators, which are programmed to detect the purposeful, highly coordinated movements of prey.

When crypsis fails, stick insects often invoke secondary defensive behaviors. Insectivorous birds usually give a tentative, investigative peck to any novel object that might be food; initial caution minimizes the possibility of injury to the beak. A pecked walkingstick responds by immediately releasing its hold on the plant and falling to the ground, where it remains motionless for a long time, perhaps the rest of the day.

Some species even jump to the ground when pecked.

If grabbed by a predator, many phasmatids become rigid. The attacker may assume that is has found a stick and drop the insect. But what if the predator arrives at a different conclusion and tries to eat the insect?

The majority of walkingsticks have yet another line of defense—glands that release distasteful or noxious chemicals. Some species regurgitate a foul liquid or leak blood from their leg joints. If a predator tastes the liquid or blood before mortally injuring the stick insect, it will likely release it. Even if the predator kills and eats a foul-tasting walkingstick, there is still a biological payoff. The predator will probably remember this unpleasant experience and avoid walkingsticks in the future. The sacrifice of one individual may spare that individual's offspring and relatives from a similar fate.

Walkingstick eggs, like those of other large insects, may be consumed by the larvae of certain tiny wasps that deposit their eggs on or in insect eggs. Some stick insects, however, have evolved a cryptic countermeasure to this threat, too. Many species produce eggs that resemble seeds, and some walkingsticks that live on only one plant species deposit eggs that look like their host's seeds. Presumably, seed mimicry makes it difficult for parasitic wasps to distinguish the eggs from the seeds.

Immature walkingsticks possess an extraordinary defensive adaptation called *autotomy*. If its leg is grabbed by a predator, a nymph can shed the leg from a joint near its body. Better to give up a leg and leave than to hang around and risk your life! This sacrifice is not as extreme as it may seem, for the nymph can regenerate its lost limb within two weeks.

No aspect of walkingstick life, not even mating behavior, has escaped investigation by evolutionary biologists eager to learn about every survival tactic. John Sivinski, a research entomologist in Florida, studied walkingsticks while a graduate student at the University of New Mexico. He had read that phasmatids mate for long periods of time. *Diapheromera veliei*, a species closely related to *D. arizonensis*, couples for 3 to 136 hours at one time, and in the extreme, a pair of *Anisomorpha buprestoides* may remain coupled for as long as 3 weeks.

Sivinski reasoned that because the transfer of sperm should require only a few minutes, protracted copulation must have a function other than fertilization. Following earlier work by Thomas Eisner, an eminent student of insect chemical defense mechanisms, Sivinski studied the predatory behavior of blue jays in the presence of both coupled and unmated walkingsticks to learn if the insects might pool their

chemical defenses and hence survive longer together than apart. His research showed no survival advantage for males, but copulating females enjoyed significantly higher survival rates over non-copulating females.

Extended copulation does have a biological advantage for males: greater reproductive success. If a male fails to remain coupled after mating, his mate may immediately seek another consort. Sperm from this subsequent mate is stored along with that of the first, reducing the number of eggs that will be fertilized by the first male. Reduced fertilizations mean fewer offspring, so males that remain paired with a single female for a long time produce more individuals carrying their genes in the next generation than will males that copulate for short periods of time.

Remember the definition of biological success? Survivors must maximize their reproduction; otherwise, survival is biologically meaningless. The 2000 living species of stick insects attest to the biological success of their designs for survival.

Termites

Robert L. Smith

Termites are morphologically uncomplicated insects, in contrast with their astonishingly complex social behavior. Superficially termites resemble and are sometimes mistaken for ants, which also exhibit social behavior. Sonoran Desert termites range in size from 4 to 11 mm (⅛ - 7/16 inches) long, not including the wings of *alates* (the winged reproductive adult forms that appear occasionally, especially during and following rains). Workers are white with small head capsules. Soldiers have enlarged head capsules, and very formidable jaws, or in one of our species, a snout-like structure.

DISTRIBUTION

There are over 40 species of termites in 10 genera widely distributed in the Sonoran Desert.

ECOLOGY

This highly successful group of social insects plays an essential ecological role in the decomposition and recycling of a nutritionally poor, highly resistant, but extremely abundant substance: *cellulose*. Cellulose is a *polysaccharide*, that is, a large number of sugar molecules linked together by tight chemical bonds to form a very long, strong chain. Cellulose is the substance that gives plants their structure and is the most abundant organic compound in the world. Wood is mostly cellulose, and so are cotton and all paper products. In the Sonoran Desert, trees, shrubs, grasses, and cactus skeletons are the primary source of cellulose, which represents more than half of all the organic material produced by photosynthesis. Cellulose

Order: Isoptera
Suborder: Apocrita
Families: Kalotermitidae, Hodotermitidae, Rhinotermitidae, Termitidae
Sonoran Desert genera: *Paraneotermes, Pterotermes, Incisitermes, Marginitermes, Zootermopsis, Heterotermes, Reticulitermes, Amitermes, Gnathamitermes and Tenuirostritermes*
Spanish name: termitos

is durable because it is a physically strong material resistant to mechanical breakdown, but more important, very few organisms produce enzymes that can chemically break it down. Among those that do produce the cellulose break-down enzyme *cellulase* are fungi and tiny animals called protozoans. Termites do not produce cellulase, but all termites contain protozoans in their guts in a mutually beneficial relationship known as *mutualism*. Termites grind up the cellulose mechanically by biting off bits and chewing them up; then the protozoans in their guts break down the chewed mass into sugars, which are readily absorbed through the termites' guts. Both the termites and their protozoans share in the nutritional benefit of these released sugars. Newly hatched termites are first inoculated with these indispensable protozoans by eating the feces of their older brothers and sisters.

The ecological importance of Sonoran Desert termites can best be understood by considering the following question: What would happen if we didn't have termites in our desert? Well, because our aridity severely limits the abundance and distribution of wood decaying fungi, without termites, we would soon be neck deep in cellulose in the form of mesquite and palo verde wood, dead grasses, cactus skeletons and dung. Eventually, few living plants would be left to produce food for animals because there would be no space for new plant seedlings to establish and no nutrients to sustain their growth.

Reticulitermes sp. alate

Reticulitermes sp. worker
(Redrawn from Linsenmaier, *Insects of the World*)

All of the space would be taken up by dry, un-recycled cellulose litter, and all of the nutrients would be tied up in this material and thus unavailable for plants in the soil. Without plants fixing carbon-producing food, most animals would disappear. So, without termites, the whole desert ecosystem as we know it would simply collapse.

Our termites partition the desert's cellulose into many ecological niches. For example, one drywood termite, *Marginitermes hubbardi*, feeds primarily on saguaro skeletons, and another very large primitive drywood termite, *Pterotermes occidentis*, is a specialist on palo verde wood. *Gnathamitermes perplexus*, the crust-building subterranean desert termite, feeds on grass, fine dry plant parts and the weathered outer surfaces of woody tissues of all kinds. *Heterotermes aureus*, the lowland subterranean termite, is an important consumer of native woods on the desert floor and also of pine.

LIFE HISTORY
In the Sonoran Desert large, well-established termite colonies of many species produce nymphs in the late spring. When these nymphs shed their external skeletons for the last time, fully-formed functional wings unfold from the wing

pads, and the resultant individuals are called *alates*. Alates are reproductively mature males and females ready and eager to start new colonies. They stay in the parent colony until conditions are optimal (usually during or after rain); then they leave the galleries of the colony, surrounded by soldier termites ready to defend their brothers and sisters against ants and other enemies as they depart. The alates then take flight.

The season and time of termite flights depend on the species, but alates from all colonies of a given species in an area fly simultaneously. Just how far alates fly is not known for any species, but it is assumed that the flights of reproductives serve to assure new colonization some distance from the home colony. The simultaneous flights also promote outbreeding by increasing the probability that reproductives from one colony will mate with members of other colonies.

Zootermopsis angustucolus soldier
(Redrawn from Wilson, *The Insect Societies*)

Amitermes sp. primary queen
(Redrawn from Wilson, *The Insect Societies*)

Soon after winged termites alight, they shed their wings by breaking them off at lines of weakness (like perforations in paper) near the point of their attachment to the body. Females may then use a chemical odor called a *pheromone* to "call" males. Males attracted to this substance may be accepted or rejected by the female. A rejected male is forced to try his luck elsewhere. The fortunate male who is accepted by a female is permitted to follow her on the ground as she runs quickly about looking for the ideal place to start a new colony. During this "tandem running" phase of courtship, the male remains within touching distance of the female until she finds the "perfect" place (in the ground or in dead wood, again depending on the species of termite) to begin a new colony. The pair then settles into a monogamous relationship and cooperative family rearing.

After mating, the queen lays a few eggs that soon hatch into tiny termite larvae. These are fed and nurtured by both the mother and father until they are large enough to begin foraging for wood and other sources of cellulose, at which time the young termites take over the work of feeding the larvae that have hatched from a second set of eggs. When the parents feed their first batch of offspring, the protozoans required to produce the enzyme needed for cellulose digestion are transferred from the mother's and father's stomachs to the larvae, and this protozoan *inoculum* is all that is required to get a culture going in the offspring so that they too, with the aid of the microbes, can digest their own cellulose.

Social Behavior

Termites can accurately be described as "tiny social cockroaches" because they evolved from a common ancestor with wood-dwelling cockroaches, to whom they are very closely related. They first appeared on earth during the age of the dinosaurs, about 100 million years ago. Termites are social in ways not unfamiliar to humans. We live together with others of our kind in complex

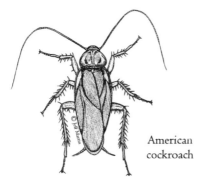

American cockroach

societies, we divide the many tasks needed to support our communities and we care for our young long after they are born. Termites likewise live in complex societies, have division of labor, and care for their young.

A well-established termite society or colony minimally consists of a king and queen, which are responsible for producing offspring: soldiers, which defend the colony against its enemies; and workers, which collect and process wood or other sources of cellulose and feed the royal couple and the soldiers, which are unable to feed themselves. Workers also care for eggs produced by the queen, and they tend to the young termite larvae that hatch from these eggs. The categories of king, queen, soldiers, and workers in a termite colony are referred to as *castes*. All of these are wingless; however, after a termite colony reaches a certain size (a few dozen to several thousand individuals, depending on the species), the colony begins to produce nymphs. These nymphs have small pads on their backs that contain developing wings. Regulation of the development of different castes in a termite colony is controlled by chemicals in the colony that are transferred from individual to individual by social feeding called *tropholaxis*. Exactly how different developmental trajectories are regulated in termite colonies remains an entomological mystery.

Knee Deep in Dung

Termites eat dead plant material and herbivore dung, thereby removing this litter from the surface of the land, permitting sunlight and moisture to reach new growth. On its own, dry cow dung decomposes very slowly. Research conducted in southwestern deserts and desert grasslands by New Mexico State University's Walt Whitford estimates that without the action of termites, cow pies would smother the land, covering 20 percent of the surface in 50 years.

Hemiptera & Homoptera

Robin Roche

Many people refer to anything small that crawls on the ground as a "bug," and indeed many insects have the word "bug" in their name, such as ladybug and light-ningbug (both are actually beetles). The Hemiptera, however, are the "true" bugs of the insect world, having distinct features that set them apart from other insect orders. *Hemiptera* means "halfwing," in reference to the unique front pair of wings, which are leathery near their base and membra-nous towards the tips. Most species hold their wings flat over their backs with the two membranous portions overlapping. This combined with a triangular structure called a *scutellum* (located between the attachment sites of the two front wings) creates an X-shaped pattern on the back of many species. True bugs have slender, beak-like mouthparts that arise from the front of the head, and are usually folded along the ventral surface of the insect except when feeding. While most hemipterans suck the sap from plants, some are predatory, sucking the body fluids of other arthropods or even the blood of vertebrates. All hemipterans undergo incomplete metamorphosis with egg, nymph, and adult stages. Nymphs look very much like diminutive wingless adults.

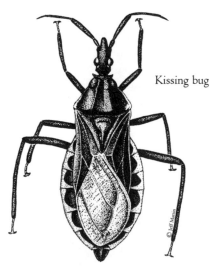

Kissing bug

The Homoptera are close relatives of the Hemiptera and also have pierc-ing-sucking mouthparts. In contrast to the Hemiptera, homopteran mouthparts arise further back on the underside of the head. Those forms that have wings have ones that are uniform in structure, hence their name, *Homoptera*, meaning "samewing." Also unlike the Hemiptera, these insects hold their wings roof-like over their backs. All are plant feeders and most have incomplete metamorphosis. Many families within this order have very strange and complex life cycles with both sexual and asexual generations, winged and wingless generations, as well as individuals with much reduced, highly specialized structures.

NATURAL HISTORY
The Hemiptera and Homoptera are large orders of insects with too many species to cover in this book. Below are examples of a few of the more commonly encountered species.

HEMIPTERA Probably everyone's least favorite representatives from this order are the conenose or kissing bugs. These blood-feeding insects in the family Reduviidae are a nuisance during the spring and summer months. We have several species (*Triatoma rubida* being the most common) whose primary hosts are the wood rat (*Neotoma* ssp.). They are most active at night during the months of May and June (a dispersal period). They occasionally venture into peoples' homes, attracted by porch lights and finding entry through holes in window and door screens. Humans are usually bitten while they are sleeping and often don't feel the actual bites. The next morning,

Giant mesquite bug

however, victims awake with large, hard, itchy welts. Some unlucky people who have been bitten several times over an extended period become "sensitized" to these bites and actually develop a dangerous allergic reaction. As the insect feeds, it injects a small amount of saliva that contains both an anesthetic and an anticoagulant. It is usually one of these two components to which the victim's immune system reacts. In Central and South America kissing bugs carry Chagas' disease (a trypanosome). The bugs defecate while feeding; the victim later scratches the bite and thereby inadvertently rubs the trypanosome (located in the feces) into the wound. Although the species in the southwestern U.S. might be able to carry the disease, their feces are not yet known to contain the parasite, but in rural Sonora below 4500 feet (1500 m) *Triatoma dimidiata* does carry the disease.

One of the more spectacular hemipteran representatives of our area is the giant mesquite bug (*Thasus gigas*). These beautiful insects can be found feeding on the fresh green stems and pods of mesquite during the early summer months. They are the largest of the true bugs in our area. Adults are brownish-black with orange-red bands on their legs and along the veins of their wings. Each of the two antennae has a round medallion-like segment. Nymphs are a gorgeous red and white and are usually found in clusters feeding together on the mesquite pods. When handled, these insects produce a potent, stinky secretion, but are otherwise harmless.

Eggs look like small brown pillows glued in rows along the stems of the mesquite. Nymphs hatch from eggs in the spring, grow to become adults, mate, lay eggs, and then die by the end of summer, leaving new eggs to overwinter until the following spring.

HOMOPTERA Among the larger and more familiar representatives of this order are the cicadas. These musical insects are prevalent during the summer months and are usually first heard during the hot dry days of June before the monsoons. Most of us are probably familiar with the song, a loud buzzing noise emanating from a tree or shrub.

These eggs hatch, and the larvae drop to the ground and dig into the soil to feed on the roots of various plants. In their underground homes, nymphs live and feed for 3 or more years before emerging as adults.

The homopteran cochineal has a fascinating natural and cultural history. Cochineal belongs to the Coccoidea, or scale insects, a bizarre group of homopterans so specialized that they often do not look like insects at all. Cochineal, in the family Dactylopiidae, feed on cactus. You may have noticed a white moldy material growing on the prickly pear cacti in your yard or in the desert. Beneath the fluffy white mass

© Jeff Martin

Cicada

Males sing to attract mates and can be very hard to locate since they usually stop singing as you approach their perch. An observant person is probably more familiar with the brown husks of exoskeleton left by nymphs when they emerge as adults. There are several species of cicadas in the Sonoran Desert region, with one of the most common species at lower elevations being the Apache cicada (*Diceroprocta apache*). Adults emerge in June, feed on plant sap, mate, and insert their eggs just under the surface of plant stems.

reside the small cochineal insects. Females are dark red, wingless, legless, and *sessile* (attached to the plant and sedentary), and look like tiny grapes. The white material covering the insects is wax that females and nymphs secrete from special abdominal glands. Males are more insect-like and resemble small pink and white gnats, each with two long tails or *caudal filaments*. Males are most prevalent during the late summer when they emerge and mate. Once the females have mated, they lay eggs that hatch into tiny 6-legged nymphs called *crawlers.* These migrate to other pads on the cactus or crawl to a pad edge, secrete thin filaments of wax, and wait for the wind to pick them up and carry them to nearby cacti. Crawlers wander until they find suitable spots on cactus pads to settle, feed, and molt. As the tiny insects molt, they lose their legs and become sessile with only their mouthparts firmly tapped into the cactus.

Females, like most other homopterans, have incomplete metamorphosis, but males pupate within a tiny silken cocoon before emerging as adults.

Cochineals' claim to fame is that their bodies contain a substance called *carminic acid* that produces a beautiful red dye. Native peoples from the Southwest, Mexico, and South America harvested these insects for the dye. Europeans discovered the dye when conquering the New World and the valuable product made many of them rich. Up to that time, most red dyes came from plant material. Cochineal, however, not only produced a superior red hue, but also withstood the effects of sun and washing much better than did the plant-derived dyes. Cochineal dye fell out of favor with the advent of synthetic dyes developed in the mid 1800s. However, it is making a comeback, as the more subtle hues of natural dyes are regaining popularity. It is also currently being used as a food and pharmaceutical dye, particularly since many of the synthetic food dyes have proven to be toxic. Look for cochineal or carmine on the labels of pink- and red-colored foods and medicines at the grocery store.

Other familiar homopterans are the aphids or plant lice. Most of us have encountered large groups of these little creatures feasting on the fleshy stems of herbaceous plants in the spring and summer. These tiny insects have very interesting and complex life cycles. Most species survive the winter as eggs and then hatch into females in the spring. These new females then begin to reproduce asexually on their host plants—which differ depending on the species of aphid—forming small colonies of clones. In many species, the colony eventually produces winged forms that fly off to a second host plant (sometimes a completely different plant species) and continue to reproduce. When fall approaches, winged forms migrate back to the original species of host plant and produce both males and females which mate and lay the overwintering eggs. An easy-to-recognize species that occurs in our area is the milkweed aphid, *Aphis nerii*. These beautiful aphids are bright yellow with black legs and antennae; they are found on various species of milkweed in the spring and summer months.

Beetles

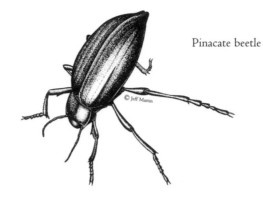

Pinacate beetle

© Jeff Martin

Arthur V. Evans

Beetles comprise the largest group of insects on Earth, representing one-quarter of all living organisms and one-third of all animals, with nearly 350,000 species grouped into more than 150 families. Beetles owe their success, in part, to an external skeleton, or *exoskeleton*, which functions as both skin and skeleton. The outer surface of the exoskeleton may be covered with spines, or hair-like structures, or coated with waxy secretions. These adornments may function as sensory transmitters of environmental information to the nervous system, or serve as additional protection from predators, abrasion, and desiccation.

The beetle head provides the housing for such delicate and primary sensory structures as the eyes and antennae. The antennae are equipped with sophisticated receptors used to detect food, locate egg-laying sites and assess temperature and humidity. Male beetles may have elaborate antennal structures to increase their powers of smell—especially those species that use specific scents, or *pheromones*, to locate mates. The mandibles are usually conspicuous and are variously modified to cut, grind, and strain foodstuffs, occasionally serving as a means of defense. The thorax is the powerhouse of the beetle body, enclosing an internal battery of muscles used to drive the legs and wings. The forewings of beetles are usually thick and leathery, protecting the delicate flight wings and the circulatory, respiratory, digestive, excretory, and reproductive systems housed within the abdomen.

Order: Coleoptera
Families: Scarabaeidae, Tenebrionidae, Cerambycidae
Sonoran Desert species: fig beetle (*Cotinis mutabilis*), pinacate beetle (*Eleodes* spp.),
   palo verde root borer (*Derobrachus geminatus*), cactus longhorn beetle (*Moneilema gigas*)
Spanish names: escarabajo, mayate, coleóptero, cochinilla (cochineal), mayate
   verde (fig beetle), torito, longicornio (cactus longhorn beetle)

LIFE HISTORY

The stages of complete metamorphosis—egg, larva, pupa, and adult—serve to adapt each beetle species to a dynamic suite of seasonal and ecological conditions. Most beetles have fairly regular life cycles, with one or more generations per year. Beetles either produce eggs singly or by the hundreds, scattering them about haphazardly, or carefully depositing each one on or near suitable larval foodstuffs. Upon hatching, the larva begins its life with a single purpose: to eat. Beetle larvae may scavenge carrion, consume animal excrement, attack roots, mine through leaves or chew their way through wood. The pupa serves as the vessel in which dramatic transformations take place, reconstructing the tissues of a non-reproductive eating machine into a precision breeding instrument that may or may not require food.

ECOLOGY

The thickened wing covers of small, compact beetles protect them from abrasion and desiccation as they move about through soil, detritus and decomposing plant materials, allowing them to hide, feed, and reproduce among the myriad of niches found in the environment. The cavity beneath the forewings serves as a site for the storage of oxygen in aquatic species, while insulating desert-dwelling species from the heat and minimizing water loss through respiration. In fact, studies with the pinacate beetle, *Eleodes armata*, have shown that the cavities beneath their forewings may be warmer than the ambient temperature, indicating that they may function as both convective cooling and heat buffering systems.

RESOURCE ECONOMISTS

Receiving little rain and exposed to often wildly oscillating fluctuations in temperature, deserts are home to a rich and dominant beetle fauna whose basic physical features are remarkably similar throughout the world. This parallel evolution is a result of the fact that beetles living in deserts have all adapted both behaviorally and morphologically to cope with the lack of water.

Many species of the beetle family Tenebrionidae are wonderfully adapted to desert life. These wingless and usually black species escape extreme temperatures by remaining buried in the sand during the heat of the day, where temperatures may be significantly lower. The pinacate bettles of the genus *Eleodes* are also called clown beetles because of their defensive stance—they stand on their heads. This action precedes the release of a foul, oily fluid from repugnatorial glands located at the tip of the abdomen, a defense that repels most predators. However, the grasshopper mouse is not deterred by the beetle's odiferous offerings—it simply grabs the beetle with its paws, forcing the tip of the abdomen into the sand, and eagerly begins to consume the tastier head and thorax, stopping just short of the ill-tasting defensive glands. A relative of the pinacate beetle, the ironclad beetle (*Asbolus verrucosus*), however, is not endowed with such a chemical defense system. Rather, when disturbed, this beetle simply collapses to one side and pretends to be dead, an act that may last from several minutes to several hours. The thick, roughened exoskeleton of the ironclad beetle, reminiscent of the impenetrable metallic exterior of battle ships, serves to conserve moisture. Further protecting this beetle from desiccation is a bluish-gray waxy secretion exuded by glands whose ducts are located at the tip of knob-like projections on the forewings. The lighter color produced by the wax also reflects ultraviolet light, helping to keep the body cool. The larvae of both *Asbolus* and *Eleodes* resemble mealworms and are sometimes encountered burrowing through the sand where they feed on plant detritus.

Beetles employ a broad array of tactics to fool their enemies, using mimicry, camouflage, and warning coloration. The cactus longhorn beetle, *Moneilema gigas*, resembles the foul smelling and unrelated *Eleodes*, both in appearance and behavior. Tests have shown that species of *Moneilema* behave in a way similar to *Eleodes* when faced with predators such as lizards, wood rats, and skunks. Although these cactus longhorn beetles can be found on the ground wandering about, their preferred spiny host, the cholla, also affords them a considerable degree of protection. Adult *Moneilema* feed on the softer, more succulent portions of the cactus. The female lays her eggs at the base of the cactus inside an earthen case. Upon hatching, the larvae bore into the roots and stems of the cactus to feed. Their boring activities above ground are conspicuous, as they expel tar-like excrement and fluid from the wounds created by their feeding activities.

Palo verde root borer

© Jeff Martin

RECYCLERS

Beetles are a major component of the "F.B.I."—fungi, bacteria, and insects—the primary agents of decomposition. Without them the planet would soon be covered with dead animals and plants. For example, a dead tree branch presents a number of feeding niches for long horned wood-boring beetles, in terms differently-sized stems and the different layers of tissue, from the bark inward. Furthermore, as a dead branch ages, there is a succession of different beetle species that work in concert to complete the recycling process. The adult palo verde root borer, *Derobrachus geminatus*, grows to more than 3 inches (76 mm) in length and may commonly be encountered during the early evening hours of summer, when it is often attracted to street lights and well-lit storefronts. Females lay their eggs just below the surface of the ground on the roots of a variety of native and non-native trees. The larvae, which may reach 5 inches (130 mm) in length, bore through the live roots of the host tree, consuming woody tissues. Microorganisms, such as bacteria, yeasts, and fungi located in the intestinal tract of these and other wood-boring beetles assist in the digestion of cellulose, whether the tissue is living or dead. These microorganisms are passed to the larvae from their mothers as they pass through a residue in the ovipositor as eggs. Upon hatching, the larvae immediately consume their own egg shells, which are laden with microorganisms that are essential for their digestion.

Another conspicuous beetle that plays an active role in the decomposition of plant materials is the green fig beetle, *Cotinis mutabilis*. The buzzing flight of this scarab, a relative of the nocturnal June bug, fills the days of summer. It is easily distinguished by its matte green back and conspicuously shiny underside. Green fig beetles feed upon soft fruits, but are also encountered around mesquite and desert broom wounds that are oozing sap. Females search for piles of organic matter in which to lay their eggs. The C-shaped grubs, which prefer to crawl on their backs and may grow up to 2 inches (50 mm) in length, are commonly encountered in compost or dung heaps. The pupal cell consists of a hardened shell of soil and larval excrement.

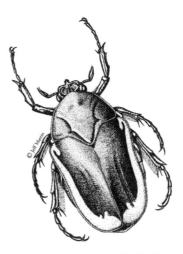

Fig beetle

Butterflies

Queen butterfly

Richard A. Bailowitz and Mark P. Sitter

utterflies are one of the most popular and easily recognized groups of insects. Together with moths, they make up one of the major insect orders or groups—Lepidoptera —which number some 160,000 recognized species worldwide. There are over 250 species of butterflies in the Sonoran Desert.

Lepidoptera comes from Greek words meaning *scaled* and *wing*. Butterflies and moths can easily be distinguished from other insects by their wings, which are covered with thousands of tiny overlapping scales, much like tiles on a roof. Each scale is one color, but collectively a butterfly's color pattern is produced by a complex mixture of differently colored scales. Butterflies are usually large, pretty, and diurnal. They are rarely pests, and consequently are well-liked by humans.

DISPERSAL AND SPECIES RICHNESS

There are a number of factors to account for the rich butterfly diversity in the Sonoran Desert. In general, as one approaches the tropics, species richness increases. Also, a varied topography means a corresponding variety of microclimates, rainfall patterns, plant distributions, and therefore butterfly distributions.

The majority of butterfly species in the Sonoran Desert are rather sedentary, occurring in fairly close proximity to their larval foodplants. But at times, for reasons not fully understood, butterflies wander. Some species move at a particular season, some nearly any time. Some species are true migrants, in that individuals push northward early in the season and southward later. However, an interesting array of taxa are influx species,

---

Order: Lepidoptera
Suborder: Rhopalocera
Families: Papilionidae, Pieridae, Lycaenidae, Riodinidae, Libytheidae,
    Nymphalidae, Hesperiidae
Sonoran Desert genera: numerous
Spanish name: mariposa

Butterfly or Moth?

Though butterflies and moths appear similar in many respects, there are some ways to distinguish between them. Generally when a butterfly lands and rests on a plant it holds its wings vertically, while moths tend to rest with their wings folded back almost horizontally. Moths have heavy, furred bodies, whereas the butterflies have more delicate, slender bodies with little hair. Butterfly antennae are thin and end with a knob at the tip. Moth antennae are often feathery and without a knob.

Color is not a reliable indicator, as some of the moths, especially the Saturnids, are beautifully colored and some butterflies, such as satyrs and mourning cloaks, have muted coloration. Also, not all moths are night fliers. Some species, such as the buck moths and the Calleta silk moth, fly by day. (You may notice the Calleta moth as it feeds on ocotillo leaves during the summer rainy season. Look for ocotillos stripped of leaves from the top down.)

Both butterflies and moths lay eggs which hatch into caterpillars. These caterpillars molt into a pupa, or resting stage. After a period of time—a few days to a season—the winged adult emerges from the pupal case. Moths tend to construct cocoons, protective silk coverings around themselves, before molting into pupas. Butterflies do not encase themselves in cocoons.

entering the Sonoran Desert yearly from other deserts, thornscrub habitats, and mountain ranges in northwestern Mexico. The strength, time of onset, and duration of the summer rainy season are thought to be responsible for the intensity of this influx phenomenon. Many of these visitors breed in the Sonoran Desert and comprise a significant or even dominant portion of the summer butterfly fauna. Several influx species, however, are on dead-end missions, there being no suitable plants to serve as larval hosts. The fact that there are both indigenous and influx species of butterflies in the Sonoran Desert accounts for the high number of species, and also delights the butterfly enthusiast.

LIFE CYCLE

The life cycle of butterflies is one of the true miracles in nature. Butterfly lives have four distinct stages: egg, caterpillar (or larva), chrysalis (or pupa), and adult. The term describing this series of distinct stages of development is *complete metamorphosis*, as distinguished from simple or *incomplete metamorphosis*, in that the animal progresses through life stages which are very similar to each other. (See the "grasshoppers" section for an example of incomplete metamorphosis.)

In most cases, butterflies produce one or more generations (broods) per year. The length of the complete life cycle varies greatly, ranging from weeks to a couple of years or more in desert adapted species. The lifespan of an adult butterfly varies as well, from merely a few days to as long as several months. After mating, butterflies *oviposit* (lay eggs), either singly or in clusters. Female butterflies typically oviposit on specific groups of related plants that will provide food for the caterpillars. The young caterpillars begin feeding and, because their skins do not expand to accommodate growth, must shed their skins several times. Each stage between molts is called an *instar*; each instar is larger than the previous one. The final molt produces a pupa, the resting stage during which the animal does not feed but undergoes the amazing transformation into a butterfly. The pupa of many butterflies hangs from a silk button called a *cremaster*. Other species' pupae are held upright by a silken girdle. Some are disguised (cryptic), being generally green or brown and resembling leaves, stems, or wood. Others are covered in thornlike tubercles or bumps. Just prior to the emergence of the butterfly, the chrysalis usually changes color. Once free of the chrysalis, the butterfly pumps fluid from its swollen body to its shrunken wings. The newly emerged butterfly then lets its wings dry and harden before taking flight in search of food and a mate.

THE IMPORTANCE OF BUTTERFLIES

Butterflies are important pollinators. They are also good indicators of the ecological quality of a habitat, as they are important components of the food chain, particularly as larvae. Few butterflies are a serious threat to economically important plants. In short, butterflies are benign, aesthetically pleasing, faunal members. In turn, the main threat to butterflies is the destruction and loss of their habitats. The channelization of

Pipevine swallowtail

© Jeff Martin

riparian areas, draining of wetlands, lowering of water tables, growth of cities, and expansion of agriculture all contribute to this habitat loss. Widespread use of pesticides may also threaten healthy butterfly populations.

THE BUTTERFLY FAMILIES

SWALLOWTAILS (Papilionidae): Swallowtails are mostly tropical and include some of the largest butterflies in the world. Adults of both sexes have 6 walking legs, take nectar readily, and often flutter their wings even while perched. One of our region's largest butterflies, with a 4 inch (10 cm) wingspread, is the giant swallowtail (*Papilio cresphontes*, also called *Heraclides cresphantes*), a brown and yellow species commonly encountered in urban areas. Its larvae, which feed mostly on the leaves of citrus, look much like fresh bird droppings. If the larvae are touched or disturbed, an unpleasant-smelling, Y-shaped orange organ called an osmeterium, is everted from just behind the head. This device and the cryptic appearance are adaptations to avoid predators and perhaps parasites.

Another impressive swallowtail is the pipevine swallowtail (*Battus philenor*), which is slightly smaller than the giant swallowtail. Pipevine swallowtails are commonly encountered March through October. The butterfly's upper surface is a dark, iridescent blue; the underside is blue with orange spots. The showy, red-orange caterpillars are poisonous due to the compounds taken from the leaves of pipevines (*Aristolochia* spp.), their larval food plant. The bright colors serve as warnings to would-be predators that the caterpillars are highly distasteful.

WHITES AND SULPHURS (Pieridae): Butterflies in this family are small- to medium-sized and most often white, yellow, or orange, with black margins. Some are among the first spring butterflies, even as early as January. Many emerge in a series of broods throughout the year. Still others are influx species during and after the summer monsoons. This family often exhibits *mud puddling* behavior wherein dozens of individuals group together on a patch of damp mud to drink and take in minerals and salts. One of the more showy species, the southern dogface (*Colias cesonia*, also called *Zeren cesonia*), has a distinctive outline of a poodle on each forewing. A common species, the cloudless sulphur (*Phoebis sennae*), exhibits strong sexual dimorphism in which the males and females have visibly different wing patterns or coloration. Males of this species are bright, clear, lemon yellow, while females are off-white with small dark markings. In addition to sexual dimorphism, some sulphur species, such as the tailed orange (*Eurema proterpia*), are seasonally dimorphic, there being short and long day-length color patterns evident. Early spring is a good time to search for orange-tips. The Sara orange-tip (*Anthocharis sara*) and desert orange-tip (*Anthocharis cethura*) both come to flowering wild mustards, plants that also serve as larval food plants for these butterflies.

GOSSAMER WING BUTTERFLIES (Lycaenidae): Lycaenidae is a large family of small butterflies. Despite their size, many are detailed with exquisite markings—truly jewels of the insect world. One of the world's smallest butterflies, the pygmy blue (*Brephidium exile*), measures little more than ½ inch (12 mm) from wingtip to wingtip, and is fairly common in our region, especially in disturbed areas. The great blue hairstreak (*Atlides halesus*) is one of the largest and most spectacular of the gossamer wings, but it is solitary and uncommon. The caterpillars feed on desert mistletoes (*Phoradendron* spp.). While there are several broods during the year, the best places and times to look for great blue hairstreaks are wherever and whenever desert broom or seep-willow (both *Baccharis* spp.) are in bloom.

METALMARKS (Riodinidae): This largely tropical family is extremely diverse in appearance and in species content. Fewer than a dozen species occur in the Sonoran Desert; most of these having metallic spotting on the ventral (underside) surface of the wings. Metalmarks have long antennae for their size and usually perch with their wings opened flat. One species, the Mormon metalmark (*Apodemia mormo*), is quite variable from one location to the next, but all populations are associated with patches of various buckwheat (*Eriogonum* species).

SNOUT BUTTERFLIES (Libytheidae): The single species of snout butterfly that occurs in the Sonoran Desert, the American snout butterfly (*Libytheana bachmanii*), can have huge population surges, particularly in the summer and fall, though in some years it is nearly absent. It has a wingspan of about 1¾ inch (44 mm) and is easily recognized by the long projecting sense organs called *palpi* (which resemble snouts) located between the antennae. The larvae feed on desert and netleaf hackberry.

BRUSH-FOOTED BUTTERFLIES (Nymphalidae): This is a very large and diverse family with most species medium-sized and generally orange and brown. Many well-known butterflies belong to this family, such as fritillaries, painted ladies, crescents, checkerspots, anglewings, admirals, longwings, and of course, monarchs. Nymphalid caterpillars feed on a wide variety of plant families. Most larvae are fiercely spined, and their pupae are usually sharply angled and adorned with silver or gold colors.

The gulf fritillary (*Agraulis vanillae*) is a large, vividly-colored, orange and black butterfly with brilliant silver spots on the underside. Gulf fritillary caterpillars feed exclusively on passion vines. Planting ornamental passion vines around the house can greatly increase the abundance of these butterflies.

Snout butterfly

One subgroup of the brushfoots, the milkweed butterflies (Danainae), is probably among the most noted and studied groups of butterflies, since it contains the monarch (*Danaus plexippus*). This well known species is a remarkably strong flier, migrating great distances every year to overwintering roosts in California and central Mexico. The monarch butterfly is not very common in the Sonoran Desert, but it can be seen regularly in late summer and early fall on its push southward. A much more common resident of the region, the queen (*Danaus gilippus*), is closely related to the monarch. The larvae of both species feed exclusively on various plants in the milkweed family (Asclepiadaceae). This family of plants is poisonous to most vertebrates, and the milkweed butterflies gain protection by ingesting its leaves.

SKIPPERS (Hesperiidae): This family is named for its rapid, skipping flight. Skippers are small- to medium-sized, differing from other butterflies in having larger bodies in proportion to the wings. They generally have broader heads and hooked antennae that continue past the clubs rather than ending at them. To some people, skippers resemble moths, being hairier and more robust and generally lacking the gaudy colors and patterns of the other butterfly families. But unlike most moths, most skippers are day-flying or crepuscular. Many skipper larvae feed on grasses. One common garden species, the fiery skipper (*Hylephila phyleus*), eats Bermuda grass and is common around desert lawns. Another common species, the funereal duskywing (*Erynnis funeralis*), is mostly dark, with white fringes on the hind wings' edges. It is regular both in towns and in desert arroyos, often in close proximity to woody legumes, the larval hosts. Several skipper species have long tail-like projections on the hindwings. The most common of these, the dorantes long-tail (*Urbanus dorantes*), is an influx species and is at times abundant in late summer and fall.

Moths

Robert A. Raguso and Mark A. Willis

With over 142,000 described species worldwide, moths are a smashing evolutionary success, second among animals only to beetles in number of species. Over 12,000 species, grouped into 65 families, are found in North America alone. The moth fauna of the Southwest is particularly rich, as it includes the northern limit of distribution for many primarily Neotropical species. Within the order Lepidoptera, moth species outnumber butterflies and skippers nearly 15 to 1, with many species left to be described, especially among the numerous "microlepidopteran" families.

Why are moths so successful? All moths undergo complete metamorphosis; that is, their life cycles progress through egg, larval, pupal, and adult stages. Thus, the typical moth lives 2 ostensibly distinct lives, filling 2 distinct ecological niches; it is born as a terrestrial, vegetarian eating machine and is "reborn" as a winged creature of the night, hell-bent on completing its reproductive cycle. Yet this is not unusual for insects. Moths share a common body plan, including a head with large compound eyes and sensitive olfactory appendages (antennae). Also, in all but the "primitive" families (Micropterigidae, etc.), moths have long, tubular mouthparts fused into a proboscis; a powerfully muscled midsection (thorax) with two pairs of scale-covered wings and three pairs of legs; and a long, segmented abdomen that includes digestive, circulatory, respiratory, and reproductive structures.

Order: Lepidoptera
Families discussed: Arctiidae (Tiger Moths), Geometridae (Inchworms), Lasiocampidae (tent caterpillars), Lymantriidae (tussock moths), Micropterigidae (mandibulate moths), Noctuidae (owlet moths), Prodoxidae (yucca moths), Saturniidae (giant silkmoths), Sphingidae (hawkmoths), Tineidae (clothes moths)
Spanish names: palomilla, mariposa de noche, polilla

Male moths can often be distinguished from females by their broader, comb-like antennae, valve-like abdominal claspers, and smaller, more slender bodies. As in beetles, moths from different families vary widely in wing venation, shape and coloration, larval and adult feeding habits and behaviors, mating systems, population structures, thermal biology, and sizes, ranging from the minute clothes moth (Tineidae) with its ¼ to ⅜ inch (7-10 mm) wingspread, to the bat-sized hawkmoths (Sphingidae) and giant silkmoths (Saturniidae). Unlike beetles, the overwhelming majority of moth species are herbivorous as larvae and adults; there are far fewer examples of carnivores, fungivores, and detritivores among moth lineages. The complex relationships between moths and their host plants may hold keys to understanding why there are so many moths.

Given the stupendous diversity of moths, and our incomplete knowledge of moth distribution and abundance, especially in the Southwest borderlands, our purpose in this section is to outline a few of the salient characteristics of moth biology and suggest a few activities by which the reader might gain an appreciation for moths as dynamic, complex organisms.

FIVE FEATURES OF MOTH BIOLOGY IN THE SOUTHWEST

1. Plants, Caterpillars, and the Arms Race

The caterpillars of most moths are highly specialized and eat only one or a few plant species. Unfortunately, moth caterpillars are infamous for the exceptional cases; the decimation of crop plants by extreme generalists such as the cabbage looper (*Trichoplusia ni*; Noctuidae), and the destruction of wool clothing and stored grains by moths in the family Tineidae. The repeated association of certain moth and butterfly lineages with specific families of host plants worldwide suggests that these relationships are ancient. A closer examination reveals complex suites of plant defenses, both chemical (terpenoids, alkaloids, phenolics, cyanide-generating compounds) and physical (hairs, spines, tough leaves, oozing resins, and latex), designed to keep caterpillars at bay. Humans owe a debt of gratitude to moths and other insects for such biochemical plant wealth, which, quite coincidentally, provides us with a pharmacopoeia of natural drugs, insecticides, flavors, and fragrances.

Caterpillars, in turn, have evolved numerous physiological and behavioral strategies to counteract these defenses, from detoxification or rapid excretion of plant toxins to avoidance of older, better defended leaves. Tobacco hornworm (*Manduca sexta*) larvae will avoid snipping the veins of tobacco leaves, thus reducing the amounts of nicotine marshalled by the plant in its defense. Some specialized caterpillars co-opt the toxins from their host plants for their own defenses, and advertise their acquired distastefulness with bright, vivid colors.

There are additional, more subtle levels to the wars between caterpillars and their host plants. When caterpillars remain undaunted by chemical or physical deterrents, plants may use extrafloral nectaries or other foodstuffs to purchase

the services of ants and wasps as caterpillar exterminators. These security
guards can be bribed, however, and
certain caterpillars do so with glandular
secretions and resume eating plant
tissues with impunity. And so on.
Caterpillars on plants are vulnerable to
many other hazards. The scents of
wounded leaves and grass, the by-
products of caterpillar foraging, are
attractive to the parasitic wasps and flies
that appropriate caterpillar tissues for
the nutriment of their own young. In
addition, caterpillars are preyed upon by
birds, wasps, and other visually foraging
predators. In order to survive, they
defend themselves by being distasteful
or covering themselves with stinging
spines, or through bluff and deceit: they
mimic leaves, twigs, galls, flower buds,
bird droppings, and even snakes.

2. Moths as Pollinators

Few people realize that the voracious
hornworm, looper and armyworm caterpillars that defoliate desert wildflowers,
crop plants and garden vegetables,
eventually become nectar-feeding adult
moths that render important pollination
services to many of the same plants.
Moth pollination is more prevalent in
the Southwest than in other regions of
North America, largely due to warm
evenings, favorable climate, and proximity to the moth-rich canyons and thornscrub of northern Mexico. Moths visit
flowers in search of nutritious rewards,
usually nectar, and transfer pollen as
a consequence of their contact with
floral structures and forging movements
between flowers. Many night-blooming
plant species, especially in grasslands
and dune areas, appear to be specialized

for moth pollination, but since most
moths feed opportunistically from a
variety of flowers, disperse widely during their lifetimes, and neither defend
territories (like hummingbirds) nor provision young with floral rewards in local
nests (as do many bees), most moth-
pollinated plants employ alternative
reproductive strategies. These include
self-pollination, recruiting other (*diurnal*,
or day-active) pollinators, or simply
waiting for the next flowering season.
Thus, moth pollination is a risky
proposition, and moth-flower mutualisms are not very exclusive.

One noteworthy exception to this
pattern is the relationship between
yucca flowers and the small, white
moths (of the genera *Tegiticula* and
Parategiticula in the family Prodoxidae)
that spend most of their lives associated with yucca plants. Yucca moths are
among the few examples of "active"
pollinators, animals that intentionally
collect pollen from anthers and apply
it to stigmatic surfaces. A female yucca
moth uses her unique mouthparts
(*tentacles*) to gather a pollen ball from
yucca anthers, then walks or flies to
another flower, deposits a number of
eggs within the flower's ovaries, and
slam-dunks the pollen ball into its stigmatic cavity. Like wasps that bury the
bodies of paralyzed spiders with their
eggs, the mother moth's pollination
services ensure that her young will have
food (developing seeds) when they
emerge as hungry caterpillars.

The yucca plant and moth are
absolutely dependent upon one another
for reproductive success, yet the terms
of their contract are usually complex.
First, the yucca plant must sacrifice a
significant percentage of its seeds as

food for the moth larvae, although limited feeding damage enhances seed germination. Second, if yucca moth females deposit too many eggs within a single flower, the plant can selectively abort that flower, effectively killing all larvae within it. Finally the yucca-moth *mutualism* (living together in such a way as to increase each other's reproductive success) is vulnerable to exploitation by cheaters: other moth species lay eggs within fertilized flowers but do not pollinate the flower.

3. Migration and Dispersal

The Southwest is an unusually good place to witness impressive directional movements of insects, especially of conspicuous desert butterflies like the snout butterflies and painted ladies. Such movements are noteworthy for the animals' prolonged flight in the same direction, at a fixed speed, just a few feet above the ground. True round-trip migration, such as that performed annually by monarch butterflies and many birds, is a relativity rare phenomenon. Most cases of mass movements by moths probably are examples of one-way dispersal, such as the northward flush of black witches (*Ascalapha odorata*) into our region from northern Mexico toward the end of the summer. Dispersal need not be limited to adults; the movements of thousands of green- and black-striped hornworms (larvae of the white-lined sphinx moth *Hyles lineata*), across the desert floor provides one of the most memorable images of the summer monsoon. There are many

potential causes of mass dispersal, such as periodic population explosions, seasonal changes in day length or humidity, and the availability of food or hostplant resources, but the actual causes are unknown in many cases. The movement of adult moths over great distances, often across political boundaries, has important implications for moths as biotic resources for plants (through pollination), predators, and parasites. With increasing fragmentation and conversion of wild habitats to agricultural lands and subdivisions, these movements also affect populations of moths and their biological interactions with plants and other animals.

4. Moths and Bats: Eavesdropping That Really Pays Off

While the age of the earliest fossil moths suggests that they shared the world with dinosaurs and flying reptiles, we probably can never know if or when moths or their ancestors abandoned daylight for a relatively predator-free night. However, with the fall of the dinosaurs and the rise of the mammals, new and deadly predators of the night skies arrived: the bats. Fast, maneuverable fliers equipped with sensitive sonar guidance systems, bats are the number one threat for night-flying moths. But moths have developed an array of sensory and behavioral strategies that enable them to avoid becoming evening snacks for a bat.

White-lined sphinx

Many night-flying moths have pairs of ears positioned on both sides of their abdomens that are tuned to exactly the sound frequencies emitted by hunting bats. These sensitive ears allow the moths to eavesdrop on the hunting cries of bats and to attempt to avoid them. Moths have two levels of escape behavior at their disposal when they hear a bat using sonar to search for food. If their bat-detecting ears inform them that a bat is on the way, but still distant, the moth turns away from the direction that the cries are coming from and leaves the area. However, if the bat gets very close before it is detected, the moth suddenly executes a series of high-speed acrobatic maneuvers, usually ending in a dive for the ground or the shelter of nearby bushes. Some moths confuse bats by emitting sounds similar to those emitted by a bat closing in on prey. Sometimes the moths can evade the hunting bats; sometimes they become dinner.

Very small moths and very large moths usually do not have bat-detecting ears. The smaller moths are too small a morsel for the bats to chase after, and many of the larger moths, such as hawkmoths and giant silkmoths, may be too large for bats to catch and eat. There is another small group of species in the tiger moth family (Arctiidae) that actually advertise their poisonous nature to hunting bats. Moths in this group are poisonous due to toxins in the plant species that their larvae eat. Diurnally-active, poisonous insects typically bear bright and conspicuous color pattern so that visually hunting predators learn to associate their horrible flavor or poison-induced sickness with the bright colors. But what do you do if you are poiso-

nous, active at night, and your major predator uses sound and not vision to identify its prey? The small group of tiger moths, which share these characteristics, make unique sounds that their potential bat predators can detect and associate with their poisonous nature. When the sensitive ears of one of these moths detect a hunting bat in the area, at first it attempts to avoid the bat. If the bat gets too close, just as the moth initiates its evasive maneuvers, it also emits a burst of high-frequency clicks that the bat detects with its very sensitive ears. Bats that have learned to associate these clicks with an unpleasant meal initiate their own evasive maneuvers and leave the moth alone.

5. Mating Systems: The Scent of a Female

It is easy to understand why the colorful and conspicuous day-active butterflies typically have mating systems that rely heavily on visual communication of species and sexual identity. In contrast, even though moths possess visual systems especially adapted for their active night life, most species identification and sexual information in moths is communicated via air-borne chemical signals known as *pheromones*. In some insects, such as the highly social bees, ants, and termites, a complex chemical "language" exists that coordinates activities in the colony and allows for group defense and the dominance of the queen. Moths and many other insects appear to have only a very imited chemical vocabulary, usually amounting to "Hey baby, I'm a fantastic guy," and "OK, I'm ready to mate."

In a large majority of the moth

species so far studied, the female moth determines when mating will occur by releasing her sex-attractant pheromones. These pheromones typically are a blend of closely-related chemical compounds, which she synthesizes in a special gland near the tip of her abdomen. The sex-attractant is unique to each particular species, and males are rarely confused into following the scent trail of the wrong species. Male moths often trace a side-to-side zigzagging flight track as they follow the wind-borne pheromone trail to its source. In most moth species studied, once the male arrives at the female's location and physical contact is made, mating proceeds almost immediately. However, in some moths, upon his arrival a male releases his own unique courtship pheromone and fans it over the female with his wings. It has been demonstrated in a few species, and suspected for others, that the female moth uses the quantity or quality of the male's pheromone to assess his "quality" as a potential mate. It is interesting to note that many of the chemical compounds identified from male pheromones are also common components of the scents of flowers. At least some male moths remember to bring a bouquet!

FIVE ACTIVE WAYS TO LEARN ABOUT MOTHS

1. Light Traps

The inexorable attraction of moths to light is axiomatic in Western culture, yet no truly satisfactory explanations for this behavior have come to light, so to speak. Nevertheless, there is no better way to gain an initial appreciation for the diversity of moths in your area than by setting out a strong ultraviolet or mercury vapor light, hung over a white sheet in an otherwise dark, wild place. Insect lights that run off automobile cigarette lighters or portable generators are available from biological supply houses; alternatively, you could run an extension cord from your porch outlet. Pull up a chair and watch the guests arrive; take notes on which species arrive at different hours of the night and at different times of year. Examine the degree of wear and tear on the wings: are the moths freshly emerged and naive, or are they time-worn and battle weary? Look at the moths' antennae and genitalia: are they all of one sex, or are equal sex ratios attracted to the lights? What do you observe when you take your light to a different habitat or elevation? Who shows up to eat the moths that are attracted to your light? Which moths never seem to get eaten? If you are interested in collecting moths, this is a good way to start.

2. Baiting

Relatively little is known about the adult feeding behaviors of most moth species. There are entire families of moths, such as the giant silkmoths (Saturniidae), tussock moths (Lymantriidae) and the familiar tent caterpillars (Lasiocampidae), that don't feed at all as adults. Others, including many hawkmoths (Sphingidae) and owlet moths (Noctuidae) are avid nectar feeders and can be important pollinators of night-blooming plants. Some of these nectar-drinkers also tipple at flowing sap or rotting fruit, resources that are exploited to a much greater extent in tropical than in temperate regions. Setting out fermented bait is an exciting way to observe these moths in action; you may attract a different fauna than you will find at light traps, and you'll have a creative way to dispose of stale beer and overripe bananas! Simply mix old beer, rotting fruit and sugar or molasses and let it brew in a warm, dark place. When the concoction becomes pungent, paint or splash it onto tree trunks along a trail at dusk, then return after dark with a lantern and a field guide. You may see large underwing moths (*Catocala* spp.)— whose camouflaged, bark-like forewings conceal colorfully banded hindwings—as well as other owlet moths, inchworm moths (Geometridae), ants, crickets, and millipedes. When you learn the species by sight, keep notes on which ones are attracted to bait at different times of year. Experiment with the kinds of bait you use, adding the proverbial "eye of newt" to the cauldron; remember that most insects are less squeamish about road kill and excrement than we are— the funkier it smells, the better!

Buck moth and larvae

3. Night-bloomers and Moth Gardens

One of the greatest thrills for a moth enthusiast is watching a large hawkmoth unfurl its 4 inch (10 cm) long proboscis to drink from a trumpet-shaped flower while hovering in place at its threshold. Hawkmoths are effective pollinators of a guild of specialized night-blooming plants throughout the Southwest, including sacred datura (*Datura wrightii*), sweet four o'clocks (*Mirabilis longiflora*), and tufted evening primrose (*Oenothera caespitosa*), which produce pale, fragrant flowers with nectar tubes as long as the moths' extended tongues. In exchange for reproductive services, these plants provide copious sucrose-rich nectar, the high-octane fuel required by hawkmoths to maintain hovering flight. The strong floral perfumes are thought to attract moths from a distance, after which the moths appear to be guided to these pale trumpets by visual cues. Plants with bunches of small tubular flowers, such as scarlet gaura (*Gaura coccinea*) and fairy duster (*Calliandra eriophylla*), attract many owlet and inchworm moths, which may spend up to 20 minutes perched on a single inflorescence, drinking leisurely from each flower. As mentioned above, yuccas are pollinated exclusively by small, satin-white yucca moths, whose frenetic mating, pollinating, and egg-laying activities are endlessly entertaining. Find populations of these and other night-blooming plants and wait until dusk, then watch as their flowers open and the moths arrive. Do flowers and their moth visitors segregate by size? Do moths stick to one type of flower, or do they sample from a buffet? Dab a few flowers of one plant with day-glo paint powder, wait until a few moths pass through, then scan all the nearby flowers with a portable ultraviolet lamp (the kind used to illuminate scorpions) to follow the moths' trails. Better still, plant your own moth garden with a variety of night-blooming, fragrant plants, and provide a nectar filling station for wayfaring moths.

4. Raising Caterpillars

In the Southwest, as in most places, chances are good that if you find an unfamiliar caterpillar, it will grow up to be a moth. This is especially true for caterpillars that mine within leaves, bear stinging spines, and spin plush silken cocoons. Modern insect science has benefited greatly from the contributions of dedicated amateur entomologists who have reared adult moths from caterpillars, and described their larval anatomies and behaviors, food plant preferences, and the timing of their life cycles. Despite the efforts of amateurs and scientists, little is known about the immature stages of most adult moths, and many await discovery.

Caterpillars can be raised in jars, terraria, or even plastic bags, as long as old leaves and grass are removed regularly and growth of mildew is prevented. When you find a new caterpillar, provide it with plenty of freshly cut leaves from the plant it was eating or walking on, along with twigs, dry leaves and soil on which to pupate. If it refuses to eat, you may have to offer it a living plant, or a buffet of different plant species from the area in which you found it. As the caterpillar grows, make drawings or photographic records of its different larval stages (instars), the pupa (cocoon) and the adult that emerges from it. Once you have reared a given species successfully, you can experiment with alternative food plants. Will your caterpillars accept leaves of another plant related to its host plant? Would they rather die (some do)? Never let caterpillars run out of food, as even temporary starvation enhances the probablility of death due to viral infection. Be persistent, as many moth life cycles involve dormancy (diapause) during one or more of the stages of development, and some caterpillars are very difficult to rear in captivity. Also, don't be discouraged if a parasitic wasp or fly emerges from your pupa; instead, photograph it or collect the specimen. Records of this kind are extremely important, as we know even less about moths' parasites then we do about moths!

5. Captive Moth Sirens

Many moth species, especially the giant silk moths and hawkmoths, feature mating systems in which newly emerged females emit volatile sex pheromones that attract males from a considerable distance. Rearing moths from caterpillars may provide you with many virgin females, and the means to conduct an interesting census of male moths in your area. (You can also obtain female moths at light traps and bait, but these may already have mated.) Place female moths into small mesh or wire cages; position the cages along a trail, near a forest clearing, or on your porch; then observe. If males of this species are about, they should find their way to the cages in short order. At what time do they arrive? How abundant are they in suburban areas, as compared with rural habitats? Do males from more than one species arrive at the cages? If you are interested in moth breeding, this is an excellent way to attract wild mates and begin the life cycle anew.

Bees

Stephen L. Buchmann

ees comprise a highly diverse group of hymenopterous insects in the Sonoran Desert region. Superficially, bees (especially the parasitic cuckoo bees) resemble some wasps, except that bees are usually hairier and more robust, and they possess specialized structures for carrying pollen back to their nests. Together with ants, bees and wasps form a natural group referred to by taxonomists as the *aculeate*, or "stinging," Hymenoptera; the stinger is called an *aculeus*. Only females sting, since the aculeus evolved from the ovipositor or egg-laying tube. Sonoran Desert bees range in size from the world's smallest bee, *Perdita minima*, which is less than .08 inches (2 mm) to carpenter bees (the genus *Xylocopa*), gentle giants that may have body lengths of almost 1½ inches (40 mm) and weigh over a gram. Our native bees burrow into the ground or create nests inside hollow, pithy, dried stems or abandoned tunnels left by wood boring beetles. All bees are herbivorous except for parasitic forms that prey on other bees.

Herbivorous bees feed on pollen, nectar, and oils offered as floral rewards by flowering plants. Most bees have solitary lifestyles in which females act alone to construct and provision nests, but there are also social forms, such as the familiar black and yellow bumblebees.

Order: Hymenoptera

Families: Andrenidae, Apidae (includes former Anthophoridae), Colletidae, Halictidae, Megachiliade, Melittidae, Oxaeidae

Sonoran Desert genera: *Agapostemon, Andrena, Anthidium, Anthophora, Ashmeadiella, Bombus, Centris, Coelioxys, Colletes, Diadasia, Epeolus, Exomalopsis, Halictus, Heteranthidium, Hylaeus, Megachile, Melecta, Melissodes, Nomadopsis, Nomia, Osmia, Panurginus, Peponapis, Perdita, Psithyrus, Sphecodes, Stelis, Svastra, Tetralonia, Triepeolus, Xenoglossa, Xenoglossodes, Xeromelecta, Xylocopa*

Spanish names: abeja (bee), jicote, abejorro (bumblebee, carpenter bee)

DISTRIBUTION

There are at least 45 genera in 7 families, and perhaps as many as 1000 species of bees distributed within the Sonoran Desert bioregion. Unlike most other groups of organisms, bees are most abundant in numbers of both species and individuals in deserts and savannahs, rather than in lowland rainforests. The region around Tucson, Arizona is thought to host more kinds of bees than anywhere else in the world, with the possible exception of some deserts in Israel. In the United States, there are about 5000 species of bees. On a global scale, there are approximately 25,000 named species, but it is likely that as many as 40,000 different species exist.

ECOLOGY

Bees, a highly successful group derived from wasps, live in almost all terrestrial habitats within our region. Except for the parasitic cuckoo bees, all female bees make their living by foraging in search of protein-rich pollen and sugary nectar from flowering plants. By moving pollen around from flower to flower and plant to plant, bees perform vital and often unappreciated roles as the most important group of pollinating animals on earth. Yet bees are not out to "help" flowers; they collect pollen and nectar in order to feed themselves and their larvae.

Of the approximately 640 flowering plant taxa growing in the Tucson Mountains near the Desert Museum, approximately 80 percent of these species have flowers adapted for and pollinated by bees. Similarly, at least 30 percent of our agricultural crops require bees to move pollen between flowers. Not only are we dependent upon these "forgotten pollinators" for over a third of our food, but for other products as well. Cotton cloth is a product that eventually results from bee pollination, and so are many beverages and medicines made from other fruits and seeds.

Africanized Bees

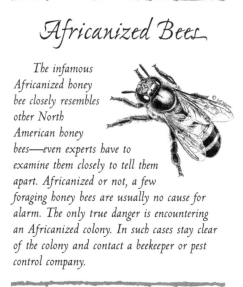

The infamous Africanized honey bee closely resembles other North American honey bees—even experts have to examine them closely to tell them apart. Africanized or not, a few foraging honey bees are usually no cause for alarm. The only true danger is encountering an Africanized colony. In such cases stay clear of the colony and contact a beekeeper or pest control company.

Without the pollination services bees provide, many plants would not produce seed-laden fruits from which the next generation of plants would grow. Without bees, there would be few or no fleshy berries or fruits to sustain birds, mammals, and other wildlife. The tunneling activity of bees aerates the soil and allows water from infrequent rains to quickly penetrate and reach plant roots; and bees' nitrogen-rich feces fertilize the soil. The bees themselves often provide food for lizards, mammals, birds, insects, spiders, and other arachnids.

In their daily quests bees harvest foodstuffs from flowers for themselves and their larvae. Pollen is a rich food source of amino acids, proteins, fatty acids, vitamins, minerals, and carbohydrates. Nectar provides the energy boost from sugars that bees need to fly. Some desert bees (*Centris*) have specialized scrapers on their legs for harvesting oils from glands on the undersides of specialized flowers in the ratany and malpighia families. These energy-rich oils are are mixed with pollen as larval food and are also used to help construct brood cells. Other bees collect small pebbles, plant hairs, or floral resins that they use as building materials. Some bees, such as mason bees in the genus *Osmia*, also require water and mud with which to construct their adobe-like nests. Leafcutter bees (*Megachile*) remove circular pieces of leaves to fashion into cell walls.

© Jeff Martin

Carpenter bee

LIFE HISTORY

The vast majority of desert bees dig burrows in the ground for their brood cells. Cells can be just a few inches deep or six feet or more down in sandy soils. Many species line their burrows or cells with waxy secretions produced by glands within their abdomens. This lining waterproofs the cells, maintains humidity, and keeps organisms like fungi from destroying the food and the developing larvae. Most bees are *solitary*, that is each female selects a site for her nest, excavates the tunnels, forms and provisions the rounded cells, and lays an egg within each one. This behavior is called "mass provisioning," since the mother bee collects and prepares at one time all the pollen and nectar food each developing larva will need to complete its life cycle from larval stages to pupa and, finally, through complete metamorphosis into a newly emerged adult. After laying an egg in each cell, the solitary female has no further contact with her progeny. Although solitary, many of our Sonoran Desert bees routinely nest with other females in very large aggregated nest sites. Among them is is our common cactus bee (*Diadasia rinconis*) which pollinates prickly pear, cholla, and saguaro cacti. Its aggregations during the spring cactus bloom may number in the hundreds of thousands of individual nests over an area the size of 2 to 3 tennis courts.

Other native desert bees don't go to the bother of excavating their own nests. Instead, they actively search out the abandoned exit holes and tunnels of wood-boring beetles (usually buprestids and cerambycids) in dead limbs or standing dead trees. These bees are known by their common names of leafcutter (the genera *Chalicodoma* and *Megachile*) and mason bees (the genus *Osmia*). Once a beetle burrow is located, these females bring back cut pieces of leaves, resins and pebbles or mud balls with which to fashion cells and their thick, protective capping plugs.

Most of our Sonoran Desert bees have but one generation per year, with adults usually emerging with the spring or summer wildflower blooms. Some species, however, have 2 or even 3 generations per year. In a typical life history, adult males and females mate soon after emergence from their natal cells. Females construct and provision nests and lay eggs, and the larvae develop rapidly underground. During cooler months the larvae usually stay in a resting condition, or *diapause*, at either the *prepupal* or the *pharate adult* stage (the pupal stage just prior to the final molt to the adult proper) until the following spring or summer when they complete their metamorphoses and emerge as adults.

Bumblebee

SOCIAL BEHAVIOR

A relatively small number of our desert bees are truly social. Bee sociality indicates that these species' evolutionary paths have diverged from those of their solitary ancestors. In the case of some sweat bees (the family Halictidae), a queen looks like other females in the colony, though she differs in having highly developed ovaries. She may or may not secrete pheromones that elicit feeding and grooming behavior in her daughters. The queen lays all the eggs within the colony. Sweat bee colonies are usually small, consisting of a few dozen or at most a few hundred individuals.

Highly social bees in our region include the introduced honey bee and the native black and yellow bumblebees in the genus *Bombus*. Bumblebee colonies are annual in nature, established by an already-inseminated queen who emerges from her winter retreat in spring and finds a suitable mouse nest or other underground cavity in which to nest. The queen is larger than her daughters, and lays all the eggs after she has produced, initially, a small brood of workers. Males and queens are produced late in the season and in the colony cycle. They mate, the males die, and the inseminated queens spend the fall and winter "hibernating" below ground until the next spring.

Farther south in Sonora, Mexico (near Alamos), extremely social bees live in colonies with many thousands of individuals—the so-called "stingless bees," in the genera *Melipona* and *Trigona*. These queens are *physogastric*, with abdomens swollen full of eggs. They are not able to fly once they begin laying eggs, and never leave the colony again. Thus, these social bees represent a still greater caste differentiation between queens and workers. Stingless bees store as much as several quarts (liters) of honey in waxen storage pots that look like clusters of grapes. Indigenous peoples who find these nests within hollow trees sometimes transport the hives back to their villages, where they tend the bees and routinely harvest their honey and beeswax.

Wasps

Justin O. Schmidt

Wasps comprise an enormous and diverse assemblage of insects ranging from the smallest known insects—tiny parasites of insect eggs—to immense cicada killers and tarantula hawks. Most wasps are predators whose young feed on other insects or arthropods, but a few groups have become vegetarians, similar to bees, and collect pollen to be fed to their larvae. Both sexes of wasps are typically strong fliers, but some species are flightless, and in others, one sex, usually the male, is an excellent flier while the other sex is flightless. Velvet "ants"—actually female wasps that superficially resemble ants—are an example of this.

Perhaps the most conspicuous and commonly seen wasps in the Sonoran Desert are the paper wasps (*Polistes*).

Paper wasps are large, (about 1 inch, or 20 to 25 mm) social wasps that build paper honeycomb nests. Paper wasps are longer, thinner, and more smooth and shiny than honey bees and have longer, narrower waists (called *petioles*) than do bees. Common paper wasps include the yellow paper wasp, whose color is true to the name; the Navajo paper wasp, which is deep chocolate brown with the end of the abdomen yellowish; and the Arizona paper wasp, which is slightly smaller and more spindle-shaped than the other 2 and is brownish-red with thin yellow cross bands on the abdomen.

Some of the most impressive insects in the Sonoran Desert are the enormous tarantula hawks (*Pepsis*). These 1 to 1¾ inch (25 to 45 mm) wasps sport brilliant gun metal, blue-

---

Order: Hymenoptera
Suborder: Apocrita
Division: Aculeata
Families: Pompilidae, Mutillidae, Sphecidae, Vespidae
Sonoran Desert genera: *Pepsis, Dasymutilla, Sphecius, Sceliphron, Polistes, Vespula*
Spanish name: avispa

black bodies carried on fiery orange wings. The desert contains more than a dozen species, some of which have jet-black wings instead of orange wings.

Velvet ants (*Dasymutilla*) are among the more colorful of all organisms in the desert. Females of these large ant-like wasps can be seen scurrying over sandy or bare soil surfaces during warmer seasons of the year. More than 3 dozen species live in the Sonoran Desert of Arizona. They range in size from tiny ⅛ inch (4 mm) species to huge 1 inch (25 mm) giants. Most are clothed in red, orange, yellow, or silver coats of hair-like *setae* (bristles) and look like moving fuzzy cotton balls. Particularly large velvet ants include the black and red *D. klugii* and Satan's velvet ant, which is black with a yellowish-white furry abdomen. The glorious velvet ant (*D. gloriosa*) is a long-haired, totally white velvet ant that looks like a creosote bush seed on legs.

Velvet ant

Cicada killers (*Sphecius grandis*) superficially resemble huge yellowjackets or hornets. Yellow with tan patches, they are 1 to 1½ inches (25 to 40 mm) long. These powerful fliers have large compound eyes.

Mud daubers (*Sceliphron caementarium*), sometimes called dirt daubers or mud wasps, are thin, 1-inch (25 mm) long black wasps with yellow legs, and long, yellow-thread waists. They are named for their habit of building mud nests under bridges or eves of houses, or in other protected areas.

LIFE HISTORY

The paper wasp is social insect whose life cycle begins as a solitary mated queen. The queen overwinters deep in rock cracks, behind peeling tar paper, or inside enclosures. In spring the queen builds a paper nest suspended from a thin stalk in a protected rock crevice, among thick vegetation such as dead fan palm leaves, or under the overhang of a man-made structure. She constructs a small cluster of paper hexagonal cells and lays an egg in each. The queen then feeds the larvae that hatch from these eggs a diet of caterpillar "meat balls." When the first young worker wasps emerge from their pupal cells, they assume the tasks of hunting caterpillars, collecting material for making papier mâché for nest expansion, and collecting water for cooling. The queen then ceases all work except egg laying. By late spring, the colonies have grown to contain 20 to 50 wasps; by late summer as many as 200 wasps may be present. At this time new queens and males are reared. After mating, the new queens imbibe nectar to fatten for the winter. By late fall, the queen mother and workers die, the nest is abandoned, and the next generation of queens goes into hibernation.

Tarantula hawks, so named for their huge size and hawk-like hunting strength, hunt tarantula spiders. During the warm months female wasps search the ground for tarantulas. Once prey is located, the tarantula hawk bites onto a leg of the tarantula and with its long, strong, sharp stinger, pierces the spider near a leg base, and injects paralyzing venom. The limp, but living, spider is dragged into an appropriate hole, sometimes the spider's own burrow, where a single egg is laid on the spider. The wasp then seals the burrow to complete her work. When the egg hatches, the larva consumes the spider and then pupates; the next spring the adult emerges to complete the life cycle. Males do not hunt, but are frequently seen visiting flowers of milkweeds, western soapberry trees, or mesquites.

Velvet ants, cicada killers, and mud daubers have life cycles similar to tarantula hawks. Wingless female velvet ants search the soil surface for burrows of wasps or bees that can serve as hosts. Male velvet ants fly above the surface looking for females. Once a female velvet ant locates a nest of a suitable host, she enters, opens a cell of a host larva that has completed feeding or has already pupated,

© Jeff Martin
Tarantula hawk

lays an egg on or near the host, and closes the cell. She does not sting or paralyze the host. Cicada killers search for cicadas, which they sting to paralyze, and then transport back to their nests dug in sand or soft soil. One to 3 cicadas are provisioned per cell for each young. Mud daubers provision their mud cells with spiders they have captured and paralyzed by stinging. The larvae of all of these wasps feed on the provisioned prey, consuming it entirely, and pupate to emerge the next year as adults.

ECOLOGY AND BIOLOGY

Most wasps are specialized hunters that track down their prey using smell and sight combined with knowledge of the habitat, activity periods, and behavior of the prey. A solitary wasp usually subdues its prey with a sting that either kills the prey or paralyzes it briefly or permanently. (Tarantulas stung by tarantula hawks can live completely paralyzed for months.) Social wasps, including paper wasps, never sting their prey. Instead, they use their powerful cutting mandibles to chew the prey into pieces to feed directly to their larvae. The venom of social wasps is used only for defense. The most serious predators of social wasps are birds, mammals, and reptiles. The sting and venom have evolved to be effective weapons against these large animals. Venoms are ideal defenses because they can be injected via the stinger directly into the assailant's body where they cause pain, toxicity, or both. The stingers of wasps, and their relatives the ants and bees, are modified ovipositors used to deliver venom rather than to lay eggs. Thus, only females can sting; males are completely harmless. Unlike honey bees, which can only sting once because the barbed sting remains in the victim's skin after stinging, thus evisceratating the bee, wasps have smaller barbs on their stings and can withdraw the stinger to sting several times. The effectiveness of wasp stings and venoms as defenses has allowed wasps to evolve bright warning colors of red, yellow, orange, or white on a black or dark background. These conspicuous warning color patterns are called *aposematic*, and their brightness usually correlates with the degree of painfulness of the sting. Potential predators and humans alike learn to avoid beautiful aposematic wasps. Failure to heed the warning often results in an excruciatingly painful, and possibly toxic, sting. Indeed, tarantula hawks deliver the most painful sting of any United States or Mexican insect, a sting that is many times more painful than that of a honey bee. Velvet ants are also true to their bright colors. They not only possess the longest stingers of stinging wasps, but deliver a powerful sting that is not soon forgotten. Although wasps can sting people, they rarely do so, and then only when they are captured, or in the case of paper wasps, when their nests are threatened or disturbed.

Paper wasp

Ants

Harvester ant

Diana E. Wheeler and Steven W. Rissing

nts are familiar creatures. Although they are small as individuals, they are social, living in cooperative colonies, and these colonies are huge. Ant colonies are made up entirely of females, and include one or more queens and many workers. All of the ants streaming in and out of a nest entrance are workers, who protect the colony, collect food, and care for the larvae. Even though the workers are females, they lack the reproductive abilities of the queen, who lives deep within the nest and does little besides produce eggs.

All ants' bodies are divided into three parts. The *head* includes the antennae that detect smells, compound eyes, and capable jaws. All legs are attached to the mid-section, called the *thorax*. The *gaster* is the last segment and contains most of the internal organs, including defensive organs, including the glands that produce formic acid. When working with ants, beware the gaster!

Ants are most closely related to wasps. Imagine a wasp without its wings—it looks like an ant. Flying ants (the new queens and males) are often mistaken for wasps.

LIFE HISTORY

Ants hatch from eggs laid by the queen and go through a series of larval stages before becoming adults, just as do butterflies. Ant larvae cannot move about on their own, however, and are completely dependent on workers for their care.

Just as individual ants go through stages of development, so do ant colonies. An adult ant colony raises reproductive (sexual) male and female forms, which can be recognized (in the desert) by the presence of wings. Reproductive females (who will become queens) look like workers but

---

Order: Hymenoptera
Family: Formicidae
Sonoran Desert genera: *Pogonomyrmex, Messor, Pheidole, Solenopsis, Myrmecocystus,*
    desert *Neivamyrmex*
Spanish name: hormegas

---

are 2 to 3 times as big; males are smaller than workers with even smaller heads and gasters. Mating occurs at species-specific times of the year. Most species fly immediately after summer rains; a few (*Messor* and some *Myrmecocystus*) fly in the winter. Reproductive individuals released within one area congregate at a mating swarm where one female may mate several times; sperm are stored for the lifetime of her colony, which may be more than 20 years!

Many desert ants (especially some *Messor, Acromyrmex, Myrmecocystus, Solenopsis, Pogonomyrmex*, and *Pheidole*), cooperate in founding colonies. Multiple new queens work together to start a small colony and raise their first workers. By working together, these queens can get underground faster (avoiding heat and predators) and produce more workers to more rapidly establish the territorial limits for their colony.

The period of colony founding—when the new queens must fly, mate, locate a good nest site, and avoid predators—is when most colonies fail. Only a few will survive. Once established, a desert ant colony may live for a decade or more. This means that the queens that establish the colonies are among the longest lived insects we know.

Long-lived, underground nests protected by thousands of ants devoted to bringing in food, offer attractive environments to other insects besides the ant-architects themselves. A variety of beetles, roaches, crickets and silverfish have evolved the ability to live in ant homes. Should we be surprised that they live in our homes as well? Perhaps the most interesting of these guests are other ant species. Some ants are "kitchen thieves" that are small and secretive.

They live in the cracks and other hide-aways of regular ant nests and eat the crumbs of food that are left out. Other ants are more insidious. They sneak into a *host nest* of a closely related species and lay eggs destined to become reproductive males and females. Workers of the host colony raise these eggs as if they were their own. When it comes time to fly and reproduce, these *social parasites* propagate their own genes, not those of the colony that reared them.

ECOLOGY

The Sonoran Desert is a great place to watch ants. They are the most abundant animal in this habitat and lack of ground cover makes them easy to see. In the desert most species of ants build underground nests that protect them from the harsh conditions. Here they can store or even grow food, find ample water, and avoid the environmental extremes of the soil's surface.

One way to look at ant diversity is to classify them by the foods they eat:

SEED-HARVESTER ANTS Many desert ants (especially *Messor, Pheidole, Pogonomyrmex*, and *Solenopsis*) harvest seeds that they use as food for their larvae. Seeds of several grasses and annual plant species are preferred; seeds of perennial plants—especially cacti—seem not to be preferred. Seeds are stored in chambers toward the top of nests where dry conditions discourage germination. Ants have interesting behaviors when learning the different types of seeds that are available to them.

Pogonomyrmex workers have large squarish heads that contain powerful muscles for crushing seeds. The workers are ½ inch (13 mm) in length and are brick red to black. They have unforgettable stings. The typical nest of many species has a prominent cleared area, with a central opening and several permanent trails radiating from it. Another ant, *Messor pergandei,* is the only species of this worldwide genus that extends into the Sonoran Desert. Workers are ¼ to ½ inches (6 to 13 mm) long and are a shiny, jet black; they do not sting.

Over a dozen species of *Pheidole* live in the Sonoran Desert. Their workers come in two distinct forms: a small "minor worker" class and a much larger "soldier" class that crushes seeds, and sometimes enemies too.

Our desert *Solenopsis* are related to the infamous "imported fire ant" that is the scourge of the southeastern United States. The desert fire ant is a natural part of the Sonoran Desert community; unfortunately, it does resemble its eastern relative in its aggressive behavior and annoying sting. Most fire ants are ⅛ inch (3mm) long with some as large as ⅓ inch (8 mm); they are shiny brick red to black.

LEAF-CUTTER ANTS Another common group of ants in the Sonoran Desert are the leaf-cutting or fungus-growing ants. *Acromyrmex* ants are related to the larger leaf-cutting ants of the tropical Americas. *Acromyrmex versicolor* is common in the Sonoran Desert. Its workers collect leaves and other plant parts to insert into fungus masses, which they grow in chambers deep within their underground nest. The fungus is completely dependent upon the ants for its care and propagation; the ants, in turn, eat a portion of the fungus as their sole source of solid food. Long columns of leaf-cutter ants search across the desert for plant matter for their fungus gardens when conditions permit in the fall and spring and on cool summer mornings; at other times, they remain underground. The fungus garden is started from a small "plug"of fungus brought by the queen from her home colony.

HONEY POT ANTS Another common food source in the desert is the liquid nectar of plants and the "juice" of other insects; both of these, however, are available only seasonally. Honey pot ants (*Myrmecocystus*) have solved this seasonal problem with specialized members of the colony that store liquid food in their engorged gasters. When other members of the colony need food, these living storage vessels share their stored reserves.

ARMY ANTS The ants described above eat seeds, fungus or nectar. Some ants prefer meat; these are the desert army ants (genus *Neivamyrmex*). These ants raid the nests of other desert ants and occasionally take other prey as well. Because they are predatory and deplete the prey in any one area, they are nomadic and move from place to place. They have no permanent nest structures, and instead tend to live in temporary quarters such as hollows under trees or kangaroo rats' nests.

Ants' ability to live in colonies and excavate deep nests where the seasonally abundant food of the desert can be stored has made them remarkably successful in the Sonoran Desert. Further, in this environment, ants are easily collected and observed; this has made them model organisms for studies of development, behavior, and ecology.

Insects and the Saguaro

Carl A. Olson

The saguaro, *Carnegiea gigantea*, is a well-known symbol of the Sonoran Desert. Watching over the scrub-like flora from its lofty height, this "sentinel of the desert" appears to be a protector of our unique landscape. It seems impervious to herbivores with its tough outer skin and latticework of spines, although packrats are known to eat spiral tunnels up the stems under extreme environmental conditions. But the saguaro is a survivor; such small indignities do it little harm.

Most insects associated with the living saguaro do little damage to the cactus. The Gila Woodpecker and Gilded Flicker, however, peck holes into the saguaro, creating new niches for themselves and for other desert dwellers. These holes heal and become the familiar "boots," nesting sites for the woodpeckers and later for other birds such as Elf Owls. Beneath some of the slightly hardened calluses of the boots, a small, horned scolytid beetle, *Cactopinus hubbardi*, sculpts galleries. This minute rhinoceros beetle is an artist-in-residence, creating mini-designs of intricate form, although unfortunately it is usually unseen by an appreciative audience.

Early in the life of a saguaro, when a nurse plant protects it and growth is slow, a jet-black, longhorn beetle, *Moneilema gigas*, may find a seedling saguaro and consume it. Normally, this beetle feeds on cholla or prickly pear cacti, but like many other of Nature's creatures, it takes advantage of the young and relatively defenseless.

Flowers act as attractants to much of the animal world, whether with beauty, odor, or food. Arizona's state flower, the saguaro blossom, is no exception—it draws bats, bees, flies and beetles to its rich smorgasbord.

If we were able to look into the flower we might be startled to see the anthers move without the aid of wind or other devices. How is this possible? Probing deep into the base of the flower, we might discover a small sap beetle, *Carpophilus longiventris*. This beetle wanders about feeding on pollen grains, then seeks a place to lay eggs. The eggs hatch quickly, and larval development is rapid since the larvae, feeding on pollen or decaying flower tissues, must leave the flower before it shrivels and drops off. The larvae pupate underground. These beetles have no effect on the developing saguaro fruit.

In the fall when life in the desert begins to slow, after the summer monsoon season has ended, secretive termites start their recycling activities.

Cactus longhorn beetle

Overnight, mud covers the base of the saguaro where barky material has formed. If the mud is scraped away, small white insects rush for cover. These are the desert-encrusting termites, *Gnathamitermes perplexus*. It is their job to recycle dead plant material, scraping off and eating the outer layers and returning them to the soil for next year's plants to feed upon. Termites are not saguaro killers, but soil enhancers.

With death the saguaro becomes really important to insects. When this stately cactus falls victim to *Erwinia*, a bacterium, or to other problems that begin the decay process, it becomes an oasis to numerous insects and other arthropods, providing food, moisture, shaded habitat or an enticement for predators seeking live food. Carve off

a piece of tough outer skin of a decaying saguaro, and you'll find the innards teeming with life. Move the whole saguaro, and spiders race from beneath it, escaping exposure to sunlight.

The sheer size of insect populations is always amazing. A small chunk of rotting saguaro (about 1 cubic foot, or 2300 cm^3) was examined at the University of Arizona; it yielded 413 individual arthropods, including adult and larval beetles, larval flies, pseudoscorpions, and mites. Compare that small portion to the size of a whole saguaro and you instantly understand why insects are this planet's dominant life form.

A closer look at the fauna in a rotting saguaro will expose an ecosystem with grazers on fungi, such as the feather-winged beetles *Acrotrichis* and *Nephanes*, and such recyclers of plant matter as the flattened, leathery syrphid fly larvae *Volucella*, the neriid cactus fly maggots *Odontoloxozus longicornis*, and numerous phytophagous mites. This habitat is no longer solid plant material, but is now quite aquatic in nature. Several hydrophilid beetles, *Agna capillata* and *Dactylosternum cacti*, may be seen swimming awkward strokes through the muck. They feed on a wide assortment of organic material, from fungus and dead plant matter to castoff exoskeletons and dead insects. What better pond is there in the desert?

And what would an ecosystem be without predators? They swarm to this bountiful table in hordes—all sizes of rove and hister beetles, each staking claim to a prey size suitable to its mandibles. Most are colored red, so an observer may readily spot these hungry terrors of the bug world as they stalk their prey through the saguaro rot. One must be careful, however, when searching through this habitat, as another great predator of the desert, the scorpion, may be lurking in a hidden recess with its pinchers and sting at the ready.

Pseudoscorpion

A tiny beast, one resembling a scorpion without a tail and appropriately called a pseudoscorpion, may be seen waving its claw-like pedipalps as it searches on top and within the saguaro for a meal. After feeding on small insect larvae or mites, this strange arachnid seeks neriid flies in the pupal stage. It lies in wait until a fly emerges into adulthood. The pseudoscorpion then latches onto the fly's leg with its pedipalps just as the fly is leaving its decomposing home.

The pseudoscorpion now has a limousine ride to the next rotting saguaro, and a chance to find a mate.

Many spiders and other arthropods also hide near the saguaro waiting for their chance to catch a snack. Arizona brown spiders, wolf spiders and other denizens of the arthropod world, such as daddy-long-legs, may be found around the fallen saguaro. Depending on the season or the region, other species may also be found playing roles in this desert drama.

This ecosystem within the saguaro shows how nature's resources, alive or dead, are always being used—nothing is wasted; everything is important to some creature.

Insects of the Living Saguaro

Class Insecta
    Order Coleoptera   beetles
        Family Scolytidae   engraver beetles
             Cactopinus hubbardi   saguaro rhinoceros beetle
        Family Cerambycidae   longhorn beetles
             Moneilema gigas   cactus longhorn beetle
        Family Nitidulidae   sap beetles
             Carpophilus longiventris   saguaro sap beetle
    Order Hymenoptera   bees, wasps, ants
        Family Apidae   bees
             Apis mellifera   honey bee
    Order Isoptera   termites
        Family Termitidae
             Gnathamitermes perplexus   desert encrusting termite

Some Arthropods Associated with the Dead Saguaro

Class Insecta
    Order Coleoptera   beetles
        Family Ptiliidae   feather-winged beetles
             Acrotrichis sp.
             Nephanes sp.
        Family Hydrophilidae   water scavenger beetles
             Dactylosternum cacti
             Agna capillata
        Family Staphylinidae   rove beetles
             Belonuchus ephippiatus
             Maseochara semivelutina
             Tachyporus grossulus
        Family Histeridae   hister beetles
             Hololepta yucateca
             Carcinops gilensis
    Order Diptera   flies
        Family Neriidae   cactus flies
             Odontoloxozus longicornus
        Family Syrphidae   hover or syrphid flies
             Volucella avida
Class Arachnida   scorpions, spiders, pseudoscorpions
    Order Scorpiones   scorpions
             Centruroides exilicauda   bark scorpion
             Vejevois spinigeris   stripe-tailed scorpion
    Order Araneae   spiders
             Loxosceles arizonica   Arizona brown spider
    Order Acarina   mites, ticks
    Order Opiliones   daddy-long-legs
    Order Pseduoscorpiones   pseudoscorpions

Aquatic Insects of the Sonoran Desert

Tom Dudley

*T*he biological diversity and abundance in natural waters of the desert can be almost overwhelming, especially when we look beyond the fish and other aquatic vertebrates to the more diverse group—invertebrates. Dozens of species often share a single site in a desert stream. The types and morphological traits of aquatic insects present are closely related to the type of stream habitat, as well as to the kinds of food energy that are available.

The abundant light in most desert streams supports the growth of algae and other plants on stream bottoms, which feed what stream ecologists call *grazer-scrapers*. Riparian plants drop leaves and other debris which, along with associated bacteria and fungi that decompose the material, sustain what are called *shredders*. Another form of food is fine organic particles, some that has passed through guts of other

Orders:	Families:
Odonata	Calopterygidae, Coenagrionidae, Libellulidae, Gomphidae, Cordulagasteridae
Ephemeroptera	Baetidae, Caenidae, Leptophlebiidae, Siphlonuridae, Tricorythidae
Plecoptera	Capniidae
Megaloptera	Corydalidae
Hemiptera	Belostomatidae, Corixidae, Gelastocoridae, Gerridae, Naucoridae, Nepidae, Notonectidae, Veliidae
Trichoptera	Glossosomatidae, Helicopsychidae, Hydropsychidae, Polycentropodidae
Lepidoptera	Pyralidae
Coleoptera	Dryopidae, Dytiscidae, Elmidae, Gyrinidae, Haliplidae, Hydrophilidae, Psephenidae
Diptera	Chironomidae, Simuliidae, Stratiomyiidae, Tabanidae, Tipulidae

animals, that *collectors* ingest. Finally, *predators*, at a higher *trophic* (feeding) level, eat all the other groups of insects, including other predators!

Feeding habits only partially explain the diversity of aquatic insects in desert streams. Different lifestages—larvae, pupae and adults of various species—are found throughout the riparian ecosystem, even far below the stream substrate. In many cases aquatic insects are the larval or nymphal stages of winged adults that will leave the stream to reproduce and disperse in the terrestrial environment.

Conditions vary from the water surface to the streambed, and from midstream to water's edge; each area has its own set of invertebrates. Their shapes, sizes, and behaviors, however, are varied, and depend upon the micro-habitat that each occupies. Each must obtain enough oxygen, which is present at lower concentrations in water than in air, especially in slow water, so most aquatic insects living below the surface have gills of some sort, but these are diverse as well. Furthermore, different flow regimes have profound effects on the populations. Riffles have rapid flow and turbulence roiling the surface, as water goes over shallow rocks. Pools are deeper, with slow-moving water and often bottoms of sand or silt; detritus collects in slower spots. Runs are deeper zones of moderate current, with combinations of sandy and rocky substrates.

Stream Riffles

In riffles, obviously, insects must be able to hold on. Mayfly nymphs of the genus *Baetis* (family Baetidae) are a little less than ⅜-inch (1 cm) long and streamlined for lessened resistance to the force of moving water, as they cling to rocks, plants, or other substrates with sharp tarsal claws. You may see nymphs of this ubiquitous mayfly swimming freely if you place a clump of the filamentous green alga *Cladaphora* (often abundant in desert stream riffles) in a pan of water. *Baetis* has a row of small gills lining the abdomen: these are sufficient in the well-oxygenated fast-flowing water and also present little surface area for drag. *Baetis* feeds on diatoms and fine organic material.

Other mayfly nymphs, such as algae-grazing Heptageniidae, are flattened; they are protected from high flows within the thin boundary layer where friction results in a zone of relatively low flow immediately adjacent to rock surfaces. Even some beetles are flattened, such as the disc-like larval "water penny" (family Psephenidae). Such flattening may also allow the water penny larvae to squeeze between crevices to feed on detritus on the underside of boulders, where they are common.

Some insects take advantage of the flowing water. The larvae of the Hydropsychidae caddisflies use silk from modified salivary glands to construct tent-like nets in fast water. The net is open upstream to capture food particles such as pieces of leaves or drifting organisms, while the insect stays hidden within its retreat. Nearly any rock that you pick up in a fast riffle will hold many of these nets, and the intricate mesh appears surprisingly delicate for such a high-stress environment.

One of the only truly aquatic lepidopterans is a moth in the family Pyralidae, which also makes a tent in somewhat slower water—a grayish-white sheet ½ to 1 inches (1-2 cm) in diameter over a rock crevice. The larva undulates slightly to circulate water through the small holes at the tent edges and past the dense gill tufts lining its abdomen.

In the very fastest water, the black-fly larva (Simuliidae) attaches itself to a rock or branch with a circular row of hooks at the end of its abdomen. From this anchor, it gathers food by extending a pair of retractable fans from its head that trap drifting organic particles. The aggressive blackfly larvae nip at neighboring insects that disrupt this food conveyor belt, thereby maintaining a small competitor-free territory that looks like a halo around the insect. Blackflies achieve impressive densities, often over 100 larvae in an area the size of your hand.

Another abundant dipteran, the midge *Rheotanytarsus* (Chironomidae), cements a ⅛-inch (4 mm) long tube to rock, and attaches silk to prongs projecting from the outer end of the tube that acts like a net, filtering particles from the fast-water current. This pronged tube looks like a dried-out hydra. The gills for both these dipterans are small, simple tube-like structures at the ends of their abdomens—adequate in these well-oxygenated sites.

Although most beetles are found in slower water, two groups inhabit riffle areas. The riffle beetle (Elmidae) is less than ⅛-inch (3 mm) long, both as a slender, tapered larva and as a chunky, longish-legged adult. Both stages clamber slowly over the substrate and feed on detritus. The adult riffle beetle has a dark oval body, usually with 4 orange or yellow spots on its wing covers, and appears hunch-backed with its head somewhat tucked into its thorax. The second beetle found in riffles, the long-toed water beetle (Dryopidae), is the only known case in which an aquatic adult insect has a terrestrial larval form; it burrows in moist sand above the stream. The ⅜-inch (1 cm) long, black adult with grooves on its upper surface spends its life slowly scavenging detritus and algae.

The largest riffle dwellers are the Corydalidae, or dobsonflies of fishing fame, which reach up to 2 inches (5 cm) long. These larvae live under stones in fast, rocky areas. Corydalids are obvious predators with huge jaws that are as big as their heads. Their bodies are soft, however, with long tube-like gills extending straight out from the sides of each abdominal segment and a pair of hooks at the tail end for holding on in the current.

Stream Pools

Insects living in pools or slow-moving water tend to be bulkier. The slow-water mayflies like *Callibaetis* or *Centroptilum* (both Baetidae), somewhat broader than their riffle-inhabiting cousins, perch well above the bottom rather than hunkering down. They flap their larger gills to create water movement. Two other common mayflies here, Caenidae and Tricorythidae, have very compact body shapes that allow them to forage for detrital food in silty areas. Fine sediments can harm sensitive respiratory surfaces, so the first pair of gills has been enlarged into a tough plate that covers the other gills. Gills on the Leptophlebiidae, on the other hand, are long and large, with double surfaces that look somewhat like a row of tuning forks extending from the abdomen.

In slow water a Polycentropodidae caddis larva makes a flimsy, tube-like net usually about ¾- to 1½-inch (2 to 4 cm) long on rock surfaces; it reaches outside this net to collect fine particles or to snatch prey. An intriguing caddis (Helicopsychidae), occasionally abundant grazing in gravelly areas, has a spiral case resembling a snail shell about ³⁄₁₆-inch (5 mm) tall, made of sand and glue from its salivary glands.

Pools provide habitat to a great many aquatic beetles. Both larval and adult stages are aquatic in the families Dytiscidae (predaceous diving beetles) and Hydrophilidae (scavenger beetles). Dytiscid adults swim to the pool's surface with the abdomen tip pointed upward, and gather air in a bubble under the *elytra* or wing-covers; hydrophilids, which clamber over pool bottoms and vegetation, also gather air, but come to the surface head first. Both use the thin bubble or *plastron*, which is held in place by extremely dense hairs or "hydrofuge," as a virtual gill, through which oxygen diffuses from the water and is taken up through the spiracles. The larvae, however, are generally wormlike, adapted for burrowing through bottom materials in search of prey. Once prey (even the occasional tadpole or small fish) is captured, the dytiscid larva inserts its sickle-like mandibles into it, injects digestive enzymes through grooves on the inner surface of its mandible, and then sucks up the partially digested contents.

Slower water is almost a predator soup; currents carry drifting organisms from fast water into slower areas, where they are vulnerable to a remarkable abundance of predators. Some interesting predators are in the order Hemiptera (true bugs). Almost all of these have sucking mouthparts and most of them occur in streams as both nymphs and adults. The water scorpion (Nepidae) perches like a praying mantis in aquatic plants or other debris, waiting for invertebrates to come near. It extends a "siphon" from the tip of its abdomen to breathe air from the surface. Toe-biters (Belostomatidae) sit among stream-edge plants, but these insects that reach up to 2 inches (5 cm) in length also crawl or swim slowly along the bottom, often preying on snails. You may see a toe-biter with its back covered with rows of pearl-like eggs. This is always a male, carrying eggs the female laid there, presumably to keep them aerated and to protect them from predators; these are among the largest eggs of all the insects. This is one of the few cases of male parental care in invertebrates!

Backswimmers (Notonectidae) and water boatmen (Corixidae) are more typically mid-water predators that capture aquatic insects swimming or drifting in the pools, and also take terrestrial insects that fall to the surface. One corixid, *Hesperocorixa*, is unique among the aquatic Hemiptera because it is an herbivore rather than a predator, grinding up (rather than sucking) parts of aquatic plants.

No predator soup would be complete without dragonflies and damselflies, the entire Order Odonata being predaceous, both as adults and nymphs. Damselfly nymphs have 3 leaf-like gills protruding from the ends of their abdomens, while dragonfly nymphs have spiny posteriors that suck oxygenated water into their rectums where the oxygen is absorbed. Many nymphs dart through the water by rapidly forcing the water out of their rectums—jet propulsion! The Libellulidae are usually brightly-patterned nymphs that crawl across the bottom searching for prey, while the Gomphidae sit and wait for their victims to stumble by. One gomphid is a flattened insect (with wide, flat antennae as well) that sits on the bottom, while another gomphid hides just under the sand, waiting for vibrations to indicate prey coming by. Quiver your hand over clean, shallow sand and watch for movement of this hidden insect. Another dragonfly (Cordulegasteridae), more common in pools in small, rocky streams, is hairy and usually covered with foreign material, camouflaged and motionless, waiting for prey.

Most damselfly nymphs are slender. The Calopterygidae perch on plants or tops of rocks on long legs; they snatch insects and small crustaceans in quiet pools. Another damselfly in the family Coenagrionidae, the only odonate common in flowing water, crawls through leaves and roots of slow runs. Its more compact shape protects it from vagaries of the current.

The face of a damselfly or dragonfly nymph appears to be covered by a toothed mask. This is actually a retractable set of mouthparts that can extend instantly up to 2 to 3 headlengths in front of the insect, then open sideways to grab a victim. You can readily pull this contraption outward to see how it functions; dragonfly nymphs are particularly sturdy.

Odonata adults cruise the streamside in search of flying insects, which they capture "on the wing" with legs that are covered with long hairs forming basket traps. Dragonflies extend broad wings outward, while damselflies' slender wings are held back over the body. Skimmers (family Libellulidae) sit on the tips of plants, darting out to catch flying insects. They also patrol stream margins to defend their feeding and reproductive territories against other dragonflies.

Odonate mating and oviposition behaviors are also complex. A male may appear to be looped awkwardly behind a female, using the claspers at his tail end to hold her "neck," while contacting the genital openings to pass sperm. Damselflies like *Argia* often fly in tandem, the blue male attached to the tan female with his claspers. To defend his genetic contribution from competing males, he stays attached while she oviposits eggs into algal mats and other soft streamside materials. There are even reports that males of some species have hooks for pulling out sperm deposited earlier in the female by another male! Many female dragonflies and some damselflies release their eggs on the wing, touching their abdomens to the water during the release, or dropping the eggs from the air like dive-bombers. The eggs drift downward, becoming sticky enough when wetted to adhere to the bottom.

Surface Film

Neuston is the collective term for organisms, such as the water striders (Gerridae, Veliidae), that live on the water surface; technically many of these airbreathers are not true aquatic insects. Most neustron prey on terrestrial insects that fall into the water.

Water striders' terminal leg segments covered with unwettable hairs allow the insects to "skate" on the surface supported by surface tension. Retractable tarsal claws, used as paddles, penetrate the surface film, giving traction without breaking through. The body weight deforms the surface film, often creating six spots of refracted light on the stream bottom below. Adding detergent to the water would break the surface tension, and the strider would fall in and drown. Smaller veliids jet forward by releasing saliva that breaks the surface film; the physical-chemical reaction pushes them forward.

Beetles of the whirly-gig family (Gyrinidae) have a split view of the neustonic feeding area: the upper portion of their eyes seeks prey floating on the surface, while the lower pair searches for prey (and scans for predators) below. Often gyrinids spin madly at the surface in groups of 30 or more.

You may see Chironomidae midges and mayflies swim up to the surface and burst out of their pupal (midges) or larval skins, rest for a moment, and then fly off to continue their life cycles This is the hatch that flyfishermen try to mimic.

Stream Margins— The Littoral Zone

The stream edge is another transition area. On the aquatic side, the creeping water bugs (Naucoridae) are highly flattened, almost disk-shaped. They crawl on sand and gravel in shallow water and have muscular front legs that make them look like weight lifters. These legs are used to hold prey while the bug sinks its beak into the hapless victim; caution—painful to you as well! The related, less painful, but drier, toad bug (Gelastocoridae) is a curious stream margin hemipteran, highly mottled and squat, with "bug-eyes"; it pounces upon insects in wet sandy-gravel sites.

Many beetles of the families Carabidae (ground beetles) and Staphylinidae (darkling beetles) are often abundant at water edges or areas drying as waters recede, where they scavenge other invertebrates left high and dry. Tiger beetles (Cicindellidae), also very common on sandy stream shores, are highly active and aggressive predators on insects near the water, or crawling/emerging onto land from the water. More often on the wing than typical beetles, tiger beetles display impressive sickle-like mandibles and beautiful metallic colors.

This is also the zone where many insects leave the water and hatch as adults. During warm months caddisflies, having gone through their pupal stages under rocks, come to the edge to break open their skins, crawl out, and dry off their wings. Fascinating enough to watch as this is, it is even more amazing to see orange-colored mites come from seemingly nowhere to climb onto an emerging caddisfly and hitch a ride—their means of dispersal. Often one sees the hollow husks (exuviae) of already-flown caddisflies, dragonflies, and damselflies still clinging near the water's edge.

Sub-Surface Creatures

Probably the most bizarre of stream habitats is the hyporheic zone, which literally means "below the flow." A variety of small insects and crustaceans dwell as many as 16 feet (5 m) under the stream bottom. Some, like the winter stoneflies (Capniidae) and a specialized chironomid midge, only come to the surface to emerge and reproduce as adults, while the crustaceans may complete their entire life cycle underground. These invertebrates survive on the bacteria and fungi which live deep within the porous sand/gravel substrate where nutrients seep into the groundwater, thus constituting a permanent community even when the surface dries totally during the dry season.

Other insects burrow just a few inches under the streambed. Crane fly larvae (Tipulidae), also called "leatherjackets" for their tough larval skin, use *peristaltic motion* (muscular wave-like contractions) to move through sand in search of food; out of water the wormlike larvae usually have swollen posterior ends. One common leatherjacket is a predator that is covered with short, golden hairs and feeds on smaller prey within the sand habitat, while another is a much "flabbier" detritivore, which feeds on decomposing organic material. Crane fly larvae pupate at stream edges to emerge as the familiar, long-legged "mosquito-hawks" (non-feeding) seen in shady corners or in your doorway.

Life Cycles

Insects that occupy the stream during their larval stage and then leave to reproduce as terrestrial adults, including mayflies, stoneflies, caddisflies, moths, dragonflies and damselflies, and the Diptera or true flies, spend almost their entire lives in the juvenile aquatic stage. Some mayflies, stoneflies, dob-

sonflies, and dipterans (for example, Chironomidae midges, Tipulidae craneflies) even have degenerated mouthparts as adults, since little or no feeding occurs in that short-lived stage. The mayfly order Ephemeroptera, in fact, has an extremely ephemeral adult stage, which lasts only 1 to a few days; some species of Baetidae mayflies don't even require a mate, but instead produce viable eggs via *parthenogenesis*.

Aquatic insects have 1 of 2 basic life cycles. Direct development (egg-nymph-adult), the more evolutionarily primitive cycle, is that of the mayflies, dragonflies and damselflies, stoneflies and the Hemiptera. These insects go through variable numbers of juvenile molts or *instars*, gradually developing mature body parts until they finally emerge as winged adults. Mayflies, some of which have as many as 30 instars, are the only insects (aquatic or terrestrial) that molt as adults, and the winged pre-adult stage or "dun" as it is called by fishermen, has smoky-colored wings. Resting on plants or other promontories near the stream, the insects soon shed this last skin to become the familiar glassine-winged "spinners." The Odonata typically crawl partly out of the water onto rocks (dragonflies) or sticks (damselflies), where they swell their bodies with water to burst the last nymphal skin before emerging as adults. The spent skins can be seen streamside from spring through fall. The eggs of baetids are often observed on the undersides of stones, where females have crawled back below the waterline to deposit them in lines that look like micro-honeycombs.

The remaining aquatic insects (caddisflies, moths, Diptera, and most beetles) have what's called "complete" development (egg-larva-pupa-adult), life cycles that include a pupal stage. Usually their cycles involve a longer aquatic larval stage and a more ephemeral terrestrial adult stage (with some exceptions as noted earlier, especially in the beetles). Pupal development lasts from less than 2 days up to 2 weeks in the larger taxa, often within a protective cocoon or chamber glued to the undersurfaces of rocks. Caddis pupae have huge sickle-shaped mandibles solely for the purpose of slicing their way out prior to emergence.

Size and Longevity

Unlike environments where colder winters restrict insect life cycles, the warm year-round temperatures in the Sonoran Desert allow many insect species to grow and reproduce throughout the year. Furthermore, warmer temperatures result in higher growth rates, so many insects can produce several generations in a year. Some caddisflies and true bugs, like the backswimmers and toe-biters can develop from egg to adult in less than 2 months and have 5 to 10 generations per year. The common mayflies, like Baetidae and Tricorythidae, and many chironomid midges, can develop in less than 2 weeks, and may have 35 generations in a year! Such development times are almost unheard of in other ecosystems, and such high production, with adult insects emerging in great abundance, partly explains why desert riparian areas are so attractive to birds and other terrestrial insectivores.

While there are many large aquatic insects in the desert, most are less than a centimeter in size, somewhat smaller than those found in streams elsewhere. This may be related to the ephemeral nature of benign conditions in the desert, because smaller animals can mature earlier, and are more likely than larger species to attain reproductive size before the next flood or drought destroys them. For example, the big blue darners (Aeshnidae), common in most of North America, are rare in the desert, probably because their 2 to 3 year life cycle makes them unsuited for an environment in which severe floods and drought occur on a frequent basis.

Survival and Repopulation

Because flash floods and summer drought can have catastrophic effects, invertebrates must have dependable means of surviving these events and repopulating the stream. Life cycles are less synchronized in desert streams than elsewhere, so at any one time there are flying adults—an "air force" ready to repopulate the aquatic environment. Summer drought may dry most of the stream, but pockets of aquatic habitat remaining in canyons or springs support residual populations that emerge and recolonize once stream waters return. Some winged aquatic adults, like beetles and true bugs, leave streams when high flows come or survive in vegetation or other protected sites. The toe-biter is one of the only insects here to produce eggs more than once; if one batch fails due to flooding or desiccation, she returns to the water to lay another batch when conditions permit. Nearly all the insects in desert streams produce large numbers of eggs. There is very little time when this extremely dynamic ecosystem is devoid of critters!

Birds

Karen Krebbs, Kenn Kaufman
& Desert Museum staff

Section Contents

The Desert Adaptations of Birds & Mammals

Peter Siminski

Have you ever wondered how animals can live in a hostile desert environment? Water, so necessary for life processes, is often scarce. Temperatures, which range from freezing to well over 100°F (38°C), make maintaining a safe body temperature a constant challenge. Add to this the catch-22 of desert survival: an organism's need for water increases as temperature rises—available water usually decreases the hotter it gets. This might sound like an impossible situation, yet, as we'll see, desert birds and mammals have developed many adaptive strategies for coping with temperature extremes and limited water.

The primary strategy for dealing with high desert temperatures is avoidance—many mammals simply avoid the high daytime temperatures by being nocturnal or crepuscular (dusk- or dawn-active). A bobcat, for instance, is typically most active at dusk and dawn;

Summer daytime temperature in the Sonoran Desert often exceeds 100°F (38°C). The primary strategy for dealing with this high temperature is to confine activity to the night or during dusk or dawn.

a javelina is never active during the day in summer, but it may be in winter. Even day-active birds are most active at the cooler dawn. Many mammals, such as ringtails or kangaroo rats, are never active during the day.

Microclimates and Burrows

Another avoidance strategy is to seek out a cool microclimate. A Cactus Wren may simply rest quietly in the shade of a jojoba; a Prairie Falcon will nest on a ledge of a cool north-facing cliff and avoid the hot south face. A cool, deep crevice in the cliff face may be the daytime refuge of a pallid bat, while a ringtail is sleeping away the day in a jumble of rocks at the base of the cliff.

Some mammals create their own microclimates. A white-throated wood rat (or pack rat) builds a den made of desert litter—cholla joints, prickly pear pads, sticks, and stones—within a clump of prickly pear cactus. It looks a little like a trash heap and may be three feet high and eight feet across. At the bottom of this pile is a series of tunnels leading to a nest of soft plant fibers. The pack rat spends its day in the soft nest, somewhat insulated from an exterior air temperature that may be 110° F (43°C), with a ground surface temperature of 160°F (71°C).

Many small mammals dig burrows in the desert soil. The burrow environment is much more moderate than is the surface temperature, which may have an annual fluctuation of between 15°F (9.5°C) and 160°F (71°C). Many desert rodents spend the entire day within the mild environment of a burrow. (A Merriam's kangaroo rat, for instance, will venture to the desert surface for less than one hour each night!) White-tailed antelope squirrels are diurnal rodents that forage for brief periods on the hot daytime desert surface. As they look for seeds, fruits, and insects, their chipmunk-sized bodies

heat up, even though their bushy tails hang like parasols over their backs. Above ground, the squirrels may often be observed pressing their bellies, with legs spread, against the cool soil—or even tile of suburban patios—in shady spots, allowing, it is presumed, their body heat to be conducted to the cool earth or tile. It has been speculated that the squirrels use the cooler earth in their burrows in a similar fashion when they retreat to them in on hot days.

Speculations made decades ago regarding the behavior of desert rodents in their burrows, and the temperature fluctuations of the rodents and their burrows during the heat of desert summers, have taken on a life of their own as "facts." So have generalizations about temperatures in burrows and pack rat nests that were based on very limited measurements at elevations and conditions far different from those of our desert extremes. The truth is, we have much to learn about these animals' temperature tolerances and their strategies to avoid overheating. Ongoing and future research assisted by modern technology will, it is hoped, provide us with more complete answers.

Large mammals do not burrow to escape the desert heat. The kit fox, however, is the exception. Unlike any other North American canid, the kit fox uses burrows year round. Burrows help it thrive in hot, dry desert valleys—an environment that is too challenging for other canids. Other large mammals, such as bighorn sheep and mule deer, seek shady spots during the day and remain inactive. Large body size actually has its advantages in the hot desert environment: a large body heats up more slowly than a small body. This phenomenon is called *thermal inertia*. It may buy enough time to get through a blistering summer day.

Heat Conduction and Radiation

Birds or mammals can conduct heat from their bodies to the environment by decreasing the insulating value of feathers or fur. On a hot day, a Curve-billed Thrasher sleeks its feathers which creates a thinner insulating layer. Coyotes lose their thick winter coats in late spring; their early summer coats are relatively thin. A bighorn sheep also sheds its winter coat in the spring— but it sheds it in stages. During the heat of June, the belly and shaded parts of the legs are shed first, providing an area from which to lose body heat; the back, however, remains covered with thick woolly fur that insulates and shades the bighorn sheep from the hot overhead sun.

Birds have some advantages over mammals in dealing with heat. The normal body temperature of birds is generally higher than that of mammals. This higher body temperature means that a Gambel's Quail, for instance, with a body temperature of 107°F (42°C), can continue to conduct heat to the air until the ambient temperature reaches 107°F. (A coyote, by comparison, has a body temperature of 102°F.) Also, by dilating the blood vessels going to its bare scaly legs, a bird can dump excess body heat to the environment. A bird's leg temperature may increase 15°F (9.5°C) after its blood vessels dilate. Thus, a hot bird sleeks its feathers and stands tall to expose its legs to the air. Mammals too have "radiators." The long ears of a

jackrabbit can transfer excess heat to the air through dilation of the blood vessels to the ear. This works best when the air temperature is below the jackrabbit's normal body temperature (104°F/40°C), or after the jackrabbit has been active.

Evaporative Cooling

The primary method for cooling down a hot bird or mammal is through evaporative cooling. As water evaporates from a surface, it cools that surface. When a coyote pants, it rhythmically moves air over the moist surfaces of the mouth, throat, and tongue. Water is evaporated and these surfaces are cooled. Abundant dilated blood vessels are near these surfaces and are cooled by them. The resulting cooled blood is then circulated throughout the body. A hot owl will flap the loose skin under its throat to move air over its mouth cavity. This is called *gular fluttering* and achieves the same result as panting. Panting and gular fluttering are energy efficient movements that produce very little heat themselves.

Brains are very sensitive to heat. In sheep, and in members or the dog and cat families, evaporative cooling of the nasal passages results in the cooling of a special network of blood vessels to the brain. The brain of an exercising dog, for instance, is cooler than the rest of its body.

Vultures use evaporative cooling in an interesting way. A vulture urinates on its legs if the daytime temperatures are over 70°F (21°C). The urine will evaporate, cooling the legs and drawing more heat from the body of the hot vulture. This is why, when the daytime

Birds lower their body temperatures during hot weather by dilating the blood vessels in their bare, scaley legs, thereby allowing body heat to escape to the environment. Vultures even urinate on their legs if the daytime temperature exceeds 70°F (21°C). This draws heat from the body through evaporative cooling. Notice the white residue on the Black Vulture's legs above.

temperatures are consistently about 70°F, a vulture's legs are white, but when the temperatures are consistently lower than 70°F, a black vulture's legs are gray and a turkey vulture's legs are red.

Water Income and Water Expense

Birds and mammals have a great need for water. Water serves as the basic transport medium for nutrients, and it is the medium for dilution and removal of body wastes. Water functions in most chemical reactions of the living process, and as we have seen, water is the body's primary coolant.

The water-budget balancing act of desert animals has been compared to balancing a bank account: there is water income and water expense. Not surprisingly, it is always a tight budget for desert animals. Stored water is generally limited to what can be placed in the gut or crop. Debts are not tolerated. A 10 to 15 percent loss in body weight due to water loss can impair an animal's ability to recover; a 20 percent loss often means death. Water loss can happen quickly on a hot day in the desert, one to two liters per hour in humans.

What are the sources of water income and water expense?

Water income can come from three sources:
- Free water (for example, a bighorn drinks at a water hole)
- Water in food (for example, a Phainopepla eats a juicy mistletoe berry)
- Oxidation water (the water produced by all animals when they metabolize food)

Water expenses can come from:
- Evaporative cooling
- Dilution and excretion of toxic body wastes
- Feces
- Eggs or milk

Some rodents, such as pocket mice and kangaroo rats, are independent of any free water —or even of moist food. The kangaroo rat is probably the best known of these. It eats primarily dry, high carbohydrate seeds; one gram of grass seed produces one-half gram of oxidation water. Seeds with much fat or high protein content are avoided: the former produce too much heat that may have to be lost through evaporative cooling; the latter require too much water for diluting waste products. A kangaroo rat can live on water produced when food is metabolized, but that is only part of its arsenal of strategies for desert survival. Additional water is available

from dry seeds which, when stored in its burrow, absorb as much as 30 percent of their weight in water from the higher humidity in the burrow. The evaporative loss from a kangaroo rat is low, as the animal has no sweat glands and little water is passively lost through its skin. Respiratory water loss is reduced by a nasal cooling system that extracts water from air as it passes through the nasal chambers as it is exhaled—a cooling system now known to be shared with other rodents and most other mammals. A kangaroo rat can produce urine twice as concentrated as sea water and feces five times drier than a lab rat's droppings. It conserves moisture further by

The kangaroo rat has a suite of physiological and behavioral adaptations to a hot, dry environment. One of these is its ability to survive without drinking water: enough water is produced through metabolizing its food to meet its needs.

being nocturnal. Finally, a kangaroo rat typically breeds only when green vegetation or insects are available to supplement its water balance.

Other rodents that do not have regular access to free water consume juicy animals and succulent plants and their fruits. Pack rats and cactus mice are good examples this feeding strategy. During June, the driest month of the year, pack rats can survive on cholla and prickly pear; cactus mice can survive on cactus fruit and insects. There are many other animals besides rodents that get most of their water from food. Elf Owls survive on katydids and scorpions. Pronghorns can survive on the water in cholla fruits. Kit foxes can satisfy their water needs with the water in their diet of kangaroo rats, mice, and rabbits, along with small amounts of vegetable material.

Other desert dwellers, such as coyotes, mule deer and bighorn sheep, require periodic free water. In fact their home ranges revolve around water holes. Such animals, including we humans, are found only where free water exists, or where it can be transported.

Humans In a Hot, Arid Environment

Humans are physiologically very good at keeping cool, but rather poor at conserving water. Sweating is the primary method of cooling the body; the evaporation of this sweat from all over the body cools the naked skin. During a really hot day in the desert, however, a human will lose as much as 12 liters (a little over 3 gallons) of water through sweat.

Humans have a special mechanism for cooling their big brain: The blood cooled by evaporation of sweat on the face and head penetrate the skull through tiny emissary veins, thus delivering freshly-cooled blood to the brain. This cranial radiator is unique among primates.

Humans' upright, two-legged stance also confers some advantages for keeping cool. When the sun is directly overhead, only the head and shoulders are in full sunlight—a four-legged animal has its entire back, shoulders and head exposed to the sun. Humans therefore gain much less radiant heat than four-legged animals. Also, by standing upright, most of the body is raised above the hot desert floor; this means that humans' rate of heat gain from the desert surface is much less than that of quadrupeds. Being upright also exposes more of the body to cool air currents, and thus body heat can be lost by convection.

Nakedness is also an advantage. Without insulating fur, heat can be lost more easily through convection, and sweat can be more easily evaporated. And that patch of thick hair on top of our heads is more than mere decoration—it shades the head and its heat-sensitive brain from the sun.

Unlike other desert mammals, humans have come up with many cultural and technological adaptations to the desert heat and aridity. Picture yourself on a typical summer day in the Sonoran Desert. What techniques and devices are you using to keep cool and hydrated?

Water Birds

Wherever there is water in arid country there will be concentrations of birds; often these will include some true waterbirds, seemingly quite out of place in the desert.

Grebes are swimmers that are more graceful underwater than in the air, but a few species do migrate across the desert. One, the Pied-billed Grebe, may show up on any small pond. A pair of Pied-billeds may remain to raise young if they find enough plant material with which to build their floating nest. Just as widespread are various members of the heron family. Our largest species, the Great Blue Heron, is sometimes seen flapping slowly over the desert, miles from water. It hunts in typical heron fashion—standing by the water's edge to spear or seize fish, frogs, and other aquatic creatures—but in a pinch it can also dine on rodents in dry fields.

Ducks are the most conspicuous water birds to reach the desert. Some lakes here may host hundreds of ducks in winter, representing up to a dozen species. Only a few of these will remain to nest in this region. While most ducks arrive here from the north, Whistling-Ducks come from the south. These odd, gangly birds — perhaps related more closely to geese than to true ducks—are typical of the tropics. Black-bellied Whistling-Ducks have increased here in recent decades, and pairs can be seen attending flotillas of downy ducklings on southern Arizona ponds.

—Kenn Kaufman

Pied-billed Grebe
(Podilymbus podiceps)

Order: Podicipediformes
Family: Podicipedidae (Grebes)
Spanish names: buzo

DISTINGUISHING FEATURES
This 12 to 15 inch, drab brown, stocky bird can be distinguished by its short, blunt bill that is encircled with a black band in summer. Its feet are set far back on body which makes travel on land awkward, but is ideal for underwater swimming. The Pied-billed Grebe is seldom seen flying.

HABITAT
Frequents ponds, lakes, and marshy areas

FEEDING
• *Diet*: Feeds on a wide variety of aquatic life including insects, spiders, tadpoles and small fish.
• *Behavior*: Forages by swimming under-water, propelled by its feet.

373

LIFE HISTORY
This bird lays 4 to 7 eggs in a floating nest that is built by both sexes. Incubation, which is performed by both parents, takes about 23 days. The young are fed by both parents and are able to swim shortly after hatching. Pied-billed Grebes are solitary birds; often only one mating pair is found on a pond or lake.

The Pied-billed Grebe has the odd habit of eating its own feathers, as well as feeding feathers to its young. It also has an interesting behavior when apprehensive or nervous: it expells air from its lungs and feathers, which causes it to slowly sink in the water; it then swims with only its head breaking the water surface.

Great Blue Heron *(Ardea herodias)*

Order: Ciconiiformes
Family: Ardeidae (Herons, Bitterns)
Spanish names: garza

DISTINGUISHING FEATURES
This 42 to 52 inch bird is the largest heron in North America. It has blue-gray wings and back, whitish underparts, and a white head with a black streak. It flies with steady, slow wingbeats; its long legs hang straight back and its long neck is tucked back on its shoulders during flight.

HABITAT
This highly adaptable bird can be found in a variety of riparian habitats including desert rivers, ponds, and marshes.

FEEDING
• *Diet:* Mostly fish but also amphibians, reptiles, insects, rodents, and birds.
• *Behavior:* The Great Blue Heron most often feeds while standing still or slowly wading in shallow water; it strikes at small fish swimming by with its spear-like bill. It is not uncommon for a heron to make a 20 or even 30 mile round trip in its quest for a good foraging site.

LIFE HISTORY
Great Blue Herons breed in colonies. The male chooses the nest site and displays to attract a female. The nest site is typically in a tree 20 to 60 feet above the ground or water, although shrubs are also used as nest sites. The female lays 3 to 5 eggs in a platform nest made of sticks. The eggs, which are incubated by both parents, hatch in 25 to 30 days. The young are fed regurgitated matter by both parents. Young are able to fly after about two months.

Herons are sometimes mistakenly called "cranes." Other than both having long legs and long necks, they are very different. Cranes are omnivores, often feeding on grain in dry fields, and although cranes may gather in huge flocks during migration and winter, they typically nest as isolated pairs, unlike the colonial Herons.

Ducks

REPRESENTATIVE SONORAN DESERT SPECIES:
Black-bellied Whistling-Duck *(Dendrocygna autumnalis)*
Green-winged Teal *(Anas crecca)*
Cinnamon Teal *(Anas cyanoptera)*
Northern Shoveler *(Anas clypeata)*
Northern Pintail *(Anas acuta)*
Redhead *(Aythya americana)*
Ruddy Duck *(Oxyura jamaicensis)*

Northern Shoveler

Order: Anseriformes
Family: Anatidae (Ducks, Geese, Swans)
Spanish names: pato (duck); cerceta aliverde (Green-winged Teal);
tilito café (Cinnamon Teal); pichichí (Whistling-Duck);
pato cabeza roja (Redhead)

DISTINGUISHING FEATURES

Black-bellied Whistling-Duck: Reddish brown with a black belly, white wing patches, and a pink bill. This beautifully marked, long-legged, long-necked duck is both sociable and noisy. *Green-winged Teal:* Very small; metallic green speculum (a patch on the secondaries). Male: gray body; chestnut head with green eye patch; vertical white mark in front of wing. Female: mottled brown.

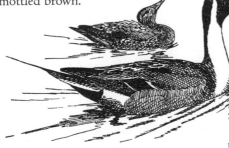

Northern Pintail

Cinnamon Teal: Male is deep rufous, female is a mottled brown. Both have a bluish wing patch.

Northern Shoveler: Often mistaken for a Mallard but can be distinguished by its large, dark, shovel-shaped bill, short neck, and pale blue wing patch. Breeding male has green head, white breast, chestnut belly and flanks, white area just before the tail. Females, juveniles, and molting males are drab and mottled. *Northern Pintail* male: slim, gray-colored body; long neck with white throat markings; brown head; long pointed tail. Female: mottled brown with a somewhat pointed tail. *Redhead* male: large, round, chestnut-red head; gray body; black breast; blue bill with dark tip. Female: pale bill with dark tip, brown back and sides. *Ruddy Duck:* Males are easily recognized from spring until late-summer by their chestnut-colored body, sky-blue bill, black cap, and white face. Females and winter males are gray-bodied, and females have a dark stripe bisecting the white cheek patch. The tail feathers of the

Ruddy Duck are often cocked up in the air. This species takes flight only reluctantly, often skittering across the surface of the water to gain speed for takeoff.

Habitat
These waterfowl frequent ponds, marshes, and lakes.

Feeding
• *Diet:* Whistling-Duck: Mostly seeds and grains. Shoveler and Teal: Seeds and aquatic plants. Redhead: Mostly insects and aquatic plants. Pintail and Ruddy Duck: Mainly seeds and insects.
• *Behavior: :* Flocks of *Black-bellied Whistling-Duck* often feed on waste grain in harvested fields, as well as seeds in grasslands and overgrown pastures; insects and other invertebrates make up a small part of their diet.

Black-bellied Whistling-Duck

Teal forage along the edge of ponds dabbling for seeds and other plant matter that are within a neck's-reach of the surface. The precocial ducklings feed on small aquatic invertebrates and vegetation.

Northern Shoveler feeds in shallow open water where large bill and comblike "teeth" are used to strain out plant and animal matter; diet varies with season; in winter diet consists mostly of aquatic plants; seldom upends or dives.

Northern Pintail are wary birds that prefer to forage in shallow water where they upend or submerge the head and neck while swimming. Some foraging is also done on land. *Redheads* forage by diving or upending, usually in shallow water. Ninety percent of their food is vegetable matter, but they will feed on snails, mollusks, and insects. *Ruddy Ducks* forage by diving and swimming under water where their broad bill collects and strains food from mud.

Life History
Except for the Whistling-Ducks and Mallards, few ducks spend the summer breeding season in the Sonoran Desert region. Most nest in the "prairie pothole region" of southern Canada and north central United States, or around lakes in the forest or tundra regions of Canada and Alaska. However, more than 20 species of ducks can be found wintering regularly in the Sonoran Desert region.

Comments
Black-bellied Whistling-Duck: United States populations are increasing, probably because of nest boxes. This species was rare in Arizona before 1949 but has since become a rather common nesting bird.

Northern Pintail: Although this species is widespread and abundant, there is some indication that numbers have been declining during the last 40 years. Drought may be responsible for dramatic reductions in nesting success.

Redhead: It appears that this species has suffered more severe population declines than most other ducks. This may be due to loss of nesting habitat.

Vultures

In cartoon fiction, vultures are
ghastly portents of death, cir-
cling ominously overhead when
explorers are lost in the desert. In
reality, vultures play a much more
wholesome role. By cleaning up the
carcasses of dead animals, they act
as the sanitation department of the
natural world.

Turkey Vulture

Vultures are perfectly designed
for living on carrion. They may have
to wait a few days in between meals, so in searching for food they burn up very
little energy, soaring for hours with scarcely a flap of their broad wings. Their
search is aided by keen eyesight and, in the case of the Turkey Vulture, a well-
developed sense of smell, unusual among birds. And although their naked wrin-
kled heads may be unattractive to human eyes, the lack of feathers on their heads
is a decided advantage when the birds are involved in the messy business of tearing open dead animals.

The vultures found in the warmer parts of Europe, Asia, and Africa are in the
same family as are hawks and eagles. The vultures of the Americas come from a
totally different background; they are now known to be more closely related to
storks. It seems remarkable that these two completely unrelated groups of
vultures should be so similar in structure and appearance. The similarities are
probably adaptations to the same feeding behavior, an example of what is called
convergent evolution.

—Kenn Kaufman

SONORAN DESERT SPECIES:
> Turkey Vulture (*Cathartes aura*)
> Black Vulture *(Coragyps atratus)*

> Order: Ciconiiformes
> Family: Cathartidae (New World Vultures)
> Other common names: buzzard (both species)
> Spanish names: zopilote, buitre (vulture); aura cabeciroja, aura común
>   (Turkey Vulture); zopilote común (Black Vulture)

Distinguishing Features

The featherless head and large dark body help to identify these 2 scavengers. In flight they can be distinguished from one another by the patterns of the underside of the wings and by the wing shape. *Turkey Vulture:* A large dark brown to black bird with a featherless red head; reddish feet; underside of wings appear two-toned; wingspan is up to 6 feet (1.8 m); an accomplished soarer. *Black Vulture:* A large shiny black bird with a dark gray, featherless head and gray feet; underside of wings have prominent white patch at the tip; tail is short and square; flies with several short wing beats and then a short glide; also soars effortlessly on hot days.

Habitat

Both vultures are found in open country.

Feeding

• *Diet:* Both are carrion eaters.
• *Behavior:* The *Turkey Vulture* hunts by soaring on thermals (rising air currents), sometimes for an hour or more with no apparent movement of the wings; food is located by smell and sight. Turkey Vultures are unusual among birds in that they have a well-developed sense of smell. The *Black Vulture* finds its food by sight; may watch for concentrations of Turkey Vultures and follow them to carrion.

Life History

Vultures use cliff faces, tree stumps, caves, and hidden areas on the ground for nesting; neither species builds nests. One to 3 whitish, blotched eggs are incubated by both parents; the young are fed by both parents by regurgitation. Turkey Vulture incubation period is 34 to 41 days. Young are capable of flight about 60 to 70 days after hatching. Black Vulture incubation period is 37 to 41 days. Young are capable of flight about 75 to 80 days after hatching.

When threatened, vultures emit a hissing sound in defense; the Black Vulture regurgitates when confronted.

These birds excrete on their legs as a means of cooling themselves. This is called *urohydrosis.*

Black Vulture

Hawks & Eagles

Harris's Hawk

Admired for their strength and hunting prowess, renowned for their keen eyesight, emblazoned on flags and national shields, the hawks and eagles are recognized worldwide. Most birds of prey that hunt by day belong to this family—more than 240 species, ranging in size from huge eagles to speedy little hawks no bigger than robins. Several kinds are familiar sights over the Sonoran Desert.

Red-tailed Hawk and Cooper's Hawk, widespread over North America, are common in the desert year-round. The red-tail, one of the typical soaring buteo hawks (*buteos* are larger hawks with broad, rounded wings and short, broad tails), is far more often seen, as it perches in the open or circles overhead, watching for rodents and other prey. Cooper's Hawk is typical of the *accipiter* group; it is a long-tailed, short-winged bird that seldom soars. Hunting near dense cover, relying on speed and surprise, Cooper's Hawk takes many birds as well as rodents.

Seen less often here are eagles and Ospreys. The Osprey is a hawk that plunges feet-first to catch fish; in the southwest it visits the larger bodies of water, but occasionally a migrant is seen flying over the open desert. Bald Eagles are also seen mostly close to water; a few pairs nest along wilder rivers in the Sonoran Desert. Golden Eagles, by contrast, are often in very dry country, where these huge predators take animals as large as jackrabbits.

Birds of prey are generally solitary, but there are exceptions. Harris's Hawk, a sharply patterned raptor of warm climates, lives in small groups. Three or more adults often care for the young in a single nest, and two or three may hunt cooperatively, actively harrying prey animals out into the open. A classic desert sight involves three or four Harris's Hawks perched on adjacent arms of the same giant saguaro.

—Kenn Kaufman

SONORAN DESERT SPECIES:
>   Osprey *(Pandion haliaetus)*
>   Bald Eagle *(Haliaeetus leucocephalus)*
>   Cooper's Hawk *(Accipiter cooperii)*
>   Red-tailed Hawk *(Buteo jamaicensis)*
>   Golden Eagle *(Aquila chrysaetos)*
>   Harris's Hawk *(Parabuteo unicinctus)*

Order: Falconiformes
Family: Accipitridae
Subfamily: Pandioninae (Osprey)
Spanish Names: gavilan pescador
(Osprey), aquila calva (Bald Eagle),
esmerejon de Cooper (Cooper's Hawk),
aquilla parda (Red-tailed Hawk), águila
real (Golden Eagle), aguilla cinchada
(Harris's Hawk).

Cooper's Hawk

DISTINGUISHING FEATURES

Osprey: Snowy white underparts and
dark brown above; head is white with
a dark brown eye stripe. In flight the
long, narrow wings have dark wrist
patches and the white belly is obvious.
Females are similar to males but usually
show a band of brown streaks across
the chest. *Bald Eagle:* A very large brown
raptor with a white head and tail; bill
is yellow; a larger head and shorter tail
than the Golden Eagle; young
resemble the adults but are
solid dark brown at first,
developing the white
head and tail over a
period of several years.
Cooper's Hawk: Medium-
sized accipiter with a long
tail rounded at the tip
and short, round
wings; adults are
blue-gray above,
and narrowly barred below with reddish-brown and white; young are more
brown with streaked underparts. *Golden
Eagle:* A very large brown raptor with
a golden head and neck; young show
white patches in the wings and a white
base to the black tail. Golden Eagles
soar with their wings slightly raised.
Red-tailed Hawk: A very common hawk
with a variety of plumages; brown

Osprey

above with broad, round wings. On
adults the upperside of the tail is typically reddish-brown, and this color
usually shows through from the underside as well. Some adults have a belly
band of dark brown streaking with
white underparts. Red-tailed Hawks in
the Southwest usually have light underparts and lack a belly band; other color
morphs may be mostly blackish,
dark brown, or cinnamon-
brown below. The
young have gray-
brown tails with
black bands.
Harris's Hawk: Dark brown
with chestnut shoulder
patches, leg feathering
and wing linings; tail is
long and black with white at the
base and tip; young are lightly streaked
below with brown.

HABITAT

Ospreys and Bald Eagles are found
near coastal areas, rivers, and lakes.
Ospreys are less common inland.
Golden Eagles can be found in moun-
tain habitats, open country, and desert.
Cooper's Hawks occur in desert,

woodlands, deciduous forests, and riparian areas. Red-tailed Hawks are found in a variety of habitats such as desert, woodlands, plains, riparian, and open areas. Harris's Hawks occur in mesquite and saguaro habitats, semi-arid woodlands, and scrub.

LIFE HISTORY

Ospreys are fish hawks. They plunge and dive with feet first into the water to catch fish. The fish is taken to a perch or nest to be eaten. There are scales with spines on the toes for grasping fish. Ospreys are the only raptors whose front talons turn backwards. Ospreys also eat frogs, turtles, rodents, and occasionally birds.

Bald Eagles prey on fish too. They will also feed on small mammals, waterfowl, seabirds, and carrion. Bald Eagles compete with Ospreys for food. Loss of habitat and the use of pesticides have affected Bald Eagles. In recent years the Arizona population has been increasing. They are still on the endangered list in our area, however.

Golden Eagles prey upon rabbits, small mammals and carrion. Pair bonds are long-term. Nests are large and bulky. The male feeds the female during incubation and chick rearing; the larger nestling usually kills the smaller. Territories are usually occupied all year.

Cooper's Hawks, members of the accipiter group, fly with rapid,

shallow wingbeats. They overtake their prey swiftly and often through dense

Golden Eagle

woods or shrubs. They feed mostly on birds and some small mammals.

Red-tailed Hawks are the most common and widespread raptors in North America. They prey mostly on rodents and are often seen perched on telephone poles watching for prey. Females often return to the same nesting territory. Red-tailed Hawk populations are stable and increasing in some areas.

Harris's Hawks are neotropical raptors that prey upon rabbits, rodents, snakes, lizards, and birds. These hawks are social and hunt in family groups. Most social groups consist of a pair and several nonbreeding helpers who assist in feeding the nestlings and defending the nest. This cooperative behavior is also used to flush and catch prey that is hiding in cover. Large family groups are observed during autumn and winter.

Bald Eagle

Caracaras & Falcons

Crested Caracara

Some of the world's most impressive fliers belong to the falcon family. Typical falcons are trim birds of prey, with long tails and angular, pointed wings, built for breathtaking speed and maneuverability in the air.

Our most familiar falcon, the American Kestrel, is also our smallest, about the size of a Mourning Dove. Kestrels nest in large cavities in trees, or in holes in saguaro cacti. Although they can put on bursts of speed when they are pursuing rodents or small birds, most of their diet consists of large insects; their fanciest flying trick is their ability to hover in one spot, on rapidly beating wings, while they scan the ground for prey.

Prairie Falcons are much larger, but surprisingly maneuverable for their size. When pursuing small birds in the open, they can twist and turn with amazing agility. Peregrine Falcons are similar in size, but rely on different hunting techniques, usually power-diving from considerable heights to take their prey by surprise. In these dives, Peregrines are thought to approach 200 miles per hour (320 km per hour). Prairie Falcons are native to the American West, while Peregrines range almost worldwide; both occur in the Sonoran Desert in small numbers at all seasons.

Classified in the same family, but far different in structure and habits, are the Caracaras. Our Crested Caracara is a broad-winged, slow-flying scavenger, often competing with vultures at road kills and other carcasses. Mainly a tropical bird, it is most common in the southern parts of the Sonoran Desert.

—Kenn Kaufman

Caracaras and Falcons

SONORAN DESERT SPECIES:
Crested Caracara *(Caracara plancus)*
American Kestrel *(Falco sparverius)*
Prairie Falcon *(Falco mexicanus)*
Peregrine Falcon *(Falco peregrinus)*

Order: Falconiformes
Families: Falconidae
Spanish names: quebrantahuesos
(Crested Caracara), gavilan pollero
(Kestrel), halcon café (Prairie Falcon),
halcon pollero (Peregrine Falcon)

American Kestrel

DISTINGUISHING FEATURES
The *Crested Caracara* has a black-brown body, white neck and throat, red-orange face, black head and long yellow legs; it has round wings and lacks the speed of falcons; its flight is direct and noisy. In flight the white base of the tail and white patch on the tips on the wings contrast with the black body. The *American Kestrel* is the smallest falcon in the United States with a wing-spread is less than 2 feet; back and tail are rust-brown; it has a black and white head pattern. Males have blue-gray wings and a more rust colored tail than females, which are brown overall, except for the lighter breast. The *Prairie Falcon* is medium-sized with a pale, sandy-brown back; its underparts are white and heavily spotted; the crown is streaked and the helmet has a thin dark mustache. In flight the Prairie Falcon shows a black patch under the base of each wing. Sexes are similar. The *Peregrine Falcon* is medium-sized to large, with a black "helmet" or "hood" covering the top of the head and extending below the eye. Adults have blue-gray

backs and barred underparts, while young birds have dark brown backs and striped underparts.

HABITAT
The Caracara inhabits arid, open country at low elevations such as desert brushlands, plains, and savannahs. It is sometimes seen at livestock or slaughter yards. The American Kestrel is found in open habitats, grasslands, deserts, and cities. The Prairie Falcon occurs in canyons, open country, grasslands, and deserts. The Peregrine Falcon can be found in deserts, mountains, and forests where tall cliffs occur. Both Peregrines and Prairie Falcons are sometimes found around tall buidings in cities.

FEEDING
• *Diet:* The *Caracara* feeds on carrion and live-caught prey such as small mammals, insects, reptiles, frogs, nestlings, weak or injured birds, and eggs. The *American Kestrel* feeds mainly on insects. The *Prairie Falcon* preys on small mammals, reptiles, insects, and

ground dwelling birds. *Peregrines* prey primarily on doves, waterfowl, shorebirds, and passerines.

• *Behavior:* The *Caracara*, whose flight is direct and steady, soar for extended periods looking for carrion. It is often found feeding with and harassing vultures.

American Kestrels take smaller and slower prey than the other falcons. During hunting they use rapid wingbeats and hover in one spot before plunging to catch their prey. In the desert, Kestrels hunt in the morning and late afternoon during summer; during winter they are active throughout the day.

Falcons have long tails and long, narrow, pointed wings designed for speed. They have tooth-like projections along the cutting edge of the mandibles that are used to kill prey quickly by severing the spinal cord with a sharp bite. The *Prairie Falcon* flies low over the ground or soars looking for prey. Its flight is swift and more maneuverable than that of the Peregrine. The *Peregrine Falcon* catches birds in flight by diving and taking them by surprise.

Peregrine Falcon

This falcon strikes its prey with its feet and returns to catch the falling bird. Pairs hunt cooperatively when not nesting.

LIFE HISTORY

The *Caracara* is adapted for walking and hunting on the ground. The bulky, loose nests are placed on the ground or in a tree and are made of twigs and sticks.

Kestrels are cavity nesters using holes in saguaros, trees, telephone poles and buildings. Eggs (4 to 5) are cared for by both of the parents. The young leave the nest in about 30 days. Kestrels have only 1 brood per year.

Both the *Prairie Falcons* and *Peregrine Falcons* nest on ledges or cliff sites. The male Peregrine Falcon usually does most of the hunting during nesting and the female broods and feeds the chicks.

Falcon populations were in decline before DDT was banned in 1972. Today Prairie Falcon populations are stable and do not appear to be suffering from the past pesticide problems. Peregrine Falcon populations have also rebounded in the United States since the ban.

Prairie Falcon

Quail

Masked Bobwhite

Traveling naturalists, accustomed to the secretive nature of quail in other habitats, are often startled to see how conspicuous Gambel's Quail can be in the Sonoran Desert.

Quail in general are plump birds, rather poor fliers, that spend almost all their time on the ground. Thus they have good reason to make themselves unobtrusive, to avoid drawing the notice of predators. Gambel's Quail are probably no less vulnerable (or tasty) than the other species, yet they behave in ways that call attention to themselves. The males call loudly from low perches; family groups go parading across the flats; coveys of two dozen or more run about clucking in the open. In the sparse plant growth of the desert, it would be impossible for Gambel's Quail to be as secretive as their relatives that live in denser cover, so perhaps shy behavior would be a non-adaptive waste of energy.

At one time, south-central Arizona had another common type of quail: the Masked Bobwhite. Unfortunately, it required not just desert, but lush desert grassland. Large herds of cattle, brought into this region before the principles of range management were well understood, eliminated most of the grasses; when the grass disappeared, so did the Masked Bobwhites. There are still captive flocks, raised from birds found in Sonora, but conservationists have faced major difficulties in trying to reintroduce these birds to the wild.

—Kenn Kaufman

REPRESENTATIVE SONORAN DESERT SPECIES:
Gambel's Quail (*Callipepla gambelii*)
Masked Bobwhite (*Colinus virginianus ridgwayi*)

Order: Galliformes
Family: Odontophoridae (New World Quail)
Spanish names: codorniz de Gambel (Gambel's Quail), codorniz común (Masked Bobwhite)

DISTINGUISHING FEATURES
Quail are terrestrial birds with short round wings, stout legs with four toes (hind toe is elevated and does not come into contact with the ground), and short, conical bills. The *Gambel's Quail* has a black top-knot that curves forward; the male only has a black throat, face and belly. Plumage is gray with white, chestnut and buff. The *Masked*

Bobwhite has no top-knot; plumage is brown, black, and buff; the male has a black face and throat with chestnut brown underparts; the female and young have cream colored underparts, face, and throat.

HABITAT

Gambel's Quail occur in mesquite habitat, desert scrub, thorn thickets, and riparian areas; often are found in habitats with water nearby. Historical habitat of the Masked Bobwhite was tall grass bordered by mesquite. Before 1880 the masked bobwhite was common in Arizona from the Baboquivari Mountains east to the Santa Cruz valley. Today, it is extinct in Arizona except for a rein-troduced popu-lation at the Buenos Aires National Wildlife Refuge.

Gambel's Quail

FEEDING

• Diet: Mostly seeds.
• Behavior: Gambel's Quail are the most arid-adapted quail. During the summer the quail are active early mornings and late afternoons when temperatures are not extreme. They avoid heat stress by resting in the shade during the hottest part of the day. Quail must either drink water daily or obtain it from their food. They can eat insects and succulent fruits of cacti to get this water. Quail also eat seeds and plants. They roost in bushes and low dense trees.

Masked Bobwhite forage in flocks (coveys), except during breeding season; they sometimes move up into shrubs and vines to forage on berries and leaves.

LIFE HISTORY

Quail are gregarious birds. In the fall and winter they often live in coveys of 20 or more individuals, but they pair off during the nesting period. They spend a lot of time on the ground in brushy areas, usually running across hot or open areas to cover. They fly short distances when startled or to avoid predators.

Gambel's Quail usually have 1 brood of 10 to 12 pale-buff eggs. The female incu-bates the eggs for 21 to 24 days. The nest is a shallow depression lined with grass, leaves, and vegetation; it is on the ground or no more than 10 feet off the ground. All eggs hatch on the same day and the precocial chicks are fully covered with down. They leave the nest soon after hatching, relying on their parents to protect them and to locate food.

Masked Bobwhite were extirpated from the United States due to the destruction of grasslands. Overgrazing and other abuses of southern Arizona resulted in the disappearance of these quail from Arizona during the late 1800s and early 1900s. Attempted reintroductions have not been very successful.

Marsh Birds

Yuma Clapper Rail

Most members of the rail family are associated with water, but not all in the same ways. "Typical" rails inhabit dense marshes, and are seldom seen as they slip among the cattails and other aquatic plants. In contrast, coots are rail relatives that act like ducks, swimming and diving freely on open water. Moorhens and Gallinules behave like either ducks or rails, sometimes swimming in the open and sometimes fading into the marsh. All of these birds have distinctive voices, but the typical rails call mostly at night, in keeping with their secretive nature.

Although they appear to be weak fliers, members of the rail family have established themselves all over the globe, even colonizing many oceanic islands. Therefore, it is hardly surprising that they manage to find the wetlands scattered through the Sonoran Desert. American Coots and Common Moorhens may appear on any marsh-edged ponds, and a couple of small species (Virginia Rail and Sora) are widespread as winter visitors. The most notable rail of the region is the Yuma Clapper Rail. This endangered subspecies is practically confined to the marshes of the lower Colorado River.

—Kenn Kaufman

Sonoran Desert species:
Yuma Clapper Rail *(Rallus longirostris yumanensis)*
American Coot *(Fulica americana)*
Common Moorhen *(Gallinula chloropus)*

Order: Gruiformes
Family: Rallidae
Spanish names: ralón (rail), gallareta (coot and moorhen)

DISTINGUISHING FEATURES

Coots and *moorhens* are both the size and shape of chickens (but behave more like ducks); they also have chicken-like bills, white in the case of the adult coot, red in the moorhen. When swimming, they pump their heads back and forth as they go. *Rails*, which are rarely seen, are henlike in appearance, narrow-bodied for running through the marsh vegetation ("thin as a rail"), and cryptically colored.

American Coot

FEEDING

• *Diet: Coots* and *moorhens* eat aquatic plants and animals. *Rails* do as well, though they forage on land, or in shallow water, slipping about through dense marsh vegetation.

Common Moorhen

HABITAT

All these birds are found associated with freshwater cattail marshes of lakes, ponds, and rivers or coastal marshes. Coots are frequently found on open water; rails seldom are.

LIFE HISTORY

All members of the rail family can swim and dive, but *Coots* are as aquatic as ducks. They have lobes along their toes, which flare out on the back stroke to give them propulsion and which fold back on the forward stroke.

Moorhens, also called Gallinules, are equally at home in the water or on shore; long toes help them walk across the water on the tops of water plants. They are less gregarious and somewhat more secretive than coots.

The *Yuma Clapper Rail* is the only Clapper Rail that occurs in freshwater marshes in the United States. Its historic habitat was the marshes of the Colorado River delta in Mexico. Most of these marshes have dried up or been destroyed through channelization, and although the creation of marshes behind dams elsewhere in the Colorado river system has extended the rail's range, it is still endangered. The only population of this subspecies in Mexico is found at the Ciénega de Santa Clara on the former flood plain of the lower Colorado River.

Shorebirds

American Avocet

Shorebirds in general do most of their foraging along the water's edge, probing in soft mud or picking at the surface in search of tiny invertebrates. They belong to several related families. The largest shorebird group is the sandpiper family (Scolopacidae); nearly two dozen species of sandpipers migrate through the Sonoran Desert, but for the most part their presence with us is fleeting, a few days' stopover as they travel between breeding grounds on Arctic tundra and wintering grounds on southern coasts. More relevant here are two long-legged waders and one plover that are with us for much of the year.

The avocets and stilts make up a small family, with only a few species world-wide. All are slim birds with long necks, thin bills, very long legs, and striking patterns. All forage in shallow water, feeding on small invertebrates. North America has one avocet and one stilt. Both have ranges which extend into the Sonoran Desert, where they seem to have benefited from human activity; most of their modern nesting sites are around the edges of artificial ponds.

Members of the plover family have distinctively short bills and short necks. When foraging they often run a few steps and then pause, and then run a few more, stopping now and then to pick up something from the ground. Although many plovers stick to typical shorebird haunts such as beaches and mudflats, a few thrive in dry fields, far from water. The Killdeer, our most familiar plover, is intermediate in its choice of habitats. Elsewhere it is common on farm fields and large lawns, but in desert regions it is usually not too far from water.

—Kenn Kaufman

Shorebirds

REPRESENTATIVE SONORAN DESERT SPECIES:
Killdeer *(Charadrius vociferus)*
Black-necked Stilt *(Himantopus mexicanus)*
American Avocet *(Recurvirostra americana)*

Order: Charadriiformes
Families: Charadriidae (Killdeer),
Recurvirostridae (Stilts and Avocets)
Spanish Names: tildío (Killdeer),
avoceta (Stilts and Avocets)

Killdeer

DISTINGUISHING FEATURES

The plover family, to which Killdeer belong, are distinguished by their pigeon-like bills; Stilts and Avocets have slender bills (curved upward in the case of Avocets) and very long legs, the Stilt's being "grotesquely long" according to The Peterson *Field Guide to Western Birds*. Their color patterns are also distinctive.

HABITAT

These shorebirds inhabit the shorelines of lakes, ponds, rivers, and seas. The Killdeer, which is common throughout the year, lives in open, irrigated farmlands or in fields far from water. The avocet and stilt occur widely in migration, but withdraw from most of our area in winter.

FEEDING

• *Diet:* Though these birds will eat aquatic vegetation or seeds, insects and tiny crustaceans make up the majority of their diet.
• *Behavior:* These birds are active feeders, often foraging in groups, often rather noisily. Their feeding behavior is fun to watch. *Killdeer* find

their food visually in the mud of fields or shores. The *Black-necked Stilt* walks about in shallow water or on the shore, picking up insects. And the *American Avocet* finds its food by feel beneath the water, using its bill to sweep in front of it as it walks along.

LIFE HISTORY

These 3 species draw attention to themselves with their loud and idiosyncratic cries. These are ground-nesting birds, whose eggs are well camouflaged and whose downy chicks can run about and find their own food shortly after hatching. Adults defend eggs or chicks with a repertoire of distraction displays. All 3 birds are good runners and strong flyers.

Black-necked Stilt

Doves

The Sonoran Desert would have a very different sound
if it were not for the doves. The cooing songs of four
species are among the classic bird voices here for much
of the year.

Mourning Doves are found throughout North
America except for the coldest regions, but in the
desert they are among the most numerous birds
year-round. A bigger relative, the White-winged Dove,
is extremely common along southwestern rivers in
summer. The rich cooing of the white-wings on
spring mornings may virtually drown out the voices of other birds. More
unobtrusive is the little Common Ground-Dove, which usually stays close to
dense thickets. Another small species, the Inca Dove, is not really a desert bird;
it is more likely to be found mincing about on lawns. Spreading north out of
Mexico, it has become one of the most familiar birds in southwestern U.S. cities.

About 300 species of doves and pigeons are found worldwide. All have short
blunt bills, stout bodies, and rather small heads. Our doves eat mostly seeds, but
tropical species may eat many small fruits as well. Dove nests are haphazard
platforms of sticks, so flimsy that the eggs or young sometimes fall through
them; as if to make up for this, the birds may make repeated nesting attempts,
raising several broods per year.

Doves love water, and it is only through their strong powers of flight that they
are able to thrive in the desert; they may fly long distances to get to reliable
sources of water. Flocks of doves hurtling overhead are a characteristic sight on
desert evenings.

—Kenn Kaufman

White-winged
Dove

Doves

SONORAN DESERT SPECIES:
White-winged Dove *(Zenaida asiatica)*
Mourning Dove *(Zenaida macroura)*
Common Ground-Dove *(Columbina passerina)*
Inca Dove *(Columbina inca)*

Order: Columbiformes
Family: Columbidae
Spanish Names: paloma pitahayera and paloma de alas blancas
(White-winged Dove), huilota común, paloma triste and tórtola coluda
(Mourning Dove), tortolita común (Inca Dove), tortolita de milpas
(Ground-dove)

DISTINGUISHING FEATURES
The *White-winged Dove* is a light brown bird with a white patch on the wing (it looks like a thin, white border when the wings are folded). Tail is round and with a short black tail; it looks slightly scaled. The adult male has a blue crown with much purple in the neck-shield area and shoulders. This is the smallest dove in the area.

Common Ground-dove

outer feathers are tipped in white. The *Mourning Dove* has a brown body, blue-gray wings, and long pointed tail. The *Inca Dove* is pale brown; rufous primaries are visible when the bird displays or flies; upper body looks scaled; slender tail with white sides. The *Common Ground-Dove* is light brown

HABITAT
The *White-winged Dove* is found in all desert habitats; most leave for the winter although pockets remain, especially in suburbs and in riparian zones. The *Mourning Dove* is found in all desert habitats throughout the year. The *Inca Dove* is most often found

Inca Dove

January to November. That fact, plus its preference for grass and weed seeds, have made the Inca Dove the most abundant bird in southwestern urban areas, after the House Sparrow.

White-winged Doves are important players in the life history of the saguaro. Along with bats, bees, and other insects, they help pollinate it as they fly from flower to flower to sip nectar. White-winged Doves also disperse saguaro seeds: they eat the fruit, then regurgitate it to their young; in the process some seed falls beneath the nest where it germinates, and the young saguaro grows in the protection of the tree.

around human settlements throughout much of the Sonoran Desert region. The *Common Ground-Dove* is found throughout the year most often in dense brushy desert or in riparian areas.

FEEDING

• *Diet:* All 4 doves are seed and fruit eaters. Doves grind seeds in their muscular stomachs (or gizzards) using sand or gravel much like internal teeth.

LIFE HISTORY

Doves are strong, fast fliers and noisy too, as they clap their wings together when they start into flight. Doves can live in deserts because they can fly long distances to find food and water. During winter they congregate, but pair off during breeding season. Dove nests look like flimsy, careless arrangements, and they can be built almost anywhere—in trees, on the ground, in hanging pots. A pair can raise several broods a year.

The *Inca Dove* has the longest breeding season of any Arizona bird:

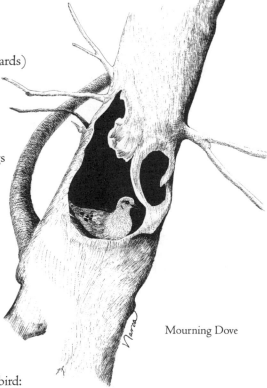

Mourning Dove

Greater Roadrunner

The most famous bird in the Sonoran Desert, without a doubt, the Roadrunner is also the most fictionalized in popular imagination. Cowboys used to tell tall tales about how Roadrunners would seek out rattlesnakes to pick fights, or would find sleeping rattlers and build fences of cactus joints around them. A later generation of Americans grew up thinking that Roadrunners were purple and cried "beep beep" as they sped about.

Greater Roadrunner

Even without such stretches or inventions, the real Roadrunner is impressive. Running in the open (and not just on roads), it reaches fifteen miles per hour. It can fly, but usually doesn't. Often it seems curiously unafraid of humans. Trotting up close to peer at us, raising and lowering its mop of a shaggy crest, flipping its long tail about expressively, it looks undeniably zany. It comes as no surprise to learn that the Roadrunner is a member of the cuckoo family.

Clownlike it may appear to human eyes, but the Roadrunner is a very effective predator. Its speed on foot is not just for show: it captures not only snakes and large insects, but also fast-running lizards, rodents, and various small birds. Gambel's Quail may pay scant attention to the Roadrunner at most seasons, but they react to it violently when they have small young, and with good reason: given an opportunity, the Roadrunner will streak in to grab a bite-sized baby quail.

—Kenn Kaufman

Greater Roadrunner (*Geococcyx californianus*)

Order: Cuculiformes
Family: Cuculidae (Cuckoos)
Spanish Names: correcaminos, churea, paisano

DISTINGUISHING FEATURES

Our largest cuckoo, this bird is characterized by a long tail, streaked appearance, frequently erected shaggy crest, and a blue and orange bare patch of skin behind the eyes. It is capable of running very rapidly across the ground (15 mph) and rarely flies. Like all cuckoos, the Roadrunner is a *zygodactyl* bird (it has 2 toes pointing forward and 2 toes backward).

HABITAT

The Roadrunner prefers open country, desert, open pinon/juniper habitat.

FEEDING

• *Diet:* Feeds upon any animal small enough for it to kill and ingest, including small birds and snakes; young are fed insects, lizards, and mice; also eats some fruits and seeds.
• *Behavior:* Hunts by walking briskly and running toward prey once it is located; also able to jump straight up in the air when small birds or flying insects are overhead. The adult uses its long tail as a rudder for maneuvering while running.

LIFE HISTORY

The pair bond in this species may be permanent; pairs are territorial all year. Courtship displays include, but are not limited to, presenting the mate with a twig or piece of grass and chasing one another.

The nest, which is constructed of twigs, is frequently found in cholla, mesquite, or palo verde. White eggs (3 to 6) are laid at intervals; if food is scarce the older, larger hatchlings will quickly seize all the food from the parents thus causing the younger, smaller ones to starve. Rarely do all nestlings reach maturity. If not enough food is available, these younger birds will be fed to the other, stronger hatchlings.

Roadrunner skin is heavily pigmented. On cool mornings, the bird positions itself with its back towards the sun and erects its feathers, thus allowing the sun to strike directly on the black skin which quickly absorbs heat energy. This makes it possible for the bird to achieve body heating without unnecessary expenditure of metabolic energy.

Owls

The Sonoran Desert at night is a very lively place. Especially in summer, there are probably more creatures abroad at midnight than at noon. Of course, most birds shun this night shift, but several species of owl are notable exceptions.

Owls are superbly equipped to hunt at night. They cannot see in total darkness— no animal can do that—but their eyes are adapted for vision under very low light conditions. Even more impressive is their sense of hearing. Studies have shown that Barn Owls can locate their prey by sound alone, in total darkness, with pinpoint accuracy. Many of the creatures that they hunt also have excellent hearing, but the owls can approach them in silence: the sound of their wingbeats is muffled by the softened edges of the larger wing feathers.

Great Horned Owls are found throughout the Americas and Barn Owls practically throughout the world, so it is

Western Screech-owl

no surprise that they can adapt to desert life. Great Horned Owls eat almost anything smaller than themselves from rabbits and skunks to snakes and insects. Barn Owls specialize on rodents and the smaller desert owls also tend to take small prey. The Ferruginous Pygmy-Owl, a tropical species that reaches its northern limit here, may hunt most at dawn and dusk, often catching songbirds. The world's smallest owl, the Elf Owl, nests in holes in saguaro cacti and ventures out at night to eat beetles and moths. Nocturnal insects are scarce in cold weather, so most Elf Owls retreat south into Mexico for the winter; other desert owls are present year-round.

—Kenn Kaufman

Owls

SONORAN DESERT SPECIES:
  Barn Owl *(Tyto alba)*
  Western Screech-Owl *(Otus kennicottii)*
  Great Horned Owl *(Bubo virginianus)*
  Ferruginous Pygmy-Owl
    (Glaucidium brasilianum)
  Elf Owl *(Micrathene whitneyi)*
  Burrowing Owl *(Athene cunicularia)*

Order: Strigiformes
Family: Strigidae (all but Barn Owl)
Family: Tytonidae (Barn Owl)
Spanish names: lechuza mono, lechuza
    común (Barn Owl), tecolotito chillon
    (Western Screech-Owl), tecolote cornudo, buho (Great Horned Owl),
    tecolote enano (Elf Owl), lechuza llanera, lechuza de ojo
    (Burrowing Owl)

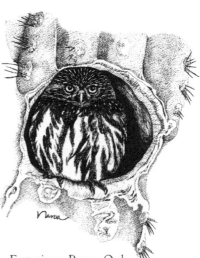

Ferruginous Pygmy-Owl

DISTINGUISHING FEATURES

Barn Owl: This is a long-legged, knock-kneed, pale, monkey-faced owl. It has no ear tufts, and the pale face resembles a heart-shaped disc. The back is golden-brown, the belly is white. The voice is a loud, rasping screech. *Western Screech-Owl:* A common bird in our area, it is small, measuring only about 8 inches (20 cm). It is gray, streaked with black and white, has conspicuous "ear" tufts and is zygo*dactylous* (2 toes point forward, 2 backward). Its call, a trill of several notes that become more rapid (like a bouncing ball), distinguishes it from the similar Whiskered Screech-Owl, which lives in oak woodlands. *Great Horned Owl:* Measuring almost 2 feet (61 cm) tall, this is our largest owl. It has a white throat, barred underside, and prominent ear tufts. *Ferruginous Pygmy-Owl:* This small

(6½ to 7 inch; 17 cm), uncommon, "earless" owl is reddish-brown with a faintly cross-barred tail. The crown has many white streaks and the underside is white with red-brown streaks. *Elf Owl:* About the size of a sparrow (5 inches tall; 13 cm), this is one of the smallest owls in the world. It has no ear tufts, and is grayish-brown with a white brow. The call consists of a variety of soft yelping notes, often running together into a high-pitched chatter. *Burrowing Owl:* About the size of a screech-owl, it is brown, spotted with tan, and lacks ear tufts; the long legs are almost featherless.

HABITAT

Barn Owl: This owl occurs throughout our area and through much of the U.S., Europe, Asia, and Africa. It is widespread but local. It is often found in conjunction with human habitation, roosting and nesting in barns, under bridges, in mine shafts, and in palm trees. It often nests in the undercuts of arroyos. *Western Screech-Owl:* The screech-owl occurs from southwestern Canada into Mexico. It is a resident of wooded areas from low desert into the mountains. *Great Horned Owl:* This owl occurs throughout the New World, except the extreme north. It is found in every habitat within our region. *Ferruginous Pygmy-Owl:* This species ranges from southern Arizona and southern Texas, south through Central and South America. The Ferruginous Pygmy-Owl is found in saguaro deserts and wooded river bottoms. In the tropics it inhabits a wide variety of wooded or semi-open habitats. *Elf Owl:* The Elf Owl occurs from western Mexico through the southwestern United States. In our region, it may be found mostly in riparian habitats or in association with the saguaro. *Burrowing Owl:* This species occurs from southwestern Canada to Tierra del Fuego, at the tip of South America. It prefers open country, prairie and desert. It is frequently seen in desert and grassland regions, standing on mounds or fence posts during the day.

Barn Owl

FEEDING

• *Diet:* The *Barn Owl* feeds on large numbers of rats and mice. *Western Screech-Owl* feeds on invertebrates and vertebrates. The *Great Horned Owl* has an extremely varied diet that includes birds, skunks, snakes, lizards, insects, and even frogs and fish. Lagomorphs (rabbits and hares) and rodents make up the bulk of the diet. The *Ferruginous Pygmy-Owl* prefers lizards and large insects, but will also take scorpions and small birds and mammals. The *Elf Owl* feeds primarily on invertebrates such as scorpions, centipedes, beetles and moths. The *Burrowing Owl* feeds on insects, rodents and small reptiles.

• *Behavior: Barn Owl:* An expert nocturnal predator, it finds its prey at night with exceptionally acute hearing and vision. The owl, like most birds of prey, ingests bone and fur when it eats its prey. However, the digestive processes of the owl are not capable of digesting bone and fur. This residual material is formed into a pellet in the stomach and regurgitated. Food habit studies of owls are easily done by examining the contents of the pellets found in or near their roosts. *Western Screech-Owl:* The streaked pattern of the owl blends well with desert shrubs while hunting. Highly nocturnal. *Great Horned Owl:* A nocturnal predator with extremely acute sound perception and night vision. The soft feathers are an adaptation to its hunting style of obtaining food. These feathers do not make any sound in flight; therefore, the bird can hear the prey and locate its position, but the prey cannot hear the owl. This owl is an extremely important predator of jackrabbits and

Elf Owl

cottontails. *Ferruginous Pygmy-Owl:* This pugnacious owl is most active at night, but it is also active during the day, especially at dawn and late afternoon.

LIFE HISTORY

Barn Owl nesting activity peaks in the spring, with 5 to 6 white eggs laid in a depression in a tree cavity, cave, mine shaft, or building. The female incubates the eggs about 33 days. The young remain in the nest for 7 to 8 weeks. The barn owl characteristically begins incubating the first egg, and while incubating it, lays additional eggs. Since the eggs are laid 1 to 2 days apart, the young hatch 1 to 2 days apart. Therefore, chicks of various ages (development stages) can be found in one nest. Often the older siblings starve out the younger. Once one dies, it is fed to the older nestlings. The young are very noisy, crying raucously when the adults feed them.

Burrowing Owls

The *Western Screech-Owl* nests in tree cavities; in our area it commonly nests in saguaro holes. The eggs are white with 4 to 5 in a clutch. They are incubated for about 26 days. The male feeds the female during incubation.

The *Great Horned Owl* begins nesting during January or February, usually in an abandoned hawk nest or on a ledge.

The white eggs are usually laid 2 or 3 to a clutch. The owls actively defend the nest territory.

The *Ferruginous Pygmy-Owl* female lays 3 or 4 white eggs in hollows of saguaros or other trees. The nest, which contains no nesting materials, may be used for many years. The incubation period is about 28 days; the male feeds the sitting female and both parents feed nestlings. The young are able to fly 27 to 30 days after hatching.

The *Elf Owl* uses old woodpecker nest cavities for its nests. The white eggs are usually laid 3 to 4 to a clutch. Starlings, which are an introduced bird from Europe, pose a threat to Elf Owls. They take over nest cavities already in use by the Elf Owls, or by other birds.

The *Burrowing Owl* nests in the used burrows of other animals, most commonly ground squirrels and prairie dogs. The white eggs number from 5 to 10, and are laid in underground nests. The babies spend most of their early life underground, but emerge before they are fully fledged to exercise their flight muscles. When disturbed in a burrow, the owl mimics a rattlesnake's rattle. The adults are probably preyed upon by other predators; the chicks are taken by snakes.

Nightjars

Nightjars are birds of mystery. Camouflaged in mottled brown and gray, they generally hide and sleep during the day, resting on the ground or on horizontal branches with their big eyes closed. At night they emerge to fly about, as silent in the air as the moths that they often capture in their wide, gaping mouths. Many nightjars are best known by, and named for, their nocturnal songs; the Whip-poor-will, which reaches the mountain forests of the southwest, is a good example.

Lesser Nighthawk

Around rocky outcrops in the desert, the lonesome cry of the Poorwill is a familiar sound on summer nights, especially when the moon is bright. Naturalists who are out at night may find Poorwills sitting on roads, and may even be able to watch them hunt when the birds flutter up from the ground to catch passing insects. Poorwills mostly disappear from our region in winter, but they are not necessarily gone: these are the only North American birds known to hibernate, and they may sleep for days or even weeks at a time.

The Nighthawks (not related to hawks at all) are the most aerial of the nightjars, longer-winged and more buoyant in flight than their relatives. They are often seen flying about at dawn or dusk, or even in full daylight. The Lesser Nighthawk of the Southwest looks very much like the Common Nighthawk, widespread in North America, but the species differ in behavior. Common Nighthawks are flamboyant birds, flying high and calling loudly; but Lesser Nighthawks tend to fly low, and they usually maintain an eerie silence, floating like ghosts over the desert at dusk.

—Kenn Kaufman

Nightjars

REPRESENTATIVE SONORAN DESERT SPECIES:
Lesser Nighthawk *(Chordeiles acutipennis)*
Common Poorwill *(Phalaenoptilus nuttallii)*

Order: Caprimulgiformes
Family: Caprimulgidae
Spanish names: tapacamino, garapena

Common
Poorwill

DISTINGUISHING FEATURES
Both birds have large eyes, tiny bills, huge gapes, and short legs. Nighthawks are larger (8-9 inches; 20-23 cm) and are identified in flight by a white wing bar and pointed wings. Poorwills (7-8½ inches; 17.5-21.5 cm) have rounded wings and no white bar.

HABITAT
The Nighthawk and Poorwill are found in all Sonoran Desert habitats. The Poorwill is more common on sparsely vegetated bajadas.

FEEDING
Both Nighthawks and Poorwills are insect-eaters, but their hunting techniques differ: the *Nighthawk* flies low, silently and gracefully, searching the sky for flying insects, and maneuvering quickly, almost like a bat. A hunting *Poorwill* sits on open ground, looking up into the sky for the backlit silhouettes of large moths or beetles. When it spots something, it flutters up, usually no higher than ten feet, and catches the insect in its mouth. Both birds are crepuscular, needing some light to hunt by. City lights may extend the activity of the more urban Nighthawk and also attract its prey. Lesser Nighthawks may also be seen until midmorning.

Poorwills like hunting by moonlight (they're *lunarphilic*) and on these nights they take over the niche of the *lunarphobic*, insect-eating bat.

LIFE HISTORY
During the day, Lesser Nighthawks and Poorwills rest on the ground or horizontally on a branch, well camouflaged by their cryptic coloration. During winter, Lesser Nighthawks migrate, and Poorwills may too. But they may also hibernate, greatly lowering body temperature, respiration, and heart rates for days, even months, at a time. This behavior is very unusual in birds—hummingbirds enter torpor, but only for one night. The first documented hibernating Poorwill was found in the Sonoran Desert, in a hollow in a rocky canyon. Its discoverer tried to find signs of life in this apparently dead bird by catching the condensation of its breath on a mirror, but failed. Ten days later, the bird still hadn't moved, but when the man touched it, the bird winked at him.

White-throated Swift

No other birds in the world are so purely creatures of the air as are the swifts. Small birds with short tails and saber-shaped wings, they speed through the sky in search of flying insects. Their small feet are not designed for perching in normal bird fashion; they can only cling to vertical surfaces, so they land only when necessary. Their nests are made primarily of their own saliva, which hardens to a substance resembling shellac (the nests of some Asian species, which separate into gelatinous shreds when soaked and cooked, are the source of bird's nest soup). Except during the season when they are raising young, or during very bad weather, swifts may spend all their waking hours in flight. Some European swifts are even thought to sleep in the air.

The only member of the family normally seen over the southwestern U.S. lowlands is the White-throated Swift. Little colonies nest in crevices in the rocks of cliffs and canyons, but flocks may be seen anywhere over the desert. Sometimes they are almost too high to see, and only their high-pitched chattering gives them away; sometimes they course low over the ground. They are perhaps best seen along the rims of canyons, where they may go hurtling past at breathtaking speeds. Some people have suggested that White-throated Swifts might exceed 200 miles per hour in level flight; although it would be hard to measure this, anyone who has seen them zoom past at close range will not find it hard to believe.

—Kenn Kaufman

White-throated Swift (*Aeronautes saxatalis*)

Order: Apodiformes
Family: Apodidae
Spanish Names: vencejo, golondrina

DISTINGUISHING FEATURES
The White-throated Swift has long, narrow, stiff wings; their black and white pattern distinguishes this species from other North American swifts.

HABITAT
The White-throated Swift roosts and nests in the crevices of cliff faces. It forages over all desert habitats in open sky.

Life History

Swifts spend most of their time in the air, foraging for flying insects. Eating, drinking, bathing, even courtship and copulation take place in the air; during mating the pair tumbles downward, sometimes for over 500 feet (250 m). Nests are built in crevices of cliffs; they are made of grass and feathers glued together with saliva. The same nest sites may be used year after year by a colony of these social birds.

Although the family name means "without feet," swifts do have feet—unusual ones, in which the 4 toes point forward. This, and the exceptionally long claws, are thought to be adaptations to clinging to vertical surfaces, such as cliff faces or nests. The legs of swifts are small and weak, as suits an animal that so rarely touches down. White-throated Swifts may be the fastest flying North American birds. Both nestling and adult White-throated Swifts can become torpid during cool weather or food shortages.

Born to Fly

Aerodynamically, flight is the triumph of lift and propulsion over gravity and drag. In flapping flight the outer wing feathers (primaries) produce the propulsion, and the inner wing feathers (secondaries) produce the lift. Lift is generated by air moving over an air foil, or wing. In addition to flapping flight, birds also glide and soar. In gliding flight, birds gradually lose altitude on steady out-stretched wings. In soaring flight, birds gain altitude on steady out-stretched wings by riding updrafts of air.

The demands and advantages of flight have produced some dramatic adaptations in birds. One of these adaptations centers around weight reduction.

Weight Reducing Adaptations in Birds:

- *Some large bones are thin and hollow*
- *Some bones are fused or reduced in size, e.g. the bones of the front limb and tail*
- *Feathers provide light strength*
- *Few skin glands*
- *No teeth or heavy jaws*
- *Air sacs*
- *Ovipary not vivipary (they lay one egg at a time)*
- *Atrophy of reproductive organs between breeding seasons*
- *Usually only one ovary*
- *Generally select high calorie, compact foods (such as seeds and insects)*
- *Rapid and efficient digestion*
- *No bladder; they excrete a dry, light-weight uric acid*

Hummingbirds

Describing hummingbirds without resorting to superlatives would be difficult, and hardly fair. This family includes the world's smallest birds, with the most brilliant iridescent colors, the fastest wingbeats, and the most amazing ability to fly up, down, sideways, and backwards. They spend their days hovering at flowers to sip nectar, feeding almost constantly to supply the sugar necessary to maintain their racing metabolism. Many people would call these the world's most fascinating birds.

Black-chinned Hummingbird

The brightest colors and most ornate patterns among hummingbirds are worn by males, and the purpose is evidently to impress females. After mating, the male takes no more part in family life. The female alone builds the nest, incubates the tiny eggs, and feeds the young. Considering the amount of energy that an individual hummingbird needs just to feed itself, it seems remarkable that the female is able to raise the young successfully all alone.

There are well over three hundred species of hummingbirds, all native to the Americas. The vast majority, not surprisingly, are found in the tropics, where flowers abound year-round. Only a handful of species reach the United States; southern Arizona hosts more than a dozen of those. Costa's Hummingbird is the only true desert hummer here, but several others live along the desert's edges. Black-chinned and Broad-billed Hummingbirds nest in streamside woods in summer, while Anna's Hummingbird, a recent invader from California, nests in the same areas (and in residential neighborhoods) in winter. Our region hosts the greatest variety of hummers in late summer, when several species are on their way south. Rufous Hummingbirds, southbound from nesting grounds in the northwest U.S., may appear in the Sonoran Desert by July, along with lesser numbers of other species, to joust for space around the blooms that follow the summer rains.

—Kenn Kaufman

Hummingbirds

REPRESENTATIVE SONORAN DESERT SPECIES:
Broad-billed Hummingbird *(Cynanthus latirostris)*
Black-chinned Hummingbird *(Archilochus alexandri)*
Anna's Hummingbird *(Calypte anna)*
Costa's Hummingbird *(Calypte costae)*
Rufous Hummingbird *(Selasphorus rufus)*

Order: Apodiformes
Family: Trochilidae
Spanish names: chuparrosa (hummingbird), chuparrosa matraquita
(Broad-billed Hummingbird), chupamirto garganti-negro (Black-
chinned Hummingbird), chupamirto cuello escarlata (Anna's
Hummingbird), chupamirto garganta violeta (Costa's Hummingbird),
chupaflor dorado (Rufous Hummingbird)

DESCRIPTION

The smallest birds in the world belong to this family. In our region, they range in length from 2½ inches to 5 inches (7 to 13 cm), and from 2 g to 10 g in weight. All in our region have long, pointed beaks for probing flowers for nectar, saber-like wings for hovering in front of flowers, a generally iridescent bronze or green dorsal surface, and primarily in males, bright, colorful throat and head patches. The iridescent throat patch is called the *gorget* (pronounced gore-JET).

Rufous Hummingbird

DISTINGUISHING FEATURES

Broad-billed Hummingbird: 4 inches (10 cm) in length; 3 g to 4½ g in weight. In poor light the males appear dark with red bills and forked tails. In good light, the male's head and breast are metallic blue or blue-green. The females are a duller grey-green, but retain the dark forked tail and have red only at the base of the lower mandible. *Black-chinned Hummingbird:* 3½ inches (9 cm) in length; 3 to 4 g in weight. This hummingbird is iridescent green above and gray below. The male has a velvet black throat, the bottom border of which iridesces violet in good light. The dark throat patch contrasts strongly with a white upper breast, giving a collared effect. The females lack dark throat patches. *Anna's Hummingbird:* 4 inches (10 cm) in length; 3½ g to 5 g in weight. These "flame-throated" hummingbirds are iridescent green above and grey below. In addition, the male's throat and forehead iridesce crimson rose in good light; in poor light, these areas appear to be velvety black. The females generally lack these iridescent rosy patches, but may

Sweating the Details

Hummingbirds "decorate" their nests to camouflage them, weaving leaves and sticks and plant fibers into the outer part of the nests. In the Desert Museum's hummingbird exhibit, however, the nest-building females frequently take thread, yarn, and even hair from visitors to weave into their nests. Some of the females are quite daring and will even land on visitors in order to pull out a particularly choice piece of yarn. I have worried that a bird might get hurt flying so close to people, but the visitors have always been very calm and gentle. They have been extemely generous also. At one point I even received a sweater in the mail, accompanied by a note explaining that one of the Costa's hummingbirds in the exhibit had taken an extreme interest in the sweater and seemed sorry to see it go when the wearer left the exhibit. "Take the sweater back to the hummingbird exhibit," the note read. "She can use it more than I can!"

—Karen Krebbs

have a few rose feathers on the throat. *Costa's Hummingbird:* 3½ inches (9 cm) in length; 2½ g to 3½ g in weight. This "flame-throated" hummingbird is iridescent green above and grayish white below. The males in good light have an iridescent amethyst purple forehead and throat. The iridescent throat patch extends into an elongated "mustache." In poor light, these patches appear velvety black. The female completely lacks these patches. *Rufous Hummingbird:* 3½ (9 cm) inches in length; 3 g to 4 g in weight. The male is cinnamon-rufous on the upper parts, tail, and lower breast and belly. In good light, the throat iridesces a metallic orange to scarlet. The female is iridescent bronze-green above and dull white below. She also has a cinnamon wash to the flanks and much rufous on all the tail feathers.

Broad-billed Hummingbird

HABITAT

Hummingbirds occupy most temperate and tropical habitats in the western hemisphere. In our region, they can be found in desert, grassland, woodland, and forest. They tend to prefer edges of dense habitats, and the scrubbier areas of open habitats.

Broad-billed Hummingbird: In Arizona and Sonora, the broad-bill is found in the lower mountain canyons and in the mesquite bosques of the larger washes and rivers. Habitats also include the thorn forest and thornscrub of southern Sonora.

Black-chinned Hummingbird: This is the "summer hummer" in southern Arizona. The bird winters in Mexico. In Arizona, the Black-chin inhabits deciduous woodland associations in low mountain canyons, desert riparian habitats, and cities. It is generally absent from desertscrub.

Anna's Hummingbird: The Anna's Hummingbird is generally considered a sedentary resident bird of the Pacific Slope, west of the Sierra Nevada Mountains, with some birds visiting southern Arizona and northern Sonora in the fall and winter. Since the 1960s the number of fall and winter visitors to Arizona has increased greatly, as has the number of breeding birds in our area. The first nesting in Arizona occurred in 1962. In Arizona, Anna's Hummingbirds are found in residential areas most commonly in association with feeders and exotic plantings. They can also be found in desertscrub and riparian woodland.

Costa's Hummingbird: This hummingbird inhabits desertscrub communities dominated by cactus, ocotillo, chuparosa, and wolfberry. It is probably our most arid-adapted hummingbird.

Rufous hummingbird: This hummingbird breeds in the northwestern United States and western Canada and winters in Mexico. It tends to migrate north through western Sonora, western Arizona, to the Pacific coast; it tends to migrate south through the Rocky Mountains, including eastern Arizona. This annual migration path forms a

Hummingbirds & Spider Webs

When we opened the Arizona-Sonora Desert Museum's hummingbird aviary in 1988, we had no idea whether or not any of the eight species of the birds on exhibit would breed and rear young. Since opening day, however, we've seen Costa's, Broad-billed, Black-chinned, Anna's, and Calliope hummingbirds nest, lay eggs, and rear young. There have been a total of 114 nests built, 186 eggs laid, 116 birds hatched, and 102 birds fledged. No other zoological institution can boast of such success.

But this success has not come without a good deal of effort on the part of the exhibit keepers and the hummingbirds—especially when it comes to nest-building. For example, in 1992 we renovated the exhibit, clearing out all the plants and expanding and replanting the new space. Within a month of the renovation, several hummingbirds began to build nests. The nests were loose and quite fragile, and even experienced nesters were having difficulty. Most of the nests fell apart and we lost several eggs that fell out and broke. We scratched our heads for days trying to figure out the problem before we finally concluded that a primary component of hummingbird nests was missing—spider webs! Hummingbirds use spider webbing as a way to bind and tie their nests together. The spiders had yet to reestablish themselves in the spanking new exhibit. I immediately went out and collected webs from around the grounds, rolling them up on twigs, which I left in the aviary. The Desert Museum's entomologist and I also collected 25 labyrinth spiders and introduced them. Within days the spiders were weaving their webs in the aviary and the birds' nests immediately improved.

—Karen Krebbs

Anna's Hummingbird

broad oval over most of western
North America. The males generally
migrate before females and immatures.
During migration in Arizona and
Sonora, the Rufous uses a wide variety
of habitats from desertscrub to
mountain meadows, wherever there
are flowers or feeders.

FEEDING
Hummingbirds are the dominant
nectarivorous birds in the western
hemisphere. Old World ecological
equivalents include the honey-eaters
of Australia and the sunbirds of
Africa. Although many of these other
birds may have specialized bill shapes
and foot types for getting nectar,
and some have bright or iridescent
plumage, none achieve the specialization
to nectarivory that the hummingbird

has. Hummingbirds also eat many small,
soft-bodied arthropods.

LIFE HISTORY
North American hummingbirds are
highly territorial, both sexes protecting
feeding territories, males protecting
courtship territories, and females
protecting nesting territories. These
territories are protected by displays,
songs, chases, or the mere presence
of the hummer on an exposed perch.

Hummingbirds are promiscuous
breeders. The male merely courts and
mates with receptive females. The
female may mate with more than one
male, but she alone builds the nest, lays
and incubates the eggs, and broods and
tends the young.

The nest is not much larger than a
jigger glass. It is typically composed of

fibrous plant down or seeds and mosses, bound together and to a branch with spider webbing. The nest may be lined with hair or feathers and decorated with leaves, bark strips, or lichens, depending on the species.

Only 2 bean-sized eggs are laid and incubated for about 2 weeks, depending on the species.

Young are altricial and are fed a mixture of nectar and small, soft-bodied arthropods like spiders and gnats. They fledge in about 3 weeks depending on the species and to some extent on the weather.

Like many animals, most hummingbirds die during their first year. After that, their life expectancy may increase to 3 or 4 years. Few live past this, although there is a record of a Broad-tailed Hummingbird living 11 years.

Broad-billed Hummingbird: In Arizona, Broad-bills nest from April through July. In Sonora, they may start earlier. The nest is saddled on a horizontal limb and is typically decorated with long strips of bark or leaves. The males have a short, buzzing, scratchy advertising song.

Hummer Facts

❖ Hummingbirds' hearts are larger in proportion to body size than those of any other warm-blooded animal.

❖ Hummers have the most rapid heart rate for a bird: up to 500 beats per minute at rest and 1260 beats per minute during activity.

❖ Their flight muscles account for 25 to 30 percent of body weight, compared to 15 to 25 percent in other strongly-flying birds.

❖ Hummers have the most rapid wing beats of birds: up to 80 beats per second.

❖ Their unique flight mechanisms allow them to hover for long periods of time, move in any direction (even backwards), and dive at over 60 miles per hour during displays.

❖ They have high body temperatures: 105° to 109°F (40.5° to 42.5°C).

❖ They have the ability to become torpid at night. (See the sidebar "Powering Down" on the next page.)

❖ Hummingbirds may consume 70 percent of their body weight, in solid food per day (8 to 12 calories) and 4 to 8 times their body weight, in water.

❖ There are over 300 species of hummingbirds. They live exclusively in the Western Hemisphere, from Alaska to the tip of South America.

Black-chinned Hummingbird: Black-chinned Hummingbirds breed in Arizona from April through July. The male's dive display consists of a swooping pendulum-like dive of about 100 feet (30 m). There is a loud whirring sound at the bottom of the pendulum that is presumably produced by the wings or tail.

Anna's Hummingbird: The Anna's Hummingbird is among the earliest of our nesting birds. Nests with young have been recorded in December in Tucson. The breeding season may extend through May. The male Anna's is the most vociferous hummingbird in our region. Its advertising song of squeaks and buzzes is louder and more "musical" than most. Both the male and female give a "war cry" consisting of a rapid series of buzzes when they chase intruders from their area. The male also exhibits a dive display in the shape of a 100 foot (30 m) "J." At the bottom of the dive, he gives a loud squeak.

Costa's hummingbird: The breeding season in Arizona and Sonora is late winter and spring. Some breeding may occur later in Baja California and in southern California. The males have a "song" consisting of one drawn-out whistle, reminiscent of a ricocheting bullet. This may be given from a prominent perch or at the bottom of its U-shaped dive display.

Rufous Hummingbird: In our region, the Rufous is known only as a migrant and a very pugnacious hummer at feeders.

COMMENTS

A related species, the Allen's Hummingbird, *Selasphorus sasin,* also occurs in our region as an uncommon migrant. The males have a greener back than the Rufous and the females are almost indistinguishable.

Powering Down

Being a hummingbird is like driving a car with a one-gallon gas tank: there is an almost constant need to refuel. Hummingbirds are often perilously close to the limits of their energy reserves. On cold nights, when the costs of keeping warm are especially high, it may be too risky for a hummingbird even to keep its engine idling.

At such times, a hummingbird bristles its feathers to let its body heat escape, and its temperature quickly approaches that of its surroundings. Its heart rate drops dramatically and it may stop breathing for minutes at a time. It appears lifeless, clinging motionlessly to its branch with its head drawn close to its body and its bill pointing sharply upward. At daybreak it revs its metabolic engines and warms itself again.

This sort of temporary hibernation is called torpor. *Hummingbirds become torpid not only to deal with fuel crises, but also to save energy for migration. And since birds lose moisture with every breath, becoming torpid also helps desert hummingbirds conserve water.*

—David W. Lazaroff,
The Secret Lives of Hummingbirds (ASDM Press, 1995)

Woodpeckers

Gila Woodpecker

We expect woodpeckers to be in the woods, so it may seem surprising that some are conspicuous in the desert. But for Gila Woodpeckers and Gilded Flickers, saguaros serve in place of trees: these woodpeckers go hitching their way up the sides of the giant cactus, and give voice to strident calls when they reach the top. The holes that they excavate for nesting sites—which may riddle the arms of some ancient saguaros—remain to serve as natural birdhouses for a variety of other birds.

Most woodpecker species feed mainly on insects, seeking them out among the irregularities of tree bark. In the desert, these birds must be more resourceful. Gila Woodpeckers eat cactus fruits, mistletoe berries, and many other items in addition to insects. Highly adaptable, they make themselves at home in southwestern U.S. cities, where they will visit hummingbird feeders and steal dog food from back porches. (They also make themselves unpopular at dawn by hammering out brash wake-up calls on metal pipes and other echoing objects.) Gilded Flickers spend much time foraging on the ground; they are among the few birds that regularly eat ants. Ladder-backed Woodpeckers, among the smaller members of this family, make their living in a more traditional woodpecker style on the trunks of mesquites along desert washes.

—Kenn Kaufman

Sonoran Desert species:

Gila Woodpecker (*Melanerpes uropygialis*)
Ladder-backed Woodpecker (*Picoides scalaris*)
Gilded Flicker (*Colaptes chrysoides*)

Order: Piciformes
Family: Picidae
Spanish names: carpintero de Gila (Gila), párajo carpintero, picapalo

DISTINGUISHING FEATURES

Gilded Flickers: Brown birds, with black barring on their backs and white rumps, visible as they fly; underside of wings and tail is golden.

Gila Woodpecker: Brown face, black and white barred back, white wing patches that are visible when in flight.

Ladder-backed Woodpecker: Black and white barred backs; face has black and white stripes.

Ladder-backed Woodpecker

HABITAT

These woodpeckers are permanent residents that are found in all desert habitats.

FEEDING

• *Diet:* As a group, woodpeckers are adapted to locating and capturing invertebrates living in the bark of trees. While these 3 species often look for insects on the side of a tree, they are also opportunistic. The flicker is often found on the ground eating ants, and all 3 eat cactus fruit.

• *Behavior:* These woodpeckers have strong head and neck muscles, and the skull is adapted to absorb the shock as the birds drive their chisel-shaped bills into the tree. The tongue is long and can be extended; its tip is bristled and sticky. Short legs, strong toes, sharp claws and stiff tail feathers keep these birds secure on the vertical surface of trees.

Gilded Flicker

LIFE HISTORY

Woodpeckers nest in cavities that they excavate. In the Sonoran Desert, Flickers and Gila Woodpeckers build nests in saguaros; the interior of the cactus provides a secure environment where the temperature is moderated year around. After the woodpeckers are finished with them, their nests are used by other birds— Elf Owls, Kestrels, Ash-throated Flycatchers, Purple Martins. According to some research, flickers and Gila Woodpeckers nest at different heights in the saguaro: flickers within 3 meters (10 feet) of the top, woodpeckers lower. The difference seems due to the fact that woodpecker nests are excavated in the outer cortex, whereas flickers need larger cavities, so their excavations go further toward the thicker center of the cactus— through the ribs and into the inner pith. Flicker bills are not adapted to heavy-duty boring: toward the top of the cactus the ribs are thinner and more easily severed. Flicker cavities may harm the saguaro, even kill it, because water is no longer transported through the vascular tissue of the severed ribs and because the excavation increases surface area/volume ratio leading to greater water and heat loss.

Tyrant Flycatchers

Northern
Beardless-Tyrannulet

Birdwatchers are sometimes driven to despair by
the challenge of telling the various flycatchers apart.
Many of the species look virtually the same. Birds
have to be able to recognize their own kind, of course, at
least during the breeding season, but the flycatchers evidently do
so mostly by voice. In the Sonoran Desert, for example, the Brown-crested and
Ash-throated Flycatchers are almost identical except for size, but their songs and
calls are different. Perhaps capitalizing on their need for vocal distinctions, many
flycatchers have "dawn songs," seldom heard later in the day.

Flycatchers in North America feed mainly on insects, and forage by watching
from an exposed perch and then sallying forth to pick flying insects out of the
air. In this they are aided by their wide, flat bills, and by the bristles on either side
of the bill. The little beardless-tyrannulet is so called because it lacks these bristles; it may have less need for them, since it often takes insects from the surfaces
of leaves.

The family name of "tyrant" flycatchers reflects the aggressive nature of some
species, which drive away much larger birds that venture too near their nests. The
kingbird group, represented here by the Western Kingbird, provides the best
example of this behavior. Seemingly more gentle are the phoebes, soft-voiced
flycatchers that often nest near houses or bridges.

Although most flycatchers are dull-colored, a striking exception is the
Vermilion Flycatcher, a relative of the phoebes, common along streams in the
desert. The male glows red and black, and in his courtship display he puffs
himself up like a ball and flutters about the sky while singing madly, in brilliant
contrast to the drabness of his relatives.

—Kenn Kaufman

Tyrant Flycatchers

Vermilion Flycatcher

REPRESENTATIVE SONORAN DESERT SPECIES:
  Northern Beardless-Tyrannulet *(Camptostoma imberbe)*
  Vermilion Flycatcher *(Pyrocephalus rubinus)*
  Black Phoebe *(Sayornis nigricans)*
  Say's Phoebe *(Sayornis saya)*
  Ash-throated Flycatcher *(Myiarchus cinerascens)*
  Brown-crested Flycatcher *(Myiarchus tyrannulus)*
  Western Kingbird *(Tyrannus verticalis)*

Order: Passeriformes
Family: Tyrannidae (Tyrant Flycatchers)
Spanish names: mosquerito copetón, mosquero (flycatcher)

DISTINGUISHING FEATURES

Northern Beardless-Tyrannulet: The smallest flycatcher in the United States, nondescript, gray-olive above and very pale gray below. The name "beardless" refers to the absence of the long rictal bristles at the base of the bill that are characteristic of most flycatchers. *Vermilion Flycatcher* males: Brilliant red head and underparts and black mask, back and wings. Females: Streaks on a white breast, brownish head, pale salmon or pink belly. *Black Phoebe:* Slate black except for white underparts; its *fee-bee* call is a typical sound along riparian areas in the Southwest. *Say's Phoebe:* Pale-gray back, dark tail, and rust-colored underparts. *Ash-throated Flycatcher:* Brown bushy head, very noticeable whitish-gray throat, light yellow belly, olive-brown back, rust-colored primaries and tail feathers. *Brown-crested Flycatcher:* Largest of its genus; olive-brown above, black bill, pale gray throat and breast, pale yellow belly; reddish tail and primaries are easily seen in flight. *Western Kingbird:* gray head, whitish-gray throat and upper breast, pale yellow underparts; the blackish tail has white margins.

HABITAT

Northern Beardless-Tyrannulet: Woodland and stream thickets; found most often in stands of mesquite or cottonwood-willow in southern Arizona. *Vermilion Flycatcher:* most commonly found near streams or ponds but also frequents grassland and desert habitats with scattered trees. *Black Phoebe:* Prefers shady areas near water including streams, ponds and walled canyons. *Say's Phoebe:* Unlike the Black Phoebe, often found in very dry open or semi-open country, far from water; typical of prairies, badlands, and ranch country. *Ash-throated Flycatcher:* Frequents desertscrub, pinyon-juniper, oak groves, creek bottoms, and dry open woodland. *Brown-crested Flycatcher:* Found in association with saguaros; also frequents river groves and other areas where trees are large enough to provide sites for cavity nesting. *Western Kingbird:* Found in semi-open country, farmland, roadsides, and in riparian vegetation.

FEEDING
• *Diet:* Mainly insects
• *Behavior:* Unlike other flycatchers which capture insects while flying, the *Northern Beardless-Tyrannulet* often moves along twigs in search of slow-moving insects. During the summer this species also forages in a more typical flycatcher manner: by flying from its perch and catching insects in its bill.

The *Vermilion Flycatcher* forages by watching from a perch and either capturing flying insects in the air or hovering and dropping to the ground.

The *Black Phoebe* is often seen perched over the water, slowly wagging its tail, and then darting out to capture an insect in the air, often just above the water's surface; occasionally takes small fish.

Because they frequent areas that lack high trees, *Say's phoebes* either perch on a low shrub and rock and dart out to capture prey, or hover over fields in search of insects in the grass.

The *Ash-throated Flycatcher* typically forages by flying out from its perch on a dead upper branch of a tree to hover and pick insects from foliage; seldom takes insects in midair; diet is mostly insects, but spiders, saguaro fruit, elderberries, desert mistletoe berries, and even small lizards, are also eaten.

The *Brown-crested Flycatcher* typically forages by flying out from its perch to hover and pick insects from foliage; also takes insects in midair or from branches or trunks of trees; will perch in shrubs and cactus to eat fruit; diet is mostly insects, but lizards are also eaten.

The *Western Kingbird* forages by watching from a perch and then flying out and either snapping up insects midair or taking them on the ground by hovering and diving; diet consists mainly of insects but also includes spiders, millipedes, berries, and fruits.

LIFE HISTORY
Northern Beardless-Tyrannulet: The baseball-sized nest is typically well camouflaged in trees or large shrubs. One to 3 white eggs have small, brown and gray dots. Nesting behavior, incubation details, and development of the young are poorly understood.

Vermilion Flycatcher: The male has a rather showy courtship display that involves fluffing up the feathers while rising vertically in the air and hovering, then swooping back to its perch or fluttering slowly to the female. The nest is built by the female in the horizontal fork of mesquite, willow or cottonwood, and other trees. The nest, which consists of twigs, grass and weeds, is lined with feathers or hairs and is often held together by spider webs. The 2 to 4 heavily spotted cream-white eggs take 14 to 15 days to incubate. Only the female incubates the eggs but both parents feed the young until they fledge at 14 to 16 days.

Black Phoebe: The male's courtship display includes fluttering in the air with rapidly repeated calls, then descending slowly. Black phoebes are solitary nesters that often return to their nest site year after year. The nest is made of mud, grass, and weeds and is usually found in

Say's Phoebe

a sheltered spot such as a cliff face or on the support strut of a bridge. Three to 6 eggs are incubated by the female and hatch in 15 to 17 days. The young, which are fed by both parents, leave the nest about 2 to 3 weeks after hatching.

Say's Phoebe: The nesting site varies greatly—it may be on a rocky ledge, under the eaves of a barn, under a bridge, or even in a well or mine shaft. The nest is made of grass, weeds, twigs, wool, moss, spider webs, and other materials. Unlike the Black Phoebe, no mud is used in its construction. Three to 7 eggs take 12 to 14 days to incubate. Both parents care for the nestlings. The young leave the nest 14 to 16 days after hatching. These birds are not shy of people and adapt well to changes to their natural landscape; populations appear to be stable.

Ash-throated Flycatcher: The nest is usually in a cavity or in an existing hole, but may also be found in a variety of other—often unlikely—places like mailboxes and exhaust pipes. The nest consists of a mass of grass, hairs, weeds, and twigs lined with softer material. Three to 7 creamy- or pinkish-white, blotched or streaked eggs are incubated by the female and hatch in about 15 days. Both parents feed the nestlings, which are ready to fly 14 to 16 days after hatching.

Brown-crested Flycatcher

Western Kingbird

This species will use nest boxes put out for bluebirds.

Brown-crested Flycatcher: This species is aggressive and conspicuous during nesting season. They arrive after most other hole-nesting birds and therefore may have to compete for nest sites. The site is in a cavity in a tree or giant cactus, usually in holes excavated by woodpeckers. Both sexes help with the construction of the nest in the cavity. Plant fibers, hair, feathers, and other debris are used. Three to 6 white to pale buff blotched eggs are incubated by the female. Young hatch in 13 to 15 days. Both parents feed the nestlings and first flight takes place 12 to 18 days after hatching.

Western Kingbird: Nest sites vary and may include tree limbs, utility poles, cliff ledges, and abandoned nests of other birds. Grass, weeds, twigs, plant fibers and softer material for the lining are used in the construction of the cup-shaped nest. Three to 5 white and heavily blotched eggs take 18 to 19 days to hatch. This spunky bird will harass hawks or other large birds that stray too close to its nest. After the young fledge, typically 18 to 19 days after hatching, it is not uncommon to see 6 or more kingbirds burst out of a tree in search of insects.

The Kingbird has adapted well to human encroachment. It has expanded its breeding range and increased its numbers during the 20th century.

Swallows

Swallows spend their days mostly in the air, often in flocks, plying the skies in search of the flying insects that make up the majority of their diet. With their graceful flight, musical voices, and sociable nature, these birds are popular with humans—and, it seems, the reverse is also true. Many kinds of swallows have prospered by learning to live alongside civilization.

Purple Martin

The biggest concentrations of swallows occur where there are swarms of flying insects, especially over water. Therefore, swallows are less numerous in the desert than in many other habitats. A few species are common here, however. Widespread but unobtrusive is the Northern Rough-winged Swallow, a brown bird that digs nesting tunnels in the vertical walls of dry arroyos and road cuts. Less sociable than most, it usually forages and nests in isolated pairs.

At one time, Cliff Swallows were quite localized in the Southwest. They required soft mud with which to build their gourd-shaped nests, and vertical or overhanging cliffs on which to place them. Modern civilization gave them an abundance of new sites: buildings close to watered lawns, and bridges over muddy creeks. Today Cliff Swallows are common in summer around many towns and roads in desert regions.

A true desert dweller is the local race of the Purple Martin, which nests in holes in saguaro cacti. Purple Martins in eastern North America today nest almost exclusively in multi-roomed birdhouses put up for them, but this habit has not yet caught on in the Southwest—our local martins are more independent of humans than are any other swallows here.

—Kenn Kaufman

Swallows

REPRESENTATIVE SONORAN DESERT SPECIES:
Purple Martin (*Progne subis*)
Northern Rough-winged Swallow (*Stelgidopteryx serripennis*)
Cliff Swallow (*Hirundo pyrrhonota*)

Order: Passeriformes
Family: Hirundinidae (Swallows)
Spanish names: golondrina (swallow)

DISTINGUISHING FEATURES
Purple Martin: Largest swallow in North America; tiny bill, forked tail. Adult male: black with some purple iridescence, deeply forked tail. Adult female: dark gray upperparts with some purple, whitish-gray underparts with some speckling. *Northern Rough-winged Swallow:* Light brown above and white underparts, small bill, forked tail, brownish throat and chest. *Cliff Swallow:* Blackish cap, light spot above tiny bill, buff-colored rump, squared-off tail, whitish belly and rust-colored throat and cheek.

Cliff Swallow

HABITAT
Purple Martins are found in semi-open country near water, saguaro forest, and woodlands. Northern Rough-winged Swallows live near water, including streams, lakes, and riverbanks; also found in desert washes. Cliff Swallows are found in a variety of open to semi-open areas, especially when these areas are near water.

FEEDING
• *Diet:* All 3 species feed on a wide variety of insects; spiders and other arthropods are also taken.
• *Behavior:* These birds do nearly all of their foraging in the air. Their short, wide bills are ideally suited for snatching insects out of the air. They are often seen flying over water because of the abundance of insects. During bad weather these birds may resort to foraging on the ground.

LIFE HISTORY
Purple Martin: It uses naturally occurring sites, such as woodpecker holes in trees or saguaros. In eastern North America, most Purple Martins today nest in multi-roomed birdhouses put up to attact them. The cup-shaped nest, which is made by both sexes, consists of grass, twigs, leaves, feathers, and often mud. Whitish eggs (3 to 8) are incubated by the female and hatch in 15 to 18 days. The young leave the nest 26 to 31 days after hatching. Purple Martin numbers have been declining; the reasons are not well

Eggs

The egg of a bird consists of several parts. The yolk is a single giant cell, the true egg or ovum, produced in the ovary of a bird. The yellow material is mostly food for the growing embryo once the egg is fertilized. The ovum is released from the ovary and picked up by the oviduct where it is fertilized by sperm from a recent mating. As the ovum travels down the oviduct, layers of material are added in an assembly line fashion. First several layers of albumen (egg "white"), then two egg membranes, then the shell, and finally coloring. Most birds will lay one egg per day until they have a complete set. This set of eggs is called a clutch.

The color of the egg is related to its need to be camouflaged. Cavity nesters usually lay all white eggs since they are out of sight. Ground nesters lay brown, gray, or olive colored eggs, the color of the surrounding vegetation. Tree nesting birds will lay white or blue eggs, splotched or speckled with brown, to help conceal them in dappled light.

understood but may be due to competition with starlings.

Northern Rough-winged Swallow: Unlike many species of swallows, this bird tends to nest singly. For nesting, the Rough-wings dig horizontal burrows in dirt stream banks, arroyo walls, or road cuts. They will also sometimes use holes in man-made structures. The bulky nest is made of weeds, twigs, and other vegetable matter. Four to 8 white eggs hatch in 12 to 16 days. Both parents feed the nestlings; young leave the nest 19 to 21 days after hatching. Because Rough-winged Swallows take advantage of man-made structures and areas of ground disturbance for nesting sites, they have actually benefited from the advance of civilization.

Cliff Swallow: This species most often uses elevated vertical surfaces with a protective overhang for nesting sites. Typical sites include cliff faces, buildings, and bridges. The gourd-shaped nest is made of mud and may be lined with feathers and possibly

vegetable matter. Old nests are sometimes reused. Three to 6 white to pale pink spotted eggs are incubated by both parents and hatch in 14 to 16 days. The young leave the nest 26 to 31 days after hatching. This species nests in colonies of up to several hundred individuals.

Rough-winged Swallow

Common Raven

Members of the corvid family, including ravens, crows, jays, and magpies, are thought to be among the most intelligent of birds. If adaptability is any sign of intelligence, then the Common Raven must rank as a superstar: it thrives from Siberia to North Africa, from arctic Alaska to the mountains of Central America. Forest, tundra, and desert are all within its realm. Technically it is classed as one of the songbirds, or perching birds, but it is larger than many hawks, and it will feed as a predator or scavenger as the opportunities arise.

In the southwestern U.S. and northern Mexico, Common Ravens range from sea level (along the Gulf of California) to the mountaintops. They can even be seen flying over desert cities, their croaking calls floating down from the sky as the big birds flap overhead. Unlike the Chihuahuan Raven—a smaller and more sociable bird that lives in grasslands along the eastern edge of our region— Common Ravens seldom travel in flocks, but members of a pair may stay together at all seasons. Their nesting sites are sometimes in large trees, where these are available, but most raven nests in the Sonoran Desert are placed on high cliffs.

—Kenn Kaufman

Common Raven (*Corvus corax*)

> Order: Passeriformes
> Family: Corvidae (Crows, Jays, Magpies)
> Spanish names: cuervo grande, cuervo holárctico

DISTINGUISHING FEATURES
A hawk-sized, shiny, black bird with a black bill and a wedge-shaped tail. The Common Raven is the largest member of the order Passeriformes (the songbirds, or perching birds).

HABITAT
This intelligent and very adaptable bird occurs from low deserts to mountains, over open desert areas to dense forests.

A Matter of Style

The folklore of more than one group of native Americans includes stories about coyote and raven interactions. I witnessed such an encounter early one morning while walking the banks of the Rillito River near Tucson. Two ravens were in the center of the dry riverbed, actively complaining. About five feet from them was a coyote, nose to the ground, casually sniffing about. Evidently located between the two groups was an object—possibly a bit of food—that all three coveted.

The ravens asserted themselves through noise and occasionally hopping about, but never getting too close to the canine. The coyote's approach to the matter was completely the opposite: he simply did not acknowledge the birds' presence. He never looked at them. He made no move in their direction. One could almost hear him mutter, "I don't see anything. Do you see anything? I don't see anything."

Eventually the coyote trotted down the riverbed, apparently oblivious to the ravens' continued harangue. It seemed as though both coyote and ravens had forgotten whatever object it was that had sparked the debate, the coyote possibly believing he was the victor in this confrontation by his superior demeanor, and the ravens believing they were the winners by raucous strength of voice.

—Peggy Larson, naturalist and author

FEEDING

- *Diet:* Omnivorous; diet may include carrion, reptiles, amphibians, bird eggs, insects, and plant matter.
- *Behavior:* Opportunistic feeders, ravens often take food wherever it can be found, including public landfills; also frequently search for nests where they feed on the eggs; sometimes hunt in pairs where one bird flushes out the prey. Their water-rich diet of carrion, insects, and eggs, along with their stocky bodies, which help them better regulate their body temperature, allow this species to cope with the desert's heat.

LIFE HISTORY

Courtship involves the male soaring, swooping and tumbling in front of the female. The pair also soar together, then perch and preen one another. Ravens mate for life.

Nesting takes place from February to May in this region. The nest, which is usually built high in a tree or on a precipice, is a platform made of large sticks, twigs and vine stems with a deep depression in the center; it is lined with grasses, animal hair and mosses. Usually 4 to 6 greenish blotched eggs are laid; incubation is by the female. The male feeds the female during the 3 week incubation period. The young are altricial; they leave the nest about 5 weeks after hatching.

Verdin

Tiny and rather plain, the Verdin seems quite unremarkable at first sight. But it is among the most character- istic birds of the desert, and it has one notable distinction: it is not closely related to any other bird in the western hemisphere. For a while it was placed, uneasily, in the same family as the chickadees, which it resembles in size and in hyperactive behavior. Scientists now believe that its closest relatives are several species of small, plain birds found in Europe, Asia, and Africa.

Verdins seem undaunted by our extremes of weather. Hardy and adaptable, they are active on the hottest days and the coldest winter mornings. Although they are most common in thick mesquite bosques, they also range out onto open flats where the plant life is sparse, and they readily move into desert cities. Insects are their main menu items, but Verdins also take nectar; among desert birds, they are second only to the hummingbirds as flower visitors, and they also come to hummingbird feeders to sip sugar water.

Except when they pair up for nesting, Verdins generally go their own way as individuals. Their nests are surprisingly large for the size of the bird; they are hollow globular masses of thorny twigs, each with an entrance low on one side. Since these birds build nests in which to sleep at night, as well as those for raising young, the nests are often easier to find than the Verdins themselves.

—Kenn Kaufman

Verdin *(Auriparus flaviceps)*

Order: Passeriformes
Family: Remizidae (Verdin and related birds)
Spanish name: verdín

DISTINGUISHING FEATURES
A small bird with gray back, white underparts, a yellow head and throat, and a red-chestnut patch at the bend of the wings. Immatures lack both the yellow and chestnut coloration of adults. This bird is easily recognized by its rather loud, rapid whistle.

HABITAT
Most common in the Sonoran Desert and mesquite bosques at lower elevations; also common in many southwestern urban areas.

FEEDING
• *Diet:* Mainly insects.
• *Behavior:* Verdins move and behave much like chickadees, flitting about singly or in pairs in search of food, often hanging upsides down to reach the underside of leaves. They are generally tolerant of people except during the nesting season. The diet consists mainly of insects but small spiders, berries, small fruits, and sometimes seeds are also taken. This bird also drinks nectar and sugar water, as people that have hummingbird feeders hung out know all too well.

LIFE HISTORY
The male may build several bulky twig nests before the female chooses one in which to lay her 3 to 6 pale green, red-brown dotted eggs. These large oval or spherical nests are placed rather far out on branches and may last many years in the dry desert environment. The entrance to the nest often faces prevailing winds, possibly an adaptation to the high temperatures of their desert habitat. Incubation takes about 10 days and the young are ready to leave the nest about 21 days after hatching, although they return to the nest at night.

Color in Birds

The color of feathers is either due to pigments deposited within the feather during growth or to structures within the feather that manipulate light; or color may be due to a combination of these two. Three common pigment types are melanins (browns, blacks, some yellows), carotenoids (some reds and yellows), and porphyrins (some reds or browns). Structural colors include blue and iridescence. Greens and violets are combinations of a pigment with a structural manipulation of light.

The functions of color in birds can generally be divided into those that will help conceal a bird (cryptic colors) or colors that will make the bird more conspicuous (phaneric colors). Patterns result in camouflage, counter shading, flash patterns, display plumages, and sexually dimorphic plumages.

Wrens

Active and inquisitive little brown birds, wrens spend their days snooping about, peering into shadows, prying with their thin bills, seeking the insects on which they feed. Often hard to see, they are easy to hear. Male wrens are inveterate singers. They are also industrious nest-builders: in some wren species, the male may build several "dummy nests" before the female chooses one, adds a soft lining to it, and lays her eggs there.

Cactus Wren

Typical of the wren family is Bewick's Wren, which hops about in the dense brush along desert washes. Usually it stays fairly low, but a male will flit to the top of an exposed stub to sing his musical trills. In more barren rocky zones, the Canyon Wren and Rock Wren manage to thrive in places where there is not enough plant life to create habitat for most birds. They play hide-and-seek among the boulders, bouncing from one rock to the next, probing into deep crevices for spiders and insects lurking there. Though the Rock Wren sings rather like a weak and uninspired mockingbird, the Canyon Wren produces a rippling, descending cascade of loud clear whistles, fitting music for the most majestic canyons of the west.

While most wrens are small, plain, secretive, musical, and solitary, the best-known wren in the Sonoran Desert breaks all those rules. Pairs or family groups of Cactus Wrens, strikingly spotted and striped, go clambering and scrambling about in the open, calling in rough scratchy voices from high perches, boldly peering in the windows of houses on the desert's edge. Its brash behavior earns the Cactus Wren the admiration of its human neighbors and it has been selected as Arizona's official State Bird.

—Kenn Kaufman

Wrens

SONORAN DESERT SPECIES:
Cactus Wren (*Campylorhynchus brunneicapillus*)
Canyon Wren (*Catherpes mexicanus*)
Rock Wren (*Salpinctes obsoletus*)
Bewick's Wren (*Thryomanes bewickii*)

Order: Passeriformes
Family: Troglodytidae (Wrens)
Other common names: Dotted Wren, White-throated Wren (Canyon Wren)
Spanish Names: matraca, saltapared (wren); matraca grande (Cactus Wren)

DISTINGUISHING FEATURES

Cactus Wren: Large with heavily spotted underparts; spots forming a dark cluster on the upper breast in the adult.
Canyon Wren: White throat and breast, overall red-brown in color. *Rock Wren:* Finely mottled gray-brown upper parts, finely streaked whitish breast, white streak over eyes. *Bewick's Wren:* Pale gray below, dull brown above, bold white eyebrow; tail relatively long, brown with white corners.

HABITAT

The *Cactus Wren* is found in deserts and arid foothills in association with cactus, yucca, mesquite, and arid brush. The *Canyon Wrens* inhabits canyons, cliffs, boulder fields and other rocky areas including man-made structures; may move to streamside habitats in winter. The *Rock Wren* is found in arid rocky slopes and canyons. The *Bewick's Wren* inhabits mesquite bosques, heavy brush along desert washes, and open woods.

FEEDING

• *Diet:* The *Cactus Wrens* eats insects, other arthropods, vegetable matter such as fruit pulp, and seeds. The three other species eat mainly insects and spiders.
• *Behavior:* The *Cactus Wren* forages on the ground and in low trees, probing crevices and ground litter and often foraging in pairs or family groups. The *Canyon Wren* forages by hopping actively in and out of cracks, crannies and dense undergrowth in canyons. The *Rock Wren* forages on the ground where it uses its bill to probe for prey; it sometimes forages within brushy vegetation or low in trees. The *Bewick's Wren* actively forages by clambering about on limbs of trees, probing into bark crevices; it sometimes feeds on the ground.

Canyon Wren

Rock Wren

Life History

The *Cactus Wren* nests from mid-March to early September. The nest is roughly the size and shape of a football, with an end opening. It is built with grasses, primarily in cholla, but also in palo verdes, acacias, and saguaros. Both parent birds construct the first breeding nest, where the female lays 3 to 5 whitish, small, brown-dotted eggs. The altricial young leave the nest at about 21 days. The male bird constructs a new nest while the female is incubating the eggs. This nest is occupied by the female as a roosting nest. The parents may raise several broods a season and these secondary nests may be used as brood nests for later clutches. Roosting nests are very important for the security of Cactus Wrens. The male cares for the young between nest-building chores while the female incubates the next clutch of eggs.

The elusive but musical *Canyon Wren* makes its small nest of twigs, grasses, and leaves in rocky crevices, cave ledges, rock piles, and other protected sites. White, red-brown dotted eggs (4 to 6) are incubated by the female and take 12 to 18 days to hatch. Both parents feed the nestlings. The young are ready to leave the nest in 15 days although they sometimes stay with the parents for several weeks longer.

The *Rock Wren* nest may be found in cracks and crevices among boulders, gopher burrows, cracks or openings in stone or adobe buildings, steep banks of washes, and other sheltered sites. The nest consists of grasses, weeds, twigs and bark lined with finer material such as feathers and hair. Brown-speckled eggs (4 to 8) are incubated by the female. The Rock Wren frequently creates a path of small rocks that lead to its nest; the purpose of this behavior is not understood.

Rock Wrens are strongly migratory,

Bewick's Wren

spending the winter in rocky lower elevations and the summer in elevations reaching 8,000 to 10,000 feet in the northern part of their range. However, the species can be found in the Sonoran Desert region during all seasons.

The *Bewick's Wren* nest is built inside almost any kind of cavity, including old woodpecker holes, natural tree hollows, or crevices in debris or buildings. Males may build several "dummy" nests, with a bulky foundation of twigs; the female chooses one and adds lining of softer material. White, brown-blotched eggs (5 to 7) are incubated by the female for about 14 days. The young leave the nest after about two weeks.

Unfriendly Embrace

As I walked to my office on a warm February afternoon I heard the familiar Chuk! Chuk! Chuk! Chuk! of a scolding Cactus Wren. I then saw the bird pursue another Cactus Wren; the second wren lacked tail feathers. The bickering and chasing continued at a frantic pace and the birds soon darted around a building. I quickly moved into a more favorable viewing position only to see the two adversaries disappear into a dense clump of jojoba. My interest piqued, I patiently waited nearby. They eventually reappeared and began a determined wrestling match less than 20 feet from me, apparently oblivious to my presence. This heated confrontation went on for some minutes until they both lay motionless, feet and wings entwined. They remained that way for over five minutes; the only indication that they were even alive was the faint movement of their tiny breasts. Finally they broke their unfriendly embrace and the tail-less bird flew to the east, its adversary in close pursuit.

This intense squabble could be explained a couple of different ways. The obvious explanation is that this was a territorial dispute between rival adult Wrens. Another possible explanation is that a parent was chasing off a youngster who was sticking around just a bit too long. With a new nesting season only a month away this second explanation seemed quite plausible.

And the missing tail feathers? They were most likely plucked out in the heat of battle—not an unprecedented occurrence. In time they will grow back.

— Steve Phillips

Black-tailed Gnatcatcher

Gnatcatchers are diminutive birds that hop and flit among foliage, flipping their long tails about in an expressive way, as they seek tiny insects. While the related Blue-gray Gnatcatcher may appear along southwestern U.S. rivers in winter, Black-tailed Gnatcatchers are true desert birds. They thrive in typical Sonoran Desert, with its relatively rich vegetation, and they also can be found on open creosote bush flats where other bird life is sparse.

Living in pairs at all seasons, the male and female Black-tailed Gnatcatchers usually forage within a few yards of each other. Perhaps this togetherness gives them a heightened need to communicate—they have a surprising variety of call-notes. They will come in close to investigate imitations of those calls, but they also seem to do some imitating of their own: some of their notes sound very much like Black-throated Sparrows and Verdins.

This Gnatcatcher and the Verdin are the two smallest resident songbirds in the desert. The two species are not related, and should not be in direct competition most of the time, but they interact at times—foraging together, or chasing each other about. This is especially true of young birds, who may be simply picking on someone their own size.

—Kenn Kaufman

Black-tailed Gnatcatcher (*Polioptila melanura*)

Order: Passeriformes
Family: Sylviidae
Spanish names: pisita colinegra, perlita colinegra

DESCRIPTION
This tiny bird measures 4½ to 5 inches (11 to 13 cm). It is gray above and whitish below. The male has a black cap during the summer that extends to the eyes. The long tail is black with white corners. Females and winter males, lacking the black cap, are difficult to distinguish from the Blue-gray Gnatcatcher. The best way to tell the

Migration

The two most significant events in the annual cycle of a bird are nesting and feather molting. In some species there is a third equally important event, migration. Migrations are generally long-distance movements made by populations of birds. They typically involve movement into or out of highly seasonal environments. The reasons for migration revolve around exploiting seasonally abundant resources and/or avoiding environments with seasonally rare resources. Yellow warblers breed in our desert riparian habitats where during the summer there is an abundance of insects. For the winter they move back to tropical America. Many grassland sparrows of the northern Great Plains winter in our desert grasslands, while their grassland home is covered in snow. Our resident American kestrel population is augmented in the winter by individuals from more northern populations. And of course there are residents that never migrate, such as the Cactus Wren, Roadrunner, and Black-tailed Gnatcatcher. There are many patterns to migration, differing from species to species.

two apart is the tail: that of the blue-gray is mostly white as viewed from below; the black-tailed is predominantly black underneath.

RANGE
The Black-tailed Gnatcatcher is a permanent resident from southeastern California and Arizona east to southern Texas and south into Mexico.

HABITAT
It is found in desert brush, dry washes, and mesquite bosques.

LIFE HISTORY
Black-tailed Gnatcatchers live in pairs all year, defending their territory and foraging in trees and low shrubs for a wide variety of small insects and some spiders. Unlike the Blue-gray Gnatcatcher, which it closely resembles, the Black-tailed Gnatcatcher rarely catches insects in midair.

The open cup nest, which is typically found in a low shrub less than 5 feet above the ground, is built by both sexes. It is constructed of a variety of materials including weeds, grass, strips of bark, spider webs and plant fibers; it is lined with finer, softer matter. Three to 5 bluish-white eggs with red-brown dots are incubated by both parents and take 14 days to hatch. Both parents feed the young, which leave the nest 10 to 15 days after hatching. Even though cowbirds often lay eggs in this species' nests, and the pair end up raising cowbird young, the Black-tailed Gnatcatcher populations seem to be holding up well.

Mockingbirds & Thrashers

Bendire's Thrasher

Songbirds truly worthy of the name are the members of the family Mimidae: the mockingbirds, catbirds, and thrashers. Some of them, especially the mockingbirds, borrow phrases from other birds—or from other sounds in their surroundings—but they work them into improvisations that are rich and musical, or at least interesting.

Although a couple of species are common in eastern gardens, and various types of mockingbirds occur throughout the Americas, the family reaches its greatest development in the arid American Southwest. There are some places where the Mockingbird and up to four species of thrashers may be found nesting in the same patch of desert. Differences in the shapes of their bills reflect differences in their feeding behavior, and probably explain how all these related birds can coexist. For example, the Crissal Thrasher, thrashing the soil with its big sickle-shaped bill, can probably root out insects that are unavailable for the short-billed Bendire's Thrasher. There are also slight differences in habitat choice; Le Conte's Thrashers, for example, are found side by side with other thrashers in some places, but their range also extends out onto barren saltbush flats where there are few other birds.

Some thrashers are secretive birds, difficult to observe, but not the Curve-billed Thrasher. Bold and inquisitive, it runs in the open and calls "whit-wheet!" from prominent perches. In parts of the Southwest it has adapted to advancing civilization almost as well as the mockingbird. Curve-billed Thrashers may even thrive in the middle of cities, as long as they can find a few cholla cacti in which to place their nests.

—Kenn Kaufman

Mockingbirds & Thrashers

REPRESENTATIVE SONORAN DESERT SPECIES:
   Northern Mockingbird *(Mimus polyglottos)*
   Le Conte's Thrasher *(Toxostoma lecontei)*
   Bendire's Thrasher *(Toxostoma bendirei)*
   Curve-billed Thrasher *(Toxostoma curvirostre)*
   Crissal Thrasher *(Toxostoma crissale)*
   Gray Thrasher *(Toxostoma cinereum)*

Order: Passeriformes
Family: Mimidae (Mockingbirds and Thrashers)
Spanish names: cenzontle norteño, cenzontle, chonte (Mockingbird);
   cuitlacoche, güitacochi (Curve-billed Thrasher)

DISTINGUISHING FEATURES
Northern Mockingbird: Grayish-white
plumage; flashes of white in the
wings and tail that show in flight.
This bird is a remarkable mimic of
other birds in its area, a characteristic
which gives it its genus and species
name; often heard singing on moonlit nights. *Le Conte's Thrasher:* Pale
gray-brown upperparts with lighter
underparts; the tail is dark dusky-
brown; slender, downcurved bill is
black; brown eyes; the palest thrasher.
Bendire's Thrasher: Short, slightly curved
bill; light grayish-brown plumage with
faint streaking on breast; yellow eyes;
voice is a clear, melodious warble with
some repetition and continuing at
length. *Curve-billed Thrasher:* Well-curved
bill; indistinct spots on breast; gray-
brown plumage; pale orange eyes; call is
a sharp "whit-wheat"; song is a musical
series of notes and phrases with little
repetition. *Crissal Thrasher:* Deeply
curved bill, unspotted breast and
underparts, dark line below the bill,
gray eyes, olive-brown upperparts and

Crissal Thrasher

lighter gray-brown underparts, reddish
undertail. *Gray Thrasher:* Slender, gray-
brown upperparts; underparts are
whitish with triangular or tear-shaped
dark spots; outer tail feathers tipped
with white; yellow eyes.

HABITAT
The *Northern Mockingbird* frequents urban
areas, ranches, densely wooded washes,
and shrub grasslands from the edge of
the desert into the lower elevations of

mountains. *Le Conte's Thrasher* prefers low, hot desert plains with scant vegetation (such as creosote bush or saltbush flats). *Bendire's Thrasher* prefers the desert scrub of the Southwest. *Curve-billed Thrasher* prefers the desert, arid brush, and shrubby woods; also found in southwestern cities as long as cholla are available to provide nesting sites. *Crissal Thrasher* frequents dense mesquite thickets along streams in the Sonoran Desert; found in dense chaparral in southwestern mountains up to about 6000 feet. *Gray Thrasher* inhabits desert scrub and mesquite; found in Baja California only.

Le Conte's Thrasher

Gray Thrasher: Mainly insects.
• *Behavior:* The *Northern Mockingbird* forages by walking or running on the ground or flying down to the ground

Gray Thrasher

FEEDING

• *Diet: Northern Mockingbird:* Insects and berries. *Le Conte's Thrasher:* Spiders, centipedes, small lizards, berries, and seeds compliment its insect diet. *Bendire's Thrasher:* Its predominantly insect diet is complemented with spiders, berries, cactus fruit, and seeds. *Curve-billed Thrasher:* Insects, fruit, berries, and seeds. *Crissal Thrasher:* Spiders, small lizards, berries, and small fruits complement its mainly insect diet.

from a perch; often opens and closes wings, a behavior which causes insects to take flight. Perches to eat berries.

Le Conte's Thrasher foraging takes place on the ground where it uses its bill to dig in the soil for food; feeding usually takes place early morning or at dusk when insects are most active.

Bendire's Thrasher forages mainly on the ground.

Curve-billed Thrasher forages almost entirely on the ground where it uses its large bill to flick aside debris and dig in the soil in search of food; berries are eaten while perched.

The *Crissal Thrasher* is seldom seen flying in the open, preferring to keep within thick streamside vegetation; forages on the ground, usually under thick brush; much of its food is found by digging in the soil or hacking the ground with its heavy bill.

The *Gray Thrasher* is often seen moving on the ground among the scrubby vegetation in search of insects or perched atop a cactus or shrub.

Curve-billed
Thrasher

LIFE HISTORY

The *Northern Mockingbird* builds a small stick nest lined with a few plant fibers in large shrubs or small trees. Here it lays 3 to 4 bluish eggs blotched with brown. These are incubated about 12 days by the female. The young, which are fed by both parents, remain in the nest for 12 to 14 days. Mockingbirds can be quite aggressive towards dogs, cats, other birds, and even humans who come within the vicinity of their nest.

Le Conte's Thrasher mating pairs, which may mate for life, remain togeth-er year-round. Cholla is a preferred nest site, although other low shrubs may be used. The bulky twig nest, which is built by both sexes, is lined with leaves, plant fibers, rootlets and sometimes with softer materials. Pale blue-green eggs (2 to 4) are incubated by both parents and hatch in about 15 days. The young leave the nest about 13 to 17 days after hatching.

The *Bendire's Thrasher* nest, which is similar but smaller than other thrasher nests, is built in cholla, yucca and various desert trees and shrubs. The female lays 3 to 5 pale green eggs spotted with brown. The young, which are fed by both parents, leave the nest about 2 weeks after hatching.

Curve-billed Thrasher nesting begins in mid-March to early April. The nest, a loose cup of thorny twigs, is built 3 to 5 feet above the ground in cholla, yucca, or mesquite. It lays 2 to 4 turquoise-colored eggs that are incubated for 12 to 15 days. The altricial young leave the nest at 14 to 18 days. Curve-billed Thrashers may tear apart Cactus Wren nests when good nesting sites are at a premium.

Crissal Thrasher pairs may remain together year-round. The nest, which is a bulky open cup constructed of twigs and lined with softer material, is typically well concealed in mesquites, willows and other dense desert vegetation. Should cowbirds lay eggs in the nest, the adults usually quickly remove them. Pale blue-green eggs (2 to 4) are incubated by both parents. They hatch in about 14 days. The young, which are fed by both parents, leave the nest 11 to 13 days after hatching.

Phainopepla

The four species of silky-flycatchers, found mainly in Central America, are not at all related to the true flycatchers; and the main items in their diet are not flies, but berries. In the southwestern United States, the silky-flycatcher known as the Phainopepla is a specialist on the berries of desert mistletoe. Few other birds in North America have such an intimate relationship with a single plant species.

Mistletoe is a parasitic plant, of course, growing on the branches of trees, and it is "planted" there through the actions of birds. When birds eat its berries, the seeds often pass unharmed through their digestive systems; if the birds' droppings happen to land on a suitable branch, the seeds may stick long enough to germinate. The Phainopepla, by specializing on the berries of desert mistletoe, is unwittingly planting its own future food supply.

In addition to eating mistletoe berries, Phainopeplas eat a variety of other small fruits, and they will fly out to catch insects in mid-air. At times they congregate by the hundreds when food is abundant, such as when elderberries are ripening along rivers at the edge of the desert. When food is scarce, they virtually disappear. Their numbers vary tremendously from season to season. As long as the mistletoe is in fruit, however, there will be at least a few Phainopeplas around. A classic winter sight in the desert is a lone Phainopepla perched atop a mesquite, its spiky crest raised, trim and alert, ready to chase away any other birds that might approach the clumps of mistletoe in the branches below it.

—Kenn Kaufman

Phainopepla *(Phainopepla nitens)*

Order: Passeriformes
Family: Ptilogonatidae
Other common names: silky-flycatcher
Spanish names: tohnehui, capulinero negro, jilguero negro

DISTINGUISHING FEATURES

The male is shiny black with a distinct crest, long tail, and red eyes; white patch on the wing is conspicuous in flight. Female and immature birds are gray.

HABITAT

This species prefers arid scrub habitats.

FEEDING

• *Diet:* Phainopeplas feed on insects and berries, especially mistletoe, on which it feeds heavily when the berries are ripe. This helps to disperse the mistletoe seed to other host trees.

LIFE HISTORY

This bird nests in early spring in mesquite brushlands, usually well up in a stout fork or horizontal branch of a tree. The smooth, slightly glossy eggs usually number 2 to 3 per clutch, and are grayish-white or pinkish, finely and profusely spotted with black, pale lavender, or gray. The eggs are incubated by both sexes (possibly the major portion by the male) for 14 to 15 days. The young are tended by both parents and leave the nest at 18 to 19 days.

The Structure of Feathers

The structure of the typical body feather of a bird consists of a central shaft and a web-like vane. The vane is composed of barbs extending laterally from the shaft. Barbules with hooks and flanges project along the length of each barb. The interlocking hooks and flanges hold the barbs together. If the barbs are pulled apart, they can quickly be re-attached with a smoothing motion. The base of the feather is buried in a follicle beneath the surface of the skin.

Bell's Vireo

A rapid-fire series of bird notes coming from the streamside mesquite thickets announces the presence of the Bell's Vireo. The short jumbled phrases sometimes rise at the end and sometimes drop, as if the bird were asking and then answering a simple question, and the whole song has a clinking quality that sounds as though the bird has a mouthful of marbles.

Tending to remain in the depths of dense thickets, Bell's Vireo is often hard to see. When it does finally allow a view, it may not make much of an impression: pale olive-gray, it lacks distinctive markings. However, it does convey a sense of energy as it hops about busily among the foliage, seeking insects. And like other species of vireos, it is a persistent singer; the males continue to sing through the heat of the day and through late summer and early fall, after most birds have fallen silent.

Vireos build compact, cup-shaped nests, often suspended by the edges in the horizontal forks of branches. Their nests are not hard to find, and they are found all too often by cowbirds, parasites that lay their eggs in the nests and leave them for the vireos to hatch and raise. This brood parasitism seems to have hurt the population levels of Bell's Vireos in some regions, including parts of the midwest and California, but so far their numbers are holding up fairly well in the Sonoran Desert.

—Kenn Kaufman

Bell's Vireo (*Vireo bellii*)

Order: Passeriformes
Family: Vireonidae (Vireos)
Other common names: Arizona vireo
Spanish name: vireo

DISTINGUISHING FEATURES

This is a nondescript little bird; its back is olive-gray and underparts are whitish with pale buff-colored sides; indistinct eye-ring and two faint wing bars.

HABITAT

Bell's Vireo are found in mesquites, desert willows, moist thickets, stream-sides, and forest edges.

FEEDING

• *Diet:* Consists mainly of insects.
• *Behavior:* Typically forages in low brush within about 10 feet of the ground; searches for insects among the foliage although it sometimes takes its prey midair, or hovers and picks it off leaves and branches.

LIFE HISTORY

The nest is situated in the fork of a horizontal branch, usually only 2 to 5 feet off the ground. In typical vireo fashion, the nest hangs from the fork and is cup-shaped with a finely-woven rim. Grasses, weeds, plant fibers, bark, leaves and spider webs are used. The inside is lined with fine grass. White eggs (3 to 5) with brown or black spots are incubated by both parents. The eggs hatch in about 14 days and the young are ready to leave the nest 11 to 12 days later. The young are fed by the parents for an additional 3 weeks.

This species is frequently para-sitized by cowbirds, which has resulted in decreasing numbers over much of its range during the last 40 years.

Hatchlings

The patterns of development of hatchlings can be divided into two general categories, pre-cocial and altricial. At hatching, precocial young are downy, their legs are developed, their eyes are open and alert, and they are able to feed themselves. Precocial young are frequently called chicks. *Altricial young are hatched nearly naked, blind, and weak; all they do is gape for food. Altricial young are called* nestlings *while they are in the nest, and* fledglings *once they fledge or leave the nest. At fledging they are nearly at adult weight. Most perching birds have altricial young. Of course there are always some species that do not exactly follow the typical patterns.*

Wood Warblers

Often referred to as the "butterflies of the bird world," wood warblers are very small, active, and colorful. They flit through the foliage from the treetops to the understory in search of small insects. More than fifty species occur in North America; almost all are migratory, and many spend the summer in northern coniferous forest and the winter in tropical rainforest.

Since these birds are lovers of trees and foliage, it is not surprising that the warbler family is under-represented in the desert. A few species, like the Yellow Warbler and the Yellow-breasted Chat, spend the summer in the dense habitats along southwestern rivers. In winter, Orange-crowned Warblers may forage quietly in the undergrowth of those same riparian thickets. Yellow-rumped Warblers, hardy little birds that arrive to spend the winter throughout the lowlands of the Southwest, are sometimes common along rivers and in dense mesquite bosques in the desert. Several other warbler species may stop through in migration. There are times in late spring and early fall when bright golden-yellow Wilson's Warblers seem to flit along every wash in the desert.

Lucy's Warbler

Of all this diverse family the only species truly adapted to the desert is Lucy's Warbler, a pallid gray-and-white sprite with touches of chestnut in the plumage. Lucy's Warblers arrive from Mexico quite early in spring, and soon the bright, simple songs of the males can be heard everywhere among the mesquite trees.

—Kenn Kaufman

Wood Warblers

Order: Passeriformes
Family: Parulidae (Wood-Warblers)
Spanish names: gusanero, chipe

DISTINGUISHING FEATURES
Orange-crowned Warbler: Olive-green upperparts, orange crown patch (rarely visible), olive-yellow underparts, broken eye ring, no wing bars; lack of conspicuous markings is a good aid in identification. *Yellow-rumped Warbler* male: bright yellow cap, throat and rump, broken eye ring, yellow patch

Yellow-rumped Warbler

at side of breast, white underparts; female: similar but duller, yellow on throat less distinct. *Wilson's Warbler:* Yellow head and underparts, olive upperparts, males have distinct black cap. *Yellow Warbler:* Overall yellow plumage, upperparts greenish-yellow. *Yellow-breasted Chat:* The largest warbler, distinct white eye rings, bright yellow throat and breast, olive-brown upperparts, thick bill. *Lucy's Warbler:* White below, pale gray above, pale face with suggestion of eye-ring; reddish brown spot on crown and patch on rump, more obvious on adult male.

HABITAT
Orange-crowned Warbler prefers riverside or shrubby vegetation, chaparral, gardens, and parks. *Yellow-rumped Warbler* inhabits streamside woodland in our region. *Wilson's Warbler* frequents thickets along streams, scrubby clearings, moist tangles. *Yellow Warbler* is found in streamside habitats, especially cottonwood-willow groves in our region. *Yellow-breasted Chat* frequents streamside thickets, meadows with tall shrubs, woodland edges. *Lucy's Warbler* prefers mesquite bosques and edges of riparian woods in desert zones.

FEEDING
• *Diet:* The Yellow, Wilson's and Lucy's Warblers feed almost entirely on insects; the other species feed primarily on insects but also eat some berries.

• *Behavior:* It is not uncommon for two or more species of warblers to exist in the same habitat and feed on the same kinds of insects. They cope with competitive pressure by foraging in slightly different ways. *Orange-crowned Warbler* forages by flitting from branch to branch or hovering in search of insects on foliage; will also take insects midair. *Yellow-rumped Warbler* forages by hovering and taking insects from foliage, taking insects midair, searching among twigs, and searching through ground litter. *Wilson's Warbler* forages by hopping about on branches, tree trunks, and on the ground; also catches insects midair. *Yellow Warbler* behavior is similar to Orange-crowned Warbler. *Yellow-breasted Chat* forages by moving about in dense vegetation; perches to eat berries; unlike other warblers it holds its food in one foot while eating. *Lucy's Warbler* hops about actively in trees and shrubs, seeking insects on the twigs and foliage.

Yellow-breasted Chat

LIFE HISTORY

These tiny, colorful, active birds are found only in the New World. Many tropical warblers are not migratory. All of ours are, but they occur here at different seasons. Orange-crowned is present mostly September to May, and Yellow-rumped mainly October to April (they winter here). Wilson's, a migrant, occurs in Spring and Fall. Lucy's and Yellow Warblers nest here, and are present mainly March to September;

Yellow-breasted Chat also nests, and is present mainly May to September.

In spite of their name, warblers are not particularly well known for the musicality of their voices. The exception, however, is the Yellow-breasted Chat. Its song consists of an odd assortment of whistles, chucks and hoots.

Most warbler nests are open cups made of leaves, twigs, weeds, and other vegetable matter with a lining of finer material. Wilson's and Orange-crowned Warbler nests are on the ground; the others make their nests in forks or on horizontal branches of trees and shrubs. Lucy's Warbler is one of only two members of the family to nest in holes in trees; it also sometimes builds its nest behind a piece of loose bark. The female is usually responsible for both nest building and incubation. The female lays 3 to 6 eggs that are usually white with brownish spots. Incubation with these species ranges from 10 to 13 days. Both parents feed the young. The young leave the nest 8 to 13 days after hatching.

The Wilson's Warbler, Yellow Warbler, and Yellow-breasted Chat are frequently parasitized by cowbirds. When this happens to the Yellow Warbler it may either build a new nest on top of the cowbird eggs or abandon the nest altogether.

Tanagers

Most tanagers are multi-colored
birds of tropical forest. There are places
in South America, in the foothills of
the Andes, where flocks of small birds
may include a rainbow palette of a
dozen species of tanagers. Only a few
species are found north of the Mexican
border. In the Sonoran Desert they have only
a marginal presence, but their bright colors
make them conspicuous when they do appear.

Summer Tanager

Summer Tanagers in eastern North America inhabit oak woods, but in the
west they are mostly streamside birds. Where desert rivers still have good stands
of cottonwoods, Summer Tanagers are common throughout the warmer months.
Since they keep to the treetops, they are not easy to see, but their crackling call
notes and lazy, burry songs are familiar sounds.

Western Tanagers nest only in coniferous forests of the high mountains and
the north, so they might seem most unexpected in the desert. Every year during
migration, however, many appear throughout the lowlands. There are days in May
and again in early fall when Western Tanagers seem to be scattered all over the
desert, and the striking yellow-and-black males look oddly out of place—which
they are. Before winter weather sets in, they will have retreated to tropical climates
with more typical tanager habitat.

—Kenn Kaufman

Tanagers

REPRESENTATIVE SONORAN DESERT SPECIES:
  Summer Tanager (*Piranga rubra*)
  Western Tanager (*Piranga ludoviciana*)

  Order: Passeriformes
  Family: Thraupidae
  Spanish names: piranga avíspera, cardenal avíspero

DISTINGUISHING FEATURES

Summer Tanager: Full-plumaged males are rose red; females and young are yellowish or olive-green; often mistaken for Cardinals, but lack crests (the only all-red non-crested bird to occur in tall riparian communities). *Western Tanager:* Adult male has a bright red head, yellow body, and black tail, back and wings; wings have white bars; female is dull green above.

HABITAT

Summer Tanagers frequent streamside cottonwood and willow groves in the southwestern U.S. and open woodlands and oak groves elsewhere. The colorful Western Tanager lives in open conifer or mixed forests of the north and high mountains. During migration it may frequent any habitat, including desert and grassland.

FEEDING

• *Diet:* Insects and small fruits.
• *Behavior:* Both species forage in the tops of trees where prey is taken from the foliage; insects also taken midair. *Summer Tanagers* may take prey from leaves while hovering; also known to raid wasp nests for larvae. *Western Tanagers* take frequent trips to flowers, possibly for both nectar and insects.

LIFE HISTORY

Summer Tanager: This species is present in southern Arizona only during the summer, when it nests in willow, cottonwood, or sycamore groves in canyons up to about 5000 feet. It usually lays 4 pale bluish-green spotted eggs in a shallow cup-shaped nest of plant fibers placed on a horizontal limb of a large tree. The female incubates the eggs for 11 to 12 days and is assisted by the male in feeding the nestlings.

Western Tanager: The nest is usually placed in the fork of an outer limb of a coniferous tree. The woven, cup-shaped nest is made of twigs, grass, bark strips, and rootlets. Bluish-green eggs (3 to 5) with brown blotches hatch in about 13 days. The young leave the nest about 14 days after hatching.

Western
Tanager

Feather Molt

Birds periodically molt their old feathers and replace them with new ones. The purposes of feather molt are to replace worn feathers, to change into or out of courtship plumage, or, maybe, to improve hygiene. The timing and frequency of molts are very important. Each species has its own pattern. One common pattern consists of a partial body molt before the breeding season to bring a bird into courtship color, then a complete molt of all the feathers, including flight feathers, after breeding and before migration. Feathers are usually molted in a gradual pattern, so that the bird does not lose its ability to fly or to protect its body from the elements.

Cardinals & Grosbeaks

Pyrrhuloxia

The Northern Cardinal is among the most popular garden birds in eastern North America, chosen as the official state bird of seven eastern states. Travelers are often surprised to discover that, in the very different surroundings of the Sonoran Desert, the same cardinal is abundant—along with several other related birds.

All the members of this group have thick bills, good for crushing hard seeds, which make up a high percentage of their diet at some seasons; most of them switch over to eating mainly insects during the nesting season. In all of our species, the males have bright colors and rich whistled songs. When they are not singing, they often hide their colors amidst dense foliage.

Very closely related to the cardinal is the Pyrrhuloxia. The male Pyrrhuloxia is mostly gray, with its red reduced to accents and highlights, but otherwise it and the cardinal are remarkably similar; some of their whistled songs are essentially identical. The two species live side by side in dense brush along desert washes, but the Pyrrhuloxia also ranges out into more open and arid places.

At the other end of the spectrum from these red birds are the Varied Bunting (mostly dark purple) and the Blue Grosbeak, mainly summer residents in the Sonoran Desert. Another species, the Black-headed Grosbeak, is mostly a denizen of oak woods in summer, but migrants show up all over the desert in spring and fall.

—Kenn Kaufman

Cardinals and Grosbeaks

REPRESENTATIVE SONORAN DESERT SPECIES
 Northern Cardinal *(Cardinalis cardinalis)*
 Pyrrhuloxia *(Cardinalis sinuatus)*
 Blue Grosbeak *(Guiraca caerulea)*
 Black-headed Grosbeak *(Pheucticus melanocephalus)*
 Varied Bunting *(Passerina versicolor)*

Order: Passeriformes
Family: Cardinalidae
Spanish names: cardenal comun, chivo (Northern Cardinal), cardenal
 pardo, cardenal torito (Pyrrhuloxia)

DISTINGUISHING FEATURES
Northern Cardinal: The male is red with a reddish bill and black face. The young males have black bills. The female is light brown or tan and red with a red bill and crest. The female is commonly mistaken for a male Pyrrhuloxia, but the Cardinal female has a red bill as opposed to the ivory bill of the Pyrrhuloxia, and the Pyrrhuloxia is grayer. *Pyrrhuloxia:* About 7½ inches (19 cm) in length, the male is a slender grayish-tan and red bird with a crest and a small stubby, almost parrot-like bill. The rose-colored breast and crest suggest the female Cardinal, but the gray back and ivory-colored bill set it apart. The female has a gray back, buff breast, a touch of red in the wings and crest and an ivory-colored bill. *Blue Grosbeak:* This bird measures 6 to 7½ inches (15 to 19 cm). The adult male is dark blue with 2 rust-colored wing bars. The female is dull brown with buff wing bars. As with other grosbeaks, this species has a heavy, conical bill. *Black-headed Grosbeak:* This grosbeak measures 6½ to 7½ inches (17 to 19 cm). The adult male has a black head, brownish-orange underparts, and black wings with white wing bars. The female has a striped head, streaked back and sides, and sparsely streaked buff-colored breast. The heavy, conical bill is pale in both sexes. *Varied Bunting:* This attractive bunting measures 4½ to 5½ inches (11 to 14 cm). The adult male has a plum or red-purple body, a blue crown, and a bright red patch on the back of the head. In poor light the bird looks black. Females and immatures are unstreaked gray-brown above and buffy below.

Northern Cardinal

HABITAT
The *Northern Cardinal* covers a large range, from southern Ontario to the Gulf states, and from the southwestern

United States to Belize and Guatemala. In Mexico, it is a resident of central and southern Baja California. It is widespread from the northern states to the Isthmus of Tehuantepec and east through the Yucatan peninsula. It prefers wood-land edges, mesquite thickets and stream edges.

Pyrrhuloxias occur from the borders of Arizona, New Mexico, and Texas southward to central Mexico and central and southern Baja California. They frequent mesquite, thorn scrub and deserts.

The *Blue Grosbeak* is found generally in the southern half of the United States south to Central America. It winters from northern Mexico to Panama. This bird inhabits brush, streamside vegetation, roadsides, and overgrown fields. In the southwest U.S., this species is most often found near water in streamside thickets or mesquite bosques.

The *Black-headed Grosbeak* is found from southern British Columbia and Saskatchewan, through the western half of the United States, south to its wintering grounds in western Mexico. This bird inhabits foothills and dense riverside wooded areas. It breeds mainly in deciduous and mixed woods such as cottonwood and willow groves, pine-oak and pinyon-juniper woodlands. In migration its habitat includes open woods, riparian areas, mesquite bosques, desert washes, and even suburban areas.

The *Varied Bunting* is found from the extreme southern parts of Arizona and New Mexico and southwestern Texas south into Mexico. Most of these birds migrate a short distance into Mexico during winter. This bird inhabits dense and thorny habitat, streamside thickets, and open desert if there is dense brush available.

FEEDING
• *Diet:* All species feed on insects, berries, and seeds; *Blue Grosbeak* adds spiders, snails, and plant parts to its diet; *Black-headed Grosbeak* includes spiders and snails. It is able to feed on Monarch butterflies in spite of the noxious chemicals they exude.

• *Behavior:* The *Blue Grosbeak* forages on the ground and from plants; it also flies out to catch insects midair or hovers and removes them from foliage. This species commonly forages in flocks during winter and migration, but not in the breeding season.

Black-headed Grosbeak forages by hovering and taking prey from the foliage of shrubs and trees and by flying out to catch insects midair.

The diet of the *Varied Bunting* is not well known, but probably consists of seeds, insects, and some berries. Insects are probably taken from leaves, and seeds from the ground. This species forages in flocks during winter.

Blue Grosbeak

LIFE HISTORY
Northern Cardinal: Nesting begins from late March to early April. The cup-shaped nest of twigs and plant stems is lined with grasses and hairs. It is located 5 to 10 feet (1.5-3 m) high in dense shrubbery or mesquites. There are usu-

ally 3 to 4 eggs that are incubated for 12 to 13 days. The altricial young leave the nest at about 10 days. Cardinals typically raise 2 to 3 broods per year.

Pyrrhuloxias begin nesting from mid March to early April. The nest is a compact cup of twigs lined with grasses and hair, located 5 to 8 feet (1.5-2.5 m) above ground in a forked branch or against the main trunk of a mesquite or other tree. The 2 to 5 eggs are incubated for 14 days and the altricial young leave the nest at 10 days.

Blue Grosbeak: The open cup-shaped nest is usually rather low in shrubs and trees. It is made of twigs, weeds, leaves, bark, and rootlets with a lining made of finer material such as fine grass and animal hair. Other items, such as pieces of fabric, string, or paper are also often incorporated into the nest. There are 3 to 5 pale blue eggs that take about 12 days to hatch. The nestlings, which are fed entirely by the female, leave the nest about 10 to 12 days after hatching. The male may begin to feed the young after they fledge, thus allowing the female to begin a second brood.

Black-headed Grosbeak: This male's courtship display involves flying above the female with wings and tail spread open and singing. The bulky, open cup-shaped nest is built by the female in trees or large shrubs. It is made of twigs, weeds, pine needles, and rootlets with a lining made of finer material such as fine grass and animal hair. Two to 5 pale greenish blue eggs with reddish brown spots are incubated by both parents. The young take about 12 to 14 days to hatch. The nestlings are able to fly about a month after hatching but remain nearby to be fed by their parents.

Varied Bunting: The nest, which is built by both parents, is a compact open cup made of grass, weeds, and plant stems lined with finer material. It is typically located in the crotches of dense shrubs or trees. Bluish-white eggs (3 to 5) are incubated by the female in 12 to 13 days. The young leave the nest after 12 days. Shortly after fledging occurs the male may take on the responsibility of feeding half the brood thus allowing the female to attempt another nesting. In Arizona the Varied Bunting's nesting is timed to summer rains.

Varied Bunting

Sparrows

Black-throated Sparrow

Most people, at the mention of sparrows, think first of the House Sparrow. That bird is familiar enough—it thrives in urbanized areas, including southwestern cities—but it belongs to a family that is native to the Old World. The sparrows that occur naturally on this continent are a much more diverse group, many with distinctive patterns and musical songs. All of them have short thick bills—designed for cracking open seeds, a major part of their winter diets—but all eat many insects as well, especially in warm weather.

The ultimate dryland sparrow is the Black-throated Sparrow, a smartly patterned bird that thrives year-round in the Sonoran Desert and also in the much sparser plant growth of the Chihuahuan Desert farther east. However, several other species are permanent residents around the edges of this habitat. Song Sparrows (of a distinctive local form, paler and redder than Song Sparrows elsewhere) are common along desert rivers where there is still healthy riparian growth. Rufous-winged Sparrows survive in desert patches that have escaped the effects of overgrazing, while Rufous-crowned Sparrows haunt rocky canyons and foothills. The towhees, big sparrows that forage by scratching actively on the ground, are represented by Abert's Towhees along lowland rivers and Canyon Towhees on drier slopes.

Winter is the season when sparrow diversity peaks, as more species move in from the north. Flocks of White-crowned Sparrows range along the brushy arroyos, flocks of Brewer's Sparrows are common on the more open flats, and several other species appear as well. Because they feed largely on seeds at this season, their numbers vary: if the summer and fall rains have produced a good crop of annual weed and grass seeds, the winter desert may be alive with sparrows.

—Kenn Kaufman

Sparrows

Order: Passeriformes
Family: Emberizidae (Buntings and their relatives)
Spanish names: gorrión, zacatero

DISTINGUISHING FEATURES

Rufous-winged Sparrow: Pale gray below, brown above with small rusty spot on shoulder; gray face with rusty stripes on crown and behind eye, 2 short black whisker stripes below bill.

Rufous-crowned Sparrow: Gray-brown above, gray below, with rusty crown and single heavy black whisker stripe below bill.

Sage Sparrow

Brewer's Sparrow: Very plain, small, and relatively long-tailed; sandy-brown above with narrow black streaks on back and crown, brown ear patch, plain pale gray below.

Black-throated Sparrow: Black throat and mask; white eyebrow and malar (chin) streak; gray crown, back, and wings; white belly; dark bill.

Sage Sparrow: White spot above the bill; white throat with black whisker; gray crown, back and wings; white on outer tail feathers; dark bill.

Song Sparrow: Brown streak extending behind eye, thick white streak below bill, brown wings with some rust, underparts white with heavy dark streaks and central breast spot.

Lincoln's Sparrow: Brown crown with central gray stripe, gray face and eyebrow, buffy breast and flanks with fine streaks, white belly.

White-crowned Sparrow: Black and white head stripes, gray face and underparts, tan back with dark streaks, pink bill.

HABITAT

Rufous-winged Sparrow: Desert grass mixed with brushes, mesquite, cholla. *Rufous-crowned Sparrow:* Grass or bushy vegetation, often near rocky slopes or outcrops. *Brewer's Sparrow:* Open desert, especially creosote flats (winter). *Black-throated Sparrow:* Found mainly in desert thornscrub and dry, open habitats such as creosote and sagebrush flats. *Sage Sparrow:* Deserts and open brushy flats (winter).

Song Sparrow: Inhabits mainly streamside and low, dense brushy areas in the Sonoran Desert region.

Lincoln's Sparrow: Areas with dense vegetation and overgrown fields (winter).

White-crowned Sparrow: Desert washes, ranches, gardens, parks, and along streams.

White-crowned Sparrow

FEEDING

• *Diet:* Sparrows in the Sonoran Desert region feed mostly on insects and seeds.

• *Behavior: Rufous-winged Sparrow* forages mainly by hopping around on the ground, occasionally taking insects midair; feeding takes place in pairs or in family groups.

Rufous-crowned Sparrow forages by walking or hopping slowly on the ground or in low bushes; usually found foraging in pairs or in family groups.

Brewer's Sparrow forages on the ground or in low vegetation usually in flocks.

Black-throated Sparrow forages by moving around on the ground or in desert trees and other vegetation; occasionally

catches insects midair.

Brewer's Sparrow

Sage Sparrow forages mainly on the ground where it is sometimes seen scratching the soil to turn up food items; forages in flocks when not nesting.

Song Sparrow forages mainly on the ground but also feeds in trees and shrubs; will come to feeders if they are in good cover.

Rufous-crowned Sparrow

Lincoln's Sparrow forages by hopping on the ground near or under denser vegetation.

White-crowned Sparrow forages mainly by hopping or running around on the ground, occasionally taking insects midair; forages in flocks when not nesting.

LIFE HISTORY

Because of their generally brownish and often nondescript coloring, identifying sparrows

Song Sparrow

can be a difficult and often unrewarding exercise for the beginning birder. But with a little persistence the birder can learn to recognize these species by their intricate markings and often engaging songs.

All of the sparrows mentioned here live near the ground or very close to it. The majority nest as isolated pairs and

Lincoln's Sparrow

aggressively defend their nesting areas by driving away intruders of their own species.

The females of most of the species mentioned here are responsible for nest building. Cup-shaped nests made of twigs and weeds are located in dense low-growing brush or cacti, or on the ground. A range of from 2 to 5 eggs, which may be plain or spotted, are most often incubated by the female only; usually both parents share in feeding the young. Rufous-winged and Rufous-crowned young leave the nest 8 to 9 days after hatching; the other species leave typically between 9 and 12 days after hatching.

Rufous-winged Sparrow

Towhees

REPRESENTATIVE SONORAN DESERT SPECIES:
    Canyon Towhee (*Pipilo fuscus*)
    Abert's Towhee (*Pipilo aberti*)
    Green-tailed Towhee (*Pipilo chlorurus*)

    Order: Passeriformes
    Family: Emberizidae
    Spanish names: vieja, viejita, ilama, toqui de Abert (Abert's Towhee),
        toqui cola verde (Green-tailed Towhee)

DISTINGUISHING FEATURES
Canyon Towhee: This is a large sparrow-like bird, about 8½ inches (21.6 cm) long. Both sexes are similar in appearance, with rusty undertail coverts and a touch of rust on the crown. It is usually found in pairs in southern Arizona and adjacent Sonora, but is not always easily seen. *Abert's Towhee:* A large (9 inches in length; 23 cm), fluffy, buff-brown bird with black feathers edging the bill. Both sexes look alike. It can be distinguished from the brown towhee by its richer brown color and black on the face. *Green-tailed Towhee:* About 6½ to 7 inches (16.5 to 18 cm) in length,

this bird has a plain olive-green back, rufous crown, a conspicuous white throat, and a gray breast. The rufous crown and the white throat are the best identifying marks. The tail is relatively long and prominent. The bill is black.

HABITAT
Canyon Towhee: A characteristic Sonoran desert bird; a common resident ground-dwelling bird, it prefers scrubby desert areas, foothills and canyons. *Abert's Towhee:* Common in southern Arizona where it is restricted to thickets near water; it frequents thickets or mesquite bosques along streams or

The female incubates the 3 or 4 bluish-green eggs for about 11 days. The young fledge in 8 to 9 days. If the nest is approached, the female, like other towhees, drops straight to the ground and runs away in mouselike fashion, probably as a distraction.

Abert's Towhee lays 3 to 4 eggs in a bulky cup-shaped nest woven of plant fibers and placed low in a tree or shrub. It is a common cowbird host; its future is threatened by cowbird impact and disappearance of desert riparian habitat.

Green-tailed Towhee builds a deep, cupped nest of twigs, grasses, and stems and lines it with small roots and hair. It is built near the ground in dense foliage. Usually 4 spotted eggs are laid.

Canyon Towhee

irrigation canals in low desert areas. *Green tailed Towhee:* Prefers arid, brushy foothills and mountain slopes, the more open pine forests, and chaparral.

FEEDING
• *Diet and Behavior:* All 3 species feed on seeds and insects found on or near the ground.

LIFE HISTORY
The *Canyon Towhee* weaves a cup-shaped nest of plant leaves and fibers, usually 6 to 15 feet high in a shrub or tree (sometimes in mistletoe clumps).

Abert's Towhee

Dark-eyed Junco (*Junco hyemalis*)

Order: Passeriformes
Family: Emberizidae

DISTINGUISHING FEATURES
This bird is 5 to 6¼ inches (13 to 16 cm) in length, with considerable geographic variation in plumage: all have white outer tail feathers, dark hood, and pale belly, but back and sides vary from gray to reddish-brown.

HABITAT
The Dark-eyed Junco inhabits the edges of woodlands or conifer forests throughout its range. In the Sonoran Desert this species occurs only in winter where it inhabits semi-open areas including roadsides, brushy spots, parks, and suburban gardens.

FEEDING
• *Diet:* Mostly seeds and insects but also eats small berries.
• *Behavior:* Forages mainly by running or hopping on the ground. Tends to feed on spilled seed under bird feeders. Also seen moving through the branches of trees or shrubs.

LIFE HISTORY
During nesting season the male sings from a high perch to defend his territory. Both birds may droop their wings and spread their tail feathers during courtship displays. The nest is usually on the ground in a well-hidden and protected spot, although it can also be located in shrubs, trees, or building overhangs. The female makes a cup-shaped nest out of grass, weeds, leaves, and twigs and lines it with finer material such as hair and grass. The 3 to 6 bluish-white eggs have blotches concentrated on the large end. Incubation by the female takes 11 to 13 days. Both parent birds feed the nestlings. The young leave the nest 9 to 13 days after hatching.

This species is migratory, although some birds in the mountains of southwestern United States are permanent residents.

Dark-eyed Junco

Blackbirds & Orioles

Great-tailed Grackle

The blackbird family is hard to characterize because it includes such diverse types: orioles, meadowlarks, grackles, cowbirds, and others. Most have at least some black in the plumage, and their other colors run to warmer tones, such as yellow, brown, and orange. All the species have sharply-pointed bills. Most are more or less omnivorous. None is adapted to extreme desert conditions, but several species make inroads to the Sonoran Desert.

Orioles in general are treetop birds, moving methodically through the foliage in search of insects, often stopping at flowers to add some nectar to their diet. In the lowlands of the Southwest, Hooded Orioles and Bullock's Orioles occur mainly as summer residents in riverside woodlands. The Hooded Oriole has a special liking for palms, however, and it may be common in desert cities where palms have been planted.

Great-tailed Grackles are recent arrivals in this region. Spreading north through Mexico, they did not reach Arizona until 1936. Even today they are closely associated with water, living near riversides, ponds, irrigated farmland, or watered lawns. Sociable birds, they nest in colonies and sleep in large communal roosts, where their cacophonous voices make them a little too conspicuous for some tastes.

Cowbirds' nesting behavior—or lack of it—makes them the most unpopular of the blackbirds and, perhaps, the most interesting. Cowbirds lay their eggs in the nests of other birds, leaving the unwitting hosts to hatch the eggs and feed the young. In many cases, only the young cowbirds survive. Cowbirds are seldom seen in natural desert areas in winter, when they mostly forage in agricultural land; but in the breeding season, Brown-headed Cowbirds and Bronzed Cowbirds infiltrate the desert (and most other habitats), seeking nests to parasitize.

—Kenn Kaufman

Blackbirds

REPRESENTATIVE SONORAN DESERT SPECIES:
   Great-tailed Grackle *(Quiscalus mexicanus)*
   Bronzed Cowbird *(Molothrus aeneus)*
   Brown-headed Cowbird *(Molothrus ater)*

Order: Passeriformes
Family: Icteridae (Blackbirds and Orioles)
Spanish names: zanate, chanate (Great-tailed Grackle); tordo (Cowbird);
tordo ojirojo, tordo mantequero (Bronzed Cowbird)

DISTINGUISHING FEATURES

Great-tailed Grackle: Very large, iridescent black plumage with a long, wide keel-shaped tail; male has yellow eyes; female is brown and much smaller. *Bronzed Cowbird:* Red eyes, conical beak; male: all black with a bronze iridescent sheen; female: gray-brown plumage without the sheen. *Brown-headed Cowbird:* Dark eyes, short, conical beak; male: brown head, black body, wings, and tail; female: grayish-brown plumage that is darker on upperparts, underparts faintly streaked, throat pale.

Bronzed
Cowbird

HABITAT

All are known to occur in open to semi-open habitat including farms, fields, river groves, thickets, and city parks.

FEEDING

• *Diet:* The grackle's varied diet includes plant matter, insects, reptiles, small fish, aquatic invertebrates, and eggs and nestlings of other birds; the two species of cowbirds feed mostly on seeds and insects.

• *Behavior:* These 3 species forage mainly by walking on the ground; the grackle also forages in trees and shrubs, especially when looking for eggs or nestlings. These birds generally feed in flocks.

LIFE HISTORY

The *Great-tailed Grackle* nests in colonies; both males and females may have more than one mate. The nest site varies but is usually in dense vegetation near water. The bulky, open, cup-shaped nest, which may be a few feet to over 20 (7 m) above the ground, is made of twigs, weeds, grass, rushes, and other available material. Bluish eggs (3-5) with brown scrawls hatch in 13 to 14 days. The young, which are fed only by the female, leave the nest about 3 weeks after hatching.

Both cowbird species are brood parasites, laying their eggs in other birds' nests where they are incubated by the host parents. In some areas this parasitic behavior has greatly diminished song bird numbers.

The Bronzed Cowbird may lay 1 pale blue-green egg per day for several weeks; the female may pierce the eggs of its host while depositing her eggs. Nestlings are fed by the host and leave the nest 10 to 12 days after hatching. Common host birds include orioles, towhees, and thrashers. Although now rather common, the Bronzed Cowbird was not recorded in Arizona until 1909.

The Brown-headed Cowbird may lay one whitish spotted egg per day for several weeks until 40 or more eggs have been laid; the female often removes the eggs of its host before depositing her own. Nestlings are fed by the host and leave the nest 10 to 11 days after hatching. Common host birds include finches, warblers, and vireos. The Brown-headed Cowbird is known to have parasitized over 200 species of birds with well over 100 species known to have successfully raised cowbird young.

Feathers

The typical feather with a shaft and vane is called a contour feather. These feathers occur on the body, the wing, and the tail. They form the outer covering of the bird providing protection from sun, wind, rain, and abrasion. The contour feathers on the wing and tail are primarily involved in promoting flight. Down feathers are frequently found underneath the body contour feathers. These feathers have a minute shaft; barbules lack hooks and flanges. The result is a very "downy" feather. Their position under the protective body contour feathers provides an air trap that facilitates heat conservation. There are several other types of feathers that have specialized functions. Semiplumes, filoplumes, bristles, and powder down are examples.

Orioles

REPRESENTATIVE SONORAN DESERT SPECIES:
  Hooded Oriole (*Icterus cucullatus*)
  Bullock's Oriole (*Icterus bullockii*)

  Order: Passeriformes
  Family: Icteridae
  Spanish names: calandria (Oriole); calandria zapotera, naranjero, bolsero (Hooded Oriole)

Hooded Oriole

Bullock's Oriole: Common and widespread in the west, this species inhabits woodland, isolated groves of trees, streamside growth (especially cottonwood), farms, ranches, city parks, and suburbs.

FEEDING
• *Diet:* Both species feed primarily on insects and take lesser quantities of fruit and nectar.
• *Behavior:* Both species forage by searching for insects in trees and shrubs; flowers are visited for nectar; feeders with sugar-water also attract orioles. The Hooded Oriole is a slow and deliberate forager, which makes it a rather easy bird to observe in the field.

LIFE HISTORY
The nest of orioles is often parasitized by cowbirds; the aggressive young cowbird usually receives the most food which eventually starves the oriole nestlings.

The *Hooded Oriole* lays 3 to 5 bluish or grayish-white, spotted eggs in a long, hanging, woven pouch that it enters from the top. The nest usually hangs in a palm, large yucca or a eucalyptus tree. Incubation is by the female and takes 12 to 14 days; both parents feed the nestlings.

The *Bullock's Oriole* lays 3 to 6 bluish-white to pale gray, spotted eggs in a hanging pouch firmly attached to a tree branch. Incubation is by the female and takes 12 to 14 days; both parents feed the nestlings. Young leave the nest about 12 to 14 days after hatching.

DISTINGUISHING FEATURES
Hooded Oriole: Sharply-pointed bill, two white wing bars. Male: black back, wings and tail and a large black bib from face to upper chest; orange head, belly, rump; white wing bars. Female: olive green above, olive-yellow below. The immature male looks like the female, but has a small black bib.
Bullock's Oriole: Sharply-pointed bill. Male: black cap, nape, eyeline, back, wings, and tail; narrow black bib; orange face, underparts and and rump; white patch on wings. Female: olive-gray upperparts, dull yellow throat and chest, whitish-gray belly.

HABITAT
Hooded Oriole: Open woods, tree plantations, palms, city parks, and suburbs; favors groups of palms for nesting, even when these trees are in cities.

Bullock's Oriole

Finches

The bird world includes few vegetarians. Even most birds adapted for eating seeds—such as the sparrows and grosbeaks, with their thick seed-crushing bills—switch over mostly to insects in summer, and feed their young on a high-protein insect diet. However, the true finches stick with plant material all year long, feeding on seeds, buds, and berries, with only the occasional insect. Their young are fed the same fare, although at first they receive their seeds in a softened, partly digested, form.

House Finch

Wild seed crops vary from season to season, so many finches are somewhat nomadic, their flocks moving around in response to changing food supplies. Lesser Goldfinches live year-round in the Sonoran Desert, but sometimes they are locally abundant. Where the flower heads of fiddlenecks (*Amsinckia* spp.) are going to seed, for example, Lesser Goldfinches may descend in flocks, bringing flashes of color and a constant musical twittering. At other times, they may be seen only as isolated couples. A pair may raise young any time from early spring to late fall, perhaps when food supplies seem favorable.

Less nomadic is the House Finch, which has adopted a winning strategy: it is adapted to living around humans. House Finches in our region may have learned to live around Native American villages centuries ago, and they found an equally good niche around farms, suburbs, and big cities. Accidentally introduced into the eastern United States about 1940, House Finches are now found from coast to coast, but in many areas they are strictly urban dwellers. Not so in the Southwest. Here they do well in the cities, but they also continue to thrive in desert canyons, where these attractive and musical little finches had lived all along.

—Kenn Kaufman

REPRESENTATIVE SONORAN DESERT SPECIES:
    House Finch (*Carpodacus mexicanus*)
    Lesser Goldfinch (*Carduelis psaltria*)

    Order: Passeriformes
    Family: Fringillidae (Finches)
    Spanish names: gorrión común, gorrión doméstico (finch)

DISTINGUISHING FEATURES

House Finch male: Brownish with shiny pink or orange-pink crown, rump, and breast; heavy streaking on the sides and back. Female: all brown, heavily streaked underparts. *Lesser Goldfinch* male: Black cap, black or green upperparts, bright yellow underparts, white wing bars on black wings. Female: greenish upperparts, yellow underparts, black wings and tail, white wing bars.

HABITAT

The *House Finch* inhabits ranches, towns, canyons and agricultural areas; found in deserts only where water is available; often seen at bird feeders, nesting on porch lights, or under eaves. The *Lesser Goldfinch* is found in open or semi-open areas where there are trees or brushy vegetation; found near water in more arid regions; also visits suburban gardens.

FEEDING

• *Diet:* Almost entirely seeds, buds, and other vegetable matter; feeds on insects to a lesser extent.
• *Behavior:* Both species forage in flocks, except during nesting season. The *House Finch* forages on the ground and in trees or shrubs; it will come to bird feeders offering seeds or sugar water. The *Lesser Goldfinch* is an active and acrobatic forager in trees, shrubs, and weeds.

Lesser Goldfinch

LIFE HISTORY

House Finch: The female builds a compact cup-shaped nest of grasses, hair, cotton, and other plant fibers; the nest is placed in a cactus, low tree, or shrub, or on a building ledge. The female incubates the 2 to 6 black and lavender dotted pale blue eggs for about 2 weeks. The male feeds the female during courtship and incubation. Both parents feed nestlings; the young fledge in 12 to 15 days. This bird may nest several times during one season.

Lesser Goldfinch: During courtship the male keeps close to the female, often perching near her and singing. The female builds an open cup-shaped nest of grasses, hair, feathers, other plant matter; the nest is typically placed in trees from 15 to 40 feet (5-13 m) above the ground. The female incubates the 3 to 6 pale bluish-white unmarked eggs for about 2 weeks. The male feeds the female during during incubation. Both parents feed nestlings; the young fledge in 11 to 13 days. In the Sonoran Desert region, breeding season is quite long, extending from February or March to September or October.

Mammals

Janet Tyburec (bats)

Pinau Merlin & Peter Siminski

Section Contents

Bats

Janet Tyburec

Bats make up one of the most diverse orders of mammals. Nearly 1000 species account for almost a quarter of the world's mammal fauna— about 1 out of every 4 mammals on our planet is a bat! Global bat diversity is due in large part to the length of time bats have been on Earth. Paleontological study indicates that bats flew through the night skies over 50 million years ago. Fossil bats are very similar to those that exist today, giving few clues about their early ancestors. Fully evolved bats lived at a time when the predecessors of the horse were small, 3-toed, fox-like animals trotting around in the Eocene swamps.

Over the millennia, bats have evolved to live on every continent except Antarctica, and in every habitat except the most extreme polar and desert regions. As a result of this radiation, bats exhibit astounding diversity. Species range in size from giants with 6-foot wingspans to Lilliputians that weigh less than a penny. Some bats are jet black and some are snow white; others have pelts from yellow, orange, and red through blue-gray. Bats employ almost every mammalian feeding strategy. Some are herbivores that visit plants and survive on flower nectar or fruit, and some even feed upon leaves. Insectivorous bats hunt flying insects or pluck stationary insects off the ground or foliage. Some bats are true carnivores that catch lizards, frogs, small rodents, birds, and even other bats; several species have evolved adaptations that allow them to fish for small minnows and aquatic invertebrates. And there are 3 species of bats that are, in a sense, parasitic; they are true vampires, surviving only on the blood of other animals.

The bat order is divided into 18 different families with all members of each family sharing similar evolutionary paths, physical features, adaptations, and sometimes even foraging strategies. Some families are large, containing hundreds of species that can be found worldwide; other families are small, with some made up of a single species that is not closely related to any other species.

Wherever bats occur, they are masters of their domains, able to navigate in the pitch darkness of deep caverns and adapted for survival in a nocturnal aerial niche like no other mammals. Bats are also cloaked in myth and superstition; as a result, misinformation about them abounds. The truth is, bats are not all blood-sucking vampires, dirty disease-carriers, or blind and likely to get tangled in hair. Worldwide, bats are efficient predators of nocturnal insects, including numerous costly agricultural pests, and they pollinate the flowers and disperse the seeds of hundreds of ecologically and economically important plants. Bats continue to astound scientists with their abilities and adaptations. Yet in North America, they rank among the most threatened land mammals, with over half of the 45 species officially listed as threatened or endangered or as candidates for such listing. It is only through information and education that we can reverse their trend toward extinction and begin to appreciate the roles bats play in healthy ecosystems.

Ghost-faced Bats

This small family of bats, with 2 genera and 8 species worldwide, is restricted to warm tropical regions of the New World (North, Central, and South America). These bats typically roost in hot, humid caves, forming colonies in the hundreds to thousands of individuals. Bats in this family have flaps and folds of flesh around their mouths, often set off by large, obvious hairs or whiskers, characteristics that give them their common family name. It is suspected that these oral structures help to focus their echolocation calls or perhaps to funnel insects into their mouths. In addition, some family members have wing membranes covering the dorsal surfaces of their fur and meeting at midline along their backs, perhaps increasing the surface area of the wing membranes and thus helping the bats to dissipate heat. Bats in this family have long, narrow wings and are fast flyers, typically *hawking* (swooping hawk-like to catch) insects near pond surfaces.

REPRESENTATIVE SONORAN DESERT SPECIES:
Peter's ghost-faced bat *(Mormoops megalophylla)*

Order: Chiroptera
Family: Mormoopidae

DESCRIPTION
The genus *Mermoops* has only 2 species, one of which, Peter's ghost-faced bat, lives in the Sonoran Desert. These bats have leaf-like folds and flaps on their faces, forming virtual megaphones, to help amplify and direct their high-frequency echolocation calls. These are medium-sized bats with wingspans of about 14½ inches (37 cm). The tails of ghost-faced bats do not extend past the tail membranes, nor are they completely enclosed by the membranes. Instead, the tails project upwards from the top surface of the membranes near the middle. Ghost-faced bats are a reddish-brown in color with similarly colored wings and membranes, though some individuals do become bleached due to high ammonia concentrations in their roosts.

RANGE
Ghost-faced bats are found from southern Texas and Arizona south throughout Mexico and into Central

A Bat Housing Shortage

Bat houses are becoming a vital conservation tool for many remaining bat populations in the United States. The little brown bat, for example, is one of the most common bats in North America. Once these bats converged in huge maternity colonies in giant old hollow trees and cavities. But, due to extensive logging practices since colonial times, especially on the East coast, few natural roosts for these bats remain. Every significant maternity colony known for this species is found in a man-made structure, such as an attic, church, or barn. As old buildings fall into disrepair or are remodeled for human comfort, bats often find themselves without homes. For these animals, specially designed bat houses are essential resources.

Bat houses should be tightly constructed with no airgaps around the roofs or sides. An open bottom allows droppings to fall free from the structure and eliminates the need for cleaning. To trap hot air and provide attractive temperature profiles throughout the cold season, the house should be at least 2 feet (60 cm) tall and about 14 inches (35 cm) wide. A "landing platform" extending beneath the bat house helps bats to gain entry, especially young bats that are not yet as adept at flying.

To protect residents from predators and to best absorb much needed solar radiation, bat houses should be placed on buildings or poles (not trees) and be free from interfering branches or other obstructions. In northern latitudes or at high elevations, bat houses should be painted dark brown to black to promote heat absorption. Even houses in warm areas should be painted tan or gray. Finally, most bats need fresh water, and those houses erected within ¼ to one mile (.4–1.6 km) of a water source are more likely to be attractive to bats.

When they conform to these simple guidelines, bat houses achieve occupancy rates similar to those for bluebird and purple martin houses. When ideally located with respect to solar exposure and habitat, bat houses experience an 80 percent occupancy rate. And they provide a much needed resource to aid in the conservation of many of North America's remaining bat species.

For more information, including a 34-page booklet complete with plans and mounting instructions, contact Bat Conservation International, PO Box 162603, Austin, TX 78746.

---

and northern South America. They are more common in arid regions, though they have been found in tropical dry forests and more humid areas.

HABITAT
These bats roost in large colonies in caves or mines and sometimes old tunnels and abandoned buildings. They forage on insects while in flight, and generally hunt for insects high above the ground instead of near foliage or water. Their narrow wings allow for strong,

fast flight and help them to travel long distances and pursue flying insects. Ghost-faced bats are not well suited for maneuvering in a cluttered environment.

LIFE HISTORY
Ghost-faced bats live where winters are not very cold. They appear to remain active year-round, neither hibernating nor migrating. Very little is known about their current distribution and habitat requirements, and much more study of their life history is needed.

New World Leaf-nosed Bats
(American leaf-nosed bats)

This is arguably the most diverse family of bats in the world, with 51 genera and 147 species. Though other families are larger, none other exhibits the variation in size, habitat, behavior, or physical traits found in the New World leaf-nosed bats.

As their common name implies, bats in this family are found only in the Americas. Members of this family are generally separated into 6 sub-families, based largely upon their common diets. Some leaf-nosed bats have adapted an insectivorous diet which leans toward carnivory, while larger members of the family take frogs, mice, birds, and other vertebrates. Other leaf-nosed bats are strictly nectar-feeders, and as such are responsible for the pollination of many tropical plants. Still others are strictly fruit-eaters, and thus essential seed dispersers. Some bats survive on either fruit or nectar during different seasons; others are specialists, feeding on just one type of fruit that is available throughout the entire year. True vampires are also grouped into this family. Only 3 species of vampire bats exist—2 specialize on the blood of birds, and 1 preys on both birds and mammals.

The diversity of leaf-nosed bats increases with proximity to the equator. Yet they can be found in all habitats from arid deserts to lush lowland tropics, and from montane deciduous forest to high-altitude cloud forests. They roost in caves and mines, cliff crevices, tree hollows, and even in human dwellings. Several species build their own roosts by biting and folding the large leaves of bananas (*Heliconia* spp.) and some palms. Among the 4 species of leaf-nosed bats that range north of the U.S.-Mexican border is an insect-eater living in the Colorado River valley; 3 are fruit- and nectar-eating bats that follow the bloom of columnar cacti of the Sonoran Desert each spring along their northward migrations. In the fall, they rely upon numerous blooming agave species to fuel their return trips south.

Lesser long-nosed bat

REPRESENTATIVE SONORAN DESERT SPECIES:
    California leaf-nosed bat *(Macrotus californicus)*
    lesser long-nosed bat *(Leptonycteris curasoae)*
    Mexican long-tongued bat *(Choeronycteris mexicana)*

    Order: Chiroptera
    Family: Phyllostomidae
    Spanish name: murciélago

DESCRIPTION

The leaf-nosed bat has a triangular fleshy growth of skin, called a noseleaf, protruding above the nose. All 3 leaf-nosed bats of the Sonoran Desert weigh between 0.4 and 0.7 ounces (12 and 20 gm), have wingspans of more than 12 inches (30 cm) and body lengths more than 2.5 inches (6 cm), and are brown in color. The California leaf-nosed bat has ears longer than 1 inch (2.5 cm). Other leaf-nosed bats have much smaller ears and long, slender snouts. The lesser long-nosed bat does not have a tail. The Mexican long-tongued bat has a tiny tail, less than 1/5 inch (.5 cm) long.

California leaf-nosed bat

RANGE

American leaf-nosed bats are found from the southwestern United States through the New World tropics. The California leaf-nosed bat lives throughout the Sonoran and Mohave deserts. The lesser long-nosed bat is found from southern Arizona and Baja California del Sur south along western Mexico to Central America, and the Mexican long-tongued bat occurs from southeastern Arizona south to southern Mexico.

HABITAT

The California leaf-nosed bat, an insectivorous species, inhabits desert scrub, roosting by day in mine shafts. It does not hibernate nor migrate, so the warm mine shafts are critical for its survival in the northern part of its range.

The other two species are nectar and pollen feeders. They are migratory

and depend on *chiropterophilous* (adapted for pollination by bats—literally, "bat loving") plants for their survival. Lesser long-nosed bats follow the flowering of cardón, organ pipe cactus, and saguaro north through the desert in the late spring. Then they shift to feeding on agaves at higher elevations during the summer, returning to southern Sonora by fall. The lesser long-nosed bat assembles in day roosts of hundreds or thousands in caves and mines.

The Mexican long-tongued bat is less common in the desert. Its primary habitat is at higher elevations where it too migrates, following the flowering of chiropterophilous plants. It roosts in small numbers, frequently at entrances to caves and mines.

LIFE HISTORY

The California leaf-nosed bat feeds on large insects that it gleans from foliage, or even from the ground. It does not alight to capture its prey, but deftly hovers above it and snags it off the substrate. It then carries its prey to an open roost such as a porch or open building to dismember, then consume it. This bat has an unusual reproductive pattern. After breeding takes place in the fall, the resulting embryos develop very slowly until March, when growth continues at a more normal rate before birth in May or June. Twins are common.

The relationship between the two

The California Leaf-nosed Bat: Uniquely Adapted for Arid Lands

Although they live in some of North America's most extreme deserts, California leaf-nosed bats have never been seen drinking and can easily live in a laboratory without water for at least 6 weeks at a time.

This species also ranks among the world's most amazing nocturnal hunters. Helicopter-like flight enables California leaf-nosed bats to pluck insects directly from foliage without even having to slow down. With their huge ears, they can detect and locate sounds as faint as a cricket's footsteps or a caterpillar's munching mandibles.

Their large eyes provide night vision on a par with our best military goggles, enabling them to spot tiny sleeping insects illuminated by nothing more than faint star-light. In fact, they prefer to rely on vision alone except when confronted by total darkness. In such situations, they simply switch to echolocation, with which they can detect objects as small as seven-thousandths of an inch (0.19 mm) in diameter. They also use special whispering signals that can be heard no more than 3 feet (1 m) away, preventing most prey from anticipating their approach until it's too late.

Their short, broad wings and slow, maneuverable flight are not suited to long-distance travel. They cannot migrate long distances to avoid cold winters, nor can they hibernate. These bats are truly desert-adapted.

nectarivorous bats and their chiropterophilous plants is not accidental. The bat receives a food reward in the form of fructose- and glucose-rich nectar and protein-rich pollen. The plant gets a pollinator that transports pollen from it to another of its own species. The bats and plants also have co-evolved features that support this relationship. The plants flower or exude nectar at night when bats are active. They attract bats with flowers that are light colored and easily seen at night and that have a strong musty or sour odor which the bats can smell in the night air. The flowers, frequently bowl-shaped and sturdy, are usually at the top of the plants or far out on limbs, so that bats can easily get to them without becoming tangled in thorny vegetation or having to slow down to navigate among branches. While getting to the large nectar reward in the flowers, a bat gets covered with pollen. The bat consumes any pollen that is not left at some other plant when it grooms itself back at the roost. Each nectarivorous bat has a long, slender snout and a long, extensible tongue with a brush-like tip for lapping nectar.

Vesper Bats (plain-nosed bats, evening bats)

Vesper bats make up the largest family of bats, and the second largest family of mammals. (Only the Muridae family, the Old World rats and mice, is larger.) More than 40 genera are commonly recognized and some 330 species are found worldwide. Bats from this family are largely insectivorous, though some have evolved long feet and sharp claws with which they fish for small minnows and fish from still ponds. Vesper bats have plain faces with simple noses largely devoid of flaps or folds. Their ears are variable in length—some are tiny, hardly protruding from their fur; others are nearly as long as their bodies.

Big brown bat

These bats live in many habitats and use every imaginable kind of roost. Species from this family are most numerous in temperate regions where the bats survive cold, snowy winters when few insects are available by hibernating deep within caves or mines that offer stable temperatures above freezing. Some hibernating species spend up to 8 months of the year sleeping underground.

Vesper bats are also among the most colorful bats, with fur that ranges from solid reds, browns, yellows, and grays to multicolored, spotted and painted. The most colorful bats are those that roost in tree foliage or cliff crevices where their markings help them blend in with bark, leaves, or areas of light and shadow.

REPRESENTATIVE SONORAN DESERT SPECIES:
> California myotis (*Myotis californicus*)
> cave myotis (*Myotis velifer*)
> Yuma myotis (*Myotis yumanensis*)
> fish-eating bat (*Myotis vivesi*)
> Townsend's big-eared bat (*Corynorhinus townsendii*)
> big brown bat (*Eptesicus fuscus*)
> spotted bat (*Euderma maculatum*)
> pallid bat (*Antrozous pallidus*)
> western pipistrelle (*Pipistrellus hesperus*)

> Order: Chiroptera
> Family: Vespertilionidae
> Spanish names: murciélago

DESCRIPTION

The *Myotis, Pipistrellus* and *Eptesicus* have small ears. A *Myotis* bat has a pointed flap of skin projecting vertically in the ear. Bats of the other two genera have rounded flaps. The *Myotis* are very difficult to differentiate. The cave myotis tend to have bare spots between their shoulder blades. The fish-eating bat has very large gaff-like feet. The *Pipistrellus* is the smallest member of the group, at 1/10 oz (3 gm). *Eptesicus*, at ½ oz (16 gm), is larger than all the *Myotis* except for the fish-eating bat. The color of *Pipistrellus* is buff, with contrasting dark black ears and wing and tail membranes. The *Eptesicus* and *Myotis* are generally brown.

Western pipistrelle

The remaining 3 species of evening bats have very large ears, more than 1 inch (2.8 cm) long. The ears of *Euderma* and *Plecotus* are rolled up around the head when the bat is at rest and then inflated with blood when the bat is active. The spotted bat is the most strikingly marked; it is pitch black with three white spots on its back and has pink ears. The pallid bat is a pale cream color or yellow. The Townsend's big-eared bat is brown.

RANGE

Most vesper bats are found throughout the Sonoran Desert, although the Yuma myotis, the cave myotis and the Townsend's big-eared bat are not found in the westernmost, driest parts of the desert. The fish-eating bat is only found along the coast of the Gulf of California. The spotted bat is widely distributed from northern Mexico to southern British Columbia, but is not abundant anywhere in its range.

HABITAT

The California myotis, cave myotis and pipistrelle forage in and above desert scrub for flying insects. The Yuma myotis forages directly over water, often within a few centimeters of the surface. The pipistrelle frequently forages while the sky is still light at dawn and dusk. The California myotis and the pipistrelle roost by day in the fissures of rocky cliffs, while the cave bat (cave myotis) roosts in caves and mines. All sometimes use buildings, particularly at night, though the Yuma myotis seem to prefer buildings during the day, too. The winter roosts of the myotis are cool caves, mines or buildings. Male pipistrelles can frequently be seen foraging on warm winter evenings in the desert, while the females are hibernating at higher, cooler elevations.

The big brown bat is usually associated with man-made structures in the desert, although it has been recorded roosting in woodpecker holes in saguaros. It forages for flying insects in the summer and migrates to higher elevations in the winter to hibernate.

Fish-eating bats catch small fish at the surface of calm coastal waters. They roost by day in the rock rubble and talus slopes on the gulf islands or in remote coastal areas. At night they rest in sea caves while consuming their catches.

Townsend's big-eared bats roost by day hanging from the open ceilings of

Spotted bat

caves or mines. They hibernate in cold (32° to 54°F; 0° to 12°C) caves, lava tubes or mines.

The pallid bat roosts by day in rock fissures in cliff walls or in tight crevices within buildings. It forages low over the ground, frequently alighting to capture prey such as sphinx moths, katydids, June beetles, and even scorpions. Its winter habits are poorly known. It is believed to migrate short distances to suitable sites for hibernation.

The spotted bat is thought to roost by day in rock fissures in high cliff walls, although it is uncommonly seen and its requirements not well known. Cliffs and water courses in arid, rocky, mountainous areas seem to be important features of spotted bat habitats. These bats are late flyers, generally emerging well after dusk. They appear to forage mainly on moths, which they intercept at great heights. They visit waterways to drink after midnight.

Pallid bat

LIFE HISTORY

The evening bats are insectivorous, with the exception of fish-eating bats. All evening bats use *echolocation* to find their prey—even the fish-eating bat. Echolocation involves emitting ultrasonic sounds (generally frequencies of 20,000 to 80,000 Hz) from the voice box and listening for the echoes that bounce off objects in the environment. A bat can determine size, shape, direction of travel and even texture of its prey from this echo information. It can even detect something as narrow as a human hair. The fish-eating bat can locate tiny fish at the surface of the water from the ripples they cause, and gaff them from the water surface with its huge clawed feet.

Generally, once an insect has been detected and found to be suitable it is captured and eaten. However, some moths use thoracic "ears" or vibration-sensitive palps to listen for the ultrasonic calls of bats, and take evasive action by flying about erratically or diving into dense shrubbery.

Moreover, some moths have a "jamming" mechanism that mimics the echoes of a bat at very close range to prey, and causes the confused bat to then break off its attack. To counter this, some bats have evolved big ears and quiet voices to sneak up on these "listening" moths. And some bats, such as spotted bats, emit low frequency sounds, lower than the moth's "ears" can hear, thus avoiding detection. This evolutionary war game is one of the finer examples of predator/prey interactions.

The reproductive pattern of evening bats usually involves mating in the fall or right before hibernation. The sperm is stored in the female's uterus until spring when ovulation and fertilization occur. Births are usually in late June and early July. Baby bats are born rump first (breech birth) while the mother hangs from her thumbs in a head-up position and catches the baby in her tail membrane. Most bats give birth to just a single offspring, but twinning is common in *Antrozous* and *Eptesicus*, especially east of the Mississippi for the latter. The females are usually in maternity colonies of a few females (California myotis) to 15,000 individuals (cave myotis). Maternity roosts are typically warm. The males during this time are usually roosting separately in cool roosts where they conserve energy and water through a daily torpor. The young stay in the roosts until they are about 6 to 8 weeks of age.

Bat Hibernation: The Tortoise & the Hare

Hibernation allows bats to survive long winters when insect prey is not readily available. While a bat hibernates, its body temperature drops to that of its surroundings and its metabolism becomes a mere fraction of its active rate.

Different species of bats use different hibernation strategies. Big brown bats hibernate in very cold areas. By lowering their body temperatures to the near-freezing temperatures of their surroundings, these bats burn stored fat reserves very slowly. But they must wake more frequently and move to more protected areas during intense cold-snaps when the temperature dips below freezing. Every time a bat wakes up from hibernation, it will burn stored fat that took from 30 to 60 days to reserve. If it wakes too often, it may starve before food is once again available in the spring. Because big brown bats are among the most hearty of our temperate bat species, they wait until the very last minute before entering hibernation, and they are among the first to emerge each spring. Big-brown bats may spend only 3 to 4 months hibernating each year.

Pipistrelles hibernate at warmer temperatures, usually deep within a vast cave where temperature fluctuations are unlikely. Because they hibernate at higher temperatures, the rate at which they burn their fat is greater than is that of the big brown bats. But, because they do not have to respond to changes in temperature, they do not have to withstand periods of activity and frequent increases in metabolism. Therefore, they burn their fat at a more steady rate all winter long. Pipistrelles may enter hibernation in September and not emerge until May, remaining asleep for 8 to 9 months each year.

In "racing against" the cold, bats remind one of the fable about the tortoise and the hare. The hare-like big brown bat has frequent spurts of energy followed by long periods of inactivity, while the pipistrelle acts more like the tortoise, keeping a slow steady march through the long winter season.

Free-tailed Bats

Free-tailed bats are found worldwide from warmer temperate regions to the tropics. Bats from this family are split into 13 genera with about 90 species. Every bat in this family has a thick tail that protrudes freely from the tail membrane for at least ⅓ of its length. These bats—high, fast flyers with long, narrow wings—are capable of making annual migrations of 500 miles (800 km) or more to overwintering grounds that are warm enough for them to feed and remain active.

Recent research on the common Brazilian free-tailed bat (*Tadarida brasiliensis*), also called Mexican free-tailed, indicates that it is an important predator of costly crop pests. In Central Texas, free-tailed bats prey on large flocks of migratory moths whose larvae attack cotton and corn. Though these bats do not range much further north than Oklahoma, they prevent hundreds of millions of moths from reaching the grain belts and bread baskets of North America, protecting farmers as far north as the Canadian border. Sadly though, Brazilian free-tailed bats have experienced some of the most precipitous declines documented for any North American bat species. In 10 key caves surveyed in Mexico, half have lost over 90 percent of their bat populations, and 4 of the remaining 5 caves have lost more than 50 percent of their populations. Similar declines have been recorded at Carlsbad Caverns in New Mexico and Eagle Creek Cave in Arizona. Protection for remaining colonies is essential for this highly beneficial species' continued survival.

REPRESENTATIVE SONORAN DESERT SPECIES:
> Mexican free-tailed bat (*Tadarida brasiliensis*)
> big free-tailed bat (*Nyctinomops macrotis*)
> pocketed free-tailed bat (*Nyctinomops femorosaccus*)
> western mastiff bat (*Eumops perotis*)
> Underwood's mastiff bat (*Eumops underwoodi*)

> Order: Chiroptera
> Family: Molossidae

DESCRIPTION

Free-tailed bats' tails extend more than one third beyond the tail membranes; most other bats have tails that are completely enclosed within the tail membranes. All free-tailed bats have very long narrow wings and forward-projecting ears. The western mastiff bat is the largest bat in the Sonoran Desert and the United States. Its wingspan is more than 20 inches (50 cm) and it weighs about 2 ounces (60 gm). The Underwood's mastiff is only slightly smaller. The American free-tail is the smallest free-tail at 0.4 ounces (12 gm) and a wingspan of 12 inches (30 cm). The big free-tail and pocketed free-tail are in between, at 0.8 ounces (25 gm) and 0.5 ounces (16 gm) respectively.

RANGE

These free-tails are found throughout the Sonoran Desert, with one exception. Underwood's mastiff occurs only in arid western Sonora to barely north of the Arizona border.

Mexican free-tailed bat

HABITAT

The free-tails are strong, fast, long-distance flyers. Mexican (Brazilian) free-tails can fly 60 miles (100 km) per hour and forage nightly as far as 50 miles (80 km) from their roosts at altitudes of up to 9800 feet (3000 m).

The *Eumops* sp. and *Nyctinomops* sp. roost in small colonies in rock fissures in high cliff faces. The *Eumops* need at least 20 feet (6 m) of vertical drop from their roosts to gain enough speed for flight. If on the ground, they need to climb up a vertical surface in order to launch into flight. The smaller pocketed free-tail sometimes roosts in caves.

The Mexican (Brazilian) free-tail roosts in large colonies, typically in caves, but sometimes in mines, buildings or bridges. Big colonies may include up to one million bats in the summer. Historically, one particular colony of 20 million Mexican free-tails existed in one cave in Arizona. Most of the Arizona colonies migrate south to Mexico in the winter. They have been recorded as migrating as far as 800 miles (500 km).

LIFE HISTORY

With the exception of the Mexican free-tail, the natural history of the Sonoran Desert free-tails is not well

known. Most seem to migrate to warmer climates in the winter. They are all insectivorous, and although moths are the specialty of the Mexican free-tail, any night-flying insect is game. Free-tails generally forage well off the ground; the western mastiff regularly forages at 100 to 200 feet (30 to 60 m) above ground level.

Western mastiff bat

A single free-tail baby bat is born during the summer. Unlike the young of most bat species, young Mexican free-tailed bats roost separately from their mothers. Babies are located in the highest reaches of the cave, where temperatures are the highest, and often in small ceiling domes where dense clusters help to trap hot air. The incubator-like conditions are essential for rapid growth and survival. In the large maternity colonies of Mexican free-tails, the mother evidently finds her own youngster among the thousands by recognizing its individual location call. The adult males roost separately during this part of the year.

Coyote & Fox

Coyote

The coyote is without a doubt the most famous desert animal, the very symbol of the west. He is prominently figured as the Trickster as well as the Wise One in Native American myths and legends. The coyote fascinates us with its intelligence and adapability. It can survive eating anything from saguaro fruit to roadkills, and is able to live in any habitat from cactus forest to the city. The coyote has expanded its range throughout the United States despite human attempts to eradicate it. The coyote is not only intelligent, curious and playful, it has very keen senses that adapt it for survival—acute hearing, excellent vision, and an extremely sensitive sense of smell.

The gray fox is the quieter relation, going about its nocturnal hunting without attracting as much attention as does the coyote. This fox doesn't dig as much as coyotes do, but it is our only canid that regularly climbs trees. It hunts in and sometimes sleeps in trees, and has even been seen napping in the arms of a saguaro.

The little kit fox inhabits the dry, open flats and is more nocturnal, so it is not often seen by people. The kit fox is also more carnivorous than the coyote or gray fox, depending on kangaroo rats for most of its diet. This fox is a great digger, and any area occupied by kit foxes will be pocked with dozens of den holes.

—Pinau Merlin

Sonoran Desert species:
  coyote *(Canis latrans)*
  gray fox *(Urocyon cinereoargenteus)*
  kit fox *(Vulpes macrotis)*

  Order: Carnivora
  Family: Canidae
  Spanish names: coyote, zorro gris (gray fox), zorrito norteña (kit fox)

Distinguishing Features

Coyote: A coyote resembles a medium-sized dog with a long, bushy black-tipped tail, big ears, and a pointy face. The fur color varies from grayish to light brown, with a buffy or white underbelly. You'll never see a fat coyote in the wild. Wiry and with long, slender legs and small feet, a desert coyote usually weighs only 15 to 25 pounds. The tracks are much smaller than those of a domestic dog of the same size.

Gray fox: The gray fox can be distinguished from the coyote by its smaller size—it weighs in at around 5 to 9 pounds. The gray fox has a grayish coat with rufous-colored hairs on the ears, neck, legs, and underside. A black stripe runs along the top of its long, bushy, black-tipped tail.

Kit fox: This is the smallest of our foxes—really only the size of a house cat. It has a large, bushy black-tipped tail, and is otherwise buff-colored. The kit fox weighs about 4 or 5 pounds (9 kg) and stands about a foot tall at the shoulder. It has large ears and a very narrow, pointy muzzle.

Kit fox

HABITAT

Coyotes are extremely adaptable and are found in all habitats, even in cities. Gray foxes prefer the rockier canyons where coyotes find it more difficult to hunt. Kit foxes inhabit the open, sparsely vegetated flats and areas with deeper soils for digging dens.

FEEDING

Diet: All three of our canines are omnivores, eating anything from road-killed carrion to cactus fruit, mesquite beans, seeds, plants, and meat. They hunt small animals such as rodents, rabbits, birds, snakes, insects—especially grasshoppers and crickets—and any injured animal they can subdue.

Kit foxes prey mainly on kangaroo rats, but also supplement these with mice, rabbits, and a very little vegetable material.

Behavior: These canids scavenge for whatever they can find, sometimes running along the roads in the morning looking for roadkills. They also actively hunt small prey. They use their keen senses of smell and hearing to find occupied burrows they can dig into after mice or ground squirrels. In addition they chase down rabbits and pounce on grasshoppers.

Gray foxes and coyotes compete for similar prey, and so coyotes will often kill gray foxes when they can catch them. When coyote populations diminish due to mange, distemper or other factors, fox populations usually increase as a result.

LIFE HISTORY

Coyotes are mostly social animals, living in small family groups. Within the larger hunting area of a coyote family is a central core area where the den sites are located. This area is scent-marked and defended, particularly during the spring and summer months. Coyotes urinate on bushes or other plants, then scrape the ground with their paws, which have scent glands. This leaves both olfactory and visual markers for other coyotes. Scats are also used to mark the territory.

Turning the Tables

Gray fox

While hiking up an old mining road in the Tucson Mountains early one April morning I caught sight of an animal moving deliberately along the sandy floor of a canyon. It was a good distance away but I could make out a large, bushy tail. It was a gray fox. In another few seconds the fox disappeared behind a cliff wall. It didn't appear to have seen me, so I decided to move up the trail and catch a better view as the fox came around the cliff. I walked on and then stood in one spot for what seemed like twenty minutes, though I suppose it was only about five. I didn't want to move again for fear the fox would reappear and see me. Suddenly, to my surprise, I noticed movement only about 40 feet (12 m) away from me down the hill. It was the fox. While I'd been staring at the canyon floor, it had climbed the hillside below me. As I watched, the fox made its way to a large, flat rock, lay down, looked around contentedly—who wouldn't on such a gorgeous spring morning!—and began a slow, careful process of licking its legs and body. I watched for quite a while before quietly moving away, not wishing to startle the fox. So often, I realized, we are the ones watched in secret by the wildlife around us. To be in this position felt like an honor.

—Steve Phillips

The breeding season is February to March with young born in April and May. Coyotes only use dens for whelping pups. They are very secretive about the den location and if it is disturbed by predators or people the mother coyote immediately moves the puppies to a safer site. Some yearling pups may stay with the parent coyotes through the next winter and help raise the new batch of puppies in the spring.

Coyotes typically sing a "good morning" wake up song around dusk as they prepare for the night's hunt. They also sing to communicate with neighbors, to keep track of family members,

after summer rains, during the full moon, and it seems, just for fun.

Gray foxes are the only canines able to climb trees (although a coyote sometimes manages to get part way up a tree with accommodating branches). They forage in trees, and can sometimes be seen sleeping up in cottonwoods or mesquites (safe from coyotes and other predators). Gray foxes breed in late winter; the pups are born in March or April. They often den in boulder piles, caves, and other natural cavities, or in mine shafts. Both parents feed the pups, but the father fox does not occupy the den with them. Instead he guards the den from a vantage point where he can watch for predators or other danger. The young foxes are able to hunt for themselves at around 4 months of age. Foxes often leave their scats in prominent places, such as on the tops of boulders, as territorial markers.

Kit Foxes can survive without free water, gaining what they need from the blood and moisture in their prey, and so can live in the most arid areas. They are the only canines to use dens year-round. Kit foxes are nocturnal—retreating to the relative coolness and humidity underground during the day probably helps them to survive in these hot, arid environments. Kit foxes change dens frequently, and there will be many mounds and holes in kit fox territory. Some dens may have as many as 7 to 10 entrance holes, making it appear that there are more kit foxes than there really are. These foxes are solitary during July through September, but females move into family dens in October to prepare for the new pups. The males join them later. The pups are born in February and March. Both parents feed them, bringing a steady diet of kangaroo rats for 5 or 6 months—then the young foxes are on their own. Signs of a den with pups include food remains, feathers, fur, and scat.

Kit fox numbers are declining due to loss of habitat and poisoning.

The Mexican Wolf

The Mexican wolf (Canis lupus *spp.* baileyi) was not a desert animal, but did inhabit the mountainous areas, woodlands, and riparian habitats of the region. The last howl of a wild Mexican wolf in Arizona was heard in the early 1970s. Despite a few unconfirmed sightings, the wolf was extirpated in Arizona; the only ones left alive were in zoos. Despite unconfirmed reports, no wild wolves have been proven to be alive in Mexico since the mid-1980s. In 1998 the U.S. Fish and Wildlife Service reintroduced the wolf into Arizona by releasing 13 captive-reared wolves into the Blue Mountains of eastern Arizona. The wolves proved very capable of surviving in the wild, successfully hunting elk, even after living their whole lives in captivity. Unfortunately, five of the wolves were shot, one was hit by a car, and one female disappeared and is presumed dead. As of this writing, others have survived and are doing well and avoiding people. Despite these setbacks, biologists feel confident that the wolf can make a comeback if the human predation stops. The Fish and Wildlife Service has increased monitoring of the wolves, stepped up law enforcement coverage to try to prevent further wolf shootings, and modified its educational programs (for example, teaching hunters how to recognize wolves).

We may again hear the howl of wolves and even find their tracks, but it will still be a rare treat indeed to actually see one of these elusive creatures in the wild.

Procyonids: Raccoons, Ringtails & Coatis

Among the most unusual, most handsomely marked, and least known of our desert animals are the procyonids—the raccoon, ringtail, and coatimundi. The raccoon is the most familiar of these, but mostly because people have seen it in other parts of the country. Many are surprised to learn that the raccoon does quite well in the Sonoran Desert, as long as there is water somewhere nearby. You can easily follow raccoons' distinctive tracks along trails leading directly to suburban desert swimming pools.

The ringtail is Arizona's state mammal, though few people have ever seen one in the wild. Large black eyes, big pink ears set on a tiny face, and a long black-and-white ringed tail that looks like a feather boa make this appealing little animal look almost cuddly, but it is really a very efficient predator. One can sometimes be seen at night as it prowls around its rocky canyon home, peering into niches and cracks where a mouse or an insect might be hiding.

Ringtail

Even though the coati is diurnal and lives in social bands of up to 30 or more animals, most people never see them, unless they make frequent visits to the oak-sycamore canyons and riparian areas coatis favor. Like the raccoon and the ringtail, coatis forage both on the ground and in trees, and are omnivores.

—Pinau Merlin

SONORAN DESERT SPECIES:
    raccoon *(Procyon lotor)*
    ringtail *(Bassariscus astutus)*
    coati, coatimundi *(Nasua nasua)*

Order: Carnivora
Family: Procyonidae
Spanish names: mapache, batepi, lavador (raccoon), cacomixtle (ringtail), chulo, coatí, solitario (coati)

DISTINGUISHING FEATURES

Raccoon: A heavy-bodied animal of about 10 to 30 pounds (4.5-13.5 kg) with a black face mask edged in white and a bushy, ringed tail. The hind foot is *plantigrade* (that is, the sole is walked on, as with humans).

Ringtail: A grayish-brown squirrel-sized animal (about 1 to 2 pounds) with a fluffy black and white ringed tail that is usually longer than its body. It has large black eyes ringed with white, big ears, and a narrow, pointy muzzle. Unlike raccoons and coatis, ringtails don't walk on the soles of their feet, one reason they are sometimes placed in their own family (Bassariscidae).

Coati: The coati is a curious-looking beast, longer than a racoon (though not as husky in the body), with a long nose and a facial mask. Its very long tail is not as distinctly ringed as are those of the raccoon and ringtail.

HABITAT

Raccoons prefer riparian habitats, and brushy and wooded areas. The ringtail lives most often in riparian canyons, especially in areas with rocky outcrops, caves, and mine shafts, and usually not in heavily wooded areas. Coatis inhabit mountain canyons with oak and sycamore in the summer, sometimes moving to lower-elevation riparian canyons or passing through desert areas in the winter.

FEEDING

• *Diet:* All three species are omnivores. The raccoon feeds on small mammals, carrion, fruit and nuts, insects, eggs, fish and aquatic insects. The ringtail favors a diet of rodents, fruit, birds, snakes, lizards, and insects. The coati eats a lot of grubs, beetles, and other invertebrates, and also fruits and nuts, rodents, eggs, snakes, lizards, and carrion.

• *Behavior:* Raccoons use their front paws to feel for food items in murky water or leaf litter.

Coati

Ringtails inspect likely niches and hiding spots in their rocky habitats, hunting for rodents, birds, centipedes, and anything else edible. They are excellent mousers, pouncing and killing with a bite to the back of the rodent's neck.

Coatis dig in the soil and leaf litter using their long claws or their noses to turn up grubs, worms, or other invertebrates. They also turn over large rocks with their front paws to search for invertebrates, lizards, and snakes.

LIFE HISTORY

Raccoons are nocturnal and usually solitary, unless they congregate at man-made food sources such as picnic areas or campgrounds. They prefer brushy, thickly vegetated habitats, but adapt very well to the many artificial ponds, lakes, and wetlands found in the suburbs and housing developments. They can eat almost anything; their dexterous paws can easily open garbage cans, so they can readily take advantage of discarded food.

In the warm desert climate, a raccoon may sleep away the day out in the open, draped over a tree branch.

Ringtails are strictly nocturnal animals, using their large eyes and keen sense of smell to locate prey. They are excellent climbers and leapers, using their long tails for balance as they negotiate steep canyon walls or trees with equal ease. The ringtails have semi-retractable claws and can rotate their hind feet 180 degrees, allowing them to descend cliffs face first. They den in niches in rock walls, boulder piles, or hollow trees. Ringtails are solitary, only pairing up for a few days of mating in April. The 2 to 4 kits are born in June. By fall the young can hunt for themselves and soon disperse. Though fierce little fighters, ringtails fall prey to Great Horned Owls, bobcats, and coyotes. When frightened, they emit a musky odor from anal scent glands.

Coatis are very social animals, living in bands of up to 20 or even 30 or more. The bands consist of females and their young. Adult males are not welcome, except during mating season, although lone males may follow a group at other times. A pregnant female leaves the group to deliver her 4 to 6 babies, rejoining the group several weeks later with her new offspring.

Coatis are diurnal, active mostly in the morning and late afternoon, then spending the night in trees or caves. As coatis forage through an area they travel with their 2 foot (.6 m) long tails held vertically.

Beach Bandits

Raccoon

We associate raccoons with woods, fields, and streams, but they also live in desert locations. On scuba-diving trips to the Sonoran shoreline along the Sea of Cortez, we've encountered raccoons living on and near a beach in the Central Gulf Coast subdivision, considered one of the driest portions of the Sonoran Desert. Only at infrequent intervals is fresh water available to these animals at this location. It appears they survive, at least in part, by the moisture derived from catching and consuming tide pool animals and by eating detritus cast up on the beach.

These raccoons are master scavengers. When humans set up camp on this beach, which is isolated and far from any permanent human habitation, the raccoons are quick to raid. At night they survey the camp from a nearby cliff, eyes shining in light reflected from lanterns or campfire. When the campers retire for the night, the raccoons enter camp and consume any groceries carelessly left about, drink water drops dripping from ice chest drains, and occasionally run across sleeping campers.

—Lane Larson, owner, Caiman Expeditions

Mustelids

The Mustelidae or weasel clan is a family known for its members' fascinating habits; it includes the badgers, skunks, otters, ferrets and weasels. In the Sonoran Desert region only the badger and the skunks are common. Both badgers and skunks are nocturnal and have *plantigrade* hind feet (plantigrades walk flat-footed instead of on their toes like *digitigrades*), giving them a distinctive waddling, shuffling gait. Four-footed plantigrades are not fast runners, but badgers and skunks have other defenses.

Spotted skunk

The badger is renowned for its power, tenacity, and irascible temperament. Few animals care to tangle with an enraged badger. One badger even attacked a tractor that inadvertently ran over its den entrance! This animal is strictly a carnivore, well adapted to digging small mammals from the ground with its 1½ inch long claws on powerful front feet. Its eyes are equipped with *nictitating membranes* (transparent third eyelids) to protect them from flying dirt while the badger is digging. Small, well-furred ears also keep dirt out while the badger tunnels. A badger is able to secrete a musky scent from its anal glands, but it can't spray this fluid like skunks can.

The skunks, on the other hand, are rather phlegmatic animals. They seem to be aware that they have the most daunting weapon for almost any situation, and show great forbearance. Skunks give plenty of warning of their intent to spray, stamping their feet, and in the case of the spotted skunk, even doing a handstand before spraying. Skunks evidently find their own odor offensive; a skunk is reluctant to spray if restricted from getting its tail out of the way. Due to the great diversity of habitats in the Sonoran Desert, all four species of skunks are present here.

—Pinau Merlin

SONORAN DESERT SPECIES:
  badger *(Taxidea taxus)*
  spotted skunk *(Spilogale gracilis)*
  striped skunk *(Mephitis mephitis)*
  hooded skunk *(Mephitis macroura)*
  hog-nosed skunk *(Conepatus mesoleucus)*

Striped skunk

Order: Carnivora
Family: Mustelidae
Spanish names: tejón (badger), zorrillo pinto (spotted skunk), zorrillo listado (striped skunk)

DISTINGUISHING FEATURES

Badger: The badger is a wide-bodied, short-legged creature of about 22 pounds. It has a distinctive white stripe running over its forehead and down its nose. Its coat is shaggy, with a yellowish brown color.

Spotted Skunk: The spotted skunk is easy to identify. Smaller than the other skunks at about 2 pounds, its bold black and white pattern resembles spots instead of stripes. The tail is black at the base and white at the tip.

Striped Skunk: There is variation in patterning among the skunks, but the striped usually has a black back with a white stripe along its sides. The tail is black with a white tip. The striped is a medium- to large-sized skunk at about 6 to 10 pounds (2.7-4.5 kg).

Hooded skunk

Hooded Skunk: The hooded skunk looks much like the striped, but has a ruff of fur around the neck, and a very long, lush white plume for a tail.

Hog-nosed Skunk: This is the least common skunk; it is all white on the top of head, back, and tail. The underparts are black. There is a bare patch of skin on this skunk's long nose.

HABITAT

Badgers inhabit the open, sandy flats with easily diggable soil, such as alluvial fans, creosote flats, farmland, grassland and even golf courses.

Skunks live in a variety of habitats from riparian canyons and wooded areas to Arizona uplands and suburbs. They prefer thick, brushy areas. The spotted skunk is most common in rocky, riparian canyons, while the hog-nose is usually found in the middle to higher elevations. None of the skunks are common in the low, dry flats.

FEEDING

• *Diet:* Badgers are carnivores, eating gophers, ground squirrels, and many other rodents, snakes, lizards, birds, eggs, rabbits, and carrion. All the skunks are omnivores and opportunistic feeders. They eat anything from beetles, grubs, and grasshoppers to rodents, birds, carrion, seeds, and fruit.

• *Behavior:* Badgers dig, dig, and dig some more. They sniff out rodents or other prey and excavate them from their burrows faster than the little creatures can dig to escape. Skunks snuffle around or dig in the ground, turning over rocks, logs, and debris looking for insects, lizards, bird eggs, and so on. They hunt for mice and also search for fruit. The hog-nosed skunk uses its long snout to turn up leaf litter as it searches for worms, grubs, and insects.

LIFE HISTORY

Badger: Badgers are nocturnal carnivores that primarily dig in the ground for burrowing animals. They are armed with formidable claws on the front feet and ferocious temperaments. Few other animals will cross a badger (at least not more than once!). Badgers are solitary, and roam widely within their home ranges. They use different dens each day, choosing whatever happens to be handy after the night's excavating. A badger sometimes makes a circuit, doubling back to check on holes it dug in the past to see if some tasty animal has moved in to take advantage of the leftover burrow.

Badgers mate in July and August, but due to delayed implantation, the badger kits are not born until February or March. Only the mother badger cares for the 2 to 5 babies. They stay with her for about 3 months; then they are on their own.

Badger

Skunks: Skunks are nocturnal creatures famous for their ability to spray a fluid so noxious that it can stop a predator in its tracks. The only major predator skunks have, aside from humans and automobiles, is the Great Horned Owl, which has almost no sense of smell. Skunks usually have enough fluid for 5 or 6 volleys of spray, which they can shoot up to about 12 feet. Their bold black and white patterns, easily seen at night, function as *aposematic* (warning) coloring, advertising the skunks' malodorous capabilities.

Skunks sometimes build their own dens, but often share the dens of other animals, particularly pack rats, or make use of other suitable sites such as brush piles, hollow logs, boulder piles, mine shafts, or underneath buildings. Their dens don't smell like skunk spray, but do have a distinctive, strong, musky odor. Although they don't hibernate, skunks gain extra weight in the fall, which tides them over the lean times during the winter. They retire underground for as many as several days at a time during cold winter storms.

The spotted skunk is the only one able to climb trees, which expands its foraging opportunities. This small skunk breeds in September and October, but delayed implantation results in the young being born in May. The other skunks all breed in the spring, with most babies born in May. The 3 to 7 kits stay with the mother through the summer, accompanying her on nocturnal hunting forays, before dispersing in the fall. Evidence of skunks in an area includes many divots in the earth and other signs of rooting, as well as scat containing berries, insect parts, and bits of fur.

River Otters

Although native river otters (Lutra canadensis sonorae) have been extirpated in Arizona, they may have been common in earlier times, especially in the Salt, Verde, and Gila rivers. Scientists aren't certain exactly what caused their disappearance; there were likely a number of factors that combined to eradicate them, including diverted water and dammed rivers.

The Arizona Game and Fish Department released 46 transplanted Louisiana otters (Lutra canadensis leutra) into the Verde River between 1981 and 1983. The transplant proved very successful, with the population now estimated to be about 100 otters. Unfortunately, with the water tables continuously dropping in the Sonoran Desert, there are not very many other permanent streams suitable for otters.

Cats

Probably no other animal evokes
the majesty and awe of wild
North America as much as does
the mountain lion. It is a mas-
sively powerful animal, yet lithe,
graceful, beautiful, and mysterious.

The lion is shy and elusive, yet we know there
are healthy populations of these desert phan-
toms because we find their tracks, scats
and the remains of kills they've made.
Most people will never see a mountain
lion in the wild, however, because it is a master of
camouflage, slipping behind a bush or rock or scrap of shadow and disappearing.

Mountain lion

Lions are superbly adapted predators, an essential part of the Sonoran Desert
ecosystem. They are designed to efficiently kill animals larger than themselves. Lions
are built with long hind legs and powerful hindquarters, which give them excellent
jumping ability and thrust for great bursts of speed. They have been known to leap
23 feet in a single bound! Huge front paws with sharp claws help to grab the prey,
while powerful jaws deliver the killing bite. Lions help maintain the health of deer
populations by culling out the weak, sick, old, or injured.

In the Sonoran Desert lions spend their days resting in thick, brushy canyons,
rocky outcrops, mine shafts, or any secluded place that provides sufficient cover.

The bobcat is both more common and more widely distributed in the Sonoran
Desert than the mountain lion. It hunts smaller prey and can more easily adapt to
marginal habitats. This small cat is solitary, avoiding other bobcats, and it is also
careful to avoid mountain lions, which will kill it, given the chance. The bobcat is
a good climber, retreating to trees for safety, but it prefers to hunt on the ground.
The bobcat is mostly nocturnal; like the lion, it is secretive and shy, usually keeping
to the more thickly vegetated areas and therefore not often seen by people.

—Pinau Merlin

SONORAN DESERT SPECIES:
    mountain lion (*Felis concolor*)
    bobcat (*Felis rufus*)

    Order: Carnivora
    Family: Felidae
    Spanish names: puma, león, león de la sierra (mountain lion),
        gato montes (bobcat)

Distinguishing Features

The *mountain lion* is the largest cat in the Sonoran Desert. A female weighs about 75 pounds, while the male can be up to 145 pounds (34 to 66 kilograms). Although the kittens have spotted coats for the first three months or so, the adult lion is a plain tawny or grayish-brown with a nearly 3 foot long, heavy, black-tipped tail. The cat itself can be over 6 feet long, with heavy legs and large feet.

The *bobcat* is much smaller, weighing only about 15 to 22 pounds (7 to 10 kilograms), though its long legs make it appear larger. It has broad cheek ruffs on the sides of its face and a very short tail that is black-tipped on the dorsal surface and white underneath. The bobcat usually has indistinct dark spots on its coat.

Bobcat

Habitat

Both cats are most common in rugged, heavily vegetated areas, but can be found in a variety of habitats, such as mountain forests, riparian canyons, and Sonoran Desert uplands. Any area that supports good prey populations will likely also have mountain lions and bobcats.

Feeding

• *Diet:* Both cats are carnivores. The mountain lion feeds primarily on deer, but also takes javelina, bighorn sheep, jack rabbits, porcupines, squirrels, and many other small mammals. The bobcat typically pursues smaller prey, mainly jack rabbits, cottontails, birds, snakes and rodents, but on occasion has been known to take down a deer!

• *Behavior:* The mountain lion may wander widely in search of deer or other prey, but when it finds something suitable, it slowly sneaks up, trying to get as close as possible before the ambush and attack. The mountain lion, though a powerful predator, has a very small heart and lungs for its size, so it lacks the endurance for long chases. A 300 yard (100 m) dash is about the best a lion can accomplish. If the deer is not caught by then, the chase is over. The lion is a very efficient predator. It often slams into the prey animal with such force that the prey is knocked off its feet. A bite to the back of the deer's neck or to its throat ends things quickly. After feeding, the cat scrapes up leaf litter or debris to cover and hide the kill, then returns each day to feed until the prey is consumed. A single adult lion without kittens usually needs to make a kill every 6 to 10 days. In the desert a pack of coyotes is sometimes able to steal the kill from a lion, harassing the cat until it gives up and retreats.

The bobcat also hunts by ambush. Sometimes a bobcat wanders in search of prey, investigating brush piles, fallen trees, or rocky areas; at other times it waits for rabbit or rodent activity, then rushes in with a pounce and a lethal bite to the neck. Because the bobcat feeds on smaller prey, it usually has to hunt every day.

LIFE HISTORY

Mountain lion: Lions breed at about 2 or 3 years of age. There is no set breeding period and kittens are produced at any time of year. A male lion's territory may overlap those of several females. He wanders continually, keeping track of the breeding status of the various females by checking urine scent markers. Normally solitary, big cats avoid each other, except when a breeding pair stays together for a few days. After a gestation period of 3 months, 3 or 4 spotted kittens are born. The kittens remain with the mother for up to a year and a half, learning how to hunt and survive. When the young cats disperse, siblings may remain together for a while to increase their hunting success before going their solitary ways. Young lions wander around searching for suitable home ranges in which to settle. Previous studies showed that these cats each had ranges of 25 square miles (65 km²) or more. Recent research has shown that lions are being crowded into ever smaller areas by human encroachment, and now several lions may occupy a much smaller area together, increasing competition, and forcing some cats into marginal areas closer to people.

Bobcat: Bobcats are also solitary, only coming together to mate in early spring. About two months later the mother bobcat seeks out a cave, a rock shelter, or even a hollow tree stump for a den and gives birth to 2 or 3 kittens. The young cats stay with the mother

---

The Story of Clyde

Many years ago Clyde, a female bobcat, was confiscated from its human owners by the Arizona Game & Fish Department and brought to the Desert Museum. She was the first bobcat to be housed in the newly completed Small Cat Habitat. But not for long. She slipped out one night and thereupon decided to be a wild, rather than a semi-tame, exhibited bobcat. Despite our efforts we could not locate her and worried about her survival. Time would prove that we were worrying needlessly.

A few weeks after Clyde's disappearance her picture appeared in a local newspaper. She had been found in the carport of a home several miles from the Museum. Game & Fish personnel had been called, and had captured and released her in the foothills north of Tucson, where over the next few years she was seen and identified three times.

Once her timing and mine were uncanny. One day driving through the area in which she had been released, I was startled to find her atop a utility pole, having been chased there by dogs. Once the dogs were driven away, she came back down the pole where she muttered and complained in a familiar fashion when I called her name, then turned tail and disappeared into the desert.

About two years later she was sighted in the same general area leading a single kitten; and another year she was spotted with not one, but two young.

— Chuck Hansen, former Desert Museum curator of mammals

until the fall, hunting on her territory until they gain proficiency, then dispersing. The bobcat's home range is only a few square miles, depending on availability of prey. If prey is scarce the cat may wander extensively. Bobcats don't usually leave kills as evidence of their presence in an area, but they do make scrapes and mark scent posts with urine, often using the same area repeatedly.

Rare Cats of the Sonoran Desert

The jaguar, jaguarundi, and ocelot reach the northern limits of their ranges in Arizona and Texas and so were probably never abundant in the United States. All three are presently listed as Endangered. Jaguars (Felis onca) have been photographed and even killed in Arizona within the last several years, but biologists believe there is no longer a breeding population of them left in the state. Jaguars are known to travel long distances, even up to 500 miles (800 km), so transient individuals may occasionally wander up from Mexico. Although water is probably not a limiting factor in the U.S. part of their range, insufficient habitat may be. Biologists studying this secretive big cat hope to learn more about its habitat preferences, as well as its habits and needs, when they can capture, radio-collar, and monitor a few animals.

Ocelots (Felis pardalis) have also been documented in Arizona, though again, loss of habitat may limit them. Ocelots prefer dense thornscrub, live oak scrub, or riparian areas with an overstory cover. In Texas, a population of 50 to 100 ocelots has been holding steady for several years, but a population that small may not have enough genetic diversity to survive indefinitely.

Jaguarundis (Felis jaguarundi) are the Sonoran Desert's mystery cats. There are a number of jaguarundi sightings in Arizona each year, but a hide or skull has never been found, nor a photograph taken of one in the wild. Even so, some biologists believe they are present in the U.S. and are residents, since these small cats can't travel long distances. Biologists studying these elusive cats in Mexico in areas where both ocelots and jaguarundis are common found it much easier to observe and catch ocelots.

Arizona Game & Fish Department biologists investigate and keep track of sightings of these rare cats, but caution that frequently juvenile mountain lions (with spots) and bobcats are confused with jaguars or ocelots, especially in poor light. Still, knowing that there could be a jaguar behind the next shrub or boulder makes the wilderness a much more exciting place!

Hooved Animals

The Sonoran Desert is home to 4 species of hooved animals—javelina, mule deer, pronghorn, and bighorn sheep. Hooves are specialized claws or toenails, adaptations for running to escape predators. Hooved animals, also called *ungulates*, are usually herbivores that socialize in herds or bands. Living in a group benefits all the individuals, since there are many sets of eyes watching for predators.

Bighorn sheep

Javelina are common throughout the Sonoran Desert region. Any area with sufficient prickly pear and other cacti can be home to a herd of javelina. They usually provide plenty of evidence of their presence, such as chewed prickly pear pads, rooted-up areas around plants, and many trails and bed grounds.

Our other three hooved desert animals are *ruminants* (even-toed, cud-chewing animals). Mule deer inhabit the lower foothills and brushy canyons in the desert, while white-tailed deer live at the higher elevations in the mountains. Mule deer depend mainly on good hearing and eyesight, camouflage, and running for defense. Their large (9 inch; 23 cm) ears move continually, checking antennae-like for any sounds of danger. The main predator of deer is the mountain lion.

Pronghorn populations are not large anywhere in Arizona. The Sonoran pronghorn is listed as endangered, but herds in a few places—the Buenos Aires Wildlife Refuge is one—are doing well.

Bighorn sheep numbers have also declined due to human encroachment and competition with domestic sheep and wild burros, but reintroduced populations are increasing in numbers. The construction of artificial waterholes in some areas has helped maintain bighorn herds.

—Pinau Merlin

SONORAN DESERT SPECIES:
   javelina, collared peccary *(Tayassu tajacu)*
   mule deer *(Odocoileus hemionus)*
   pronghorn antelope *(Antilocapra americana)*
   bighorn sheep *(Ovis canadensis)*

Order: Artiodactyla
Family: Tayassuidae (javelina), Cervidae (deer), Antilocarpidae (pronghorn), Bovidae (bighorn, and, formerly, pronghorn)
Spanish names: jabalí, cochi javelín (javelina), venado buro (mule deer), berrendo (pronghorn), borrego cimarrón, chivo cimarrón (bighorn)

DISTINGUISHING FEATURES

Javelina: The javelina is a medium-dog-sized animal of about 40 to 55 pounds (18-25 kg), that somewhat resembles a small wild boar. It has a coarse salt and pepper pelage, short legs, and a distinctly pig-like snout. The head appears to join the shoulders with very little neck. A collar of lighter-colored hair rings the neck and shoulders. Both males and females have

Pronghorn antelope

long, sharp canines, the "javelins" for which they are named.

Mule deer: This is a brownish-grey deer with big ears like a mule's, and a white rump patch; its small white tail has a black tip. Only the buck grows antlers, which are shed each spring and regrown during the summer and fall. The main beam of the antlers diverges into 2 branches, each bearing 2 or more tines.

Pronghorn antelope: The pronghorn is a medium-deer-sized animal with long, thin legs, a large white rump patch, and white on the sides of its face. Both males and females have forked or pronged horns that are shed each year, except for a small inner core.

Bighorn sheep: The desert bighorn is a heavy-bodied, gray-brown, deer-sized animal with a large white rump patch. Both males and females have horns, but the males' are much larger, growing into a curled spiral shape over the course of several years.

HABITAT

Javelina are common throughout the Arizona Upland. They favor saguaro-palo verde forests, and grasslands with mixed shrubs and cacti, but make use of other habitats as well.

Mule deer are widespread throughout the desert areas wherever there is sufficient vegetation for browse and cover. They make elevational migrations in the foresummer and winter.

Pronghorn are grassland animals, preferring open areas with rolling hills and a few shrubs.

Bighorn sheep are found in rugged, rocky desert mountain ranges and some canyons. They prefer precipitous slopes and cliffs where they can be safe from predators.

FEEDING

• *Diet:* These hooved animals are all herbivores. Javelina eat roots, tubers, seeds, mesquite beans, green vegetation, cactus fruits, agaves, and prickly pear pads. When the opportunity presents itself, they will also eat dead birds or rodents. Mule deer browse on mesquite leaves and beans, catclaw, jojoba, buckbrush, and fairy duster, and graze on a variety of forbes and grasses. Pronghorn feed on grasses and forbes but also browse on globemallow, brickellia, sagebrush,

No Way Out

I was walking up a deep, dry arroyo bottom north of Phoenix a few years ago. As I rounded a corner, I came face to face with a doe and two fawns, having unintentionally "chased" them up the narrow canyon. Behind the deer the arroyo ended abruptly in a vertical rock wall at least fifteen feet high. The doe trotted toward me a few steps, judged that the cut was too narrow to safely pass by me, then turned and bolted back toward the rock wall. I stepped to one side of the arroyo, hoping she would pass by, but instead she ran towards the wall and easily ascended it. I watched her disappear in amazement, then moved further aside to encourage the fawns to pass back down the arroyo rather than attempt such a foolhardy exit. But they had other plans. In turn, they bolted for the wall and deftly surmounted it. I walked over to the wall, looked up, and judged that no matter how much I might want to follow those deer, there was no way I could do so without running the risk of serious bodily harm. I was forced to retrace my steps and find an easier way to move upstream. As I walked down the arroyo I realized that this was truly mule deer habitat; I was but a visitor.

—Bob Scarborough

Mule deer

rabbitbrush, and other shrubs. Bighorn sheep eat many different grasses as well as mesquite leaves and beans, desert lavender, fairy duster, desert ironwood, palo verde, globemallow, cactus fruits and agave.

• Behavior: Javelina meander in loose groups, feeding as they move through an area. They dig up roots and bulbs with their sharp hooves or with their snouts. They eat prickly pear cacti, spines and all, by tearing off bites with their large canines. Because they don't have sharp cutting teeth, much fibrous material is left on the prickly pear. Javelina chew as they walk, so bits and pieces fall from their mouths,

sometimes leaving a short trail.

Mule deer usually eat early in the mornings and in the evenings, and are sometimes active on moonlit nights. They both browse and graze.

Bighorn sheep feed early in the mornings and again late in the afternoons. A sheep sometimes kicks the top off a barrel cactus to reach the succulent inside, or butts its horns against a saguaro to get at the tender flesh.

LIFE HISTORY

Javelina live in groups of 2 to 20 animals, the average being about 8 to 12. Each group defends a territory of about 700 to 800 acres, the size and boundaries varying in different seasons and different years; the territories include bed grounds and feeding areas, but they may overlap at critical resources, especially watering holes. An older, experienced sow leads the herd, determining when to bed down, feed, or go to water. Javelina have no defined breeding season; the babies, usually twins, can be born in any month. Not many predators other than a mountain lion will attack an adult javelina, but the babies are also prey for coyotes, bobcats, and other animals.

Javelina have poor vision, relying instead on their sense of smell. (See sidebar, next page.)

The *mule deer* breeding season occurs in December and January, but the one or two fawns are not born until the next summer rainy season, about seven months later. The does often move higher on the mountains into more rugged terrain to drop their fawns, so as to avoid coyote predation. The little fawns are up and able to run within a few days after birth. They usually remain with the mother until the next year. Most deer spend their lives within a 1 or 2 square mile home range.

Pronghorn antelope are well adapted to living in open areas of little cover. Instead of hiding from predators, they rely on their large eyes and exceptionally good vision to spot predators from up to 4 miles away! They are the fastest runners in North America, attaining speeds of 35 to 40 miles per hour. Pronghorn run with their mouths open in order to gulp extra oxygen to supply their large hearts.

Bighorn sheep live in and around the most inaccessible steep canyon walls and rugged terrain. Their defense is to retreat to these hard-to-reach spots where predators cannot follow.

Bighorn feed with their herds early in the morning, bedding down to rest near each other in shallow caves or thick brush, while they chew their cuds. Activity resumes in the late afternoon. Breeding occurs during August and September. Rams butt heads to establish dominance and the right to breed, and the cracking sound of their butting heads can be heard a mile away.

Bighorn: Adapted for Desert Living

Bighorn sheep inhabit some of the driest mountain ranges in Arizona—they are superbly adapted to life in these arid environments. During winter months when dew is available and plants contain more moisture, the sheep can last for several months without drinking free water. In the intense heat of summer when most green plants have dried up, they seek out barrel cactus, chain fruit cholla fruits, and other cactus fruits for their water content. They avoid the sun by resting in the shade or in caves and shallow overhangs, but they can withstand body temperatures of up to 107°F (42°C)! (Normal body temperature for bighorn sheep is 101 to 102°F; 38 to 39°C.) Enlarged stomach compartments can store water to last for several days, allowing the sheep to go 2 or 3 days without a drink. They can then drink up to 20 percent of their body weight (up to 2 gallons; 7.5 liters) in just a few minutes at a waterhole.

Peccary Perfume

Usually hidden beneath the bristly hairs near the rear of a javelina's back is a small organ producing an oily liquid with a strong and unmistakable odor. This musky perfume helps maintain order in the busy javelina social life.

Javelina usually stay with the same herds throughout their lives, traveling, feeding, bedding down, even playing together. From time to time pairs of animals stand nose-to-tail and rub their heads across each other's scent glands. As a result of this frequent contact, every javelina herd develops its own distinctive group scent.

Javelina sniff each other to recognize their herd-mates. As the herd moves through the desert it's enveloped by its communal smell, helping members to stay in contact with the group. (They also listen for the sounds of the other animals' grunting, chewing, and moving about.)

Odor also helps keep different herds apart. Herds mark the boundaries of their territories by rubbing scent glands across rocks and trees.

— David W. Lazaroff,
Arizona-Sonora Desert Museum Book of Answers (ASDM Press, 1998)

Javelina, or collared peccary

The ewes move off alone to drop their lambs in February, rejoining the group about a week later with the new lamb. Lambs imprint on the spot where they were born, and try to return there when they are ready to give birth.

Both males and females have horns.

The ewe's horns are narrow and only grow about 12 inches (30 cm) long (about a half curl). The male's horns are broad and massive and eventually curl in nearly a full spiral. The ram's horns may weigh as much as 40 pounds (18 kg)!

Shrews

Shrews, which are insectivores rather than rodents, are intriguing little animals. They are voracious nocturnal predators with a high metabolic rates—in fact so high that they must hunt and eat frequently or face the real possibility of quickly dying of starvation. When sleeping, however, shrews sometimes conserve energy by falling into a torpor, which lowers their body temperature.

Scientists have discovered that some shrew species use echolocation to find their prey, giving the animals an added advantage on their hunting forays.

Though shrews are seldom seen, they are actually fairly common in many habitats, including arid regions of the Southwest.

—Pinau Merlin

SONORAN DESERT SPECIES:
    desert shrew *(Notiosorex crawfordi)*
    Order: Insectivora
    Family: Soricidae

DISTINGUISHING FEATURES
The desert shrew vies with the pipistrelle bat as the smallest desert mammal. An adult desert shrew weighs from 3 grams (the weight of a penny) to 5 grams (the weight of a nickel). It has short legs, a pointed snout, tiny eyes, pale grey fur, and a short tail. The body is a little over 2 inches (5 cm) long.

RANGE
The desert shrew is found throughout the southwestern United States and northern Mexico.

HABITAT
Desert shrews do not appear to be restricted to any particular habitat, so long as there is sufficient cover. They are often found in packrat houses, or under dead agaves, old logs, or other debris.

FEEDING
• *Diet:* The desert shrew eats a variety of arthropods as well as lizards and even small mice. It eats scorpions and seems to be immune to their sting.
• *Behavior:* Shrews store food items in their nests. To keep food fresh, shrews crush their victims' heads and bite off their legs, but do not kill them. Shrews eat up to 75 percent of their body weight in food every day.

LIFE HISTORY
The desert shrew prowls and searches for its prey at night, so it often falls victim to nocturnal hunters such as owls and snakes. The shrew has scent glands on its sides, however, that emit a strong musky odor; this scent appears to make it less palatable to mammalian predators. Even so, most shrews probably only survive a year or so in the wild.

The desert shrew is well adapted to desert life, surviving without free water and obtaining what moisture it needs from its prey. It also conserves water by concentrating its urine and remaining in its nest during the heat of the day.

Rabbits and Hares

Of all the desert-dwelling mammals, the desert cottontail is probably the one you will see most frequently. Preyed upon by everything from snakes to coyotes to owls, most cottontails are killed within their first year. These rabbits have few defenses other than good eyesight, good hearing, and the ability to flee quickly. They compensate for heavy losses by reproducing at a prodigious rate. Female cottontails can breed at 3 months of age and have multiple litters in a year. Young stay at the nest for only about 2 weeks before venturing off. These reinforcements make for a fairly constant supply of cottontails.

Desert cottontail

Jack "rabbits" —they are actually hares— live in open areas with little cover; they rely on exceptional speed and great leaping ability to evade predators, but they also suffer predative losses.

In the field, size is an easy way to tell cottontails apart from jackrabbits. Cottontails are small, 1 to 2 pound (.45 to .9 kg) animals, while jacks are quite large, weighing up to 10 pounds (4.5 kg) and standing just under 2 feet (.6m) tall. Cottontail babies (true rabbits) are born blind, naked, and helpless; but jackrabbit young (like all young of true hares) are born furred and with their eyes open; they can move around just a few hours after birth.

—Pinau Merlin

SONORAN DESERT SPECIES:
    desert cottontail (*Sylvilagus audubonii*)
    antelope jackrabbit (*Lepus alleni*)
    black-tailed jackrabbit (*Lepus californicus*)

Order: Lagomorpha
Family: Leporidae
Spanish names: conejo (rabbit), conejito, conejo (cottontail),
    liebre (jackrabbit)

DISTINGUISHING FEATURES

Although the desert cottontail resembles most other cottontails, its ears are much larger. This grey rabbit with rufous nape and white tail weighs around 2 pounds (900 grams). It is the only cottontail in the Sonoran Desert.

The antelope jackrabbit is one of the largest hares in North America, weighing 9 to 10 pounds (4.5 kg). This jackrabbit's huge ears are edged in white. The large eyes are placed high and towards the back of its slightly flattened head, allowing it to see nearly 360 degrees as it watches for predators. The antelope jackrabbit is so named because it has a patch of white fur on its flanks that it can flash on one side or the other as it zigs and zags, running from a predator, much as the pronghorn antelope does.

The brownish black-tailed jackrabbit is smaller than the antelope jack, at about 8 pounds (3.6 kg). Its ears and the top of the tail are tipped in black.

HABITAT

The desert cottontail is found throughout the Sonoran Desert, especially in thick, brushy habitat with plenty of hiding places. The antelope jackrabbit inhabits the drier areas of the desert, including creosote bush flats, mesquite grassland, and cactus plains into and beyond southern Sonora. It prefers open places with sparse grasses where it can see predators and flee if need be. The black-tailed jack is also found in open, flat places, though not in habitats as dry as the antelope jack can tolerate. Its range does not extend into southernmost Sonora.

FEEDING

• *Diet:* The rabbits and hares are herbivores, feeding on grasses, forbs, mesquite leaves and beans, and cacti (for moisture). Twigs nipped off by jacks have clean, slanted cuts, while ends bitten by cottontails have a rougher, nibbled appearance. (Twigs browsed by deer look pinched off.)

• *Behavior:* Cottontails stay within about a 400 yard (366 m) home range, foraging on almost anything green. Resins or chemicals in some plants deter browsing by most animals, but in drought years even these will not stop rabbits.

Black-tailed jackrabbit

Moisture from cacti and other plants fills most of their water needs, but they readily drink water if it is available.

Black-tailed jackrabbits may travel up to several miles a night to find suitable food, returning to their home ranges each day. The large antelope jacks need more food than the black-taileds and can consume ¼ pound (118 g) of food per day.

LIFE HISTORY

Cottontails are primarily crepuscular and nocturnal, but during milder temperatures may be active during the day as well. They usually rest during the heat of the day in "forms"—shallow depressions under grasses or brush, or in brush piles. They also use burrows of ground squirrels, skunks, packrats or badgers, when available. Cottontails breed throughout the spring and summer months. They usually have at least 2 litters per year. About a month after parents mate, 2 to 4 baby rabbits are born. The mother rabbit excavates a nest a few inches deep and lines it with grasses and fur, covering the babies with another layer. Mothers mostly stay away from the nest so as not to alert predators, returning only a few times each night to nurse the babies. In 2 weeks the little rabbits leave the nest. At this age they are very timid and cautious, never venturing away from the safety of a bush or shrub. Rabbits live about 2 years, if they're lucky.

Antelope jackrabbits are also nocturnal, resting in the shade of a cactus or other plant during the day. They don't escape the heat by digging or borrowing burrows, and seldom get to drink water, so they must conserve energy and moisture by restricting their activities to the cooler hours of the day.

When a jack suspects a predator is nearby it often stands on hind legs to get a better view, then crouches down and freezes. If the predator gets too close the jack springs up, leaping away in 15 foot (4.6 m) bounds. After a few leaps the jack jumps even higher as it sails over bushes and other obstacles and tries to spot the predator. These hares can run up to 35 miles per hour (56 km/h) for up to ½ mile (.8 km)!

The habits of *black-tailed jackrabbits* are similar to those of the antelope jack, although black-taileds seldom freeze to escape predation, instead usually zigzagging rapidly away.

Jackrabbits breed throughout the year. Courtship is dramatic with chases, charges, leaps over each other, and sprayed urine. Six weeks later, 1 or 2 baby jacks weighing about ½ pound (230 g) each are born already furred and with their eyes open. The youngsters are *precocial*—able to move about shortly after birth—but may stay with the mother for several months. Jacks are very social. Large groups of 25 or more congregate, especially on moonlit nights.

Ground Squirrels

The ground squirrels belong to the rodent order—small, gnawing, mammals that many predators depend on for food.

Since we lack large trees in most of the Sonoran Desert, it's not too surprising that our squirrels are ground dwellers. (Actually the rock squirrels and round-taileds climb quite well, and often forage in mesquites when the new leaves are budding out in the spring.) They all dig burrows to live in and retreat to for safety, but spend days on the surface when the temperature is moderate, foraging and sunning. These little creatures are well designed for digging in the dirt. All three squirrels discussed below have good claws and small ears set lower on their heads than those of tree squirrels. The round-tailed squirrel also has sleek, short fur.

All three of our squirels are diurnal; only Harris' antelope squirrel is active all year. The round-tailed ground squirrel hibernates in winter in most of its range and estivates during the summer drought. The rock squirrel retreats to its burrow during cold winter periods, though scientists doubt that it actually hibernates.

—Pinau Merlin

SONORAN DESERT SPECIES:
Harris' antelope squirrel (*Ammospermophilus harrisii*)
round-tailed ground squirrel (*Spermophilus tereticaudus*)
rock squirrel (*Spermophilus variegatus*)

Order: Rodentia
Family: Sciuridae
Spanish names: ardilla (squirrel), chichimoco (Harris' antelope squirrel),
   juancito (round-tailed ground squirrel)

DISTINGUISHING FEATURES

The little *Harris' antelope squirrel* is often mistaken for a chipmunk; chipmunks, however, are higher elevation animals, while Harris' antelope squirrel is a creature of the rocky deserts. The Harris' antelope squirrel has a white stripe on its side, but not on its face, and a bushy black tail that it often carries arched over the back. The underbelly is white.

The *round-tailed ground squirrel* is a social animal. Although it resembles a tiny prairie dog, and shares some of its habits, the two animals are not related. The round-tailed ground squirrel is light beige colored with a long, black tipped tail. It weighs only 6 or 7 ounces (170-200 g).

The *rock squirrel* looks like a typical tree squirrel, but is a ground dweller. It is the largest of the ground squirrels, weighing up to 1½ pounds (.7 kg). The rock squirrel has speckled greyish-brown fur and a long bushy tail.

HABITAT

Harris' antelope squirrels prefer the rockier habitats of the desert, though their ranges often overlap those of the round-tailed ground squirrels. Round-tailed ground squirrels live mostly on lower alluvial fans or open, flat areas of valleys. They need deeper soils where they can dig their 3 foot deep burrows. Rock Squrrels are found in many habitats throughout the region, even at high elevations in the mountains, but they are absent from the driest areas of southwestern Arizona. They are most commonly in rocky outcrops, boulder piles, or canyon walls, but they are very adaptable and make use of suburban lots, tree roots, and many other places.

Rock squirrel

FEEDING

• *Diet:* The Harris' antelope squirrel feeds less on green vegetation and more on fruits of cholla, prickly pear and barrel cacti, seeds, mesquite beans, insects, and occasionally, mice. The round-tailed ground squirrel depends on succulent green vegetation, such as new spring wildflowers, cactus flowers and fruit, mesquite leaves, grasses, and ocotillo flowers, but it eats seeds as well. It also will take advantage of carrion, including roadkill of its own species. The rock squirrel is an omnivore, feeding on seeds, mesquite beans and buds, insects, eggs, birds, carrion, and a variety of fruits, including the fruit of barrel cactus and prickly pear.

• *Behavior:* Harris' antelope squirrels run around the desert sniffing out seeds in the ground and digging them up. Many shallow divots in the dirt are an indicator of their activity. These ground squirrels also climb up barrel cacti to get the fruit, despite the spines.

Round-taileds usually don't have to venture far from their burrows, finding enough grass seed, cacti, and vegetation nearby to satisfy their needs. They intersperse bouts of feeding with periods of sunning or relaxing in the shade of bushes.

Rock squirrels forage for food on the ground of their home areas, but can also climb trees very well. They often climb into mesquites, willows, and ocotillos when they are first leafing out, to feed on the tender new growth. They also climb flower stalks of agaves to feed on the tender tips. In addition they hunt and kill small birds and rodents.

LIFE HISTORY

The *Harris' antelope squirrel* is active year-round. Evidence of Harris' antelope squirrels in an area usually includes several 2 inch (5 cm) diameter holes under a bush or cactus or among rocks, and food remains, such as bits of cactus fruits nearby. This squirrel can be active even during the midday heat of summer. It holds its tail arched over its back; this shades the animal, keeping it cooler. During hot weather the squirrel seeks a cooler, shaded spot and lies down, spreading all legs out (it is frequently observed doing this on shaded tile patios in desert suburbs) presumably to dump heat from its body.

Round-tailed ground squirrels are social, living in small colonies. They hibernate through the winter months, emerging in early February to take advantage of the new spring growth to regain the weight they lost over the winter. Round-taileds breed shortly after coming out of hibernation; the average of 6 to 7 young are born in the middle of March or April. By May the youngsters accompany the mother to the surface. Mother squirrels emerge first in the morning, checking the area for predators, then call the youngsters out. The young come spilling out for several hours of wrestling, playing, and feeding, then the whole family retires to the burrow until late in the afternoon when temperatures again start to cool. Round-taileds often stand on their hind legs trying to get a better view as they watch for their many predators. Because they're very dependent on succulent vegetation for moisture, these squirrels estivate for a few weeks during the summer drought, until the summer rainy season again brings new growth.

Harris' antelope squirrel

Rock squirrels are dormant during parts of the cold winter months. During this time they gather and store food and retire to their burrows, but occasionally do come out on warm days. They become active in the spring, and can be seen in the mornings sunning on high rocky perches where they can keep a lookout for hawks, roadrunners, coyotes, snakes, other predators. They give whistle-like warning calls.

Rock squirrels mate early in the spring. Young squirrels are born in March; there may be second litter in August or September. Rock squirrels can be colonial or solitary.

When a rock squirrel encounters a snake, it stamps its feet and waves its tail from side to side while facing the snake. It also tries to push sand or dirt in the snake's face with its front paws.

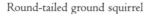
Round-tailed ground squirrel

Pocket Gophers

The *fossorial* (burrowing) pocket gopher is a rarely seen animal, since it spends almost its entire life underground in its extensive tunnels. Only the many mounds of dirt on the surface show where these animals are active. The mounds have no visible holes, because the gophers plug them from underneath. Occasionally the gopher opens a hole to allow some air exchange in the tunnel, or to let the tunnels dry out after heavy rains.

These little animals are active year round, always digging and extending their tunnel systems. They perform a valuable service in turning and aerating the soil. Many other creatures such as rabbits, ground squirrels, mice, skunks, snakes, lizards, and toads use gopher holes and tunnels.

—Pinau Merlin

SONORAN DESERT SPECIES:
  Botta's pocket gopher *(Thomomys bottae)*

  Order: Rodentia
  Family: Geomyidae
  Spanish Names: tuza, topo

DISTINGUISHING FEATURES

This heavy-bodied animal is about 9½ inches (24 cm) long and weighs 6 to 8 ounces (170-225 g). It has very small ears and eyes, a short naked tail and large forelimbs with long claws. The lips close behind the large incisor teeth, so that the teeth are always visible. This gopher ranges in color from pale gray or white to almost black.

HABITAT

Pocket gophers are found throughout the Sonoran Desert region where there are easily dug soils, such as those in riparian areas, washes, farms, mesquite bosques and golf courses. They are found at all elevations, but not usually in the hard caliche soils of the desert.

FEEDING

• *Diet:* These animals are vegetarians, eating roots, tubers, grasses, green plants, and prickly pears.
• *Behavior:* Pocket gophers are very shy and timid, seldom leaving their underground tunnel systems. They prefer to pull plants down into the tunnel from below. They also store food for lean times in chambers off the main tunnel.

LIFE HISTORY

Although male gophers leave their tunnels to seek mates, and 2-month-old youngsters leave home to establish their own territories, pocket gophers spend most of their time underground in their own tunnels. A pocket gopher only occasionally ventures to the surface for a tasty plant, or to dump a load of dirt. It rarely moves far from the tunnel entrance, and if startled, quickly retreats backward down the hole. A sensitive tail helps the animal feel its way as it runs backward through its 90 to 200 foot (27-61 m) long tunnel. Side chambers are used for food storage, latrines or nest chambers.

Botta's pocket gopher

The gopher's teeth are continuously growing (9 to 14 inches/23 to 35 cm a year!) and must be kept trimmed by constant gnawing. Both the teeth and the long claws are used for digging, and the gopher's lips close behind its teeth so that dirt doesn't get in its mouth while it digs. Gophers are solitary, only getting together for mating once or perhaps twice a year, with 2 to 6 young born 19 days later. These youngsters will be sexually mature adults in 3 months.

Heteromyidae: Kangaroo Rats & Pocket Mice

The heteromyids are a group of rodents consisting of kangaroo rats and pocket mice. Despite their names, they are neither rats nor mice; and in spite of their mouse-like appearance, they are not closely related to any other species of North American rodent.

Kangaroo rats and pocket mice are all nocturnal, burrowing animals with external fur-lined cheek pouches for storing and transporting the seeds that are their primary food. They are all well adapted to living in arid environments since most of them never need to drink water. They also have efficient kidneys that can conserve precious fluids by concentrating the urine.

Because there are many of these little rodents and they are closely related to each other, each species has evolved with different foraging times and places, which minimizes competition. Bailey's pocket mouse, for example, climbs up into desert wash vegetation to find seeds and berries still on the plants, while the desert pocket mouse hunts along the ground in washes and open areas for seeds. Merriam's kangaroo rat, a creature of open, creosote flats, tends to dash from one clump of bushes to the next, overlooking seeds out in the open spaces, leaving those for other mice to find. In this way many species of heteromyid mice and rats can share the same environment.

—Pinau Merlin

SONORAN DESERT SPECIES:
>   little pocket mouse *(Perognathus longimembris)*
>   Arizona pocket mouse *(Perognathus amplus)*
>   long-tailed pocket mouse *(Perognathus formosus)*
>   rock pocket mouse *(Chaetodipus intermedius)*
>   desert pocket mouse *(Chaetodipus penicillatus)*
>   Bailey's pocket mouse *(Chaetodipus baileyi)*
>   banner-tailed kangaroo rat *(Dipodomys spectabilis)*
>   Merriam's kangaroo rat *(Dipodomys merriami)*
>   desert kangaroo rat *(Dipodomys deserti)*
>   Baja California kangaroo rat *(Dipodomys peninsularis)*

Order: Rodentia
Family: Heteromyidae
Spanish names: ratón (pocket mouse), rata canguro (kangaroo rat)

DISTINGUISHING FEATURES

All three of these kangaroo rats have long tails and large hind feet with 4 toes. (Some other species have 5 toes.) They also have large heads with big eyes, small ears and external, fur-lined cheek pouches.

Arizona pocket mouse

The *banner-tailed kangaroo rat* is one of the largest kangaroo rats, weighing up to 4½ ounces (128 g). It is light brown or buff-colored with a white underside. The tail is black towards the end, but has a white tip.

Merriam's kangaroo rat is only about half the size of the banner-tailed rat; its tail is dark at the tip.

The *desert kangaroo rat* is similar to the banner-tailed in size; it also has a white tip on the tail, but its fur is more yellowish.

The *Arizona pocket mouse* is a small mouse with a thinly-furred tail that does not have a brushy tip. This mouse has tan- to orange-colored fur that is softer than that of many of the other species of pocket mice.

The *desert pocket mouse* has buff to brownish, coarse fur, and a white underside. The tufted tail is long, and has a crest of fur forming a ridge along its length.

Bailey's pocket mouse is the largest of the pocket mice, weighing up to 1⅜ ounces (39 g). It has a long, tufted tail and coarse, grayish fur, though the color can vary considerably.

The *rock pocket mouse* is medium-sized (⅜ to ⅝ ounce; 11-18 g), with rough brownish-gray fur and spine-like hairs on its rump. The long tail is tufted, with a ridge of fur running along its top.

HABITAT

The banner-tailed kangaroo rat lives in open desertscrub, creosote bush flats, open grasslands and sandy places. It favors a sparse covering of grasses, interspersed with a few mesquite trees and cacti.

Merriam's kangaroo rat is the most common and widespread kangaroo rat in the Sonoran Desert. It inhabits washes, open grassland, and sandy soils of the desertscrub.

The desert kangaroo rat is found in the driest parts of the Sonoran Desert, among sand dunes and creosote flats.

The Arizona pocket mouse and the desert pocket mouse both inhabit the sandy, open desert with sparse vegetation of grasses, mesquites, creosote bushes, and a few cacti.

Bailey's pocket mouse is found on rocky slopes and areas with boulders and rocks mixed in among the cacti and shrubs.

Bailey's pocket mouse

The rock pocket mouse also prefers rocky slopes in the desert, lava flows, and gravelly soils. It may use crevices and cracks in rocky ledges as retreats.

FEEDING

• *Diet:* All of these kangaroo rats are primarily seed eaters, feeding especially on mesquite beans and the seeds of native grasses. Merriam's kangaroo rats eat a few more insects and a little green vegetation as well. The pocket mice are also primarily granivorous (seed eating), most often eating mesquite beans and the seeds of grasses, creosote bushes, and weeds. They may also eat some insects and a very little vegetation. Bailey's pocket mice are adapted to eat jojoba nuts, which are toxic to most other mice.

Banner-tailed kangaroo rat

• *Behavior:* Kangaroo rats forage around the desert at night, sniffing out buried seeds, collecting as many as they can by stuffing them into their cheek pouches. These rodents forage in fairly large areas, as they hop along with their bipedal gait. Banner-tailed and desert kangaroo rats store piles of seeds in their burrows, where the seeds can absorb up to 30 percent more moisture from the humid air. Merriam's doesn't store seeds in burrows, but stores them instead in shallow surface caches scattered around its home range.

The pocket mice also search for seeds, but less widely than do the kangaroo rats. They carry the seeds in their cheek pouches for transport back to the burrow, where they are cached.

LIFE HISTORY

The Sonoran Desert kangaroo rats are remarkable in not needing to drink water, even though their diet is almost entirely composed of dry seeds. They survive almost entirely on water metabolized from the seeds they eat. The rats prefer seeds high in carbohydrates, from which they can produce a half gram of water from each gram of seed eaten. They even bathe without water, keeping their shiny fur clean by taking dust baths in the sand.

Kangaroo rats are adapted to a bipedal gait with large, strong hind legs and big feet. They can jump up to 9 feet (2.75 m) in one bound, an effective aid in escaping predators. The long tail is used as a counterbalance while the rat hops and leaps about.

These rodents have very keen hearing and are able to detect an owl's silent approach and the movement of snakes.

Pocket mice are also well adapted to arid desert life. They seldom drink, and can conserve water in a number of ways. They spend the days underground in the burrow, where in summer the humidity is higher and the temperature lower than above ground. The entrance hole is usually plugged to keep the moisture from escaping to the dry air above. The kidneys of these rodents concentrate the urine to a viscous consistency, reducing water loss. When temperatures become extreme, some pocket mice go into a torpor, or *estivate*. These animals are solitary and defend small territories, often fighting when they encounter each other.

Convergent Evolution

There are many reasons why one kind of living thing
may look like another. Usually it's simply because
they're related. But sometimes an organism
mimics another's appearance, thereby deceiving
an unsuspecting third party. And sometimes
living things come to resemble each other by
adapting in similar ways to similar habitats or
ways of life—a process called convergent
evolution.

Merriam's kangaroo rat

Kangaroo rats are such perfect
examples of desert adaptation that it's
perhaps not surprising nature has
arrived at similar designs more
than once—a wonderful example of convergent evolution. There are small jumping rodents
very much like kangaroo rats in Old World deserts, including the jerboas and gerbils of
Africa and Asia and certain hopping mice in Australia. (And in case you're wondering,
there are such things as "rat kangaroos" Down Under—but they're something completely different!)

— David W. Lazaroff, adapted from
Arizona-Sonora Desert Museum Book of Answers (ASDM Press, 1998)

Muridae: Mice & Rats

The rats and mice are among the most sucessful mammal groups on Earth. They are adaptable creatures that can inhabit almost any environment. The Sonoran Desert, with its great diversity of habitats, is blessed with an abundance and a wide variety of these fascinating creatures. Here we have predatory grasshopper mice that hunt and kill other mice, and packrat builders that construct houses of sticks and debris up to 2 or 3 feet (1 m) high and 8 feet (2.4 m) wide. The desert woodrat lives in the most *xeric* (arid or dry) environments, while Merriam's mouse fills a niche in mesquite bosques.

Arizona cotton rat

These little rodents are at the bottom of the vertebrate food chain, preyed upon by everything from coyotes and snakes to hawks and bobcats. In response, they breed prolifically, with some species, like the cotton rats, able to produce 8 to 10 litters a year. Populations still fluctuate with drought and predation, but the mice and rats are able to respond to good conditions by rapidly rebuilding their numbers.

All rodents, including the mice and rats, are gnawers. Their teeth are ever-growing and must be kept trimmed down by constant gnawing. A layer of hard orange enamel covers the front surface of the teeth. The rest of the tooth is softer and wears down quicker than the enamel as the rodent gnaws, thus creating a chisel-like shape to the front teeth that is unique to the rodent family.

—Pinau Merlin

SONORAN DESERT SPECIES:

    cactus mouse *(Peromyscus eremicus)*
    Merriam's mouse *(Peromyscus merriami)*
    canyon mouse *(Peromyscus crinitus)*
    southern grasshopper mouse *(Onychomys torridus)*
    Arizona cotton rat *(Sigmodon arizonae)*
    white-throated woodrat, or packrat *(Neotoma albigula)*
    desert woodrat *(Neotoma lepida)*

    Order: Rodentia
    Family: Muridae
    Spanish names: rata (rat), ratón (mouse), rata de campo (woodrat)

DISTINGUISHING FEATURES

The three *Peromyscus* species are similar in appearance, with long tails and big ears. The *cactus mouse* has a sparsely furred tail with a slight tuft at the tip; its pelage is brownish to cinnamon colored with a white underside. *Merriam's mouse* is gray with a white underside. The *canyon mouse* has long silky fur, with the color varying to blend in with the rock substrate on which it lives.

The *southern grasshopper mouse* is a medium sized gray-brown- to cinnamon-colored mouse with a short, white-tipped tail.

Rats are larger than mice. The *Arizona cotton rat* is a thick-bodied rodent of medium length, with a thinly-furred tail and small ears and eyes. Its pelage is brown interspersed with black hairs.

The *white-throated woodrat* (perhaps better known as *packrat*) is medium-sized (up to 1 pound; .45 kg), with big ears and eyes and a short tail. Hairs on this animal's throat are white to the bases, while they are gray or dark at the base on the *desert woodrat*. The desert woodrat is smaller than the white-throated.

Cactus mouse

Southern grasshopper mouse

HABITAT

The cactus mouse lives in the desertscrub and grassland areas throughout the Sonoran Desert. Merriam's mouse is only found in the low desert mesquite bosques in south-central Arizona. The canyon mouse inhabits the canyons, rocky slopes and cliffs of northern and western Arizona. The southern grasshopper mouse lives in grassland and desertscrub communities in southern and western Arizona.

Cotton rats are usually seen in grassy areas near streams and ponds and around irrigated fields. The white-throated woodrat is found throughout most of Arizona and Sonora in a variety of habitats, especially in areas with mixed cacti. It can live in very arid environments as long as prickly pear or cholla are available.

The desert woodrat lives on desert floors or rocky slopes; its range includes most of the western Sonoran Desert, extending south into Baja California.

FEEDING

• *Diet:* The cactus mouse, canyon mouse and Merriam's mouse all eat seeds, mesquite beans and leaves, and to a lesser extent, green vegetation, and insects. Grasshopper mice are predators, hunting insects, beetles, grasshoppers, and scorpions, but they also hunt and kill other mice. Cotton rats eat mostly green plants and grasses. woodrats eat mesquite beans, palo verde seeds, green plants, and cacti, particularly prickly pears and chollas, which provide them with moisture as well as food. Occasionally they eat insects or other meat.

• *Behavior:* The *Peromyscus* are all skilled climbers of the cacti, trees, and cliffs in their various habitats, using their long tails for balance and support as they move about at night in search of seeds and invertebrate prey. During the day they remain in burrows in clumps of cacti, in the ground, or among rocks. Cotton rats construct little tunnels or runways in grasses and weeds where they can forage in safety.

The grasshopper mouse is an efficient predator, killing other mice with a bite to the back of the neck, and biting the stingers off scorpions before consuming them. Pinacate beetles emit a toxic spray from their rear ends, deterring most predators, but grasshopper mice catch them and shove the defensive ends of the beetles into the sand, then bite off the good parts, leaving beetle bottoms embedded in the sand.

White-throated woodrat

woodrats forage at night, eating food and carrying some items back to the house to store for later use (as is the case with mesquite beans), or to incorporate them (especially cactus parts) right into the house structure.

LIFE HISTORY
The *cactus mouse* often climbs around in vegetation and brush, searching for seeds and fruits to eat. It may nest in wood piles or rock piles, or use the abandoned burrows of other animals. Although this mouse needs less water than many others, and is desert-adapted, it may estivate or go into a torpor in the summer when resources for food and moisture are not available.

Merriam's mouse also climbs the mesquite trees and shrubs of its habitat in search of food. The canyon mouse doesn't burrow, instead placing its grassy nest in a natural cavity or rocky crevice. The fur color of this mouse usually matches the color of the cliffs and rock walls it haunts.

Grasshopper mice not only hunt prey, they also have lifestyles reminiscent of canid hunters. They form family packs with both parents feeding and caring for the young and teaching them how to hunt. They defend a territory, but will range widely in search of food. The grasshopper mouse even vocalizes like a tiny wolf, standing on its hind legs, throwing back its head and howling. These eerie, high pitched calls may be used to communicate with other family members.

Cotton rats are active day or night and are often found in dense populations. A high density of these rats usually attracts predators, which then reduce their numbers.

woodrats are famous for their houses made of sticks, cactus parts, animal dung and debris, usually tucked in around a prickly pear cactus, under a mesquite tree or hackberry bush, or among boulders. The house acts as insulation for the nest, which is underneath but close to the ground surface. The spiny cactus parts may also offer some protection from coyotes digging up the nest. Several entrance holes allow the packrat a quick escape should a snake come visiting. Packrats are solitary, with only one rodent per household, unless a female has young.

Fishes

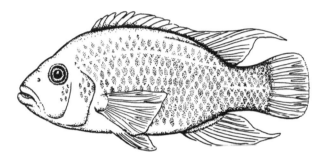

Section Contents

Fishes in the Desert

Craig Ivanyi

Traveling around the Sonoran Desert today, especially the northern portion, one would never suspect that it is home to more than 100 species of freshwater fishes. Though there isn't much free-standing water today, as recently as 100 years ago several of the dry washes were true rivers. The presence of these bodies of water resulted in an incredible array of animals in the region, everything from two inch (5 cm) long topminnows to six foot (1.8 m) long squawfish.

Even today Arizona boasts an impressive 30 or so native species of fish and at least twice as many introduced species. Unfortunately, most of the natives are disappearing with some of them virtually gone. Of Arizona's native species, at least two-thirds are listed as Threatened or Endangered by state or United States federal agencies. A myriad of factors has resulted in their precarious position, including exotic competitors (and predators), irrigation diversion, dams, and overuse of available water by our burgeoning human population. These factors have led to fragmentation and reduction of habitat available to native species. The status of fishes in Mexico is less certain because there hasn't been as much study of them.

Colorado River Fishes

Originating high in the Rocky Mountains, the Colorado River drains seven North American states. The watershed encompasses 242,000 square miles (625,000 km²) before it reaches (on a good day) the Gulf of California, some 10,000 feet (3000 m) lower in elevation. The river has a fascinating history of human use over the last 10,000 years that includes fishing, farming, mining, and even steamboat transport! Today it is of major commercial importance, providing electric power, water for

511

irrigated crops and for 20 million people, of which only around 2.5 million actually live in the river basin.

By the time the river reaches the gulf, the water has been claimed by farmers, power companies, Indian nations, and both the U.S. and Mexican governments. Combined, the capacity of agricultural and municipal diversions actually exceeds the average annual flow of the river through Lee's Ferry (in northern Arizona).

In the last seven decades, a series of massive dams has been created to regulate the flow of the Colorado River. Since dams create physical barriers to aquatic wildlife and severely alter the characteristics of the river's flow, natural seasonal changes in flow and water temperature are virtually eliminated. Instead, flow is regulated by electric power demands. Water for power generation is often drawn from the bottom of the reservoir. Waters from the bottom tend to be cold and temperatures are fairly constant. At times, closed dams can shut off all downstream flow.

Many of the introduced fishes evolved in quieter waters, and could be considered "preadapted" for life in reservoirs above dams. Whereas native fishes tend to do poorly or leave areas where great physical and chemical modifications have occurred, many non-natives have flourished. Thus, human modification and stabilization of rivers and construction of reservoirs set the stage for establishment of an exotic fish fauna.

Habitat changes and species introductions have resulted in declining populations or extinction of native Colorado River fishes. The Colorado River today seems more akin to a man-made plumbing system than a wild, naturally flowing river.

The Colorado River basin contains more endemic freshwater fish species than any other river basin in North America. Some are large animals remarkably well-adapted to the swiftly moving waters of the Colorado. Adaptations common to the three so-called big river species listed below are reduced or embedded scales (to reduce friction) and a bizarre hump occurring just behind the head (apparently an adaptation that helps the fish maintain its position in swift currents by pushing it down toward the river bottom).

REPRESENTATIVE SPECIES:
  bonytail chub
    (*Gila elegans*)
  Colorado River squawfish
    (*Ptychocheilus lucius*)
  razorback sucker
    (*Xyrauchen texanus*)

Stream Fishes
of Intermediate Elevations

Many streams and rivers in lower elevation coniferous forests, woodlands, and desert grasslands have also been altered. These channels are largely unchanged as they pass through the deep canyons of the headwaters. But, as they reach more gentle terrain and flow through broad valleys, streams have been altered dramatically by human activities. Many streams and their associated riparian areas have disappeared due to irrigation diversion and channelization. The remaining streams, however, still support large populations of native fishes.

REPRESENTATIVE SPECIES:
   loach minnow
      (*Rhinichthys cobitis*)
   longfin dace
      (*Agosia chrysogaster*)
   Yaqui chub
      (*Gila purpurea*)
   Sonora chub
      (*Gila ditaenia*)
   desert sucker
      (*Catostomus clarkii;*
       also *Pantosteus clarkii*)
   Yaqui catfish
      (*Ictalurus pricei*)

Intermediate- to
Low-desert Fishes

Few desert streams remain in Arizona. Most surface water sources have been diverted and subterranean supplies have been cut off by pumping. In most valleys, the water table has been lowered by more than 160 feet (50 m), and in some places, it has been lowered by 1000 feet (300 m)! Some of the fishes occupying intermediate elevations have been forced to migrate upstream by decreasing water tables. Streams below 2000 feet (600 m) in elevation remain only if they receive reliable spring inflows (or sometimes sewage effluent), and many flow only through man-made channels. Other desert streams flow only during times of high precipitation.

REPRESENTATIVE SPECIES:
   Gila topminnow
      (*Poeciliopsis occidentalis*)
   desert pupfish
      (*Cyprinodon macularius*)
   Sinaloan cichlid
      (*Herichthys beani;* also
       Cichlasoma beani, Heros beani)

bonytail chub *(Gila elegans)*

Order: Cypriniformes
Family: Cyprinidae (minnows)
Spanish name: charalito elegante

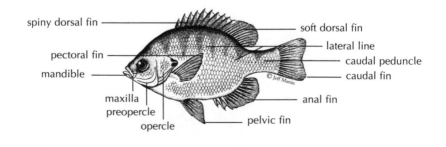

DESCRIPTION

With a streamlined
body, terminal mouth, and a
thin, pencil-like caudal peduncle,
the bonytail chub can grow to over
2 feet (60 cm) long. As with many
desert fishes, its coloration tends to
be dark above and lighter below. This
color pattern may serve to camouflage
the fish thereby facilitating prey
capture and reducing susceptibility
to predation. Breeding males have red
fin bases.

RANGE

Bonytail chub once occurred in
Arizona, Colorado, Utah, Wyoming,
New Mexico and California. They
have suffered severe reductions in their
range and are an Endangered species
in the U.S. They persist in the Green
River of Utah and maybe in the larger
Colorado River impoundments.

HABITAT

Bonytail chub have been reported to
occupy swiftly moving water but seem
to prefer flowing pools and backwaters
with rocky or muddy bottoms. Though
primarily restricted to rocky canyons
today, they historically were abundant
in the wide, downstream parts of rivers.

LIFE HISTORY

Adult bonytail chubs feed mainly on
terrestrial insects, plant debris, and
algae. The young usually eat aquatic
insects. Almost nothing is known
about reproductive habits. They are
presumed to spawn in mid-summer
and in some lake situations may
hybridize with both humpback and
Colorado chubs.

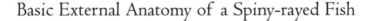

Basic External Anatomy of a Spiny-rayed Fish

spiny dorsal fin

pectoral fin

mandible

maxilla
preopercle

opercle

soft dorsal fin

lateral line

caudal peduncle

caudal fin

anal fin

pelvic fin

© Jeff Martin

Colorado River squawfish *(Ptychocheilus lucius)*

Order: Cypriniformes
Family: Cyprinidae (minnows)
Other Common Names: salmon
Spanish name: charalote

DISTINGUISHING FEATURES

The Colorado River squawfish, growing to 6 feet (180 cm) and over 80 pounds (36 kg) in weight, is the largest minnow in North America. It has a long, conical head with a very large, horizontal, terminal mouth that extends to or beyond the eyes. It tends to have a narrow caudal peduncle and a deeply forked caudal fin. Color above is tan, olive, or brownish, and yellow to whitish below. The sides of the body will often have some scales with a metallic sheen.

HABITAT

Squawfish are found in swiftly moving water.

LIFE HISTORY

Adult squawfish feed on other fish and smaller birds or mammals that light on or fall into the water. The young feed on aquatic insects and crustaceans.

Squawfish spawn in river channels over gravel beds. Apparently they reproduce in late spring to mid summer when water temperatures exceed 21° C (70° F) and water levels drop.

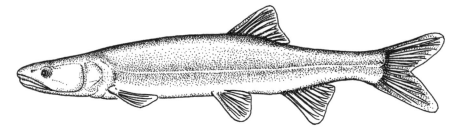

RANGE

The Colorado River squawfish occurs in Arizona, Colorado, Utah, Wyoming, California, and a few streams in Nevada. The squawfish has been extirpated from the southern part of its range. This fish has suffered severe reductions in its range and is listed as Endangered in the United States.

Below dams these necessary temperature and water fluctuations usually don't occur. Water released from dams is cold and constant in temperature and the amount of flow seldom drops at the right time to stimulate spawning. At times, no water is released below dams. These conditions severely restrict where fish can survive.

Though now virtually extinct in Arizona, historically these fish were so common that they were removed from irrigation canals and ditches with pitchforks and used to fertilize crops.

RECOVERY EFFORTS

Efforts to recover the Colorado squawfish, bonytail chub, razorback sucker, and other native fishes have been largely unsuccessful in Arizona. Federal and state agencies and private organizations produce millions of fingerlings (juvenile fish about as long as a person's finger) that are released into previously and currently occupied habitats. However, habitats have been severely modified and exotic species often prey upon these small, juvenile fish. Currently, native species that are to be released in the wild are being raised to larger sizes before being released. It is hoped that larger size will discourage predation by exotics. Additionally, in some locations, flow regimes below dams have been modified in recent years to provide more of the natural cues necessary to stimulate spawning by these species. Fish ladders have been installed in some locations to help migrating fish pass around dams and other obstructions. It will take years before the success or failure of these efforts can be assessed.

That's a Mouthful!

I was down at the Desert Museum's beaver pond exhibit one spring morning feeding crickets to squawfish. As I tossed the crickets towards these 3½ foot (1 m) long fishes, they'd breach the surface and snatch the tiny morsels in their huge mouths.

Some Mallards, realizing there was free food to be had, watched nearby. One of them finally swooped in to nab a cricket. Unfortunately, a squawfish, with its eye on the same cricket, broke the surface and grabbed the duck's head in its mouth. The water exploded with beating wings and flailing fins as both startled parties fled in opposite directions. I let a few minutes pass and then threw some more crickets into the pond. The squawfish eyed them suspiciously, peering up out of the water for signs of dangerous ducks before gingerly taking the insects.

This incident really demonstrated to me that fish are very much in tune with their surroundings. And that they can be quick learners!

—Ken Wintin, Desert Museum keeper

razorback sucker (*Xyrauchen texanus*)

Order: Cypriniformes
Family: Catostomidae (suckers)
Other Common Names: buffalofish, buffalo
Spanish name: matalote jorobado

DISTINGUISHING FEATURES

The razorback sucker is our largest native sucker, perhaps reaching 36 inches (91 cm) in length and 11 pounds (5 kg) in weight. Adults have subterminal mouths with large fleshy lips and sharp, blade-like keels running down their backs. This keel extends from behind the head to the dorsal fin. The young of this species are virtually indistinguishable from those of other suckers. Colors vary from olive to light brown laterally and yellow or white below.

RANGE

The razorback sucker is a fish of the Colorado River drainage, occuring in Arizona, Colorado, Utah, Wyoming, California, and a few streams in Nevada. It is currently found only north of the Grand Canyon, except for relict populations in Lake Mead, Mohave, and Havasu. This fish has suffered severe reductions in its range and is listed by U.S. agencies as Endangered.

HABITAT

Razorback suckers seem to prefer flowing pools and backwaters. They are most often found in sandy areas and undercut banks. Historically they were reported to occupy areas with strong, uniform currents.

LIFE HISTORY

Razorback suckers feed primarily on algae and fly larvae, as well as planktonic crustaceans. Spawning has been observed in some human-made impoundments. Generally, a group of 2 to 12 male suckers attend to one female, the males nudging the female's genital region with their heads and the front part of their keels. As the female settles to the bottom, she is flanked by at least one male on each side. Vibrations ensue and end with strong convulsions; many males then fertilize the eggs.

COMMENTS

Several agencies have been involved in a recovery effort for this species. Millions of larvae and juvenile fish have been stocked in current and former habitat, but there is little evidence of recovery.

loach minnow *(Rhinichthys cobitis)*

Order: Cypriniformes
Family: Cyprinidae (minnows)
Spanish name: charalito adornado

DISTINGUISHING FEATURES
This is a small minnow usually less than 2½ inches (6.3 cm) long. The eyes are situated somewhat dorsally. The species is typically olive-brown with black specks and blotches dorsally, and a black spot in the middle of a white bar at the base of the caudal fin. Breeding males develop brilliant orange to blood-red patches on the lips, fins (bases mainly), and sometimes on the body. Females are fairly drab but can develop yellow coloration on the fins and body.

RANGE
This minnow is found mainly in the Upper Gila River system of New Mexico and Arizona and in the San Pedro River of Arizona and northern Sonora. It is uncommon in Arizona but locally abundant in New Mexico. It is protected in the U.S. as a Threatened species.

HABITAT
Loach minnows live in shallow riffles, usually less than 6 inches (15 cm) deep. Generally the substrate in these riffles is coarse gravel. Loach minnows tend to be secretive bottom dwellers most commonly associated with beds of filamentous algae.

LIFE HISTORY
The loach minnow feeds almost entirely on mayfly and blackfly larvae. It spawns from March through September. Males may be territorial and defend nests, which are placed downstream of stones in riffles.

This species has a greatly reduced air bladder, which means swimming above the streambed is energetically expensive. Therefore, it spends its life on the bottom. In many respects the ecology of the loach minnow resembles that of darters. Loach minnows live only 2 or 3 years.

longfin dace *(Agosia chrysogaster;* formerly *Rhinichthys chrysogaster)*

Order: Cypriniformes
Family: Cyprinidae (minnows)
Spanish name: charalito aleta larga

DISTINGUISHING FEATURES

Longfin dace can be up to 4 inches (10 cm) long; it has a rounded snout and a slightly subterminal mouth with a small barbel at each corner. Coloration is dark gray above with a dark band along silver sides and creamy white or light yellow below. There is a black spot at the base of the caudal fin. In adult males paired fins are yellow at their bases while large females have an elongated lower lobe of the anal fin. In breeding condition males develop nuptial tubercles on the fins and head which superficially resemble the common disease of aquarium fish known as "ich."

RANGE

Longfin dace are our most common native fish and have the widest distribution of any species in the region. This species is found primarily in the Gila and Bill Williams river systems (Colorado River drainage) of Arizona and New Mexico. It is also common in the Pacific drainages of western Mexico. This fish is most commonly found at intermediate elevations, occurring from the Arizona Uplands to low-desert streams.

HABITAT

Longfin dace are found in relatively shallow, sandy, gravelly, to rocky streams. They are frequently found in flowing pools. This species can survive in a dry stream by living in moist algal mats. In these mats they encounter high temperatures, extreme temperature fluctuations, and low dissolved oxygen—conditions that would kill most other fish. Young longfin dace are known to be active when water temperature is as high as 33° C (92° F).

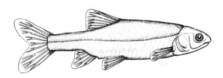

LIFE HISTORY

An omnivore, the longfin dace feeds on whatever is available. It spawns throughout the year over relatively soft, sandy bottoms. Eggs are laid in nest pits and hatch in a few days. After hatching the young tend to select slower moving waters along the edge of the stream, whereas juveniles and adults move into the smooth runs of the main channel.

Yaqui chub (*Gila purpurea*)

Order: Cypriniformes
Family: Cyprinidae (minnows)
Spanish name: charalito Yaqui

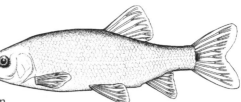

DISTINGUISHING FEATURES
Yaqui chubs can be up to 5½ inches (14 cm) long. Body color varies from steel blue to yellow-brown. The base of the caudal fin has a triangular black wedge, and the anal fin has 7 rays. Males turn blue-gray when they are in breeding condition.

RANGE
The Yaqui chub is found principally in Sonora, Mexico in the Yaqui, Sonora and Matape rivers. This Endangered Species naturally enters the United States only in San Bernardino Creek, but it has been introduced into Leslie Creek in southeast Arizona.

HABITAT
Prefers quiet pools of headwaters, creeks, and marshes.

LIFE HISTORY
The Yaqui chub is omnivorous, feeding on algae, terrestrial and aquatic insects, and arachnids. It generally reproduces from spring to early summer.

COMMENT
The Yaqui chub is listed as Endangered by the U.S. Fish and Wildlife Service. Apparently all naturally occurring populations in Arizona were lost because the waters in which they occurred dried up. Prior to the loss of the last natural population in Arizona, individuals were translocated into Leslie Creek at the south end of the Swisshelm Mountains, where they still persist.

Where Have All the Natives Gone?

Today in Arizona there are three times as many species of introduced or non-native fishes as there are native species. The reasons for this are numerous and complicated. Man's increased demands for water to provide electrical power have resulted in the drying up of some waterways, as well as in the creation of man-made dams and channels that inevitably alter the habitat of native species of all kinds. Those species whose specific requirements cannot be met in this altered environment face the probability of extinction within that area. In addition, large numbers of non-native fishes have been introduced in Arizona streams and lakes for game purposes. The native species are often preyed upon or outcompeted for food and territory by introduced species. Loss or alteration of habitat is equally detrimental to often highly adapted native fishes.

Sonora chub *(Gila ditaenia)*

Order: Cypriniformes
Family: Cyprinidae (minnows)
Spanish name: charalito sonorense

DESCRIPTION

In the United States, the Sonora chub is generally smaller than 8 inches (20 cm), but in Mexico it may grow up to 10 inches (25 cm) long. The Sonora chub has a somewhat chubby body. The upper half of the body is generally dark with two black lateral bands above and below the lateral line; the lower half of the body is much lighter. When fish are in breeding condition the bases of the anal and paired fins are red, and part of the belly may be orange.

RANGE

The Sonora chub is only found in the Rio de la Concepción drainage of northern Mexico. Sycamore Creek, near Nogales, forms the headwaters of this drainage and is the only place in Arizona where you can find this fish.

HABITAT

This fish tends to inhabit shaded pools with undercut banks which, during droughts, may be the only areas where water is present. Even in times of high stream flow, chubs seem to stay in these same pools.

LIFE HISTORY

Sonora chubs are omnivorous, feeding on aquatic and terrestrial insects and algae. They spawn in early spring. These remarkable fish are capable of surviving in small habitats under extremely difficult environmental conditions. At times this species persists in wet algal mats with almost no free water.

desert sucker *(Catostomus clarkii;* also called *Pantosteus clarkii)*

Order: Cypriniformes
Family: Catostomidae (suckers)
Spanish name: matalote del desierto

DISTINGUISHING FEATURES

The desert sucker grows to 13 inches (33 cm). Its color varies from green to silver or tan above and silver to yellow below. During the spawning season breeding males develop a striped pattern consisting of 1 or 2 light lateral stripes on a darker background. This sucker has a subterminal mouth with enlarged cartilaginous lobes behind the lower lip. The front of the upper lip lacks *papillae* (small, flap-like growths).

RANGE

This common fish of the Lower Colorado River drainage downstream of the Grand Canyon also lives in the Virgin River of Utah, Arizona, and Nevada, the Bill Williams River of Arizona, and the Gila River drainage of New Mexico, Arizona, and northern Sonora, Mexico.

HABITAT

The desert sucker occupies small and medium rivers. Young fish occupy backwaters, while adults are found in pools with undercut banks during the day and riffles at night.

LIFE HISTORY

While the desert sucker is omnivorous, it prefers diatoms and algae that cover rocks. It uses the cartilaginous sheaths on the jaws to scrape these food items from the rocks. Spawning occurs in the spring over riffles and generally involves

Pattern variations in the desert sucker

1 female and 2 or more males. The female creates a depression in the gravel with lateral movements of her body. She then deposits eggs which are fertilized by the males. The eggs are buried in the loose gravels, and hatch in a few days.

Yaqui catfish *(Ictalurus pricei)*

Order: Siluriformes
Family: Ictaluridae (catfishes)
Spanish name: bagre del Yaqui

DISTINGUISHING FEATURES
Yaqui catfish reach almost 2 feet (60 cm) in length. Externally they are similar to the channel catfish (*Ictalurus punctatus*) though they have shorter pectoral spines. This species has 4 pairs of barbels (whiskers) along the side and the rear of the mouth. It has an adipose (soft, rayless) fin and spines in the dorsal and pectoral fin origins and in the pectoral fins. Coloration is silver-tan to goldish above, transitioning to silver or silver-white below.

RANGE
This is a fish of the Río Yaqui and Río Casas Grandes drainages of north-western Mexico and extreme southeastern Arizona. It is found mainly in small or medium rivers from intermediate to lower elevations. A rare fish in Arizona due to limited range, it is protected in the United States as a Threatened Species. It is still relatively common and unprotected in Mexico.

HABITAT
The Yaqui catfish prefers quiet water over sand-rock substrates.

LIFE HISTORY
Almost no information is available on this species, but its life history is presumed to be similar to that of the channel catfish. Catfish tend to be nocturnal and feed upon aquatic invertebrates while young. When they reach 4 inches (10 cm) in length they begin to also feed on algae and small to medium sized fishes. They spawn from April through early June. An egg mass is laid in a protected depression in the substrate. Eggs are generally guarded by the male. While guarding eggs, males are quite aggressive and bite readily. Though there are at least 7 species of catfish in Arizona, the Yaqui catfish is the only one native to the state.

Gila topminnow *(Poeciliopsis occidentalis)*

Order: Cypriniformes
Family: Poeciliidae (live-bearers)
Spanish name: charalito

DISTINGUISHING FEATURES

This is a very small fish measuring only 2¼ inches (5.7 cm). Males are smaller than females and have a very long *gonopodium* (over ⅓ of the body length), which is an anal fin modified into a sexual organ for internal fertilization. Color is light olive-tan to silver with a small black spot at the back of the dorsal fin. Large males are black with an orange gonopodium base. Males have slender bodies with small eyes.

RANGE

Gila topminnows were once found in the Gila River basin of Arizona, New Mexico, and northern Mexico from intermediate to lower elevations. At one time Gila topminnows were among the most common fishes in the low desert of the United States; they are now endangered. They are rare in Arizona and have been extirpated from New Mexico.

HABITAT

This species prefers shallow, warm, slow-moving waters of creeks and small or medium rivers.

LIFE HISTORY

Gila topminnows feed upon vegetation and amphipod crustaceans as well as aquatic insect larvae. They breed mostly from January to August, though some pregnant females may be found throughout the year. Topminnows have internal fertilization and are live-bearers; they are the only native fish in Arizona to bear live young. Broods range from 1 to 15. Usually females are carrying 2 broods simultaneously, one more developed than the other.

COMMENTS

This fish was once very common; it is also very prolific. The Desert Museum obtained 85 Gila topminnows for breeding and in less than one year this number had swelled to 3000. Given these reproductive capabilities it is hard to believe that this species has almost completely disappeared. The main cause of its disappearance is predation by a close relative, the mosquitofish (*Gambusia affinis*), which was introduced from the Mississippi River basin for mosquito control. Mosquitofish primarily eat young topminnows and may compete with them for critical resources. Mosquitofish have been known to replace a population of topminnows in one season.

desert pupfish (*Cyprinodon macularius*)

Order: Cypriniformes
Family: Cyprinodontidae (pupfishes)
Spanish name: cachorrito del desierto

DISTINGUISHING FEATURES
Males reach a length of almost 3 inches (76 mm); females are two-thirds that length. Breeding males become a beautiful, iridescent blue, while females, young, and non-breeding males tend to be drab whitish-tan with dark vertical bars (parr marks). Breeding males develop lemon-yellow or orange caudal peduncles and fins.

RANGE
Originally this species was found in the Lower Colorado River drainage in Arizona, California, and northern Mexico. Now this endangered species is restricted to a few parts of its former range.

HABITAT
Considered a fish of desert oases, this species of pupfish is found in springs, marshes, lakes, and creek pools. It prefers sandy substrates.

LIFE HISTORY
Pupfish are omnivorous, feeding on aquatic invertebrates and plants.

Breeding occurs mainly in the spring and the summer. Breeding males are fiercely territorial, defending their turf against invasion by other males. Females that are ready to spawn will enter the male's territory. After spawning the male guards the nest. Periodically a subordinate male may sneak into his territory while the dominant male is busy chasing away other fish. The subordinate may spawn with the female and then retreat to a safe place. The dominant male may therefore inadvertently wind up protecting eggs fertilized by another male.

This species is remarkably adaptable to rapid changes in temperature and salinity. It has been known to survive in fresh as well as salty water (with greater salinity than sea water) and to endure water temperatures over 100°F (38°C) or below 50°F (10°C).

Pupfish are endangered due to habitat destruction and fragmentation. They have been, and continue to be, preyed upon and displaced by non-native fishes.

Sinaloan cichlid *(Herichthys beani; also Cichlasoma beani, Heros beani)*

Order: Perciformes
Family: Cichlidae (cichlids)
Other common name: green guapote
Spanish names: mojarra sinaloense, mojarra verde

DISTINGUISHING FEATURES

The Sinaloan cichlid is a large fish, up to 10 or more inches (25 cm) long. The species is similar in form to sunfish, with co-joined spiny and soft portions of the dorsal fins. Coloration varies depending upon breeding condition. Non-breeding animals are usually a mixture of subdued yellow and black. Breeding females are much smaller than males and develop striking yellow tops and upper sides with jet black lower sides and belly. Breeding males generally have yellow scales with black spots in the centers of the scales.

RANGE

Native to the Río Yaqui basin, mainly along the western side of northern Mexico, these cichlids have a tropical distribution, with the lower Río Yaqui being their northern limit.

HABITAT

The Sinaloan cichlid tends to prefer quieter waters of medium to large rivers, generally in rocky or gravelly areas.

LIFE HISTORY

This cichlid is an omnivore that feeds on detritus, aquatic vegetation, invertebrates, and other fishes. It is capable of spawning throughout the year. Breeding pairs aggressively defend the adhesive eggs, which are laid on medium sized rocks. They also defend the young. One parent remains with the young until they are able to fend for themselves.

COMMENTS

In spite of an extremely pugnacious nature, this species may be declining from competition with and predation by introduced species such as other cichlids and sunfishes.

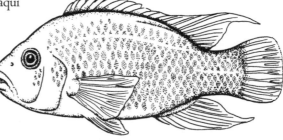

References

Minchley, W. L. *Fishes of Arizona.* Phoenix: Arizona Game & Fish Department, 1973.

Page, Lawrence M. and Brooks M. Burr. *A Field Guide to Freshwater Fishes: North America North of Mexico.* Boston: Houghton Mifflin, 1991.

Rinne, John N. and W. L. Minckley. *Native Fishes of Arid Lands: A Dwindling Resource of the Desert Southwest.* Ft. Collins, CO: USDA Forest Service General Technical Report RM-206, 1991.

Tellman, Barbara, R. Yarde and M. Wallace. *Arizona's Changing Rivers: How People Have Affected the Rivers.* Tucson: Water Resources Research Center, College of Agriculture, Univ. of Arizona.

Reptiles & Amphibians

Adaptations of Desert Amphibians & Reptiles

Thomas R. Van Devender

mphibians and reptiles have
many different adaptations that
allow them to live in deserts,
avoiding extremes in aridity, heat, or
cold. The animals may be active only in
certain seasons and at favorable times of
the day. Many use the environment to
actively regulate their body temperatures, preventing lethal extremes. And
some are well adapted to the surfaces
they live on—with modified appendages
for burrowing or the capacity to run on,
dive into, swim in or sidewind across
loose sand.

Before vertebrate animals adapted to
specific terrestrial habitats, such as
deserts, they first had to adapt to living
on land. The primary adaptations to life
on land occurred in the Paleozoic 400
to 360 mya (million years ago) with the
evolution of amphibians. Amphibians,
a name derived from the Greek word
amphibios (a being with a double life), live
in fresh water as larvae and can move
onto land as adults. In the amphibian's

metamorphosis from larva to adult, one
can read the story of its evolution from
lung fish: the larva uses gills to breathe
and openings along its lateral line to
sense its environment; in the adult these
are lost, and lungs, limbs and digits
develop. Aquatic larvae and thin permeable skin vulnerable to water loss and
sunlight prevent amphibians from
entirely living on land and limit their
radiation into arid habitats. Although
early amphibians had lumbered ashore
in search of insects, vertebrates didn't
finally leave the water until later in the
Paleozoic when the first reptiles evolved
waterproof skin and an egg with membranes (amnion, chorion) to protect
embryos from desiccation.

The evolutionary radiations of
modern amphibians and reptiles, as well
as of modern mammals and birds,
began as the dinosaurs declined in the
late Cretaceous (98-65 mya). Most general adaptations to aridity evolved in the
dry seasons of tropical deciduous

forests from the Eocene (about 45 mya) through the middle Miocene (15 mya), long before the deserts of North America came into being. The adaptations of Sonoran Desert endemics likely evolved in tropical deciduous forests or thornscrub. The uplift of the Sierra Madre Occidental by 15 mya changed weather patterns. Preadapted reptiles thrived as increasing aridity formed the Sonoran Desert by the late Miocene (8 mya)

Desert environments present great difficulties to amphibians. Tiger salamanders and lowland leopard frogs enter the desert only near permanent ponds, streams or springs. Tiger salamanders often become *neotenic* (retaining their larval forms) even reproducing as larvae, and only rarely metamorphosing into terrestrial adults.

The Sonoran Desert toad, desert spadefoot, northern casque-headed treefrog and others survive in the desert because of their abilities to excavate burrows as much as three feet deep where they spend nine or ten months at a time. Spadefoots and the northern casque-headed treefrogs have hardened areas, called spades, on their hind feet with which to dig. To prevent water loss in the burrows, spadefoots secrete a semipermeable membrane that thickens their skins, while the casque-headed treefrog forms a cellophane-like cocoon by shedding outer layers of skin. Spadefoots have a high tolerance for their own urea, since they do not excrete while in their burrows.

The ultimate challenge for desert amphibians is to reproduce in the temporary pools produced by highly sporadic and localized summer thunderstorms. Most breeding occurs at night with females attracted to calling males. The desert spadefoots evolved an accelerated development rate—from egg to toadlet in less than two weeks! In southeastern California, where summer rainfall is less dependable, spadefoots emerge during the first storm, travel to ponds, call and breed, and gorge on lipid-rich, swarming termites, often in single night. The adults may have only enough fat reserves to survive for a year without feeding.

To meet the challenge of life in temporary pools, the spadefoot matures at break-neck speed: egg to toadlet in under two weeks!

Primitive reptiles were able to radiate into drier habitats than amphibians because of the amniote egg with a leathery or hard shell, and because of their relatively impermeable skin with scales. Populations no longer were concentrated near water sources and embryos developed directly into small adults at hatching.

Since reptiles have thin skin with little insulation and most do not produce heat internally to fuel their metabolisms, adaptations to regulate body temperature (*thermoregulation*) are very important. Thermoregulation is possible because of complex relationships between body temperature, physiological processes (chemical reactions, hormone production, etc.) and behavior. Activity patterns change with the seasons, from midday in spring and fall to early morning and late afternoon in summer. Nocturnal reptiles such as the banded gecko and most snakes passively exchange heat with the air and soil. In contrast, diurnal lizards absorb heat by basking in the sun. Relatively uniform body temperatures are maintained in a number of ways: through the timing of daily activities, by shuttling in and out of shade and changing body orientation to the sun (*insolation*), by adjusting contact with the surface to regulate heat transfer (*conduction*), by changing color (dark skin absorbs energy faster), and so on. Additionally, some desert reptiles can tolerate quite high body temperatures; the desert iguana's active range, for example, is 100 to 108°F (38-42°C).

During times of environmental stress, desert reptiles spend long periods of inactivity in burrows, often borrowed from those dug by rodents or other mammals. During hibernation in winter and estivation in summer, animals in burrows have greatly reduced metabolic processes. They live on water and nutrients stored in their bodies, while wastes accumulate to potentially-toxic levels. Desert tortoises, for example, have a large urinary bladder that can store over 40 percent of the tortoise's body weight in water, urea, uric acid and nitrogenous wastes for months until they are able to drink. Urates are separated from water and can be eliminated in solid form, freeing water and ions to be reabsorbed. During extended droughts while the tortoises are inactive, they can reabsorb minerals from their shells to use in their metabolic processes. The giant Isla San Esteban and spiny chuckwallas on islands in the Gulf of California have a pair of lateral lymph sacs in the sides of their bodies that allow them to store extracellular fluid. Chuckwallas and Gila monsters, as well as the barefoot and western banded geckos store water in fatty tissue in their tails.

Species in the Lower Colorado River Valley of Arizona and California and the Gran Desierto of northwestern Sonora have a number of specializations for living in loose windblown sand. Sidewinders have evolved with an unusual form of locomotion where the body contacts the surface at only two points as it lurches along. The flat-tailed horned lizard and the Baja California legless lizard (a snakelike burrowing lizard about the size of a lead pencil, restricted to a small area on the western coast of Baja California) have lost the sand collecting external ear openings present in most lizards. Several species, including the legless lizard, banded sand snake, and shovel-nosed snake, have small eyes, narrow heads, counter-sunk lower jaws, and very smooth scales— adaptations to swimming and breathing in loose sand. The fringe-toed lizard has pointed, fringe-like scales on the elongated toes of its hind feet to give it traction as it runs across dune surfaces. The wedge-shaped head, nasal valves, ringed eyelids, scaly ear flaps, and fine body scales allow this lizard to escape predators by diving and burrowing into sand.

Thus, amphibians and reptiles use a variety of mechanisms not merely to survive extreme heat and aridity but actually to thrive in hot, dry deserts. Virtually all of these adaptations were inherited from tropical ancestors before the late Miocene formation of the Sonoran Desert.

Reptile & Amphibian Accounts

Craig Ivanyi, Janice Perry,

Thomas R. Van Devender (desert tortoise)

and Howard Lawler (desert tortoise)

Herpetologists (people who study amphibians and reptiles) from other parts of the United States have long been intrigued by the unique and rich assemblage of reptiles and amphibians (herpetofauna) of the Sonoran Desert. This diversity is not surprising considering the complex landscapes of mountains and valleys with rocky to sandy soils, the variable climate with two distinct rainy seasons, and the diverse vegetation. But the diversity of reptiles and amphibians reflects evolutionary origins intertwined not only with the history of the Sonoran Desert itself, but also with other major biotic provinces such as the grasslands of the mid-continent to the east and the "New World" tropics to the south.

In the following section, accounts are given for the species commonly encountered in the Arizona Upland subdivision of the Sonoran Desert. Other species of special interest that are seen less often or occur in other habitats in the region will also be covered.

Section Contents

Couch's spadefoot (*Scaphiopus couchi*)

Order: Salientia
Family: Pelobatidae (spadefoots)
Other common name: spadefoot toad
Spanish names: sapo con espuelas

DISTINGUISHING FEATURES

Couch's spadefoot is a 3 inch (8 cm), smooth-skinned, greenish, yellowish, or olive spadefoot with irregular blotches or spots of black, brown, or dark green. The belly is white and without markings. At the base of each hind foot is a dark, sickle-shaped keratinous "spade," hence the name spadefoot. The width of the eyelids is approximately the same as the distance between the eyes. The pupils are vertical.

RANGE

In the southwestern United States, Couch's spadefoots range from southeastern California through southern Arizona and southern New Mexico. They continue into all of Texas, except the extreme north and east, and northward to southwestern Oklahoma. In Mexico, this frog is distributed along eastern Baja California and on the western and eastern coasts of mainland Mexico south to Nayarit and southern San Luis Potosi.

HABITAT

Couch's spadefoots do well in extremely xeric (dry) conditions in areas with sandy, well-drained soils often occupied by creosote bush and mesquite trees. They are also found in short grass prairies and grasslands, cultivated lands, and along desert roadways during summer thunderstorms.

Though often called spadefoot toads, spadefoots are not true toads and should therefore simply be called spadefoots.

LIFE HISTORY

During summer monsoons, the spadefoot is well-known for emerging from its subterranean estivation to breed in the temporary ponds created by the heavy runoff. Interestingly, the cue for adult emergence during these summer thunderstorms is not moisture, but rather low frequency sound or vibration, most likely caused by rainfall or thunder. Upon emergence, males begin calling to attract females. Their calls sound like the bleating of sheep or goats. One female may lay as many as 3000 eggs. Once the eggs are laid, they must hatch quickly into tadpoles before these shallow pools disappear. And hatch quickly they do—at water temperatures of 86°F (30°C) eggs hatch in 15 hours! Tadpoles must also metamorphose quickly—2 weeks on average, sometimes as little as 9 days—into froglets before the ponds dry up. In this exacting atmosphere very few eggs make it to young frogs.

Using the spade on the hind foot, spadefoots can quickly bury themselves in loose, sandy soil. Adult spadefoots burrow into the ground to avoid heat and desiccation, but recently metamorphosed spadefoots may be seen during and immediately after the rainy season in any moist place—under vegetation, former ponds, or moist soil. During this time young spadefoots need to eat enough food to survive the unfavorable living conditions above the surface of the ground. After eating as much as possible, they too burrow beneath the surface. Breeding may not occur in years with insufficient rainfall. Preying primarily upon beetles, grasshoppers, katydids, ants, spiders, and termites, a spadefoot can consume enough food in one meal to last an entire year!

COMMENTS

Couch's spadefoots have a skin secretion that may cause allergic reactions in some humans. Cuts and scratches may become painful, and sneezing and discharge from the eyes and nostrils may also result from the handling of this amphibian.

Sonoran Desert toad (*Bufo alvarius*)

Order: Salientia
Family: Bufonidae (true toads)
Other common names: Colorado River toad
Spanish name: sapo grande

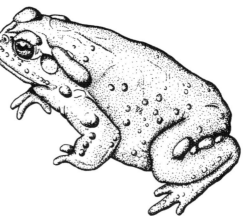

DISTINGUISHING FEATURES

At 7 inches (18 cm) or more this is one of the largest toads native to North America. Adults have a uniformly green to greenish-gray dorsum (topside of the body) and creamy white venter (underside). Large white turbercles, or "warts," are found at the angle of the jaw, but aside from the large parotoid glands and a few large lumps on the hind legs, this species has relatively smooth skin. Recently metamorphosed toadlets will be tan to green with orange or red spots on the dorsum. Unlike other male toads in our region, male Sonoran Desert toads do not have dark throats; males develop darkened, thick callosities (calluses) on the inside of the thumbs of the forelimbs during the breeding season.

An Open Door Policy

Colorado River toads are impressive amphibians, the largest toads in the Sonoran Desert, with an equally impressive diet of insects, including the large palo verde wood borer beetles. Each summer we are regularly visited inside our home by some of these toads. One in particular visited us for five consecutive years.

The welcome sighting of this toad always indicated to us that the summer rains were due to commence almost immediately. He made his appearance in the house by entering through the flap-type dog door, after which he traveled about the house consuming any insects he found along the way. Eventually he arrived in the sun porch off our living room, where he plopped himself in the water of a fountain in the corner of that room. We are not sure, but believe that he hibernated one winter in the soil of a flowerbed adjacent to the fountain.

It has been several years since we last saw this, our favorite toad, but others of his species visit us each summer.

—Bill Woodin, ASDM Director Emeritus & Beth Woodin, ASDM trustee

RANGE

Found from Central Arizona to southwestern New Mexico and Sinaloa, Mexico; historically entered southeastern California, though it has not been seen there since the 1970s.

HABITAT

This toad is common in the Sonoran Desert. It occurs in a variety of habitats including creosote bush desertscrub, grasslands up into oak-pine woodlands, and thornscrub and tropical deciduous forest in Mexico.

LIFE HISTORY

Sonoran Desert toads feed upon a variety of insects throughout their lives. Adults eat primarily beetles, although large individuals will occasionally eat small vertebrates including other toads. Sonoran Desert toads are active from late May to September, though principally during the summer rainy season. They are nocturnal during the hot summer months. The male's call is weak, sounding somewhat like a ferryboat whistle. Eggs are laid in temporary rainpools and permanent ponds. Larvae metamorphose after 6 to 10 weeks. This species lives at least 10 years, and perhaps as many as 20 years.

COMMENTS

Sonoran Desert toads have extremely potent, defensive toxins that are released from several glands (primarily the paratoids) in the skin. Animals that harass this species generally are intoxicated through the mouth, nose, or eyes. Dog owners should be cautious: the toxins are strong enough to kill full grown dogs that pick up or mouth the toads. Symptoms of intoxication are excessive salivation, irregular heartbeat and gait, and pawing at the mouth. If a dog displays any of these symptoms, use a garden hose to rinse its mouth from back to front and consult a veterinarian.

Sonoran green toad (*Bufo retiformis*)

Order: Salientia
Family: Bufonidae (true toads)
Other common names: reticulated toad
Spanish name: sapo

DISTINGUISHING FEATURES

The Sonoran green toad is small, reaching only 2¼ inches (57 mm) in length. Brightly colored, this toad is green to greenish-yellow with reticulations (net-like lines) or spots of black or brown on the back and legs, and numerous small, black-tipped warts on the back and sides. The underside is white with an occasional speck or two of black. The parotoid glands are large. Males have a dark throat.

RANGE

Endemic to the Sonoran Desert region, this toad is found from south-central Arizona to west central Sonora.

HABITAT

The Sonoran green toad is found along washes in mesquite grasslands and creosote bush flats between 500 and 1500 feet (150 to 450 m).

A Time for Toads

Toad and spadefoot activity is highly correlated with the monsoon season. Some species may be active as early as late spring while others will be out only after summer rains. If it is cool enough, desert amphibians may occasionally be active during the day. However, most species are primarily active at night when one often hears the strange calls of males from quite a distance. Some sound like bleating sheep, others chirp, snore, or wheeze; some make almost no sound at all.

LIFE HISTORY

Once the summer rains begin, males move into grasses around temporary rainwater pools and washes and begin to call. The call lasts a few seconds and sounds like a combination buzz and whistle. Hatchlings are only ⅛ inch (3.5mm) in length—smaller than a pea!

COMMENTS

Male Sonoran green toads, like all *Bufo* species, have rudimentary ovaries that can become functional if the testes are damaged or removed.

red spotted toad (*Bufo punctatus*)

Order: Salientia
Family: Bufonidae (true toads)
Spanish name: sapo

DISTINGUISHING FEATURES

This small, up to 3 inch (76 mm) long toad has round parotoid glands, a characteristic which distinguishes it from other toad species in the region. It tends to be whitish when found in association with limestone, light tan to red around volcanic rocks, to brown above, with scattered reddish tubercles (raised bumps); the underside is creamy white. Males have dark throats and single vocal sacs. The body and head are dorsoventrally compressed, giving this toad a flattened appearance.

RANGE

This toad is found from southern Nevada to southwestern Kansas, south to Hidalgo, Mexico, and throughout Baja California. It occurs from below sea level up to 7000 feet (1980 m).

HABITAT

A riparian inhabitant, this species is commonly encountered in and around rocky streams and arroyos. Its flattened body allows it to wedge into narrow rock crevices.

LIFE HISTORY

The red spotted toad is insectivorous. It breeds mainly after summer rains in quiet pools. The call of the male is a high-pitched musical trill, which may be confused with the sound of a cricket. This is the only toad species native to our region that lays its eggs singly. Tadpoles metamorphose in 6 to 8 weeks. This species is nocturnal through the hot summer months, but may be active in the morning or late afternoon when temperatures are cool enough.

canyon treefrog (*Hyla arenicolor*)

Order: Salientia
Family: Hylidae (treefrogs)
Spanish name: rana

DISTINGUISHING FEATURES
This is a small treefrog with highly
variable pattern and color. It grows to
2¼ inches (56 mm) with a ground
color of cream to brown and irregular
bars, blotches, and spots of olive to
brown. Its color matches its substrate
extremely well. Large adhesive toe pads
are present on all 4 feet. Adult males
have dusky (darkened) or yellow
throats, whereas females have white to
cream colored throats (which match
the underside).

RANGE
Barely entering Colorado, this treefrog
is mainly found in southern Utah,
western New Mexico, southwest Texas,
all but western Arizona, and northern
Mexico at elevations up to 9800 feet
(2990 m).

HABITAT
This species is largely restricted to riparian areas in rocky canyons. It is typically
found along streams among medium
to large boulders from desert to desert
grassland and into oak-pine forests.
The canyon treefrog can operate at
cooler temperatures than many frogs;
it avoids cold surface temperatures by
retreating underground.

LIFE HISTORY
The canyon treefrog eats insects of
various kinds. It breeds in July and
August during summer rains as well as
in spring. The abrupt, explosive call of
males attracts females to breeding sites;
males then mount females and spawning may begin. Eggs are laid in a large
mass that floats on the surface of the
water.

COMMENTS
This is an extremely well camouflaged
species that usually does not move
until a potential predator is almost on
top of it.

northern casque-headed frog (*Pternohyla fodiens*)

Order: Salientia
Family: Hylidae (treefrogs)
Other common names: lowland burrowing treefrog
Spanish name: rana, sapito

DISTINGUISHING FEATURES
This treefrog reaches 2½ inches (63 mm) in length and has a tan to light brown dorsum with a dark brown network of blotches and bars and a creamy white underbelly. The toes are slightly webbed and the toe pads are small for a treefrog. The name "casque," which means helmet-shaped, is given to this frog because the skin of the head is fused to the skull; there is a fold of skin at the back of the head. Males have dark throats and a double vocal sac.

RANGE
This toad barely enters the United States in south central Arizona but it is common in western Mexico from Sonora to Michoacán. Occurs from sea level to 4900 feet (1490 m).

HABITAT
This species inhabits desertscrub to thornscrub. In Sonora this species is more typically found in riparian areas.

LIFE HISTORY
The northern casque-headed frog is insectivorous, terrestrial or fossorial (burrowing), and nocturnal. It breeds from June to September during the summer rainy season. The male's call is an explosive, hoarse "wauk-wauk-wauk."

COMMENTS
After burrowing underground, this species sheds several layers of its epidermis that form a virtual "cocoon" around the entire body and probably reduce dehydration in the dry season. In addition, this species uses its head to block the opening to its burrow. This reduces water loss from the frog's body, and may protect it from some predators.

leopard frog (*Rana* spp.)

Order: Salientia
Family: Ranidae (true frogs)
Spanish name: rana

DISTINGUISHING FEATURES
At least 9 species of leopard frogs and several close relatives are found in the Sonoran Desert region. Differences among species are small and often indistinguishable to the human eye. All species are fairly large with pointed snouts, webbed hind toes, long, powerful hind limbs, and large external eardrums. Colors vary from tan to green or brown with irregular patterns intermittent streams from near sea level to 7900 feet (2410 m).

HABITAT
These frogs inhabit permanent and intermittent streams, irrigation canals, and some ponds.

Vanishing Frogs

*There appears to be a world-wide decline in amphibian populations. In our region, leopard frogs are much scarcer than in years past. A close relative, the Tarahumara frog (*Rana tarahumarae*), has disappeared from Arizona, the only place in the United States where it occurred. No one knows for sure why amphibian numbers are decreasing, but surely there is more than one cause. Water pollution, acid rain, ozone depletion and excessive ultra-violet radiation, habitat destruction, and introduced species (such a bullfrogs in Arizona) are being examined as potential causes for the crisis. Unfortunately natural populations of amphibians have not been subjects of intensive, long-term study. We therefore rarely know about normal population fluctuations and cannot easily determine if what we are seeing today is an overall trend or just a temporary aberration.*

composed of spots, bars, and blotches of darker green, brown or black.

RANGE
These frogs occur from coast to coast and from Canada through Mexico. They are found throughout the Sonoran Desert along permanent and

LIFE HISTORY
Leopard frogs are insectivorous and piscivorous (fish-eating). Highly aquatic, they often jump into water from the streambank to avoid capture. Leopard frogs may breed year-round, and tadpoles may take more than a year to metamorphose. Tadpoles get very large.

western box turtle (*Terrapene ornata*)

Order: Testudines
Family: Emydidae (subaquatic turtles)
Other common names: yellow box turtle
Spanish name: tortuga de caja

DISTINGUISHING FEATURES

The words *Terrapene ornata* translate
to "ornate (patterned) small turtle,"
which aptly describes this 6 inch
(15 cm) long, colorful, terrestrial
turtle. Both the high-domed *carapace*
(upper part of the shell) and hinged
plastron (lower part) are black or brown
with radiating yellow lines and dots.
The male has an orange to reddish iris,
reddish spots on the foreleg, and an
inwardly curved nail on each hind foot.
The shells of older animals often lose
their pattern, becoming a uniform pale
green or straw color.

RANGE

This turtle is found as far north and east
as South Dakota, Michigan, and Indiana,
south through southeast Arizona, New
Mexico and Texas into northern Mexico.

HABITAT

The western box turtle of grasslands,
found in treeless plains to gentle hills
with grass or low bushes and sandy
soils, though occasionally it is found in
desert habitats. It tends to create shallow burrows in loose soils; it will also
use mammal burrows and bannertail
kangaroo rat mounds. These burrows
are used to avoid temperature extremes
and reduce desiccation.

LIFE HISTORY

The western box turtle is omnivorous,
feeding on insects (especially beetles),
berries, leaves, fruits, and sometimes
carrion. It reproduces from March to
November, laying 2 to 8 eggs per
clutch. Breeding strongly correlates
with rainfall.

COMMENTS

Box turtles can be either timid or
pugnacious, retreating into their tough
shells or defending themselves with
strong beaks.

Sonoran mud turtle (*Kinosternon sonoriense*)

Order: Testudines
Family: Kinosternidae (mud and musk turtles)
Spanish name: tortuga del agua

DISTINGUISHING FEATURES

The Sonoran mud turtle is medium sized (up to 6½ inches, 16.5 cm) with a smooth, high-domed *carapace* (upper part of the shell). The name *Kinosternon* means moveable chest, which refers to the hinged *plastron* (lower part of shell). The Sonoran mud turtle can be distinguished from the yellow mud turtle by noting its non-enlarged ninth marginal *scute* (indicated by the arrow on the illustration). The same scute on a yellow mud turtle is usually taller than it is wide. (Scutes are the large, hornified plates that cover the shell.) This turtle generally has a uniform light brown or yellowish-brown shell. The shell is often partially covered with algae. The head and neck have light and dark marks; there are several barbels (fleshy projections) located under the chin. Males have tails with a hooked tip and 2 groups of thickened scales on the inner surfaces of the hind legs.

RANGE

This species is widespread from central Arizona to Durango, Mexico and southeastern California to west Texas up to 6700 feet (2042 m).

HABITAT

A stream dweller, the Sonoran mud turtle is usually found in springs, creeks, ponds, and intermittent streams. Though occasionally found in desert and grassland areas, this turtle usually inhabits oak to pinyon-juniper woodlands or pine-fir forest.

LIFE HISTORY

Although the Sonoran mud turtle favors invertebrate prey such as crustaceans, insects, and worms, it also eats some fish and frogs. It is also known to scavenge. Females lay 2 to 9 eggs from May to September.

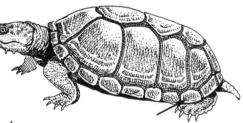

desert tortoise (*Gopherus agassizii;* also *Xerobates agassizii*)

Order: Testudines
Family: Testudinidae (tortoises)
Spanish names: tortuga del desierto, tortuga de los cerros, tortuga del monte

DISTINGUISHING FEATURES

Tortoises differ from other turtles in having cylindrical and elephantine hindlegs and short, broad, club-shaped feet. The genus *Gopherus* also has flattened forelimbs for digging. The adult desert tortoise can measure up to 14 inches (35.5 cm) in length; the hatchlings are only about 2 to 2½ inches (5 to 6.5 cm) long. The carapace is brown to gray and rounded. The yellowish plastron is not hinged and is connected to the carapace at the sides. The male's plastron is concave at the posterior to accommodate the rounded carapace of the female during copulation. The female's plastron is flat. Like all turtles it is toothless; the large tongue helps push food back in the mouth. There are no visible ears.

DISTRIBUTION

The desert tortoise lumbering over rocks on a hillside among foothills palo verdes and saguaros is the modern representative of a lineage that began 50 million years ago in early Tertiary tropical forests long before the Sonoran Desert existed. Today, the desert tortoise occurs from tropical areas in northern Sinaloa and southern Sonora northward through the Sonoran Desert in central Sonora and western Arizona to the Mohave Desert in southeastern California, southwestern Utah, and southern Nevada. The climates in these areas range from tropical in southern Sonora to temperate in southern Nevada. Tortoises inhabit vegetation ranging from tropical deciduous forest and thornscrub in the south to various Sonoran and Mohave desert communities in the north, to desert grassland in Arizona.

ECOLOGY

The ecology of desert tortoises in the Mohave Desert differs in important details from that of its Sonoran Desert and tropical relatives. Tortoises in Nevada and Utah typically live in relatively flat valley bottoms and construct extensive burrows up to 35 feet (11 m) in length that are often shared with other tortoises. These tortoises excavate their own dens, allowing the populations to reach the highest known densities for their species. Tortoises in the Sonoran Desert in Arizona live on rocky, boulder-strewn hillsides and are mostly solitary. Burrows are shallow and casually made; they also make use of existing crevices or depressions. Populations of these tortoises are limited by available natural shelter sites.

Desert tortoises are well-adapted to withstand the extended dry periods typical of deserts. Although they extract much of their water from the plants they eat, tortoises drink prodigiously from temporary rain pools. They have large urinary bladders that can store over 40 percent of their weight in water and urinary wastes. Urea is precipitated as solid uric acid in the bladder, freeing additional water and useful ions. During periods of inactivity in winter (hibernation) or summer (estivation), metabolic rates, digestion, and water loss from defecation and urination are greatly reduced. As soon as fresh water is available, the solid urates are eliminated from the bladder. Well-hydrated tortoises are able to eat dried plants and store fats in the body. Dehydrated tortoises are physiologically stressed and cannot digest dry plant foods.

Tortoises are generally active in times when water stress is reduced (early morning or evening); they can be diurnal or crepuscular depending on temperature and season. Mohave tortoises are mostly active in the spring months from February to May. Sonoran tortoises are primarily active in the summer monsoon months from July to October. Estivation during the hottest, driest parts of the summer helps conserve water because burrows or rockshelters are relatively cool and relative humidity of up to 40 percent reduces evaporation. Without extensive burrows it is unlikely that tortoises could survive the dry Mohave Desert summers. The common defensive behavior of emptying the bladder when molested or handled can have serious consequences in drier periods.

Desert tortoises are herbivores, although occasionally carrion, insects, rocks, bones, and soil are ingested. The intestines are elongated and can digest the cellulose tied up in plant fiber. In the Mohave Desert, fresh spring annuals are the primary food plants. In the Sonoran Desert, tortoises eat mostly grasses, globemallows, desert vine (*Janusia gracilis*), and many other plants. Spring annuals are eaten fresh in April and dry from May to October. In August and September, Sonoran tortoises supplement their diets with summer annuals, especially spurges (*Euphorbia* spp.) and verdolaga (*Portulaca oleracea*), and prickly pear (*Opuntia* spp.) fruits.

The social behavior of tortoises is relatively straightforward—males fight all adult males and court all adult females they encounter. Mating occurs throughout the summer, although females lay eggs in late June or early July in the Sonoran Desert. Mohave Desert tortoises typically lay a second clutch of eggs at the end of the summer. Retention of viable sperm in the cloaca of the female for at least 2 years is an excellent survival strategy in non-colonial animals: it ensures fertilization during extended droughts when males are less active or populations crash. The eggs contain all of the water and nutrients necessary for complete development of the hatchlings. Hard egg shells retard desiccation. The normal incubation period of about 90 days in the summer can be much longer later in the year as temperatures fall.

Tortoises are long lived. In the Sonoran Desert, free-ranging adults live about 35 or 40 years and spend their lives within a few miles of where they hatch. Sexual maturity is attained at 13 to 15 years, at which point the carapace is about 7 inches (18 cm) long. Growth is controlled by environmental conditions and amount and quality of food. Captives with unlimited water and high protein diets grow and mature faster than wild animals. Ages can be estimated using the annual rings on the epidermal scutes, although tortoises may not grow in especially severe years, and the rings are worn off in older animals. Few hatchlings survive to maturity except during a series of unusually wet years.

In the Mohave Desert, populations have declined in response to habitat destruction, a debilitating upper respiratory disease, and predation of young by ravens. In 1990, this population was designated as Threatened by the United States Fish and Wildlife Service. In Arizona, most populations of Sonoran tortoises studied are healthy. However, there are serious threats to desert tortoise populations in Arizona and Sonora as grasses, especially the exotic red brome (*Bromus rubens*) and buffelgrass (*Pennisetum ciliare*), introduce fire into vulnerable Sonoran desertscrub and thornscrub habitats.

No matter where they live, desert tortoises are protected. They cannot be collected, killed, transported, bought, sold, bartered, imported, or exported from Arizona without authorization by the Arizona Game and Fish Department. Tortoises held in captivity before protection was implemented, and their progeny, may be possessed legally. Captive tortoises should not be released into the wild for a variety of reasons: they may transmit disease to the wild population; they may disrupt natural population levels and/or social structure; they may not survive due to maladaptation resulting from long-term captivity and captive diets.

For these reasons, the Desert Museum has established a tortoise adoption program which places unwanted or surplus captive tortoises with qualified private custodians. This reduces incentive to collect them from the wild and helps to satisfy some people's interest in having them. Applicants for the program are carefully screened for attitude and aptitude. The custodian must comply with ASDM standards of housing, care and diet before receiving a tortoise, which remains the property of the State of Arizona. The program is not a pet program, but rather a captive tortoise "recycling" effort to insure that tortoises already in captivity are provided the best life possible.

western banded gecko (*Coleonyx variegatus*)

Order: Squamata
Family: Gekkonidae (gekkos)
Spanish name: salamanquesa de franjas

DISTINGUISHING FEATURES
This delicate-looking lizard seldom exceeds 3 inches in length, excluding the tail. It has moveable eyelids and large eyes with vertical pupils. The small body scales are granular and soft; the toes are slender. There is a constriction at the base of the otherwise bulky tail. The tail is about as long as the body with indistinct rings. Between the pairs of legs are dark brown crossbars on a pale yellow, pink, tan, or cream background. The eyelids are edged in white. The head and body are mottled with light brown. The belly is somewhat translucent. Males have prominent spurs on either side of the body at the base of the tail.

RANGE
The western banded gecko occurs in the Mohave and Sonoran deserts.

HABITAT
The western banded gecko is found in open arid deserts and desert grassland, in canyons and on hillsides. It is usually associated with rocks or other shelters, but is also is found in sandy arroyos and dunes.

LIFE HISTORY
Active principally at night, western banded geckos can be seen crossing roads during the summer. It has been suggested that their gait and carriage mimics that of the scorpions of the genus *Hadrurus* that share the same habitat. If disturbed, the gecko will wave its tail to divert attention of a would-be predator away from its head and body. The tail has specialized fracture planes that allow it to easily break off. Blood vessels close off rapidly to prevent much blood loss and the writhing tail is left behind. This may allow the lizard to escape predation; its tail is very rapidly regrown. However, the regenerated tail consists of cartilaginous material that lacks fracture planes; it is also shorter than the original and has different color patterns and scales.

The tail of the gecko serves as more than a way of escaping predation. It also stores food and water that the animal uses during lean times, including winter dormancy. Regrowing the tail is energetically expensive and the loss of the tail can put the survival of the lizard in jeopardy—especially if it was lost just before the onset of winter.

Banded geckos feed on a variety of invertebrates including beetles, spiders, grasshoppers, sowbugs, termites, and solpugids. Two eggs are laid in late spring, with females laying 2 clutches a year. After 6 weeks, the eggs hatch into 1 inch (2.5 cm) long lizards.

COMMENT

This small lizard is often mistaken for a young Gila monster due to the simi- larity of the pattern. A banded gecko can be distinguished from a Gila monster by its small size (young Gila monsters are around 6 inches; 15 cm) and the lack of bead-like scales on the back. Banded geckos may emit a squeak when captured.

whiptails (*Cnemidophorus* spp.)

Order: Squamata
Family: Teiidae (whiptails)
Other common name: racerunner
Spanish name: huico

DISTINGUISHING FEATURES

Whiptails are long, slender lizards with pointed snouts and extremely long tails. Snout-vent lengths range from 2¾ inches (69 mm) to 5½ inches (137 mm) among the various species of whiptails. Giant spotted whiptails (*C. burti*), with tails longer than their bodies, can have a total length of over a foot. Color tends to be tan, olive, or brown with lighter stripes and/or spots of yellow or white. Male western whiptails (*C. tigris*) may have very dark forelimbs, throats and upper torsos. Whiptails have large, square belly scales arranged lengthwise and transverse rows. The scales on the upper part of the body are very small and granular in appearance.

Western whiptail

RANGE

Whiptails are found throughout the Sonoran Desert region from sea level up to 8000 feet (2440 m).

HABITAT

These lizards occupy low desertscrub through grasslands, woodlands and pine forests. They are often found under rocks or nosing around leaf litter.

LIFE HISTORY

Whiptails feed on a variety of terres- trial invertebrates and occasionally on smaller lizards. Most species repro- duce sexually and lay 1 or more clutches of 1 to 6 eggs in late spring or early summer. However, in Arizona approximately 60 percent of whiptail species are *parthenogenetic*, meaning that they reproduce asexually. These species, such as the Sonoran spotted whiptail (*C. sonorae*), consist entirely of genetically identical females that lay unfertilized eggs, creating a popula- tion of clones. Oddly enough, many of the behaviors exhibited by sexually reproducing species are expressed by

Lizard Displays

Most Sonoran Desert lizards use a variety of behaviors in a purely social context. Displays to establish dominance or territory or to aid in courtship are common. Though many species are similar in their behavioral repertoires, each species' behaviors are unique to that species. They may differ only slightly, such as with head-bobs, where the number of or intensity of them may be the only difference. Then again, some are vastly different. The open-mouthed gape and vertical extension of the body and throat of collared lizards, which serve as a challenge from one male to another, appear quite different from the four-legged push-ups accompanied by the display of the brightly colored dewlap of tree lizards, though both serve the same purpose.

Some of these displays are used only within a species, while others may be used between species. Aggressive male collared lizards will display to establish dominance over just about any other lizard they can intimidate. Often, agonistic (aggressive) behaviors are ritualized display between males. These behaviors serve to reduce physical contact and the potential for injury to either animal.

An inventory of a species' repertoire may include gaping, lunging, chasing, biting, inflation of the body or throat, head bobs, and push-ups, as well as subtler shudders of the body. Some of these actions will be used only between males, while others facilitate courtship. What may superficially resemble agonistic push-ups may actually be a subtler shudder of the fore-body, which a male uses to court a female.

Taking a closer look at how lizards interact with one another can be fascinating entertainment, as well as a way to give the observer a better understanding of how animals that cannot speak actually communicate desire, intent, and need to each other.

---

these parthenogenetic lizards. Females will engage in pseudocopulation and mount and bite other females. Apparently this triggers hormonal changes necessary for ovulation and egg laying. Eggs typically take 60 to 75 days to hatch regardless of reproductive style.

COMMENTS

Lizards' behavioral habits vary considerably. Some are day active. Others, such as most geckos, are nocturnal.

Some are sit-and-wait predators, while others are active foragers. Whiptails fit the latter description in the extreme. Beyond actively foraging, they forage intensely for prey, often at a frenetic pace.

When they are on the move under plants or through leaf-litter, their jerky, start-stop movements create unmistakable and unique sounds. Someone familiar with whiptail lizards can often locate and identify a whiptail from auditory cues alone.

Gila monster (*Heloderma suspectum*)

Order: Squamata
Family: Helodermatidae (venomous beaded lizards)
Spanish name: escorpión

DISTINGUISHING FEATURES

The Gila monster is a large, heavy-bodied lizard reaching a little over 1½ feet in length. The head is large, with small, beady eyes; the tail is short and fat. The family name Helodermatidae means "warty skin," referring to the beaded look of the dorsal scales, due to the presence of osteoderms (small bones) under the scales. The lizard is bright pink and black, usually in a reticulated pattern, but in a banded pattern in some populations.

RANGE

The bulk of this lizard's range is in western and southern Arizona, continuing to southern Sonora, Mexico, but it can also be found in extreme southeastern California, southern Nevada, extreme south-western Utah, and southwestern New Mexico.

HABITAT

The Gila monster is most commonly found in mountain foothills dominated by saguaros and palo verde trees. It also uses washes that extend down into valleys. It may use burrows dug by other animals, or construct burrows of its own.

LIFE HISTORY

Gila monsters prey on newborn rodents, rabbits, and hares, though ground nesting birds and lizards, as well as eggs from birds, lizards, snakes and tortoises are also eaten. Young Gila monsters may consume as much as 50 percent of their body weight in one feeding, while adults are capable of consuming 35 percent of their body weight in a single feeding. They are active mainly during the day from

March through November, and may be seen basking at the entrances to their shelters in winter and early spring. Hibernation takes place from the end of November through February. Some sources estimate they spend up to 98 percent of their time in their subterranean shelters. Generally an animal occupies two burrows over the course of a year, one from autumn through early spring and another during the warmer spring and summer months. The latter burrow is usually in or near a bajada, while a higher elevation, foothill burrow is used when cooler temperatures arrive. Little is known about reproduction in the wild. An average of 5 eggs, but as many as 12, may be laid in late summer. In southern

---

Dr. Ward's Prescription

Dr. Ward, of Phoenix, an old practitioner in the valley, says: "I have never been called to attend a case of Gila monster bite, and I don't want to be. I think a man who is fool enough to get bitten by a Gila monster ought to die. The creature is so sluggish and slow of movement that the victim of its bite is compelled to help largely in order to get bitten."

—*Arizona Graphic*
September 23, 1899

Arizona, Gila monsters breed in May and June, with eggs laid in late June through mid August. The eggs incubate and develop from fall to early spring; young appear the following April through June. There is no other known egg-laying lizard in North America where eggs over-winter and hatch the following year.

COMMENTS
Gila monsters are one of only two venomous lizards known to occur in the world. The other, the beaded lizard (*Heloderma horridum*), is found in southern Sonora and further south

in thornscrub and tropical deciduous forest.

Venom is produced in glands in the lower jaw and expressed along grooved teeth as the animal bites. Once the lizard bites, it generally holds on and chews more of the venom into its victim. Though the bite is rarely life-threatening to humans, it may cause pain, edema, bleeding, nausea and vomiting. A Gila monster's venom is believed to be a defensive weapon. The animal probably does not need venom to subdue its defenseless prey and the intense pain caused by the venom readily causes a predator to change its mind. Before biting, the lizard will hiss, gape, and back away from its would-be attacker. If these efforts fail, it will bite with amazing speed. Gila monsters should not be handled!

Reptiles, especially venomous ones, are often poorly understood and greatly feared. With the Gila monster this combination has led to misinformation and the generation of many myths. Interesting but untrue are stories about how the Gila monster is venomous because it lacks an anus and "all that stuff went bad in there." Or about how "once they bite down, they can't let go until sundown," or "if one bites you, don't worry, it has to turn upside down to get the venom in you." In 1952 the Gila monster became the first venomous animal in North America to be afforded legal protection; it is therefore illegal to collect, kill, or sell them in Arizona. Though it is an animal with a fairly large range, it has a spotty distribution primarily clumped around mountain ranges.

desert iguana (*Dipsosaurus dorsalis*)

Order: Squamata
Family: Iguanidae (iguanid lizards)
Spanish names: porohui, lagartijo

DISTINGUISHING FEATURES

This medium-sized lizard with a blunt head and long tail reaches a length of 16 inches (40 cm), including the tail. It is pale gray or whitish with a tan or brown reticulated pattern on the back and sides. Down the center of the back is a row of slightly enlarged, keeled scales. Rows of brown spots are on the tail, which is as long as the body from snout to vent. The belly is pale. During the breeding season, in both sexes, the sides become pinkish.

RANGE

This heat-loving, desert dweller occurs in southeastern California, southern Nevada, southwestern Utah, and western and south-central Arizona in the United States. The Mexican distribution includes eastern and southern Baja California, northwestern Mexico, and some of the Gulf of California islands. The range of the desert iguana is largely contained within the range of the creosote bush.

HABITAT

This lizard is most common in dry, sandy areas dominated by creosote bush. It can also be found in rocky streambeds up to 4000 feet (1200 m). In the southern portion of its range this lizard lives in arid subtropical scrub and tropical deciduous forest associations.

LIFE HISTORY

Desert iguanas emerge from hibernation in mid-March. Breeding occurs in April and May with 2 to 10 eggs laid from late May to early July. Hatchlings emerge late July through August. Desert iguanas may lay 2 clutches of eggs a year under the right conditions. This lizard is extremely tolerant of high temperatures and can be seen active during mid-day even in the hottest summers.

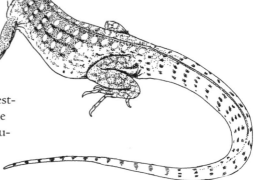

Primarily herbivorous, the desert iguana crawls into the branches of creosote bushes and other shrubs to feed upon the leaves and flowers. Additionally, it eats insects, carrion, and fecal pellets of its own species, which aids in the digestion of plant cellulose by establishing the proper gut fauna.

Studies have shown that secretions from the femoral pores (on the underside of the thighs of the rear legs) fluoresce and that the lizard has vision in these wavelengths. So, besides using these secretions as scent markings, they may also serve as visual indicators of a desert iguana's presence.

Though desert iguanas seem to prefer open, relatively flat habitat, they rely heavily on the creosote bush in these areas for a number of needs. This plant provides some of the lizard's diet (flowers), and the lizard burrows around and under the plant's roots to avoid thermal extremes and predation.

chuckwalla (*Sauromalus obesus*)

Order: Squamata
Family: Iguanidae (iguanid lizards)
Spanish name: iguana

DISTINGUISHING FEATURES
The chuckwalla is a large, bulky lizard reaching nearly 16 inches (40 cm) with folds of loose skin on the sides of its body. The color varies between sexes and with the age of the individual. Adult males have black heads and forelimbs; their trunks may be black, red, orange, gray, or yellow. Females and juveniles may have gray or yellow banding.

On the inside of the male's thigh are well-developed femoral pores, which are small openings that allow secretions to be exuded. These secretions are thought to be a way of marking areas.

RANGE
A resident of southwestern deserts in the United States and Mexico, the chuckwalla is found in southeastern California, southern Nevada, southwestern Utah, western Arizona, eastern Baja California, and northwestern Mexico.

HABITAT
Strictly a rock dweller, the chuckwalla is found in rocky outcrops, lava flows, and rocky hillsides of the Great Basin, Mohave and Sonoran deserts.

LIFE HISTORY
This herbivorous lizard emerges from hibernation in mid to late February, although it may be seen in rock crevices close to the surface on any warm winter day. During the active season, it emerges in the early morning to bask in the sun. It is active in temperatures as high as 102°F (39°C). When disturbed it seeks shelter in rock crevices and gulps air, wedging itself in a crack, thus making it extremely difficult for predators to extract it. During summer an average of 6 eggs are laid; hatchlings emerge in late September. The chuckwalla feeds mainly on annuals, but also eats perennials; it will consume insects on occasion.

If food resources are abundant, large male chuckwallas become territorial during certain parts of the year. Below the "tyrant" male, the other males will set up a dominance hierarchy based on size. When food is scarce, no territoriality is exhibited and some males form a hierarchy centered around food resources rather than the size of the animal. Often in these lean times, reproduction will not occur.

common collared lizard (*Crotaphytus collaris*)

Order: Squamata
Family: Iguanidae (iguanid lizards)
Other common names: mountain boomer
Spanish name: lagartija de collar

DISTINGUISHING FEATURES
This is a medium-sized lizard reaching nearly 10 inches (25 cm) in total length; males are larger than females. The head is large. Two black collars around the neck give the lizard its name. The small body scales have a ground color of tan, bright green, olive, brown, bluish or yellowish with many light spots and dark crossbands. The belly is whitish. Juvenile collared lizards have distinct banding that slowly fades as the animal matures. The adult male is usually very green with dark spots on the throat. Adult females are only slightly green. In breeding season spots and bars of bright red or orange appear on the sides of the female's body and neck indicating that she is pregnant; these fade after egg deposition. Color varies among the different populations throughout the range.

RANGE
The collared lizard is widespread throughout the western United States. In the Sonoran Desert, it is found in

Arizona, southeastern California, and northern Mexico, including eastern Baja California.

HABITAT
Collared lizards are found in rocky areas of a variety of habitats: pinyon-juniper, sagebrush, desertscrub, and desert grassland. They are usually in areas with open vegetation.

LIFE HISTORY
Collared lizards are capable of running swiftly on their hind legs, the body held off the ground at a 45° angle, with tail and forelimbs raised. THe stride is up to 3 times the length of the body. They do not lose their tails easily, as they are useful in maintaining

balance as the lizards sprint on hind legs. Speed facilitates the capture of prey by these visually oriented lizards. They have large heads with strong jaw muscles that allow them to get a powerful grip on large prey such as lizards. Though fairly bold, if confronted by a predator, collared lizards quickly dive into rock crevices to avoid being eaten.

These lizards often sit on large rocks basking in the sun and looking out for other individuals or food. Males are highly territorial and have stereotypical head-bobbing and push-up displays. Collared lizards primarily eat grasshoppers, but also eat other insects as well as lizards, including their own species. In early summer females lay 1 to 13 eggs; they are capable of reproducing more than once a year. Hatchlings emerge in late summer and early fall.

spiny lizards

desert spiny lizard (*Sceloporus magister*)
Clark spiny lizard (*Sceloporus clarkii*)

Order: Squamata
Family: Iguanidae (iguanid lizards)
Spanish name: cachora

DISTINGUISHING FEATURES
These are medium to large lizards with snout-vent lengths ranging from 2½ to 5½ inches (63 to 138 mm). These robust lizards have keeled, pointed scales. Background color is usually subdued gray, tan, or blue with a striking wide, purple stripe down the back and single yellow scales scattered on the sides (*S. magister*), or scattered turquoise scales mixed with tan and brown on the back and sides (*S. clarkii*). Both species have a dark collar under or around the neck; males have vivid blue throats and under-bellies. Females develop orange to red heads during the breeding season.

RANGE
S. magister occurs in 6 western states including almost all of Arizona; it occurs east to Texas and south to Sinaloa, Mexico; it is found from sea level to 5000 feet (1520 m). *S. clarkii* is found in central to southeastern Arizona, southwestern New Mexico, and south to northern Jalisco, Mexico, from sea level to around 6000 feet (1830 m).

Desert spiny lizard

Meeting One's Match

Collared lizards can be quite pugnacious. In fact we had a male in one of our enclosures that routinely visited all other similar-sized lizards, just to make sure they knew exactly who was in charge.

When we introduced a male desert spiny lizard to this enclosure, the collared lizard quickly ran over to assert himself. However once he was within a few inches of the spiny lizard he seemed to realize how large this newcomer was. The two lizards took positions next to each other, bodies parallel, then sized up one another with sidelong glances. Eventually they took off in opposite directions. Apparently neither felt superior enough to press the issue.

—Craig Ivanyi

HABITAT

S. magister and *S. clarkii* overlap in their use of arid to semiarid regions, lower mountain slopes, and subtropical thornscrub. *S. magister*, primarily an inhabitant of desertscrub and thornscrub, is found mainly on the ground in rocks and, less frequently, in trees. *S. clarkii* prefers trees, but also inhabits rocky areas with large boulders; it ranges from rocky Sonoran desertscrub into oak woodland. This species also occurs in tropical deciduous forest and oak-pine forest in Chihuahua, Mexico.

Clark spiny lizard

LIFE HISTORY

These two species are insectivorous. Both lay 4 to 24 eggs in the summer (into early fall for *S. clarkii*) which take 60 to 75 days to hatch.

COMMENTS

Like many other lizards, spiny lizards exhibit *metachromatism*, which is color change as a function of temperature. When it is cooler, colors are much darker than when the temperature is high. Darker colors increase the amount of heat absorbed from the sun and lighter colors reflect solar radiation.

tree lizard (*Urosaurus ornatus*)

Order: Squamata
Family: Iguanidae (iguanid lizards)
Spanish name: cachora, lagartija

DISTINGUISHING FEATURES

A small, up to 2¼ inch (56 mm), black, dark brown, tan, or gray lizard, often with a rusty area at the base of the tail. The ground color of this slim-bodied lizard is broken with a dusky pattern of blotches and/or crossbars. There are two bands of enlarged scales down the middle of the back, separated by a strip of smaller scales. Adult males have bright blue or blue-green belly patches that have a metallic sheen. The color of the throat varies from yellow to green or blue-green. The throat of females can be white, orange, or yellow.

RANGE

This species occurs from southwestern Wyoming to southern Sinaloa and northern Coahila, Mexico, and from the Colorado River east to central Texas. It is found from sea level to 9000 feet (2770 m).

HABITAT

This arboreal lizard most commonly lives in riparian zones in mesquite, alder and cottonwood, but it also is found on non-riparian oak, pine, and juniper. The tree lizard is also found on some non-native trees such as eucalyptus and tamarisk, and in some treeless areas; it is often very abundant on granite boulder piles. Color and pattern serve it well in avoiding detection by would-be predators.

LIFE HISTORY

The tree lizard eats insects and spiders. It reproduces 1 to 6 times per year, laying 2 to 13 eggs per clutch from March through August.

COMMENTS

Secondary sexual traits, such as the brightly colored throat fold (*dewlap*) of the tree lizard, are often considered to play a part in mate selection (as well as in male-male competition). The theory is that females choose males based on physical or behavioral traits (dewlap color, body size, frequency of and intensity of social display or courtship) that they equate with "fitness" (superior reproductive success). Recent research on tree lizards, however, has failed to support this theory: female tree lizards did not seem to prefer one throat color over another, nor did they necessarily select the largest or most vigorous males. Instead, females may first select a suitable territory and then select a male whose range overlaps their territory.

horned lizards (*Phrynosoma* spp.)

Order: Squamata
Family: Iguanidae (iguanid lizards)
Other common names: horny toad
Spanish name: camaleón

DISTINGUISHING FEATURES

Up to 10 species of horned lizards occur in the Sonoran Desert region, from the 2¾ inch (69 mm) long round-tailed horned lizard (*P. modestum*) to the 5 inch (127 mm) long Texas horned lizard (*P. cornutum*). With squat, flat, toad-like bodies (Phrynosoma means "toad-body") and thorn-like projections at the rear of their heads, horned lizards are easily distinguished from other lizards. The projections differ in size and arrangement from one species to another. Along the sides of the body, fringe like scales occur in one row, two parallel rows, or they may be absent. Males have enlarged post-anal scales, and during the breeding season, a swollen tail base.

RANGE

Horned lizards are found throughout the Sonoran Desert region from near sea level up to 11,300 feet (3440 m). Some species are widespread, such as the round-tailed and Texas horned lizards which occur in several U.S. and Mexican states, while the flat-tailed horned lizard (*P. mcalli*) is restricted to southwestern Arizona, extreme southeastern California, a small part of northeastern Baja California and the upper neck of northwestern Sonora, Mexico.

HABITAT

Horned lizards are found in extremely diverse habitats. The flat-tailed horned lizard occurs in areas of fine sand, while the short-horned lizard (*P. douglassii*) is found in shortgrass prairie all the way up into spruce-fir forest. The most common species in the Arizona Upland subdivision is the regal horned lizard (*P. solare*), which frequents rocky or gravelly habitats of arid to semiarid plains, hills and lower mountain slopes.

LIFE HISTORY

The diet of some horned lizards consists of specific insects, while other species are more catholic in their tastes. Not only does *P. solare* prefer ants, it has a strong preference for harvester ants, which may make up to 90 percent of its diet. As diets go, ants are low return items because so much of their body consists of indigestible chitin. Thus, the regal horned lizard must eat a great number of ants to meet its nutritional needs. This diet requires space, which is why the stomach of the regal horned lizard may represent up to 13 percent of its body mass.

Ant-eating horned lizards usually capture their prey with their sticky tongues rather than grabbing it with their jaws. In addition, they have modified skeletal morphologies, such as shorter teeth and reduced diameter of the bones of the lower mandible.

Horned lizards are no exception to the general rule that lizards are not attracted to dead insects as food—the ants must be alive and moving for the lizard to show interest in them as prey. Harvester ants can bite and have a potent venom, but apparently this has little effect on the esophagus or stomach of the lizard. However, when faced with swarming ants the lizard will make a hasty retreat, for these little invertebrates can kill an adult horned lizard.

Most species of horned lizards lay eggs between May and August, with clutches ranging from 3 to 45 depending on species. Even with such high numbers of eggs only around 2 from each clutch will reach sexual maturity. The short-horned lizard bears live young. This is considered an adaptation to living at higher elevations, where eggs may be at risk due to low temperatures, and egg development might be slowed considerably.

COMMENTS

The body form and armor of the horned lizard cost it speed and mobility, but they confer great advantages as well. Small animals, such as snakes, have more difficulty with a horned lizard's wide, thorny body than with a smooth, slender lizard. In fact, when confronted by a snake, a horned lizard will continually present the largest part of its body to the snake. Some horned lizards are difficult to distinguish from

rocks; thus they avoid detection by would-be predators. In response to a threat, a horned lizard may play dead, or it may run away and then suddenly turn around to face its attacker, hissing or vibrating its tail in leaf litter. Several species can rupture small capillaries around their eyes and squirt a bloody solution at would-be predators. These

Regal horned lizard

fluids, beyond coming as a surprise, can be irritating to the mucous membranes of some predators.

The large, flat body surface of the horned lizard also works well as a solar collecting panel: at cooler temperatures, the lizard will orient its body to maximize the amount of exposure to the sun. When it gets too hot, the horned lizard will burrow into loose soil. Initially, the lizard uses the scales on the front edge of its lower jaw to literally cut into the earth as it vibrates its head into the ground. Then it will shake and shimmy its body into the soil until almost none of it is above the surface.

"sand" lizards

fringe-toed lizard (*Uma notata* and *U. scoparia*)
greater earless lizard (*Cophosaurus texanus*)
lesser earless lizard (*Holbrookia maculata*)
zebra-tailed lizard (*Callisaurus draconoides*)

Order: Squamata
Family: Iguanidae (iguanid lizards)
Spanish names: lagartija de las dunas (fringe-toed lizard), lagartija
    (greater earless lizard), lagartija (lesser earless lizard), perrita
    (zebra-tailed lizard)

DISTINGUISHING FEATURES

This is a group of small- to medium-sized lizards with wedge-shaped heads and countersunk jaws, smooth granular scales, and dark crescents or bars on the sides of their bellies. Color is usually sandy tan to salmon with irregular spots, blotches, or bars of darker tan to brown. *Uma* spp. tend to have extensive networks of light spots with dark centers. Aside from the fringe-toed lizard, the species are easily confused. Their defining characteristics are outlined below.

RANGE

Fringe-toed lizards are restricted to southeastern California, southwestern Arizona, northeastern Baja California, and northwestern Sonora, Mexico. The greater earless lizard is found in southeastern Arizona; southern New Mexico; central, west, and south Texas into Mexico at elevations up to 5600 feet (1700 m). The lesser earless lizard occurs in 5 central to western states, including eastern Arizona, south to Guanajuato, Mexico at elevations ranging from sea level to 7000 feet (2130 m). The zebra-tailed lizard ranges from Nevada and southeastern California through the western half of Arizona south throughout much of Baja California, Sonora and Sinaloa, Mexico at elevations ranging from sea level to 5000 feet (1520 m).

Species	Ear Openings	Belly Marks	Tail Marks
fringe-toed	present	1 large blotch	bars present
greater earless	absent	2 crescents behind midbody	bars present (underside)
lesser earless	absent	2 crescents midbody	bars absent
zebra-tailed	present	2 crescents at or in front of midbody	bars present (tail often banded)

HABITAT

Fringe-toed lizards occur only in low desert areas having fine, loose, sandy substrate. The greater

Greater earless lizard

earless lizard seems to prefer rocky bajadas and canyons in upland desert areas with mesquite, ocotillo, palo verde, and occasional creosote bush, in sandy or gravelly soils; it often rests on large rocks. The lesser earless lizard is more of a habitat generalist, occurring in desert grassland, Sonoran desertscrub, pinyon-juniper woodlands, thornscrub, and tropical deciduous forest; it is usually found in open areas with loamy soils. The zebra-tailed lizard is most commonly encountered in canyon bottoms, washes, desert pavement, and hardpan, where plant growth is minimal and there are wide or long stretches of open sandy soil.

Lesser earless lizard

LIFE HISTORY

All of these lizards are insectivorous, though some may also eat small lizards. They breed from spring to summer laying between 1 to 15 eggs, with the fringe-toed lizard usually laying fewer eggs than the other species. Eggs take 60 to 75 days to hatch.

Females of several lizard species exhibit color changes during breeding season. The "sand" lizards, for instance, develop bright orange or red areas on the different parts of the body. In some species this indicates readiness for breeding, while in others it is a signal to males that the female has already mated and is *gravid* (pregnant).

COMMENTS

The fringe-toed lizard with its projecting toe scales, countersunk lower jaw, overlapping eye scales, and nasal valves is ideally suited to its sand dune habitat. Often it will dive into the sand to escape predators or extreme heat.

The zebra-tailed lizard has the peculiar habit of wagging its curled tail, which may serve to visually distract predators by drawing attention away from the lizard's body and head. If a predator seizes the tail it easily detaches, a process known as *autotomy*. The tail has built-in fracture planes in the vertebrae to help it readily break off. The lizard grows back a cartilaginous

Zebra-tailed lizard

replacement, which is shorter and has a different appearance than the original. Many species of lizards have this ability.

western blind snake (*Leptotyphlops humilis*)

Order: Squamata
Family: Leptotyphlopidae (blind snakes)
Other common names: worm snake
Spanish names: culebra

DISTINGUISHING FEATURES

This small snake, which is about as thick as a pencil lead on average, reaches a maximum length of 16 inches (40 cm). It is usually pale brown, pink, purplish, or beige with a silver sheen. The hard, shiny scales on the underside are similar in appearance to those on the back, except lighter in color. The head and tail are blunt, with a spine at the tail tip. Vestigial eyes appear as spots under the head scales. Teeth are lacking from the upper jaw.

RANGE

In the northern part of its range, the western blind snake occurs from southern California to western Texas. Continuing south into Mexico, it is found in all of Baja California and western and north-central Mexico.

HABITAT

Found from below sea level to 5000 feet (1500 m), the western blind snake prefers moist, loose soils suitable for burrowing. This may include the sandy washes or canyon bottoms of mountain brushy areas or desert grasslands.

LIFE HISTORY

By spending most of its life underground, the western blind snake has no need for visual acuity. While not entirely blind, it does have vestigial eyes thought to be capable of seeing light only. If disturbed, it will writhe and wiggle its tail to focus attention here instead of on the head. Preyed upon by a wide variety of animals, including birds, mammals, snakes, fish and even spiders, the western blind snake is a specialist in its culinary desires: ants and termites along with their eggs, pupae, and larvae. Millipedes and centipedes are also occasionally eaten. When searching for food, a western blind snake will hunt until it finds an ant pheromone trail and follow it back to the nest to consume the residents. The smooth, tightly overlapping scales provide protection against the bites and stings of ants. This small serpent shares a feature with the much larger boas and pythons—the remains of a pelvic girdle and femur, complete with a tiny spur! A secretive, nocturnal snake, the blind snake lays up to 7 eggs in mid summer.

rosy boa (*Lichanura trivirgata*)

Order: Squamata
Family: Boidae (boas)
Spanish names: corua del desierto

DESCRIPTION

This heavy-bodied, medium-sized snake reaches a length of 3½ feet (107 cm). The eyes have cat-like vertical pupils; the long, pointed head is barely distinguishable from the neck. Shades of beige, slate blue, or rose on the back are adorned with 3 brown or gray stripes that run the length of the body. The light, creamy or gray underside has flecking of gray or black. Males have a spur on either side of the anal opening.

RANGE

The rosy boa is restricted to southern California, southwestern Arizona, and northwestern Mexico.

HABITAT

The rosy boa inhabits desert foothills with rocks and boulders from sea level to 4500 feet (1400 m). Although this snake prefers rocky terrain where crevices make safe homes, it occasionally is spotted in shrublands or chaparral without rocky areas, although this is unusual.

LIFE HISTORY

Active principally at night, the rosy boa eats birds and mammals. Boas and pythons, unlike most snakes, have a vestigial pelvic girdle complete with rudimentary femur bones. These can be seen as anal spurs located on either side of the anal opening. More pronounced in the male, these shiny, almost metallic looking spurs are used in courtship to gently stroke the female. Mating takes place in spring; females give birth to as many as twelve young in late fall. The young are about 1 foot (90 cm) long at birth.

Boa Locomotion

Concertina locomotion, used by heavy-bodied snakes such as rosy boas, involves using a posterior loop to push the anterior body forward.

garter snake (*Thamnophis* spp.)

Order: Squamata
Family: Colubridae (colubrid snakes)
Spanish name: culebra de agua

DISTINGUISHING FEATURES
Garter snakes in the Sonoran Desert region are slender with a maximum length of 3½ feet (106 cm). Most species have light-colored stripes on the top and sides of an otherwise olive-green or dark body. There are many different kinds of garter snakes, but only two will be discussed here. They are the black-necked garter snake (*Thamnophis cyrtopsis*) and the checkered garter snake (*Thamnophis marcianus*). They can be distinguished from each other by the position of the side stripe: this stripe is confined to the second and third scale rows on *T. cyrtopsis* and the third scale row on *T. marcianus*. Also the black-necked garter snake has 2 large black blotches behind the head, while the checkered garter snake has large, squarish, dark blotches in a checkered pattern on its body.

RANGE
Both species are found throughout the Sonoran Desert. Additionally, *T. cyrtopsis* ranges from southeastern Utah to Guatamala and from central Texas to central and southern Arizona. Isolated populations occur in the Hualapai Mountains, Burro Creek, and Ajo mountains in western Arizona.
T. marcianus is found in southwestern

Kansas south to Zacatecas and northern Veracruz, Mexico, and from east-central Texas to south central Arizona and east-central Sonora, Mexico. A disjunct population occurs around the juncture of California, Arizona, northern Baja California, and Sonora, Mexico.

Checkered garter snake

HABITAT
Both species are semi-aquatic, generally found in or near bodies of water ranging from streams to canals, ponds, and cattle tanks. The black-necked garter snake is found from desert through mixed conifer forest, as well as tropical habitats in Mexico. The checkered garter snake usually inhabits lowland river systems, ponds, springs, streams, rivers, and irrigation ditches in arid and semiarid regions. Occasionally it is found in oak-pine woodlands.

LIFE HISTORY
Both species feed on aquatic or semi-aquatic prey, including fish, frogs, toads, tadpoles, worms, salamanders, and crustaceans. Lizards, small mammals, and birds may also be eaten. Garter snakes are live-bearers, generally giving birth to 6 to 18 young in the summer. If disturbed, garter snakes will bite, defecate, and emit foul-smelling musk to deter the intruder.

COMMENTS
Garter snake habitat is disappearing throughout the Sonoran Desert due to habitat destruction and fragmentation. Garter snake numbers have also declined due to competition with and predation by introduced species such as bullfrogs and sunfish. These introduced animals eat small snakes as well as the fishes and tadpoles which are food for garter snakes.

common kingsnake (*Lampropeltis getulas*)

Order: Squamata
Family: Colubridae (colubrid snakes)
Other common names: desert kingsnake, black kingsnake, California kingsnake
Spanish names: culebra

DISTINGUISHING FEATURES
A long, slender, harmless snake, the common kingsnake reaches lengths of approximately 3½ feet (100 cm) in Arizona, although rare specimens reach 6 feet (180 cm). In most of the Sonoran Desert it is a dark brown or black snake with narrower bands of yellow, white, or cream going around the body, widening on the belly. However, there is considerable variation in pattern throughout the range. In some parts of the range (southern Arizona) the common kingsnake is entirely dark with no light bands, while in other areas the bands degenerate into a speckled appearance. In California, this species can have a single stripe that runs from behind the head to the tail. The scales are smooth and glossy in appearance.

RANGE
This is a widespread species, found coast-to-coast in the United States from southern New Jersey to Southern Florida, and west through California and southern Oregon. In Mexico this species is found through much of Baja California and mainland Mexico, south to northern Sinaloa, San Luis Potosí and northern Tamaulipas.

HABITAT

A habitat generalist, the common kingsnake has been found in deserts, riparian areas, woodlands, forests, and farmland from sea level to 7000 feet (2100 m).

LIFE HISTORY

Active in early morning and late afternoon when the weather is mild, the common kingsnake becomes nocturnal with the onset of extreme heat in summer. An opportunistic feeder, it will eat many things including lizards, birds, mammals, frogs, bird eggs, and snakes (including rattlesnakes). When a rattlesnake encounters a kingsnake, or its odor, it will not rattle. Instead it slightly elevates the front portion of its body in a line horizontal to the ground, making itself appear large, a behavior known as *body bridging*. Once seen, body bridging is easily distinguished from the more frequently seen defensive coil of all rattlesnakes. Body bridging does not deter a hungry kingsnake. The kingsnake, highly tolerant of and perhaps immune to the rattlesnake's venom, will bite the rattlesnake behind the head; coils of the kingsnake's body are then thrown around the rattlesnake, constricting it until death results. Sometimes the kingsnake will begin consuming the rattlesnake before it is dead.

When confronted, the common kingsnake may hiss, strike, and rattle its tail or hide its head in coils of its body, releasing a foul-smelling musk. Five to 17 eggs are laid in mid summer, hatching occurs in late summer and early fall.

longnose snake (*Rhinocheilus lecontei*)

Order: Squamata
Family: Colubridae (colubrid snakes)
Spanish names: coralillo

DISTINGUISHING FEATURES

This slender snake reaches lengths of slightly over 3 feet (90 cm). While there is considerable variation in pattern and color, generally this snake is banded or blotched with black, white, and usually red; red may be entirely absent in some individuals. The belly is cream or yellowish with some dark spotting on the sides. The lower jaw is countersunk. This snake is easily confused with the venomous coral snake due to the similarity in color banding pattern. The longnose snake can

be distinguished from the coral snake by its long nose, its light colored flecking on the nose (within dark fields) and by body bands that do not completely encircle the body. Additionally, color sequencing is generally different.

RANGE
The longnose snake is found throughout the southwestern United States, in northern Baja California, and south to San Luis Potosí and southern Taumalipas in Mexico.

HABITAT
The longnose snake is found in desertscrub, grassland, chaparral, and tropical habitats below 5400 feet (1650 m). It prefers brushy, rocky areas.

LIFE HISTORY
When disturbed, the longnose snake writhes and twists its body, vibrates its tail, and defecates feces and blood from its anal opening. An excellent burrower if the soil is sandy, it can also retreat under rocks or into rock crevices or rodent burrows, if necessary. In warmer parts of its range, the longnose snake may lay 2 clutches of 4 to 11 eggs per year, while in cooler portions of its range only 1 egg clutch per year is laid. Active primarily at night, it feeds on many things including lizards, lizard eggs, small snakes, small mammals, and sometimes birds.

coachwhip (*Masticophis flagellum*)

Order: Squamata
Family: Colubridae (colubrid snakes)
Other common name: red racer
Spanish names: chirrionero, alicantre

DESCRIPTION
This long, slender snake reaches lengths of 3 to 8½ feet (90-260 cm) long. Quite variable in color, it can be tan, gray, pink, black, reddish-brown, or any combination of these colors. Broad crossbars may be present. The scales are smooth and the eyes large; the head is distinct from the body. Unlike the adults, young may have obvious dark brown or black blotches or bands on a light brown background. This snake receives its name from the braided appearance of its scales which resemble the whip used by stagecoach drivers in earlier days.

RANGE
This is a wide-ranging species occurring throughout the southwestern United States south through Baja California and Mexico (except the Sierra Madre).

HABITAT
This common snake is found in deserts, prairies, grasslands, woodlands, thornscrub, and even cultivated lands. It is found from below sea level

to 7700 feet (2350 m) in hilly or flat lands, rocky or sandy soils.

LIFE HISTORY
Active during the morning and late afternoon, the coachwhip is often seen crossing roads. A speedy snake, it has been clocked at 3.6 miles per hour. The coachwhip is a nervous snake and may retreat into rocks or rodent burrows when threatened, but it is just as likely to approach an intruder hissing, striking, and possibly shaking its tail; it will bite if handled. During summer 4 to 20 eggs are laid, hatching 44 to 88 days later. Young and adults feed on mammals, birds, bird and reptile eggs, lizards, snakes, carrion, and insects; the prey is seized and swallowed without being killed.

Sonoran whipsnake (*Masticophis bilineatus*)

Order: Squamata
Family: Colubridae
Spanish names: alicantre, chirrionero

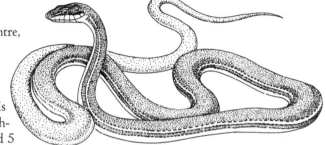

DESCRIPTION
The Sonoran whipsnake is a long, slender snake reaching lengths between 2 and 5 feet (60 to 150 cm). The head is broad with large eyes. It is olive or bluish-gray fading to yellow towards the tail, with 2 or 3 light-colored stripes on each side. The belly is cream fading to pale yellow towards the tail.

RANGE
Chiefly a Mexican species, the Sonoran whipsnake enters the United States in extreme southwestern New Mexico and southeastern to west-central Arizona. In Mexico, it is found south to Oaxaca.

HABITAT
The Sonoran whipsnake is most commonly found in rocky canyons, riparian areas, foothills and mountains with dense vegetation in elevations up to 6100 feet (1800 m). However, this snake is also found in open creosote bush flats. It is equally at home on the ground or in low, shrubby vegetation. Look for the Sonoran whipsnake in areas with ocotillos, saguaros, and palo verdes, but also in chaparral, cottonwood, juniper, and pine-oak forests.

LIFE HISTORY
This fast-moving snake is primarily active in the early morning from March to October. When hunting, the Sonoran whipsnake may elevate its head off the ground and scan the surrounding area for possible prey. When a bird, lizard or frog is spotted, the snake will seize and swallow the prey without killing it first. Six to 13 eggs are laid in June and July.

The Sonoran whipsnake bites readily if handled.

western shovelnose snake (*Chionactis occipitalis*)

Order: Squamata
Family: Colubridae (colubrid snakes)
Spanish names: culebra, coralillo

DISTINGUISHING FEATURES

The western shovelnose snake is small, reaching 10 to 17 inches (25 to 43 cm). As the name implies, the snout is flattened and shovel-shaped. The lower jaw is deeply inset. The dark brown or black bands may be saddle-like or may encircle the body. The basic ground color is cream, whitish, or yellow. Red or orange saddles may or may not be present between the dark saddles. The scales are smooth.

RANGE

Strictly a desert dweller, this snake is restricted to southeastern California, southern Nevada, south-western and central Arizona, northeastern Baja California, and northwestern Sonora, Mexico.

HABITAT

This snake is found in loose sandy areas such as washes, dunes, sandy flats and rocky hillsides that have sandy areas between the rocks. Vegetation is usually sparse and may include creosote bushes, grasses, cacti, and mesquite.

LIFE HISTORY

The underset lower jaw, muscular body, smooth scales, and shovel-shaped nose make this snake very good at "swimming" in the sand. During the heat of the day it is usually submerged beneath the surface, emerging at night to hunt for food. It feeds on numerous kinds of insects (including their larval stages) as well as spiders, scorpions, centipedes, and moths. If disturbed, the western shovelnose snake may try to bury itself in the sand, climb into low-lying vegetation or form a ball with its coils, hiding its head deep in the coils. As many as 9 eggs are laid in the summer.

western hognose snake (*Heterodon nasicus*)

Order: Squamata
Family: Colubridae (colubrid snakes)
Spanish names: culebra

DISTINGUISHING FEATURES

This squat, heavy-bodied snake reaches a maximum length of 3 feet (90 cm), but 2 feet (60 cm) is more typical. Most noticeable on the western hognose snake is the strongly upturned, pointed snout. Dark blotches extend down the pale brown or yellowish back from behind the head to the tail, with 2 rows of smaller, alternating blotches on the sides. The belly is heavily pigmented, with solid black pigmentation underneath the tail.

RANGE

The western hognose snake ranges from south-central Canada, south to southeast Arizona, New Mexico, and Texas, southward into San Luis Potosí, Mexico.

HABITAT

This snake prefers scrubby, flat prairie areas with loose, sandy soil suitable for burrowing.

LIFE HISTORY

The western hognose snake uses its upturned snout to burrow through the earth in search of toads, its principal food. Other items eaten include frogs, lizards, mice, birds, snakes and reptile eggs. Not dangerous to man, the western hognose snake uses a slightly toxic saliva to help subdue its prey. The venom flows down enlarged rear teeth. As many as 39 eggs are laid in the early summer, hatching in as little as 50 days.

Master Bluffer

The western hognose snake has one of the most elaborate bluff behaviors in the snake world. When threatened, the snake flattens the skin on its neck giving it a hooded appearance. It then takes a huge breath, inflating its body dramatically, and releases the air with a loud hissing noise. The snake may strike at the intruder, but the mouth is closed. (It is difficult to get a hognose snake to bite in self-defense.) Occasionally, if the snake is not left alone, it will go into convulsion-like motions, turning over on its back, thrashing its head from side to side, and pretending to die. During this death feign, the mouth is open and the tongue sticks limply out. The snake may even bleed from the mouth or the anal opening and expel feces, although this behavior is more common with the eastern hognose of the southeastern United States. When the snake is picked up, it is limp. If it is turned belly down, it quickly flips over. After a few minutes, the snake lifts its head and, if it perceives no threat, quickly slithers away.

western patchnose snake (*Salvadora hexalepis*)

Order: Squamata
Family: Colubridae (colubrid snakes)
Spanish names: culebra

DISTINGUISHING FEATURES
This slender, docile snake reaches 1½ to 3½ feet (107 cm) in length. Most noticeable is the large, patch-like rostral scale on the end of the nose. A wide yellow or beige stripe with a dark border runs down the center of the back; one dark stripe runs down each side. Occasionally the stripes are broken or obscured by crossbars. The belly is pale, sometimes faintly orange. Males have keeled scales at the base of the tail and above the anal opening.

RANGE
This snake is found in the southwestern United States, northwestern Mexico, and Baja California. It is also found on Isla Tiburón and Isla San José in the Gulf of Mexico.

HABITAT
The patchnose snake is found in sandy soils or rocky areas in lowland desert with open creosote bush flats or desertscrub. It is also found in grasslands to the lower slopes of mountains with chaparral, and in pinyon-juniper woodlands as high as 7000 feet (2100 m).

LIFE HISTORY
Active in the daytime year round in warmer climates, this snake is crepuscular in the heat of the summer. In milder climates it may be active from early April to early November. The enlarged rostral scale is useful for burrowing in both loose sandy areas or rocky areas in search of its food: lizards, grasshoppers, small mammals, and reptile eggs. While the western patchnose snake does not constrict its prey, it does throw loops of its body on top of the prey to subdue it. It locates reptile eggs by scent, using its nose to unearth them. Much like the whipsnake, it moves quickly on the ground, and may climb into the lower branches of vegetation. If picked up it will thrash wildly. During the summer, it lays 4 to 10 eggs. Eleven inch long hatchlings emerge in late summer.

Gopher snake (description on page 573)

gopher snake (*Pituophis melanoleucus*)

Order: Squamata
Family: Colubridae (colubrid snakes)
Other common names: bullsnake
Spanish names: víbora sorda

DISTINGUISHING FEATURES

Large and heavy-bodied, the gopher snake is reported to reach 9 feet (275 cm) in length, but 4 feet (120 cm) is more common. On its back are 33 to 66 light- to dark-brown or reddish blotches on a ground color of yellow, straw, tan or cream. Smaller blotches are located on the animal's sides. A dark stripe runs from in front of the eye to the angle of the jaw. The underside is creamy or yellow, often with dark spots. The scales on the back are strongly keeled, becoming smoother on the sides.

RANGE

This is one of the most widespread snakes in North America. Its range extends from the Atlantic to Pacific oceans, as far north as southern Canada, and as far south as Veracruz and southern Sinaloa, Mexico, including Baja California.

HABITAT

A habitat generalist, the gopher snake is found in deserts, prairies, woodlands, brushlands, coniferous forests, and even cultivated lands. These biomes can be rocky, sandy, sparsely or heavily vegetated, and range from below sea level to over 9000 feet (2700 m).

LIFE HISTORY

When disturbed, the gopher snake will rise to a striking position, flatten its head into a triangular shape, hiss loudly and shake its tail at the intruder. These defensive behaviors, along with its body markings, frequently cause the gopher snake to be mistaken for a rattlesnake. The tapered tail, the absence of a rattle, the lack of a facial pit, and the round pupils all distinguish the gopher snake from the rattlesnake. The gopher snake is active mainly during the day, except in extreme heat when it ventures out at night. It is a good climber. A constrictor, it consumes mostly mammals, although birds and their eggs are also eaten. During the summer 2 to 24 eggs are laid which hatch in the fall.

Male gopher snakes engage in ritualistic combats during the spring mating season. The combatants remain on the ground, entwined from tail to neck. Each tries to maintain its head and body position, although occasionally they will exert so much force that they roll. Hissing frequently, they rarely bite one another. Presumably the combat ritual is a means of determining the sexual fitness of a male, for usually only the victor will copulate afterward.

night snake (*Hypsiglena torquata*)

Order: Squamata
Family: Colubridae (colubrid snakes)
Spanish names: culebra

DISTINGUISHING FEATURES

This small 12 to 26 inch (30 to 66 cm) snake has a triangular-shaped head, a dark eyestripe, and elliptical pupils; it is often mistaken for a young rattlesnake. The night snake is easily distinguished from the latter by a tail that tapers to a point and the absence of a rattle. The night snake has a pair of large, dark brown or black blotches on the neck immediately behind the head. It is pale gray, light brown, or beige with dark gray, brown, or black blotches on the back and sides; the underside is white.

RANGE

The range of this snake extends from British Columbia through the western United States and south to Guerrero, Mexico.

HABITAT

A habitat generalist, the night snake is found in rocky areas of grassland, chaparral, desertscrub, woodland, moist mountain meadows, and thornscrub from sea level to 8700 feet (2650 m).

LIFE HISTORY

Seldom encountered during the day, this nocturnal snake is often seen crossing roads at night. The night snake preys upon lizards, small snakes, frogs, salamanders, and small mice, which it subdues with its mild venom; this venom poses no threat to humans. Young night snakes feed upon insects. If disturbed, the night snake raises its head and weaves, hisses, and flattens its neck in threat. Two to 9 eggs are laid spring through summer.

lyre snake (*Trimorphodon biscutatus*)

Order: Squamata
Family: Colubridae (colubrid snakes)
Spanish names: culebra

DISTINGUISHING FEATURES
Named for the V-shaped "lyre" on its head, this medium-sized snake reaches a length of nearly 4 feet (1.2 m). The broad head with narrow neck gives the lyre snake a triangular-shaped head. Dark brown saddles reside on a light brown to light gray back. The underside is creamy-white or yellow with scattered brown spots. The scales are smooth; the pupil is vertical.

RANGE
The northern part of the lyre snake's range is in southeastern Nevada and southwestern Utah; the range continues south through western Mexico to Costa Rica.

HABITAT
The lyre snake lives mainly in the lower rocky canyons and arroyos of hills and mountains from sea level to 7400 feet (2300 m). A rock dweller, it wedges itself in the many crevices and fissures that are abundant in rocky areas. This snake is an occasional resident of flat lands.

LIFE HISTORY
This mildly venomous, rear-fanged snake feeds primarily on lizards, but also eats birds and bats. Constriction may be used to subdue prey since the hemorrhagic venom is not very effective on birds and mammals. Primarily nocturnal, it is seldom active during the day. Easily alarmed, the lyre snake will raise its body off the ground, shake its tail, hiss, and strike and bite the intruder if not left alone. This behavior, along with the body pattern, triangular head, and elliptical pupils, sometimes causes the lyre snake to be mistaken for a rattlesnake. As many as 20 eggs are laid during the summer.

Arizona coral snake (*Micruroides euryxanthus*)

Order: Squamata
Family: Elapidae (fixed front-
    fang venomous snakes)
Other common names:
    western coral snake
Spanish name: corallilo

DISTINGUISHING FEATURES
The Arizona coral snake is a slender, small snake reaching only 13 to 21 inches (33-53 cm) in length. It is brightly colored with broad alternating bands of red and black separated by narrower bands of bright white or yellow. The bands completely encircle the body, but are paler on the belly. The head is black to behind the eyes. The snout is blunt.

RANGE
The Arizona coral snake is found in central and southern Arizona, extreme southwestern New Mexico and southward to Sinaloa in western Mexico.

HABITAT
This snake occupies arid and semiarid regions in many different habitat types including thornscrub, desert-scrub, woodland, grassland and farmland. It is found in the plains and lower mountain slopes from sea level to 5800 feet (1768 m); often found in rocky areas.

LIFE HISTORY
Carnivorous, as are all snakes, the Arizona coral snake specializes in feeding primarily on blind and black-headed snakes. Occasionally it eats lizards or other small, smooth-scaled snakes. A secretive snake, it usually emerges after sundown, and may remain active well into the night. It is also frequently active during the day after rains or if the sky is overcast. If disturbed it will bury its head in its coils, elevate and wave its tightly coiled tail, and evert its anal lining, making a popping sound. Two to 3 eggs are laid during the summer.

COMMENTS
The venom of this snake is similar to that of the cobra. However, due to the small size of the snake (less venom), smaller mouth, and small fangs (less effective means of delivery), the venom does not pose as much danger to humans as that of rattlesnakes. As with any venomous reptile, medical attention should be sought in the event of a bite.

Many people use a rhyme to remember the coral snake: "Red touch yellow, harmful fellow." Unfortunately, this rhyme does not always work in our region (and many parts of the western hemisphere). We have several non-venomous snakes in our region that have red bands touching yellow bands. The best way to identify a coral snake is by: 1) a very blunt head that is black to behind the eyes, and 2) bands that completely encircle the body, along with the yellow or white bands occurring on both sides of the red bands.

Rattlesnakes

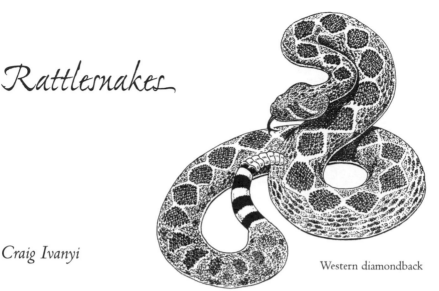

Western diamondback

Craig Ivanyi

Almost everyone is fascinated by rattlesnakes. The fascination is most often caused by fear of these animals, which are legendarily perceived to be aggressive and deadly. Although their threat to humans is grossly exaggerated, these snakes are fascinating for another compelling reason: rattlesnakes are among the most highly specialized organisms on the planet.

Their venom—in fact a toxic saliva—is among the most complex substances known: a mixture of enzymes unique to pit vipers that destroys blood or paralyzes nerves. And the delivery system is equally amazing—the snakes' fangs are movable hypodermic syringes. Rattlesnakes are also among the few animal groups with dual visual systems. In addition to their eyes, they have sensory organs in their upper jaws which can actually "see" infrared images. They can detect the heat from a candle flame 30 feet (9 m) away. These animals merit admiration more than fear.

While other areas have larger rattlesnakes, the Sonoran Desert region is blessed (no, that isn't a misprint) with more species of rattlesnakes than is any other region in the world—and many of them will bite large beasts like us if we are perceived to be a threat. Rather than seeking to eradicate these animals, we have more to gain by finding ways for all of us to coexist. Why? Because these animals are the natural predators of a suite of other animals (e.g., mice and rats) that can cause plant damage, carry diseases, and so on. Besides, trying to kill rattlesnakes actually puts us at greater risk than does leaving them alone.

Knowledge of "who" these creatures are—that is, what they do for a living, where they live and when they are active—will help us coexist without harm to either snakes or humans.

Most rattlesnakes avoid contact with humans. They tend to avoid wide open spaces that offer little protection from predators, so they usually spend

Mohave
rattlesnake

to a venom gland (located toward the back of the head on the upper side) by a venom duct. Baby rattlesnakes (born live from mid-summer to early autumn) arrive fully equipped with teeth and toxins!

Rattlesnake venom functions primarily to help the animals feed, facilitating capture and partial digestion of prey. It is not believed to have originated as a defensive weapon to avoid predation or molestation (though it can be very effective in this regard).

Approximately 20 percent of defensive strikes are "dry;" that is, no venom is injected. This has led to the common belief that rattlesnakes can "choose" not to envenomate, although it is unknown why rattlesnakes elect to control venom injection in this defensive context.

their time in and under low-growing shrubs, natural and artificial debris, rocks and the like.

They are most active in the warmer times of the year—spring through early fall—and many of them are nocturnal during the summer months. When favorable temperatures occur, many rattlesnakes are marginally active even during the winter. You are most likely to see them when the air temperature is between 70° and 90°F (21° to 32°C), regardless of the time of day—be it June or January.

Venom

Rattlesnakes may be among nature's best examples of efficiency and economy of motion. The use of venom to capture prey conserves considerable energy that would otherwise be needed to capture and subdue an animal.

Rattlesnakes have very long, curved fangs that lie parallel to the jaw line when not in use. When rattlesnakes strike, the fangs are rotated by muscular contraction to an erect position, somewhat perpendicular to the jaw line. The fangs are hollow, which allows the snake to inject venom through the tooth into its victim. Each fang is connected

Rattlesnake skull showing fang position when mouth is closed (left) and during strike (right).

Snake venoms are extremely complex; there is no "standard" rattlesnake venom. They consist of combinations of proteins that range from hemotoxins, which break down cells and tissues, to anticoagulants and neurotoxins that may cause circulatory arrest or respiratory paralysis.

Heat Vision

Rattlesnakes and other pit vipers have remarkable heat-sensing pits. Located behind each nostril, below a straight line that would directly connect the nostril to the eye, is a *loreal pit* (called this because it is a depression in the loreal scale). These pits are highly effective in detecting differences in temperature even several yards away. At short ranges within a foot or so, minute differences (of perhaps fractions of a degree) may be perceived.

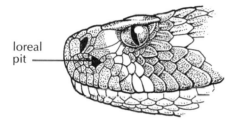

loreal pit

Heat given off by an animal creates a heat image; therefore, rattlesnakes have "heat vision." The heat images are integrated with visual ones in the brain. This type of vision is helpful for nocturnal predators, for it enables them to hunt effectively even in total darkness. It may also help distinguish predator from prey, allowing rattlesnakes to determine whether they are at risk themselves. Larger, non-prey animals give off larger heat images, signaling the snakes to avoid potential encounters with these animals.

If It Rattles . . .

The best way to distinguish rattlesnakes from other kinds of snakes is by the rattle. Most other characteristics are either too subjective or inconsistent, or else they require one to get too close to the animal for safety. Even at birth, rattlesnakes have the first segment of a rattle, which is called a "prebutton." The prebutton is lost the first time the snake sheds its skin and is replaced by a button. Each shedding episode that follows adds another segment to the rattle, and rattlesnakes may shed one to four or more times each year. Only when there is more than one segment can the rattle produce sound. There isn't anything inside the rattle; the various segments merely bump against each other to produce the sound. The tremendous amount of sound produced results from movement of the rattle back and forth 60 or more times per second!

The rattle is composed of fibrous protein called keratin—as are your fingernails—and segments can easily break off. Because segments are added at various rates, and because they break off, it is not possible to tell a snake's age by counting the segments.

In addition to the change in structure from small scales to large rattle segments, rattlesnake tails look stumpy, almost as though their tips had been snipped off. Other snakes have tails that come to smooth tapering points, much like pencil points.

Motherly Love

Few reptiles exhibit parental care of their young. However, there are notable exceptions —mother American alligators may watch over their young for periods of up to four years. Without evidence to the contrary, scientists take a conservative approach and assume that parental care is absent in most reptiles.

Sometimes scientists may make premature judgments based on incomplete information. For years it was assumed that rattlesnakes do not care for their young after birth. Yet, Dr. Harry Greene of the University of California, Berkeley, has found that female black-tailed rattlesnakes stay with their young until they complete their first shed, about a week's time. Similar observations have been made by herpetologist Hugh McCrystal with other species of rattlesnakes such as the rock rattlesnake. In the future, we may find that most rattlesnake species exhibit parental care.

Human Bites and Treatment

Rattlesnakes do not always rattle before they strike, nor must they coil before they strike. If they feel threatened, they may coil, strike, retreat or do nothing at all. Each snake is a unique individual and responds accordingly. In spite of the snakes' lethal potential, fewer than I percent of the people bitten in the United States by venomous snakes die. Many bites to adult humans are the result of human provocation..

If you are bitten, remain calm and get to a hospital as soon as possible. Most first-aid treatments suggested many years ago are no longer recommended.

Important Note: Do not attempt to capture or kill the snake. In modern medical facilities, all rattlesnake bites receive the same antivenin.

Rattlesnake Rules

Here are a few simple rules that will help keep you from having an unexpected, and potentially dangerous, encounter with a rattlesnake.

1. Identify everything before you pick it up.
2. Don't touch anything that can hurt you. If you don't know if it can hurt you, don't touch it.
3. Always look under things before picking them up, and whenever possible, before stepping on or around them.
4. Look under things from a distance (use a tool to lift, then look).
5. Always use a flashlight when you are out at night.
6. Do not pick up a "dead" rattlesnake. It may not be dead, and even if it is, it may still bite (reflexively) and envenomate.
7. Don't walk barefoot or in open-toed shoes in the desert.

Guidelines for Removing a Rattlesnake from Your Yard or Home

Purpose

There are two reasons for attempting to remove (rather than kill) a rattlesnake from your property:
- Human or animal safety.
- Snake safety. Why? An injured snake is a dangerous one!

Mindset

To successfully remove a rattlesnake you need the right technique as well as the right mental and emotional state. You must remain calm when dealing with venomous animals. You must be in control of yourself to be in control of the animal.

Equipment

- Tongs

  We recommend 42-inch tongs for dealing with a range of wild-caught animals.

- Container

  A specially designed, hinged, double-lidded wooden box, with a latching top is best. Single-lid boxes or large trash cans with secure lids can also be used. Be sure to check equipment periodically and immediately prior to use. You don't want to find out that something is missing or broken when you need to capture an animal.

Tiger rattlesnake

Technique

- Be sure to have an escape route should you need it.
- Keep others away from the area (an assistant is very helpful for this).
- Approach the snake cautiously, preferably at an angle rather than head on.

If you encounter a venomous snake on the trail or in your yard, back away from it and give it a chance to move away. In the event of a snakebite, do not wait to see if symptoms appear—go immediately to the nearest medical facility for treatment.

- Maintain a safe distance. A snake may be able to strike a distance of up to two-thirds or more of its body length. If you stay more than one body length away you will be quite safe.
- Set up the container.
- Gently put the tongs around the animal. It is best to grasp the snake somewhere slightly before or at mid-body. Farther forward or backward risks injury to the snake and makes the creature more difficult to manipulate, which increases the risk to you.
- Gently apply enough pressure to slow or stop the animal from moving through the tongs. Too little pressure and the animal escapes, too much and it may be injured.
- Put the lid on the container without putting your fingers, or any other part of your body, on the underside of the lid or in a place where the animal can strike them.
- Secure the lid. You can use masking tape or bungee cords to secure a trash can lid.
- Relocate the animal, preferably a short distance from where it was captured, in an area where the snake doesn't pose a threat to people or their pets.

Rattlesnakes

REPRESENTATIVE SONORAN DESERT SPECIES:
   western diamondback *(Crotalus atrox)*
   Mohave rattlesnake *(Crotalus scutulatus)*
   tiger rattlesnake *(Crotalus tigris)*
   blacktail rattlesnake *(Crotalus molossus)*
   sidewinder *(Crotalus cerastes)*

Order: Squamata
Family: Viperidae (moveable front-fang venomous snakes)
Spanish names: víbora de cascabel (rattlesnake), víbora de cuernitos
   (sidewinder)

DISTINGUISHING FEATURES

Western diamondback: Dark, diamond-shaped or hexagonal blotches along the center of the back, light eye stripe from eye to upper lip, bold black and white tail banding, small scales on head.
Mohave rattlesnake: Difficult to distinguish from western diamondback; tail generally has narrower black bands than white, 2 to 3 enlarged scales on top of the head between the eyes.
Tiger rattlesnake: Small head and large rattle, 35 to 52 distinct closely-spaced crossbands on back and sides.
Blacktail rattlesnake: Black tail and snout.
Sidewinder: Horn-like projection over each eye.

HABITAT

The *western diamondback* is a generalist which can be found in diverse habitats from below sea level to 6500 feet (2000 m). The *Mohave rattlesnake* prefers open areas with grasses, creosote bush, palo verde, mesquite and cactus; most common at lower elevations; also common in desert grasslands of southeast Arizona; usually not present in rocky areas or areas with heavy vegeta-

Blacktail rattlesnake

tion. The *tiger rattlesnake* is strictly a Sonoran Desert region species; most common in very rocky canyons and foothills or arid desert mountains up to 4800 feet (1460 m); usually restricted to cactus and mesquite of the rocky foothills; seldom encountered in flat, sandy areas devoid of rocks. The blacktail is primarily a mountain snake found in pinyon-oak woodland or coniferous forests up to 9600 feet (2900 m) near rocky areas; also resides in saguaro-covered desert uplands. The *Sidewinder* is a common resident of sand dunes and other loose, sandy areas where vegetation is sparse and composed primarily of creosote bush; rarely seen in rocky areas.

Mohave Rattler

The Mohave rattlesnake may be the most dangerous venomous snake in the Sonoran Desert. Quick to go on the defensive, the Mohave has very toxic venom that has caused human fatalities. Venom toxicity varies among different populations. The seriousness of a bite from this rattlesnake, as from any rattlesnake, depends on many factors, including, but not limited to, the amount of venom injected and the health and size of the victim. A person bitten by a Mohave rattlesnake should seek medical attention immediately.

FEEDING
• *Diet:* Rodents make up the majority of the diet for all of these snakes. Birds, lizards and other small animals are also taken.

LIFE HISTORY
Western diamondback: This snake is active at night during the warm months and during the day in spring and fall; it returns to rocky cliffs for a winter hibernation period, but may exit to bask in the sun on warm days.

Prior to copulation in the spring, male diamondbacks (as well as males of at least some other rattlesnake species) perform well-documented, ritualized "combat dances." When two males encounter each other they raise their bodies off the ground—as much as one-third of their lengths. Belly-to-belly, they begin an intense wrestling contest. Occasionally one snake or the other falls to the ground, only to rise up to continue the contest anew. This wrestling match may continue for thirty minutes or more. At some point, one snake finally gives up and crawls away, often with the victor in hot pursuit. Victors have even been observed climbing into shrubs several feet off the ground, apparently to make sure the loser does not try to return to the females. There are occasions when a third male is present. He does not join the duo at battle, but instead copulates with the females while the other two males are battling. Biologists have termed this the "sneaky male strategy." The inseminated female will give birth to as many as 23, 9- to 14-inch-long (23-36 cm) young in the late summer. Young diamondbacks feed on rodents, and adults also eat rabbits and ground-dwelling birds.

Mohave rattlesnake: The Mohave is active primarily at night from February to November. Unlike most rattlesnakes, which usually hibernate in larger groups, the Mohave hibernates singly or in pairs or trios in rodent burrows. Courtship and copulation occur in the spring or, occasionally, in the fall. Following fall copulation, sperm may be viably retained for several months, resulting in births during the next year's warmer seasons. As many as 13, 9-inch-long (23 cm) young are usually born in late summer and early fall. Rodents comprise the bulk of the diet.

Sidewinding

While many snakes use the method of locomotion called "sidewinding," the sidewinder is particularly adept at it. Sidewinding is a method of locomotion adapted for areas with loose, hot, sandy soils where traction is difficult. To sidewind, the snake throws a loop of its body forward and then pulls itself up on the loop. As the loop is thrown forward, the head and neck of the snake push down into the sand. In this way much of the snake's body is held up off the hot surface. To the observer, the sidewinding snake appears to be going sideways with respect to the direction in which the body points. While sidewinding is the primary form of locomotion for a sidewinder, it can also use all other methods of locomotion characteristic of snakes. (See rosy boa account, page 564.)

Direction of Movement

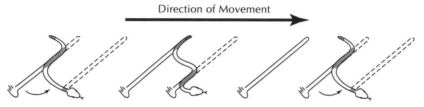

(Only the shaded portions of the snake are in contact with the surface. Solid lines indicate contact with the surface and dashed lines indicate future impressions.)

Tiger rattlesnake: While this snake is active from spring through late fall, its peak of activity correlates with the summer monsoons. Though not rare, it is rarely seen; it is primarily nocturnal. Mating occurs in April, with 4 to 6, 9-inch-long (23 cm) young born late June through September. The tiger rattlesnake fang is proportionately shorter than that of other rattlesnakes; the venom is strong. Tiger rattlesnakes eat lizards and rodents; juveniles generally favor lizards more than do adults. Though these are small rattlesnakes, they have been known to eat fairly large prey, including kangaroo rats, packrats, and even spiny lizards!

Blacktail rattlesnake: This snake often climbs into the lower branches of trees and shrubs several feet off the ground to bask in the sun or to feed on birds. It also readily eats mammals and lizards. Born in mid-summer, the

Sidewinder

young number 3 to 16, about 1 foot long (30 cm). Like many rattlesnakes, the blacktail is not aggressive nor easily alarmed and may not rattle if approached.

Sidewinder: Where it is warm throughout the year, the sidewinder is active year-round. During the summer the sidewinder seeks shelter during the day, retreating to animal burrows or burying itself in the sand under the shade of a creosote bush. It ventures out at night to eat lizards, small snakes, birds and mammals. During cooler seasons sidewinders may be diurnal (active during the day) or crepuscular (active at dusk or dawn). In areas where hibernation does occur, the sidewinder hibernates singly in a rodent burrow, or occasionally in a desert tortoise burrow. Copulation occurs in the spring, and 5 to 18 young, 6½ to 8 inches (16-20 cm) long, are born in late summer and fall.

References

Brown, David E. and Neil B. Carmony. *Gila Monster: Facts and Folklore of America's Aztec Lizard.* Silver City, NM: High-Lonesome Books, 1991.

Campbell, Jonathan A. and William W. Lamar. *The Venomous Reptiles of Latin America.* New York: Cornell University, 1989.

Ernst, Carl H. *Venomous Reptiles of North America.* Washington, DC: Smithsonian Institution Press, 1992.

Ernst, Carl H. and Roger W. Barbour. *Turtles of the World.* Washington, DC: Smithsonian Institution Press, 1989.

Greene, Harry W. *Snakes: The Evolution of Mystery in Nature.* Berkeley: University of California Press, 1997.

Klauber, Laurence M. *Rattlesnakes: Their Habits, Life Histories and Influence on Mankind.* Berkeley: University of California Press, 1997.

Stebbins, Robert C. *Field Guide to Western Reptiles and Amphibians.* Boston: Houghton Mifflin, 1985.

Stebbins, Robert C. and Nathan W. Cohen. *A Natural History of Amphibians.* Princeton, NJ: Princeton University Press, 1997.

Wright, John W. and Laurie J. Vitt, eds. *Biology of Whiptail Lizards: A Genus Cnemidophorus.* Norman, OK: Oklahoma Museum of Natural History, 1993.

Afterword

Carol Cochran

This volume has its genesis in the *Arizona-Sonora Desert Museum Docent Notebook*, a compilation of natural history information that has long guided the training and development of the Museum's world-renowned volunteer interpreters. The acclaim of the *Docent Notebook* led the Museum to undertake this ambitious book project.

The *Docent Notebook* was begun almost thirty years ago by the Museum's first education curator, Doris Evans, who researched and wrote fact sheets about desert plants and animals and assembled them into a notebook. Through the efforts of many Desert Museum staffers, it evolved into a four-inch-thick binder containing accounts of the plants and animals most likely to be encountered by visitors to the desert or the Museum, as well as essays on the non-living processes of the desert, such as geology and climate. It centered on the needs of desert visitors, answering questions they might ask, offering information to make their desert experience richer and more enjoyable.

The present volume grew from the desire to make the Notebook more complete and accessible. In so doing, the *Docent Notebook* has metamorphosed into something more scientifically rigorous and broader in subject matter. Most importantly, the present volume tells stories—stories of interrelationships and adaptations. The docents learned these stories from curator lectures, or they put them together on their own using the pages of the *Docent Notebook* as strands in their narratives. Now these stories are found in the pages of this volume, available to anyone curious about the intricate connections that sustain life in the Sonoran Desert.

Every copy of the *Docent Notebook* I've ever seen has been dog-eared, full of scribbled notes, and stuffed with extra information, evidence of active dialog with the content. There could be no happier fate for this work than to become as well-used, and as well-loved, as it precursor.

Glossary

Note: These definitions supplement explanations and parenthetical definitions found in the text. Refer to the note at the top of the index (page 597) for information on how to locate additional definitions in this book.

altricial Helpless when born or hatched and dependent upon parents for food and care for a period of time. Compare precocial.

angiosperms Flowering vascular plants with seeds enclosed in the ovaries (as compared to gymnosperms, such as conifers, which have "naked" seeds).

annual plant A plant which completes its life cycle from seed germination, through reproduction, to death within one year.

autotomy The self-amputation of a limb or body part in order to escape predation; e.g., a western banded gecko can shed its tail or an immature walkingstick its leg when threatened, then regenerate the excised part, although not perfectly.

biotic community The plants and animals living and interacting with each other and with their physical environment in a given locality, such as in a salt marsh or in a bursage-creosote community.

caldera A landscape feature that is a broad basin-like depression formed by a volcanic explosion or the collapse of a volcanic cone.

chaparral A plant assembly characterized by dense growth of resinous shrubs that readily resprout from roots and seeds following fires—typical of a Mediterranean climate with mostly winter rain.

chiropterophilous plant A plant having features that are attractive to bats, such as pale flowers with musky scents, abundant nectar produced at night and pollen containing amino acids important to bat nutrition; literally, "bat loving."

cienega (sometimes cienaga) Southwestern U.S. and Spanish term referring to a swamp or marsh, especially one fed by springs. (First "e" bears an accent mark in Spanish.)

cohort A group of individuals of the same species or other taxa which start life at approximately the same time, e.g., saguaro cacti that become established during a given period or a population of butterflies that emerge from pupas at the same time. (From military usage.)

commensalism Symbiosis in which one organism benefits at no cost (or benefit) to the other.

community See biotic community.

complete metamorphosis An insect life cycle which includes four distinct morphological stages—egg, larva, pupa and adult.

composite plant A dicotyledenous plant in the family Asteraceae (Compositae); many species bear individual flowers tightly clustered in disk-like or daisy-like heads.

convergent evolution The independent development over time of similar structures by unrelated (or distantly related) species in response to similar environmental factors or stresses, e.g., spines, leaflessness and succulence in both cacti and many African euphorbias.

crepuscular (Zool.) Active in the dim twilight of dusk or dawn; compare nocturnal and diurnal.

cryptic colors Coloration which makes a given species difficult to see in its habitat or which makes it resemble an inedible object such as a pebble.

cryptobiotic soil (also cryptogamic or microphytic crust) The fragile, crusty top layer of many desert soils characterized by the growth of lichens, algae, blue-green algae (cyanobacteria), liverworts, or mosses, in combination or singly.

deciduous plant A plant that drops its leaves annually in response to a dry or cold period.

dicotyledon (dicot) A plant in the larger of two angiosperm classes—the Dicotyledones. The first leaves that emerge from dicot seeds are in a pair. Compare monocotyledon.

dimorphism The state of having two forms which differ in appearance, shape, or size within the same species.

dioecious Having only anther-bearing "male" or carpel-bearing "female" flowers on a given plant. Compare monoecious.

diurnal Active during the day.

dormancy The temporary suspension of biological activity, such as growth or movement.

ecology The study of interrelationships among plants, animals and their physical environments.

ecosystem The combined living organisms and non-living physical components of the environment and their relationships in any given area, e.g., a small mud flat or the biosphere; through these energy moves and minerals are recycled.

endemic Growing or living exclusively within a particular region or locality.

ephemeral (Bot.) A plant that germinates, reproduces and dies over a very short period, usually within a few weeks to several months; pertaining to such plants.

estivation (Zool.) A period of torpor or dormancy in response to heat or dryness, especially in summer.

family A taxonomic group consisting of genetically related genera.

fledgling The avian life stage in which young birds begin to fly and are able to leave the nest.

flowering plant See angiosperm.

forb A broadleafed flowering plant (as distinguished from a grass or sedge).

fossorial Digging or burrowing.

glochid A readily detachable barbed bristle, usually tiny and hairlike, peculiar to opuntioid cacti.

granivorous Feeding upon grains, seeds.

gravid Pregnant with fetus or eggs.

herbivorous Feeding upon plants or parts of plants.

hibernation (Zool.) A period of dormancy or inactivity in response to cold or drought during which metabolic processes are significantly reduced and body temperature may be greatly lowered.

hybrid Offspring which is the product of cross-fertilization between two different species, subspecies or varieties.

incomplete metamorphosis A life cycle of insects in which the larva hatching from the egg resembles the adult stage but lacks wings; lacking a pupal stage, it changes to the (usually) winged adult form over the course of several molts.

insolation Exposure to the sun.

intar An arthropod in any given stage between molts; or any given stage between arthropod molts.

lagomorph A rabbit or hare.

larva An immature feeding stage of some animals which differs from the adult form.

mesic Of or pertaining to a habitat that has at least a moderate amount of mois-ture, such as a riparian area or cienega, or adapted to such a habitat. Compare xeric.

metamorphosis The physical (morphological) transformation undergone by various animals from one life stage to the next. See also complete and incomplete metamorphosis.

mimicry (Zool.) The characteristic of resembling something unpalatable, distasteful, or poisonous or otherwise dangerous.

monocotyledon (monocot) A plant in the smaller of two angiosperm classes—Monocotyledones. A single leaf emerges first from the seed. Compare dicotyledon.

monoecious Having separate anther-bearing "male" and carpel-bearing "female" flowers on the same plant. Compare dioecius.

nectarivorous Feeding on flower nectar.

nestling Young altricial bird incapable of flight and dependent on parent for food.

niche The role or function of a particular species in its community or ecosystem.

nocturnal Active at night.

nymph An immature stage in the incomplete metamorphosis of some arthropods which resembles the adult form. Also a butterfly in the genus *Satyrus*.

omnivore An animal which feeds on both animal and plant material.

orographic Pertaining to mountain-building or orogeny.

oviparous Laying eggs which hatch after leaving the mother's body.

ovoviviparous Producing eggs which hatch inside the mother's body so that the young are born with neither a placental attachment nor a shell.

parotoid glands Conspicuous glands behind the ears of toads which secrete poisonous or toxic substances. (These resemble parotid salivary glands which have a similar location.)

pedipalps Leglike appendages in arachnids just behind the mouthparts used in maneuvering prey and during mating.

parasitism Symbiosis in which one organism benefits at the expense of the other.

petiole The stalk of a leaf which attaches to the stem (Bot.). The narrow part of the body, or waist, connecting the abdomen and thorax in a wasp (Zool.)

pinna (pl. pinnae) The primary subdivision of a pinnately compound leaf, a leaflet which may bear other leaflets (pinnules).

playa The floor of a desert basin with interior drainage that becomes a shallow lake after heavy rains.

poison A substance which causes illness or death when ingested (e.g., plant parts or secretions from toad glands).

precocial Born or hatched furred or feathered, with eyes open and able to move away from the birthplace or next relatively soon.

pupa The non-feeding, nonmobile stage in insects between the last larval form and the adult stage, during which the insect undergoes anatomical changes.

sexual reproduction In both plants and animals the process in which two cells with nuclei---one from each of two parents--- fuse and the resulting offspring inherits a set of chromosomes from each parent, resulting in a unique individual. Compare parthenogenesis, apomixis, and vegetative reproduction (see listings in the Index).

species richness The number of species in a given area.

spine A hard sharp-pointed modified leaf or part of a leaf.

substrate The surface on which an organism rests or to which it is attached, e.g., a stream bottom, soil surface or leaf.

symbiosis A state in which two different species live together in intimate association; the term is often used to describe relationships that are mutually beneficial, but it also includes other relationships. See commensalism, mutualism and parasitism.

taxon A taxonomic group of any rank or size.

taxonomy The science of classification of plants and animals.

torpor A state of dormancy or inactivity.

venom A substance which may contain hemotoxins and/or neurotoxins and causes pain, paralysis and/or death when injected by mouthparts (including fangs) or stinger.

xeric Of or pertaining to a dry habitat such as a desert, or adapted to such a dry habitat. Compare mesic.

Index

Note: Bold-faced page numbers indicate terms defined or explained in the text.

Highway 286, 20, 28
Hilaria belangeri, 267, 271
Himantopus mexicanus, 389, 390
Hirundinidae; *Hirundo pyrrhonota*, 417, 418, 419
Hispanic-Americans, 119
Hispanics, 110-11, 120
Histeridae, 356
Hodotermitidae, 316
Hohokam, 106-9, 119, 156, 158-59
Holbrookia maculata, 23, 561, 562
Holly, desert (*Atriplex hymenelytra*), 220
Holocene, 65-66, 67, 68, 88, 92, 102, 273
Hololena hola, 299
Hololepta yucateca, 356
Homoptera, 290, 320-21, 322-23. *See also*
  Aphids; Cicadas; Leafhoppers
Hoover Dam, 106
Hopbush (*Dodonaea viscosa*, also *D. angustifolia, D. viscosa* var. *angustifolia*), 22, 253, 273
Hordeum murinum, 275; *pusillum*, 268, 273
Hornaday, William T.: *Campfires on Desert and Lava*, 203
Horned lizards (*Phrynosoma* spp.), 559-60;
  flat-tailed (*douglassii*), 531, 559; regal (*solare*), 26, 559(fig.); round-tailed (*modestum*), 559;
  Texas (*cornutum*), 559
Hornworm; black striped (*Hyles lineata*), 336;
  tobacco (*Manduca sexta*), 259, 334
Horse latitude deserts, 11, 13
Horse latitudes, 10-11
Horses, 67, 68, 80, 110
Horsetail (*Equisitum* spp.), 80
Horticulture, 160, 167, 193, 256
Host plants, 251-52, 330, 334-35
Huachuca Mountains, 33, 36, 121-22
Huizache. See Acacia constrica
Huizapol (*Ambrosia dumosa*), 173-74
Humans, 125, 166, 265, 372, 580; agave use
  by, 157-59; and desert pavement, 98-99;
  impacts of, 106-17; Pleistocene predation by,
  67, 68, 80; recreational vehicles and, 99-100;
  and saguaros, 192-93
Hummingbirds, 31, 122, 124, 404-10; Allen's
  (*Selasphorus sasin*), 26, 410; Anna's (*Calypte anna*), 19, 21, 28, 404, 405-6, 407, 410;
  Black-chinned (*Archilochus alexandri*), 23, 404, 405, 406-7, 410; Blue-throated, 34; Broad-billed (*Cynanthus latirostris*), 23, 404, 405, 406, 409; Broad-tailed, 34; Costa's (*Calypte costae*), 21, 28, 404, 405, 406, 407, 410;
  Magnificent, 23; and ocotillos, 119, 243; as
  pollinators, 157, 178, 259; pollination syndrome for, 145-46; Rufous (*Selasphorus rufus*), 21, 26, 404, 405, 406, 407-8, 410

Humphrey, Robert, 241
Humulus lupulus, 253
Hunting of the Snark, The (Carroll), 241
Huntington Botanic Gardens, 199
Hurricanes, Pacific, 16, 45, 241
Hyacinths, desert (*Dichelostemma pulchella*), 22, 246
Hybridization, 212, 233, 526
Hydrology, 79
Hydrophilidae, 354, 356, 357, 360
Hydropsychidae, 357, 358
Hyena (*Chasmoporthetes johnstoni*), 67
Hyla arenicolor, 34, 540
Hylaeus spp., 341
Hylephila phyleus, 332
Hyles lineata, 336
Hylidae, 540-41
Hymenoptera, 290, 341, 345, 349, 356.
  See also Ants; Bees; Wasps
Hyphae, 99
Hyporheic zone, 362-63
Hypsiglena torquata, 574

Ice ages. *See* Pleistocene
Ictalurus pricei, 513, 523; *punctatus*, 523
Icteria virens, 438, 439, 440
Icteridae, 453, 455-56
Icterus bullockii, 23, 453, 455, 456; *cucullatus*, 453, 455, 456
Idria columnaris. See Fouquieria columnaris
Iguanidae (iguanas), 64, 107, 553-54; desert
  (*Dipsosaurus dorsalis*), 23, 531, 553-54; *Iguana iguana*, 65; *Pumilia novaceki*, 65; spiny tailed,
  120, 557
Imperial Valley (Calif.), 72, 122
Incense, 177, 180
Inchworms, 333
Incienso (*Encelia farinosa*), 176
Incisitermes spp., 316
India: grasses from, 277
Indicator plants, 12, 13, 14
Influx species, **328-29**
Infrared, **58**, 59
Inocullum, 318
Insara covilleae, 263, 310, 312
Insecta (insects), 39, 210, 290, 408; aquatic,
  357-61, 362-64; and flight, 285-86; and
  plant toxins, 137, 312; as pollinators, 157,
  161; predation on, 396, 398, 413, 415, 418,
  432, 461; and saguaros, 353-55; social,
  318-19; on surface film, 361-62; *See also by order; family; genus; species*
Insectivora, insectivores, 461, 492. *See also various species*

Contributors

RICHARD BAILOWITZ is the author of numerous publications about lepidoptera including *Butterflies of Southeastern Arizona* and *70 Common Butterflies of the Southwest*. He is a high school mathematics teacher in Tucson.

STEPHEN BUCHMANN is an entomologist with the Carl Hayden Bee Research Center and associate professor at the University of Arizona. His research on conservation biology involves native Sonoran Desert bees and flowering plants.

CAROL COCHRAN was director of education at the Desert Museum from 1987 to 1997. Her interest was, and remains, developing ways to encourage public understanding and enjoyment of science and the natural world.

GOGGY DAVIDOWITZ, a research associate in the Department of Ecology and Evolutionary Biology at the University of Arizona, studies the relationship between environmental variability and developmental regulation of life history and morphological traits.

MARK A. DIMMITT came to the Desert Museum as curator of botany in 1979 and has been director of natural history since 1997. One of his central interests is in interpreting ecological relationships to the public. He also propagates several groups of arid-adapted plants.

TOM DUDLEY is a lecturer at the University of California at Berkeley. He conducts research on stream ecology and riparian conservation throughout the Southwest.

ARTHUR V. EVANS is the Insect Zoo director at the Natural History Museum of Los Angeles County. Co-author of *An Inordinate Fondness of Beetles* and an authority on scarab beetles, he has conducted extensive field work in southern Arizona for 25 years.

ROSEANNE and JONATHAN HANSON are native Arizona naturalists and authors. Together they have written numerous award-winning nature and outdoor books, including *Southern Arizona Nature Almanac*. They work as naturalists in the Chiricahua Mountains of southeastern Arizona.

MRILL INGRAM has written on a range of environmental and scientific topics and has focused on issues of climate and environmental change. She coordinated the Desert Museum's Forgotten Pollinator campaign from 1996 - 1998.

CRAIG IVANYI is collections manager of herpetology at the Desert Museum. His interests are in management of captive and wild reptiles and amphibians as well as informal education for the general public about Sonoran Desert herpetofauna.

KENN KAUFMAN, a longtime Tucson resident, has been obsessed with birds since childhood. He is field editor for *Audubon Magazine* and author of serveral books, including *Lives of North American Birds*.

KAREN KREBBS, collections manager of mammalogy and ornithology, has worked at the Desert Museum since 1987. She has done extensive research on hummingbirds in captivity and has published five papers on the subject.

HOWARD LAWLER was the curator of herpetology and ichthyology at the Desert Museum from 1981 to 1996. He has managed captive wildlife collections and related programs for over 25 years.

DAVID W. LAZAROFF, an independent naturalist, writer, and photographer, is the author of two Desert Museum books. He enjoys serving as a naturalist-in-residence at Tucson elementary schools.

RENEE LIZOTTE, a specialist on spiders and other arachnids, is the keeper of invertebrates at the Desert Museum.

JOSEPH R. McAULIFFE is a research ecologist with the Desert Botanical Garden in Phoenix. In 1995 he received the W.S. Cooper award from the Ecological Society of America for his studies on plant ecology, soils, and geomorphology in the Sonoran Desert.

PINAU MERLIN is a Tucson-based naturalist, writer, photographer, and artist specializing in the Sonoran Desert region. She is the author of *A Field Guide to Desert Holes*.

GARY PAUL NABHAN is director of conservation biology at the Desert Museum, and author of 12 books on natural history. He has been conducting field research on desert ecology ever since he first worked for the Museum in 1976.

CARL A. OLSON has been curator of the entomology research collection at the University of Arizona for 23 years, and has been active educating the public about the wonders of insect life in many forums.

JANICE PERRY, a keeper at the Desert Museum, has been in zoo herpetology for over 18 years. She has written several articles on herpetology and is currently studying the genetics of beaded lizards and the relocation of rattlesnakes.

STEVE PRCHAL, a 45-year resident of Tucson, worked at the Desert Museum for 16 years before founding the research and education organization Sonoran Arthropod Studies Institute in 1986.

ROBERT RAGUSO has studied hawkmoth pollination at the Desert Museum and the University of Arizona since 1996. He is now an assistant professor of biology at the University of South Carolina.

STEVEN RISSING is professor of biology at Arizona State University. His research concerns the evolution of cooperation in founding ant colonies. He is also interested in the teaching of introductory college biology courses and writes on issues concering the teaching of evolutionary biology.

ROBIN ROCHE, as program coordinator for the Center for Insect Science Education Outreach at the University of Arizona, has been sharing her enthusiasm and life-long fascination with arthropods with K-12 teachers and their students since 1994.

ROBERT SCARBOROUGH, staff geologist at the Desert Museum and life-long Southwesterner, enjoys exploring the vast canyons of time and sharing his discoveries with museum members and visitors. He is currently writing a book on the region's geology.

JUSTIN SCHMIDT is an evolutionary biologist and natural historian specializing in predator and prey relationships of insects and arthropods. He has been associated with Southwestern Biological Institute in Tucson since 1986.

THOMAS SHERIDAN is curator of ethnohistory and head of the Ethnology Division at the Arizona State Museum, University of Arizona. He has written or edited numerous books and articles on the ethnology, ethnohistory, and political ecology of the southwest U.S. and northern Mexico.

D. PETER SIMINSKI is the director of living collections at the Arizona-Sonora Desert Museum. He has been involved with conservation and education in the Sonoran Desert region for 16 years.

MARK P. SITTER is a horticulturist at the Desert Museum specializing in cacti and succulents. He has been studying entomology with particular interest in the natural history of lepidoptera and coleoptera since 1972.

ROBERT L. SMITH has taught entomology and conducted research on aquatic insects and insect behavior in the entomology department of the University of Arizona since 1977. He has written two books and numerous scientific papers and popular articles on entomological topics.

BARBARA TERKANIAN is collections manager of invertebrate zoology and ichthyology at the Desert Museum, a position she has held for two years. She has studied sexual selection, parasitism, and wing coloration in butterflies.

JANET TYBUREC is the director of education programs for Bat Conservation International in Texas. She has been intimately involved in the structure and execution of training workshops for educators and wildlife biologists.

THOMAS VAN DEVENDER, senior research scientist at the Desert Museum, has reconstructed the environmental history and evolution of the Sonoran Desert region using fossils, inventoried floras in desert and tropical areas, and studied the ecology and the conservation of the desert tortoise.

DIANA WHEELER, a professor and specialist in ant biology at the University of Arizona, has been studying in general ants for 20 years and Arizona ants for the last 10 years.

MARK WILLIS has been conducting a research program on the olfactory control of moth behavior at the University of Arizona since 1990. He has studied chemical communication and the mating systems of moths in both laboratory and field settings for the last 20 years.

Editors

STEVEN J. PHILLIPS joined the Desert Museum in 1994 as its first publications manager. Before coming to the museum, Steve owned and directed a book publishing company in Denver, Colorado. His professional background includes historical and prehistoric archaeology, cultural and natural resource management, technical and scientific writing, and graphic design. Steve is the author of an award-winning book on historic architecture.

PATRICIA WENTWORTH COMUS is a research associate at the Desert Botanical Garden in Phoenix. Pat's varied fieldwork and research background ranges from the study of the territorial and mating behavior of sympatric long-tailed brush and tree lizards to the study of the flora of the Lower Santa Cruz Valley and the soils on which they grow. She has experience in book and periodical editing, proofreading, and fact-checking.

Photo Credits

Arizona Historical Society/Tucson #40579: 116 (top); Arizona State Museum, Helga Teiwes, photographer, ASM #78719: 108; Arizona-Sonora Desert Museum: 113; Paul & Shirley Burquist: 371; Harry Casey: 98; Mark Dimmitt: 131, 135, 138, 141; Kyle House: 89; Peter Kresan: 84; Joseph McAuliffe: 92, 93, 95, 103; C. Allan Morgan: 370; Tad Nichols: 46; Steven Phillips: 40, 116 (bottom); Robert Scarborough: 77, 81, 83, 85; Lewis W. Walker: 367. All of the color photographs were taken by Desert Museum staff and volunteers, with the exception of plate 9, which was taken by former Desert Museum board president Peter Kresan.

Illustration Credits

All illustrations are copyrighted by the artist unless otherwise noted

Narca Moore-Craig 365, 378, 379, 381, 382, 384, 385, 386, 391, 392, 393, 394, 396, 397, 398 (bottom), 399, 400, 401, 404, 405, 406, 408, 412, 413, 414, 415, 416, 420, 422, 425, 426, 427, 428, 430, 431, 432, 436, 438, 439, 441, 443, 446, 447, 448, 449 (left), 450, 451, 452, 453, 456, 457, 458

Jeffrey L. Martin 281, 292, 293, 297, 298, 300, 301, 302, 304, 305, 306, 307, 311, 313, 317, 318, 319, 320, 321, 322, 324, 326, 327, 328, 330, 332, 336, 339, 342, 343, 344, 346, 347, 348, 349, 354, 355, 435, 514 (bottom)

Randy Babb 26, 527, 538, 539, 543, 544, 545, 548, 549, 551, 553, 554, 555, 556, 557, 558, 560, 562, 563, 564, 565, 566, 567, 568, 569, 570, 573, 574, 575, 576, 577, 578 (top), 579, 581, 582, 584 (bottom)

Brian Wignall 127, 148 (left), 149, 158, 159 (top), 160, 162, 163, 164, 169, 176, 194, 195, 196, 203, 204, 205, 221, 233, 240, 245, 246, 247 (right), 248, 250, 255, 259, 265, 266, 277, 278

Rachel Taylor 464, 472 (top), 502, 503, 505, 506 (top), 509, 515, 516, 517, 519, 520, 523, 535, 536, 540, 541, 542

Deborah Reade 11, 12, 15, 30, 42, 52, 53, 55, 57, 58, 73, 76, 82, 89, 91, 96, 584 (top), plate 1

Ken Wintin 514 (top), 518, 522

Pamela Ensign 459, 473

Kim Duffek 303, 461

Illustrations on pages 177, 220, 223, and 254 reprinted by permission of the University of California Press, from *A Flora of Southern California* by Philip Muntz, 1974.

Sandy Truett illustrations on pages 465, 467, 468, 469, 472 (bottom), 474, 475, 477, 478, 479, 480, 481, 482, 483, 484, 487, 488, 489, 491, 493, 494, 497, 498, 500, 504, 506 (bottom), and 507 reprinted by permission of the University of Arizona Press, from *Mammals of the Southwest* by E. Lendell Cockrum, 1982.

Lucretia Breazeale Hamilton illustrations on pages 172 (left), 175 (right), 179, 222, 225, 239, 251, 258, 261, and 276 (right) reprinted by permission of the University of Arizona Press, from *An Illustrated Guide to Arizona Weeds* by Kittie F. Parker, 1972.

Lucretia Breazeale Hamilton illustrations on pages 159, 172 (left), 173, 175 (left), 178, 181, 224, 225, 226, 229, 230, 231, 232, 234, 235, 236, 242, 244, 247, 253, 257, 260, and 262 reprinted by permission of the University of Arizona Press, from *Trees and Shrubs of the Southwestern Deserts* by Lyman Benson and Robert A. Darrow, 1981.

Lucretia Breazeale Hamilton illustrations on pages 268 (bottom), 272, 273, 274, and 279 reprinted by permission of the University of Arizona Press, from *Arizona Range Grasses* by Robert R. Humphrey, 1970.